3 0250 01070 6773

D0554945

DATE DUE

DEC -6 1989			
DEC - 7 1989			

f
QR
111
E22
985

Ecology and Management of Soilborne Plant Pathogens

Proceedings of Section 5 of the Fourth International Congress of Plant Pathology

University of Melbourne, Melbourne, Australia, 17–24 August 1983

EDITED BY
C. A. Parker, A. D. Rovira, K. J. Moore, and P. T. W. Wong

AND

Proceedings of the First International Workshop on Take-all of Cereals

Victorian Crops Research Institute, Horsham, Victoria, Australia
10–11 August 1983

EDITED BY
J. F. Kollmorgen

The American Phytopathological Society
St. Paul, Minnesota, U.S.A.

Publisher's note: This book includes the proceedings of the most recent of an ongoing series of international symposia on soilborne plant pathogens. The symposia have produced five volumes; the preceding four are *Ecology of Soil-borne Plant Pathogens: Prelude to Biological Control* (K. F. Baker and W. C. Snyder, eds., University of California Press, 1965); *Root Diseases and Soil-Borne Pathogens* (T. A. Toussoun, R. V. Bega, and P. E. Nelson, eds., University of California Press, 1970); *Biology and Control of Soil-Borne Plant Pathogens* (G. W. Bruehl, ed., American Phytopathological Society, 1975); and *Soil-Borne Plant Pathogens* (B. Schippers and W. Gams, eds., Academic Press, 1979). These symposia now coincide with the International Congresses of Plant Pathology, held every 5 years.

Cover: *Phytophthora cinnamomi* sporangium with three remaining zoospores. Micrograph by N. Malajczuk, CSIRO Division of Forest Research, Wembley, Western Australia, 6014. © N. Malajczuk; used by permission.

Library of Congress Catalog Card Number: 85-71116
International Standard Book Number: 0-89054-066-7

© 1985 by The American Phytopathological Society

All rights reserved.
No part of this book may be reproduced in any form by photocopy, microfilm, retrieval system, or any other means, without written permission from the publisher.

Printed in the United States of America

The American Phytopathological Society
3340 Pilot Knob Road
St. Paul, Minnesota 55121, U.S.A.

Preface

The Fourth International Congress of Plant Pathology, held in Melbourne in August 1983, provided a rare opportunity for many overseas scientists working on soilborne plant pathogens to visit Australia and for us Australian scientists to meet them on our own ground.

Because of its isolation, Australia's geography is little known overseas. Our climates range from wet tropical and monsoonal conditions in the north, through warm temperate and temperate in the east, to cool temperate in the far south of Victoria and in Tasmania. Westward are very large areas of arid and desert lands in which no agriculture and little grazing is practiced. In southern South Australia and the southwest corner of Western Australia, a Mediterranean climate prevails, with hot, dry summers and cool, moist winters.

The soils of Australia range from the black and red loams of the tropics to the more leached soils of the higher rainfall and cooler regions of the south to the sands and stony deserts of the arid center. Deeply and extremely weathered lateritic and bauxite materials are widespread, especially in the western third of the continent. Soils are commonly coarse textured and infertile, deficient in many of the trace elements as well as in major nutrients such as phosphorus.

Early agricultural development in the 1800s by European settlers naturally took place on the better soils of the temperate and tropical regions of eastern and southern Australia. Following World War II, agricultural development extended onto the poorer soils, and dramatic successes were achieved in identifying and remedying major and minor element deficiencies and in raising fertility through legume-based ley pastures. Agriculture in Australia now faces the problems of developing from this base toward farming systems that are economically viable, stable, and sustainable—often in a harsh, unpredictable climate.

Australia enjoys some unique "firsts" in root pathology. Within 14 years of the first European settlers arriving in South Australia in 1836, a disease appeared in wheat crops that was described as follows:

> The hitherto fertile fields of South Australia have been visited by a demon-like enemy to agriculture, which is already beginning to fill the minds of farmers with despair. Small patches were observed here and there, where the crops died off not long after their appearance above the ground. These patches increased in number and size and were taken possession of by weeds. At first, very little notice was taken of those patches, but when they increased and spread year by year attention was attracted to them. The people, very appropriately, called the mysterious disease "take-all"; but no one hardly ever thought at that time that this pest would be able to destroy the prosperity of the whole district, and even to endanger the very existence of the colony as an agricultural country.

These words were written in South Australia by Dr. Carl Mucke in 1870 in an essay called "The Take-all—the Corn Disease of Australia," 20 years before the take-all fungus was identified by Prillieux and Delacroix in France and 34 years before the first identification of the disease in Victoria, Australia, by McAlpine. The problem was so serious that the government of South Australia set up a commission in 1868 to investigate "diseases in cereals" that were reported to the government. At that time there was a great deal of speculation as to the cause of the dreaded take-all disease. Among the so-called remedies, however, were treatments that are advocated today: application of salt, phosphorus, potassium, nitrogen, and silicates, along with stable manure. Even at such an early stage of development of the colony of South Australia, it was recognized that our soils were low in nutrients and were depleted rapidly by continuous cropping, which led to a worsening of the take-all disease.

With the improvement of soil fertility by the application of fertilizers and growth of legumes, the disease now takes a different form. It affects the plant at a later stage to cause premature ripening—a disease that South Australian farmers call "Haydie." Despite our increasing knowledge of the take-all fungus and the disease it causes, take-all is still considered the major root disease of wheat in Australia and of major importance in most wheat-growing areas of the world.

In view of this long history of the disease in Australia, it is especially appropriate that the First International Workshop on Take-all should be held in this country. Dr. J. F. Kollmorgen worked diligently to organize the workshop and to produce the Proceedings that are published in this volume. Most of the world's experts on take-all attended the workshop, and this section of the book records their combined knowledge and the gaps in knowledge that are now the focus of their research.

Australia is also unique in that large areas of native forests have suffered the onslaught of another soilborne root disease—this time caused by *Phytophthora cinnamomi*. For many years, areas of *Eucalyptus marginata* (jarrah) forests of Western Australia were dying of a mysterious complaint known as "jarrah dieback," which was thought to be associated with waterlogging and impeded drainage. This disease was sometimes associated with the development of roads through the forest, but at first no one linked this occurrence with the use of gravel as a road-building material; this gravel was often taken from affected parts of the forest where the commercial jarrah trees had already died. When *P. cinnamomi* was identified by Zentmyer, Podger, and Doepel as the cause of jarrah dieback in 1965, it was still difficult for plant pathologists to convince others that their earthmoving and other operations that disturbed the soil were contributing to the spread of the disease. It was necessary, they argued, to introduce quarantine measures before shifting heavy earthmoving and tree-logging equipment from one part of the forest to another. However, such a quarantine system, with sterilization of implements, is now standard procedure in the Western Australian jarrah forest, and it has been effective in slowing the spread of the disease.

These are but two classic examples of how root pathogens have affected the welfare and landscape of Australia. The visit of Dr. Kenneth Baker to work with Australian plant pathologists led to one of the first successes in the biological control of soilborne root disease by the introduction of suppressive bacteria into soil. Thus it is gratifying to find some 15 papers on the biological control of soilborne plant disease in these proceedings. We find that research in the United States, France, and Japan is unraveling some of the mechanisms involved in biological control, and research in Israel and the United States is demonstrating that *Trichoderma* spp. can be used as a biocontrol agent against *Rhizoctonia* and other soilborne root diseases.

The extension of biological control outside the traditional world of soil microbiology has revealed the likely role of soil animals such as mycophagous amoebae, which attack the hyphae of the take-all fungus, and microarthropods, which feed on fungal hyphae and spores; this information has introduced new dimensions into our concepts of biological control at this Congress. May we now hope for more frequent and vigorous collaboration between soil zoologists and plant pathologists?

The ultimate control of soilborne root diseases will depend on our understanding of the ecology and epidemiology of the pathogens in the field and the influence of soil, climate, plant nutrition, and plant defense mechanisms on root diseases. The importance with which root pathologists view such areas of research is shown by the 34 papers on these topics in this book, which reveal that soil and crop management strategies are being developed to control some root diseases.

Of these strategies, solarization must be the most exciting. The success of this technique, which leaves no toxic residues in the soil, has been spectacular; its further development for high-value crops offers a significant breakthrough in controlling certain soilborne root diseases.

The development of conservation tillage practices (no-till, reduced tillage, minimum tillage, or direct drilling) to reduce erosion, sustain soils, and reduce cropping costs by the use of herbicides to replace the traditional plow creates a different environment for soil microorganisms and root pathogens. This is leading to unforeseen problems with some root diseases and is becoming an important research area as we move toward more efficient, long-term, sustainable agricultural systems.

The complexity of the biological environment in soil and the problems arising from the use of fungicides to remove particular components of the soil biota on a broad-acre basis lead many root pathologists to dream of the day when resistance to plant pathogens can be provided within the plant. Plant selection and breeding have been successful in developing resistant lines for many root diseases, but many problems remain—including those caused by the diversity of strains within many root pathogens and the interactions of soil and seasonal conditions with the resistance or tolerance of plants to root disease.

The active participation in this Congress of scientists from Japan working in the areas of ecology, epidemiology, and biological control of root diseases was very welcome. These scientists are the bridge between the Fourth International Congress of Plant Pathology held in Australia and the next Congress, to be held in Japan in 1988.

We hope these Proceedings will inspire more scientists to enter the difficult area of root disease research, bringing papers to the next Congress that will report further successes in the quest to control root diseases and thus to increase world food production.

Authors of papers designated "Invited Paper" on the Contents page were invited to present an overview of their topic rather than a specialized research paper. Topics and speakers were chosen to ensure adequate coverage of the field of soilborne diseases from many different countries. For this reason, five papers were drawn from Section 3 (Mycology) and three from Symposium 5 (Effect of New Cropping Practices on Plant Diseases in Temperate Countries) of the Congress.

We thank all the participants and organizers who helped make Section 5, the section of the Congress devoted primarily to soilborne plant pathogens, a success. We thank Drs. K. J. Moore, P. T. W. Wong, and J. F. Kollmorgen, who have served as our coeditors in the task of preparing this publication. We are also grateful to Dr. Thor Kommedahl and the American Phytopathological Society for support in publishing these proceedings and to its staff for helping to bring the publication through the many stages from manuscript to book.

C. A. Parker and A. D. Rovira
Senior Editors

Contents

Proceedings of Section 5 of the Fourth International Congress of Plant Pathology

Part II. Colonization of Roots and Rhizosphere

Part III. Biological Control

Part IV. Plant Responses and Resistance

Part V. Influence of Soil, Environment, and Nutrition

Part VI. Management Practices, Chemical Control, and Solarization

Proceedings of the First International Workshop on Take-all of Cereals

"Graveyard" site in a Western Australian jarrah (*Eucalyptus marginata*) forest, resulting from root infection by *Phytophthora cinnamomi*. This scene demonstrates the damage that a microscopic soilborne plant pathogen can do in a valuable hardwood forest. (Photograph by N. Malajczuk, CSIRO Division of Forest Research, Wembley, Western Australia, 6014. ©N. Malajczuk; used by permission.)

Proceedings of Section 5 of the Fourth International Congress of Plant Pathology

University of Melbourne, Melbourne, Australia
17–24 August 1983

The Study of Soilborne Plant Pathogens: Changing Outlook or More of the Same?

D. HORNBY, Rothamsted Experimental Station, Harpenden, Herts. AL5 2JQ, U.K.

Since the Third International Congress of Plant Pathology in 1978, many researchers have found themselves working in an increasingly competitive environment, under pressure to pursue objectives with more immediate practical relevance than hitherto. The recent economic recession has much to do with this pressure, and it is natural that such experiences affect views of the present and the future.

Three months after the Fourth International Congress of Plant Pathology in Melbourne, the 10th International Congress of Plant Protection took place at Brighton, England. The proceedings of the Brighton Congress (2) provide a broad, contemporary background against which to consider the subject of soilborne plant pathogens. They underline the need to sustain the agricultural growth necessary to achieve food security for humankind and, consequently, to overcome undulations in agricultural production that are attributed mainly to weather fluctuations, pests, and diseases. There is evidence that production is becoming more unstable, with disturbing regional differences. Production must, however, increase by 50–60% by the end of the century to keep up with the needs of an expanding population for food and materials.

The view that the global economic climate is "bottoming out of recession" might be grounds for cautious optimism, but overproduction in developed countries and millions of hungry people in developing countries are realities that present plant pathology with different sets of problems. In developed countries, agriculture is partly shaped by factors other than economic ones, and overproduction must be solved by developing alternative systems; whereas the priority in developing countries is to encourage food and fuel production by those most in need. The agrochemical industry believes that it is adapting to these changing circumstances and that it can make a major contribution to world food production, but the promise of crop protection research also encompasses nonchemical procedures and biotechnology. A decade of small improvements in yields of major food and fiber crops has renewed interest in better definition and reduction of losses as a means to increase crop yields. This interest has highlighted the need for multidisciplinary cooperation and the holistic "systems approach." What we do in our science is unlikely at all times to accord with these views, but at least we are given pointers about where the future may lead. Therefore, as an introduction to the specialist sections of this volume, I offer a few observations that seem relevant to any assessment of the present and predictions for the future.

Looking back over the five international symposia on Factors Determining the Behavior of Plant Pathogens in Soil, a period of 20 years, one cannot help but be struck by certain long-running themes. The first symposium addressed seven major topics (1): soil microorganisms; soil environment; root and rhizosphere; pathogenesis and resistance; mechanisms of antagonism; soil inoculum; and interactions between soil, microorganisms, and plants. Except perhaps for the first, all the topics expanded and evolved in the ensuing years. Successive symposia dealt with root and rhizosphere (1), root exudates (30), root environment (3), root habitat (26), and, most recently (this volume), colonization of roots and the rhizosphere. Resistance has been a recurring theme (1,3; this volume): mechanisms of antagonism (1) grew into microbial antagonism (3) and then, supported by allied themes such as crop residues and amendments (3), bloomed into a popular, main feature, biological control (26; this volume). Soil inoculum reappeared through the years under population dynamics (30), quantification (26), or epidemiology (this volume). Other examples can be given, but the point is made: there seem to have been no sudden leaps forward and no major change in direction.

Indeed, in that time agriculture and the problems we confront have changed more than our preoccupations. Quite often, outside influences have dictated new topics in these symposia. Contributions on soil and crop management (26) acknowledged changes in farming systems, and those on side effects of pesticides (26) and integrated pest management were, in part, a response to increasing awareness of, and concern over, the excessive use of agrochemicals. Increasing skills, improved techniques, and growing sophistication have thus far enabled us to respond to most of these challenges with few conceptual upheavals.

At international congresses of plant pathology, the content of the section on soilborne plant pathogens (Section 5 in the two most recent congresses) is very much circumscribed by a tradition of symposia dominated by microbial ecology and diseases caused by fungi. This now-familiar but somewhat artificial demarcation has meant that other considerations that could easily be grouped under soilborne plant pathogens *sensu lato* are treated elsewhere. At Melbourne, there were joint sessions on the Rhizoctonia disease complex with the mycology section, and on soilborne diseases with Section 8 (Epidemiology and Crop Loss Assessment). In future congresses, the preservation of these traditional barriers should be questioned further. Soilborne viruses are an example of a study that has been pursued outside the traditional limits of Section 5, except perhaps for occasional interest in fungal vectors. It is a research field that has grown slowly over the years and has recently received a fillip in Europe from increasing concern about barley yellow mosaic virus, transmitted by *Polymyxa graminis* (19), and rhizomania in sugar beet (14,32), caused by beet necrotic yellow vein virus, which is transmitted by *P. betae*. After several years of observing colleagues in soil microbiology grappling with problems of mycorrhizae that have parallels in the field of root-infecting pathogens, I can only wonder at the persistence of these traditional demarcations.

The soil is an extremely difficult medium in which to study disease, and this is reflected in the small number of new techniques that are actually improvements or innovations. Greaves and Malkomes (8), reviewing the effects of herbicides on soil microflora, concluded that too little is known of soil microbial ecology and that considerable basic research is required because it is the only way information essential to applied scientists can be obtained. In their specific field, the following familiar topics were identified as warranting special attention: nutrient cycling, decomposition of crop residues, rhizosphere studies, soilborne plant pathogens, microbial antagonism, and soil structure. The failure of some attempts at biological control has been ascribed to inadequate knowledge of the biocontrol microorganisms (12). Here, therefore, is a strong argument in favor of closer links between soil microbiology and plant pathology (18). Whether it is finding ways to measure soil suppressiveness (11), assessing root disease and its effects, or trying to quantify soilborne inoculum, it is a sobering fact that often only one or two well-used, less-than-perfect methods are available. It has not helped that some of our most cherished concepts, such as inoculum potential and the root surface, either tend to defy quantification or have been wide of the mark. Electron microscope studies of roots in the last few years (5) have revealed a complex, convoluted interface with the soil that must surely refute the idea of a rhizoplane. Sometimes when the end is clear, the means are not, as when many years ago it was suggested that the considerable problem of assessing root disease could be tackled by estimating the amount of functional tissue remaining, rather than assessing diseased tissue. Although there have been attempts to do this, often by indirect measurements of root activity (e.g., silicon in wheat glumes and soil moisture content under diseased and healthy crops [25]), it has made little impact and most root disease is still assessed by traditional methods. Perhaps the very difficulty of attaining this goal makes us view it as an elusive panacea. It is worth remembering, however, that aboveground assessments of late blight of potatoes by visual grading and measurements of healthy and diseased plant areas each revealed details of the epidemic difficult to measure by the other methods, although healthy haulm area was the assessment that reflected the integrated influences of factors affecting the crop and disease (21). With the need to continue basic studies in order to improve or develop new techniques, we should not overlook the benefits of reexamining old ideas when new technology becomes available. For example, the lure of seeing the pathogen operating in situ is one that has attracted researchers for many decades, but now minirhizotrons, improved borescopes, and video recording systems offer new prospects for plant pathologists (23), although difficulties of fewer than expected root intersections in the top 20 cm of soil have still to be resolved (31).

What concerns me most at this time is the problem of premature publicity and its uncritical acceptance, from which the field of biological control has been suffering more than most. As Ruppel et al (22) pointed out, "The literature is replete with reported successes with biocontrol of soil-borne pathogens by *Trichoderma* spp. in glasshouse experiments, but effective and economical control of disease in the field has not been achieved." Such statements contrast starkly with the razzle-dazzle that has surrounded biological control recently, and they remind us of its two sides. The promising, optimistic side is typified by the glasshouse environment and high-input, high-value crops, which with integrated crop management afford unique opportunities to put many biological control mechanisms into practice (12). Modern glasshouse growing systems are based on media such as peat or synthetic fiber products or on hydroponics, and fewer crops are grown in true soil (10); as a result, disease problems have been more responsive to this kind of control. The less successful and less publicized side of biological control is the experience of Ruppel et al (22) and many others working on arable crops, which tends to be swept aside in the full flood of biological control euphoria. Biological control still receives serious attention in many quarters and it is likely to have an increased role to play in plant protection, but too many sensational, ill-considered, or hollow claims will destroy the support and good standing it needs if it is to prosper. The problem is one facet of a general problem that besets agricultural research, as glimpsed in Pearce's (20) remarks that a weakness of some research institutes is that they conduct experiments on their own farms and then announce the results without giving much thought to generalization.

Identifying topics, other than continuing themes, that are going to claim attention at the next congress is not easy. Before the Munich symposium in 1978, minor pathogens (24) was a minor subject; but following an explosion of interest in plant-growth-promoting rhizobacteria (PGPR) (27), it has expanded and flourished. Taxa that provide PGPR are just as likely to contain organisms that under certain conditions have detrimental effects. For example, *Pseudomonas* is a genus currently providing PGPR and biocontrol agents, but some fluorescent pseudomonads, physiologically and biochemically similar to plant-growth-promoting isolates, extensively colonize and damage roots of nontarget plants affected by herbicides (9). Bioassays of such pseudomonads from roots of wheat sprayed with mecoprop (phenoxy group of herbicides) revealed (Greaves, *personal communication*) that about 40% stimulated root growth, 40% inhibited it, and 20% had no effect. Although these proportions stayed about the same in subsequent tests, many isolates had a different effect in each bioassay. The capacity of these organisms to lose their favorable characteristics or even to assume harmful properties may well become a major concern in the future if PGPR and biological control are to become more generally useful in agriculture. Mycostasis, a major topic in earlier symposia, has been overshadowed by new interests such as suppressive soils. This compartmentalization and apparent novelty may be misleading us, however, and in future it may be profitable to pay more attention to similarities and relationships among these phenomena, a point that has

4

already been made by Lockwood (16). Mathematical modeling of natural phenomena is a comparatively recent occupation in soil microbiology that now appears to be entering a log-phase period of expansion (29). Models are simplifications of reality and may be descriptive, predictive, or conceptual; their appeal is often as an aid to organizing and disciplining work and avoiding inappropriate experimentation (15). There are models for decomposition, nutrient cycling, microbial abundance in the rhizosphere, and soilborne plant pathogens (29). Most models of soilborne plant pathogens have been deterministic or simple probability models, and Gilligan (6) concluded that such modeling has perhaps generated more contention than consensus. He was of the opinion that stochastic models (models that invoke probability theory to describe changes in time and/or space) will play an important role in the future, but I think it is worth suggesting that they may be less valuable for predictive purposes. Even with airborne plant pathogens, where epidemiological studies are usually considered to be more advanced, most work on disease in populations of plants has considered increases with time. There are few mathematical descriptions of disease gradients in space and no analytical models to describe disease progress in time and space (13).

Other areas where we await progress concern phloem-translocated fungicides and the genetic manipulation of soil organisms. The requirements for phloem motility are becoming known (4), and the advent of fungicides with such properties would introduce a whole new dimension to the control of root disease. Progress in the transfer and manipulation of genes in ways other than sexual crossing to improve crop plants has been dramatic. We have seen plants transformed by microbial (e.g., *Agrobacterium tumefaciens*) plasmid vectors, but we have yet to see these techniques making an impact in the field of soilborne plant pathogens. However, if it is hoped that the manufacture of natural biological pesticides such as *Bacillus thuringiensis* will be improved by genetic engineering technology (7), it seems reasonable also to hope that the organisms eventually intended for biological control in soil would ultimately be susceptible to such improvement. More general-purpose biological control, capable of operating over a wider range of environmental conditions, may also be possible and might even be combined with PGPR. According to Jarvis (12), there is remarkably little integrated pest management, which is a situation that overlooks the realities of crop production; but it is increasingly advocated where problems have arisen from overreliance on pesticides, as in perennial plantation and orchard crops (32). It is also becoming clear that biological control agents and PGPR require extensive testing in the field, and we are likely to see the development of more appropriate experimental designs and procedures for this purpose.

Most disease problems nowadays concern endemic diseases that insidiously reduce crops over a wide area by a small amount. We are thus entering an era when better experiments and sound analyses will be essential (20). "Better experiments" is not always just a matter of ensuring that the work meets with the statistician's approval: it also means that it should be sound in all respects. In London in 1980, the Federation of British Plant Pathologists celebrated 100 years of Koch's

postulates with a meeting entitled Plant Disease Etiology. There I suggested that the etiology of disease is often taken for granted in experimental studies and that full and rigorous use of Koch's postulates has been the exception rather than the rule. The need to pay careful attention to their proper use is as important today as it ever was. When all the requirements cannot be fulfilled, as with ill-defined, poor growth phenomena with no diagnostic symptoms, then conclusive proof of the cause is likely to be lacking (28). The proposition (17) that Koch's postulates can be used as rules of proof for suppressive entities is appealing, but it needs further examination (11).

Throughout its history, there can have been no time when the study of soilborne plant pathogens was devoid of challenges, fashions, and excitements. The contributions to this volume and the number of scientists engaged in studies of soil organisms today testify to the health of the subject. It is not the time for proselytizing or making fervent promises about great things to come; after all, achievements come from painstaking work over many years, or arise in unsuspected ways that none but an oracle could foresee. As far as our transactions at the international level are concerned, the choice is becoming increasingly clear: we can either go the way of rigid tradition or make a start in breaking down some of the outmoded divisions that have kept our section in its narrow confines (the "more of the same" or "changing outlook" of our title). In 1978, at the Munich congress, it was proposed that the International Society for Plant Pathology have a committee for soilborne plant pathogens. When this eventually came to pass, the new ISPP Committee for Soil-borne Plant Pathogens saw itself primarily as an advisory body for local organizers of international congresses. In other areas, the committee has been underused, and in the future it could become a means by which individual plant pathologists may enter this debate, either directly or through their national societies.

LITERATURE CITED

1. Baker, K. F., and Snyder, W. C., eds. 1965. Ecology of Soil-borne Plant Pathogens. University of California Press, Berkeley. 571 pp.
2. British Crop Protection Council. 1983. Plant Protection for Human Welfare. 3 vols. Proc. Int. Congr. Plant Prot., 10th. Croydon. 1,228 pp.
3. Bruehl, G. W., ed. 1975. Biology and Control of Soil-Borne Plant Pathogens. American Phytopathological Society, St. Paul, MN. 216 pp.
4. Butcher, D. N., Chamberlain, K., White, J. C., Briggs, G. G., Bromilow, R. H., Chen, Q. F., Evans, A. A., and Rigitano, R. 1984. Uptake and movement of chemicals in plants. Direct measurement of mobility. Rep. Rothamsted Exp. Stn. 1983, Part 1, 102-103.
5. Foster, R. C., Rovira, A. D., and Cock, T. W. 1983. Ultrastructure of the Root-Soil Interface. American Phytopathological Society, St. Paul, MN. 157 pp.
6. Gilligan, C. A. 1983. Modeling of soilborne pathogens. Annu. Rev. Phytopathol. 21:45-64.
7. Glick, J. L., Peirce, M. V., Anderson, D. M., Vaslet, A., and Hsiao, H. 1983. Utilization of genetically engineered microorganisms for the manufacture of agricultural products. Pages 67-87 in: Genetic Engineering: Applications to Agriculture. L. D. Owens, ed. Beltsville Symposia in Agricultural Research 7. Granada Publishing, London. 327 pp.
8. Greaves, M. P., and Malkomes, H. P. 1980. Effects of soil

microflora. Pages 223-253 in: Interactions Between Herbicides and the Soil. R. J. Hance, ed. Academic Press, London. 349 pp.

9. Greaves, M. P., and Sargent, J. A. 1985. Herbicide-induced microbial invasion of plant roots. Weed Sci. In press.

10. Hockenhull, J. 1982. Biological control of root diseases of glasshouse crops in Scandinavia. Chron. Hortic. 22:48.

11. Hornby, D. 1983. Suppressive soils. Annu. Rev. Phytopathol. 21:65-85.

12. Jarvis, W. R. 1983. Progress in the biological control of plant diseases. Proc. Int. Congr. Plant Prot., 10th, 3:1095-1105. British Crop Protection Council, Croydon.

13. Jeger, M. J., Jones, D. G., and Griffiths, E. 1983. Disease spread of nonspecialised fungal pathogens from inoculated point sources in intraspecific mixed stands of cereal cultivars. Ann. Appl. Biol. 102:237-244.

14. Koch, F., ed. 1982. Die Rizomania der Zuckersube. Proc. Int. Inst. Sugar Beet Res. (IISBR), Brussels, pp. 211-238.

15. Kranz, J., and Royle, D. J. 1978. Perspectives in mathematical modelling of plant disease epidemics. Pages 111-120 in: Plant Disease Epidemiology. P. R. Scott and A. Bainbridge, eds. Blackwell, Oxford. 329 pp.

16. Lockwood, J. L. 1979. Soil mycostasis: Concluding remarks. Pages 121-129 in: Soil-borne Plant Pathogens. B. Schippers and W. Gams, eds. Academic Press, London.

17. Lui, S. D., and Baker, R. 1980. Mechanism of biological control in soil suppressive to *Rhizoctonia solani*. Phytopathology 70:404-412.

18. Lynch, J. M. 1983. Soil Biotechnology. Blackwell Scientific Publications, Oxford. 191 pp.

19. Macfarlane, I., and Adams, M. J. 1984. Barley yellow mosaic virus (BaYMV). Rep. Rothamsted Exp. Stn. 1983, Part 1, 120-121.

20. Pearce, S. C. 1983. The Agricultural Field Experiment. John Wiley, Chichester. 335 pp.

21. Rotem, J., Kranz, J., and Bashi, E. 1983. Measurement of healthy and diseased haulm area for assessing late blight epidemics in potatoes. Plant Pathol. 32:109-115.

22. Ruppel, E. G., Baker, R., Harman, G. E., Hubbard, J. P., Hecker, R. J., and Chet, I. 1983. Field tests of *Trichoderma harzianum* Rifai aggr. as a biocontrol agent of seedling disease in several crops and Rhizoctonia root rot of sugar beet. Crop Prot. 2:399-408.

23. Rush, C. M., Upchurch, D. R., and Gerik, T. J. 1984. In situ observations of *Phymatotrichum omnivorum* with a borescope mini-rhizotron system. Phytopathology 74:104-105.

24. Salt, G. A. 1979. The increasing interest in "minor pathogens." Pages 289-312 in: Soil-borne Plant Pathogens. B. Schippers and W. Gams, eds. Academic Press, London.

25. Salt, G. A., Ellis, F., Howse, K. R., and Brown, G. 1975. Nutrient and water up-take by healthy and diseased wheat roots. Field observations. Rep. Rothamsted Exp. Stn. 1974, Part 1, 218-219.

26. Schippers, B., and Gams, W., eds. 1979. Soil-borne Plant Pathogens. Academic Press, London. 686 pp.

27. Schroth, M. N., and Hancock, J. 1981. Selected topics in biological control. Annu. Rev. Microbiol. 35:453-476.

28. Sewell, G. W. F. 1981. Effects of *Pythium* species on the growth of apple and their possible causal role in apple replant disease. Ann. Appl. Biol. 97:31-42.

29. Smith, O. L. 1982. Soil Microbiology: A Model of Decomposition and Nutrient Cycling. CRC Press, Boca Raton, FL. 273 pp.

30. Toussoun, T. A., Bega, R. V., and Nelson, P. E., eds. 1970. Root Diseases and Soil-borne Pathogens. University of California Press, Los Angeles. 252 pp.

31. Upchurch, D. R., and Ritchie, J. T. 1983. Root observations using a video recording system in mini-rhizotrons. Agron. J. 75:1009-1015.

32. Wildbolz, T. 1983. Integrated pest management in perennial plantation and orchard crops. Proc. Int. Congr. Plant Prot., 10th, 3:965-970. British Crop Protection Council, Croydon.

Part I
Ecology and Epidemiology

Characteristics of Trends in Disease Caused by Soilborne Pathogens with Spring Barley Monoculture

P. C. CUNNINGHAM, An Foras Taluntais, Oak Park Research Centre, Carlow, Ireland

Disease levels caused by soilborne cereal pathogens of temperate climates have rarely been studied over extended periods of winter wheat monoculture, and even less with spring sowings. Short- and medium-term investigations have revealed that with *Pseudocercosporella herpotrichoides* (Fron) Dei. severe eyespot levels are reached within 2 or 3 years following even long-term absence of cereals (3,6,8). Take-all infections caused by *Gaeumannomyces graminis* (Sacc.) von Arx & Olivier var. *tritici* Walker increased until between the third and seventh crops (3,4,12) and diminished subsequently. The reduction in take-all level after its peak was quantified by Glynne (5) at Rothamsted, where it was later considered in a more practical context with the phenomenon becoming known as take-all decline (13).

Long-term disease trends require intensive study to assess pathogen behavior in equilibrium with a newly established soil environment as influenced by monocereal cropping and by annual seasonal environment. In this study, eyespot and take-all trends are examined at two centers over an 18-year period with maximum standardization of growing conditions. At one of these centers, the influence of different crop husbandry practices maintained on the same plots over 14 years was studied, which enabled comparison of agronomic with seasonal variation in disease levels.

MATERIALS AND METHODS

Monoculture cropping of spring barley was carried out from 1965 to 1982 at two centers (Oak Park, County Carlow, and Clonroche, County Wexford) about 50 miles apart. The soil at Clonroche was a medium- to heavy-textured, well-drained, brown earth derived from Ordovician shale with surface texture a medium to heavy loam containing 25% clay. The Oak Park site had a light- to medium-textured, imperfectly drained, gley soil derived from limestone gravels with a light to medium surface horizon having 16% clay and a water table that rose close to the surface during periods of high rainfall. Whereas at Oak Park the barley sequence was preceded by 3 years of pasture, the cropping history at Clonroche was wheat in 1962, sugar beet in 1963, and wheat in 1964. Disease data over the 18-year period were obtained at each site from fourfold replicated plots, each 1/30 ha in area, in which crop management practices that might influence disease were standardized. Crops were sown each year with a similar type malting barley cultivar (Hunter or Emma) at 140 kg/ha between 15 March and 15 April. Annual and perennial weeds were controlled. Plots received 70 kg of N/ha annually, and fungicides were not used.

Agronomic variables, studied at Clonroche over a 14-year period (1965–1978) in which individual treatments were repeated on the same plots each year, consisted of undersowing the barley crop with ryegrass each spring; autumn sowing of mustard on rotavated stubble immediately following harvesting of the previous crop, and plowing it under in December; and control consisting of conventional spring plowing. Each husbandry treatment was carried out at 70 and at 35 kg of N/ha, and all plots received 28 kg of P/ha and 55 kg of K/ha per year. The experiment was laid down as six randomized treatments with fourfold replication.

Plots were sampled in the second week of July by taking eight 30-cm samples of corn drill row from each, using the method of Glynne (7) to ensure representative sampling. Disease diagnosis was confirmed initially by isolating the pathogens from a proportion of stems and roots. Leaf sheaths were removed and individual straws with their roots were pressure washed before grading in one of four categories of take-all severity: less than 25%, 25–50%, 50–75%, and more than 75% diseased roots. By allocating values of 1, 2, 3, and 4, respectively, to these categories, a disease index was calculated by expressing the aggregate as a percentage of the maximum possible grade. A similar disease index was calculated for eyespot based on two categories of infection severity: straws with lesions on up to or more than 50% of stem circumference were assigned values of 1 and 2, respectively. A statistical analysis of disease data as influenced by agronomic treatments at Clonroche was done for individual years as well as for the 14 seasons combined.

RESULTS

At the Clonroche site take-all reached a peak of 80% disease in 1966 and subsequently gave rise to a series of minor peaks and troughs (Fig. 1). The minor peaks that occurred in 1968, 1969, 1972, 1976, 1978, and 1982 had less than 40% disease. Fluctuations in eyespot followed take-all trends most years, with high levels concurring in 1968, 1969, 1972, 1976, and 1982 and troughs in 1967, 1970, 1975, and 1977. There was no agreement between disease levels in 1974 and 1981.

Take-all trends at Oak Park reached a peak with a little more than 40% disease in 1967 and 1968, and subsequent levels were uniformly low with the exceptions of 1972 and 1981 (Fig. 2). By contrast, eyespot levels were high and varied enormously between seasons, from more than 85% in 1972 to less than 5% in 1975. Along with concurrence of peaks in 1972 there was similarity between trends for low levels of the two diseases on this site from 1975 to 1980, with a subsequent increase.

Effect of agronomic treatments on take-all at Clonroche (Table 1) was significant in 1965, 1967, 1968, 1969, and 1974. Differential effect of cultural treatments on take-all levels was not pronounced, although it made the main contribution to significant differences in 1968 and 1969 (Fig. 3). Influence of nitrogen regime was also small, but it would have been the main contributor to significance in 1965 (Fig. 4). While the analysis of variance showed that treatment effect was not significant over all seasons, the effect of season was (P=0.05). Treatment by year interaction was very highly significant (P=0.001). However, seasonal variances differed markedly and χ^2 (=87.4) was significant for nonhomogeneous variances from year to year, having an invalidating effect on comparisons over years; this was also indicated by the error mean square values.

As with take-all, season also had a much more profound influence than agronomic treatment on eyespot (Figs. 4 and 5). Effect of treatment on eyespot level within season was highly significant in 1968 and 1969 and was also significant in 1965, 1971, 1972, and 1973 (Table 2). Whereas nitrogen did not influence eyespot consistently to any extent, Fig. 5 suggests that greater disease levels with mustard sown on rotavated stubble was the main contributor to significance in 1968, 1969, 1972, and 1973. The F value for treatment effect on eyespot over the years was not significant, whereas that for years and interaction of treatment by years (P=0.001) was very highly so. Again, while there was highly significant nonhomogeneous variance between years (χ^2=125), it is unlikely to affect the significance of the pronounced seasonal effects on eyespot levels.

DISCUSSION

The concurrence of take-all peaks at Clonroche and Oak Park (Figs. 1 and 2) could not be expected because the Clonroche barley sequence followed a tillage rotation containing two cereals in the previous 3 years and the Oak Park sequence was preceded by 3 years of grass. The peak in the second crop at Clonroche in 1966 was followed by a major decline in disease the following year, whereas the less pronounced peak at Oak Park was maintained for 2 years followed by a gradual decline in disease over the following three seasons. A further interesting phase in the take-all trend at Oak Park was the gradual disease diminution from 1974 almost to the point of extinction in 1979, only to increase subsequently in the 1980s. In fact, until 1979–1980, take-all curves at both sites suggest continuously diminishing disease levels from the take-all peak onwards. Hornby and Henden (9) found a similar

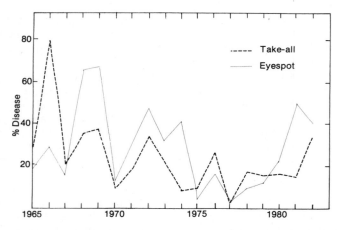

Fig. 1. Take-all and eyespot trends with spring barley monoculture over 18 seasons at Clonroche, County Wexford.

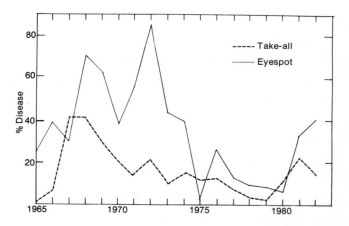

Fig. 2. Take-all and eyespot trends with spring barley monoculture over 18 seasons at Oak Park, County Carlow.

Table 1. Effect of agronomic treatment on mean percentage of take-all on continuous spring barley at Clonroche

	35 kg of N/ha			70 kg of N/ha			Variance error mean square[a]
Year	Ryegrass undersown	Mustard sown on rotavated stubble	Control	Ryegrass undersown	Mustard sown on rotavated stubble	Control	
1965	32.03	38.08	41.34	24.68	26.52	25.53	58.14*
1966	72.76	73.75	79.76	67.82	80.23	79.13	77.51
1967	20.74	24.69	24.75	12.74	19.46	14.45	16.99**
1968	21.52	34.39	28.52	19.69	35.37	19.44	59.93*
1969	24.99	50.33	40.37	16.05	37.43	29.78	91.81**
1970	10.38	15.03	11.57	7.99	13.98	9.71	17.35
1971	18.97	24.13	23.90	12.97	18.39	16.79	36.45
1972	44.06	44.60	38.66	38.22	34.11	42.04	93.24
1973	31.59	32.82	23.71	22.12	20.13	22.60	39.67
1974	11.05	13.00	5.98	5.43	8.53	6.25	11.73*
1975	5.41	6.50	11.11	11.41	9.34	18.39	160.05
1976	31.57	33.11	31.94	37.71	31.45	36.14	111.33
1977	1.95	2.66	4.20	1.51	1.28	2.71	2.14
1978	11.24	21.55	12.70	15.46	16.98	15.39	29.16
Mean	24.02	29.62	27.04	20.99	25.23	24.17	

[a]* = $P \leqslant 0.05$; ** = $P \leqslant 0.01$.

resurgence in take-all in soil infectivity bioassays in 1981 and 1982 in spring barley monoculture work in the east of England, which suggests a common, overriding seasonal influence on this disease even during the accepted "decline" phase.

It is obvious that take-all was more effectively buffered on the Oak Park site throughout the sequence. However, such suppressiveness was not confined to the period following the peak of the disease, and the effect seemed to prevail in the first 5 years of the sequence as well as later. There was more than 50% greater incidence of take-all on the Clonroche site over the 18-year period. Disease levels in the minor peaks during the "take-all decline" phase on the Clonroche site, although not particularly injurious to barley, would have caused appreciable yield losses if wheat were grown, which is both more susceptible (1,5,10) and less tolerant (2).

Although Oak Park had less take-all than Clonroche, it had about 20% more eyespot over the 18 years. This may be related to the fact that before 1975 crops tended to be more vigorous at Oak Park, which would favor sporulation and infection by the eyespot pathogen. Annual eyespot levels were much greater on both sites before 1975 than after. The similarity between eyespot trends at the two centers was very close in most years.

The chief features of this common pattern were a steep increase from 1967 to a plateau in 1968–1969, followed by a decline in 1970 and a further peak in 1972. A major trough occurred on both sites in 1975, followed by a small increase in 1976 and lower levels for a few years before increasing to moderately high levels in 1981 and 1982. Similar weather conditions during the months of April and May, which have a pronounced influence on this disease in spring cereals (*unpublished data*), probably account for such similarity because the sites do not differ appreciably in rainfall and temperature. The generally similar pattern of take-all and eyespot trends at the two centers also suggests an influence of common season weather variables on the two diseases, particularly after the early years of a cereal sequence.

The fact that there was no effect of agronomic treatment on take-all in 9 years out of 14 confirms the difficulty of obtaining a response from cultural treatments with this disease. Interaction between treatment and year is effectively illustrated by the high take-all levels with mustard sown on stubble relative to the other treatments in 1967–1970 by contrast with 1976, when the mustard-sown plots had least take-all (Fig. 3). An explanation for the relatively high take-all levels in the mustard-sown plots in 1968 and 1969 was found by

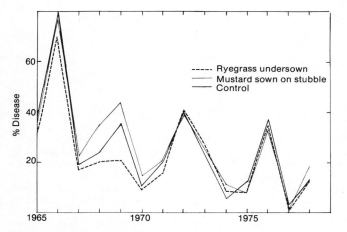

Fig. 3. Influence of cultural treatments on take-all levels in spring barley over a 14-year period at Clonroche, County Wexford.

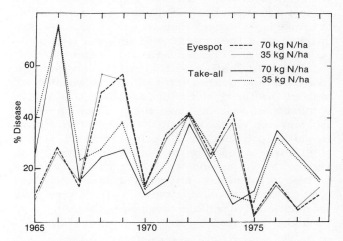

Fig. 4. Effect of level of applied nitrogen on the incidence of take-all and eyespot of spring barley over 14 seasons at Clonroche, County Wexford.

Table 2. Effect of agronomic treatment on mean percentage of eyespot on continuous spring barley at Clonroche

	35 kg of N/ha			70 kg of N/ha			Variance error mean square[a]
Year	Ryegrass undersown	Mustard sown on rotavated stubble	Control	Ryegrass undersown	Mustard sown on rotavated stubble	Control	
1965	16.93	14.96	17.11	17.63	19.18	21.44	5.63*
1966	23.49	26.36	29.31	27.45	29.08	29.57	60.02
1967	16.55	12.74	17.92	10.45	15.38	12.80	13.34
1968	47.71	54.43	46.59	49.60	65.21	56.08	41.98**
1969	55.97	67.14	46.66	46.69	67.09	49.39	26.39***
1970	14.45	12.54	13.59	14.58	12.32	15.08	6.30
1971	24.89	34.78	35.31	32.79	30.71	39.53	32.72*
1972	41.64	47.69	34.01	47.64	47.65	32.05	70.42*
1973	22.13	33.38	17.25	27.34	31.36	19.04	53.94*
1974	38.09	45.60	31.25	45.38	40.56	38.93	1.97
1975	1.77	2.56	2.15	3.03	3.59	2.38	1.06
1976	10.49	16.10	15.13	13.54	15.61	17.13	11.98
1977	5.96	4.31	5.43	5.06	2.80	5.19	5.25
1978	13.45	15.10	12.09	8.98	9.15	12.91	25.72
Mean	23.82	27.69	23.13	25.01	27.83	25.11	

[a]* = $P \leqslant 0.05$, ** = $P \leqslant 0.01$, and *** = $P \leqslant 0.001$.

observations that volunteer barley growth under the mustard canopy in the autumn was very severely infected by *G. graminis* var. *tritici*. This condition, which arose through harvesting losses of grain, probably gave rise to more inoculum in the following crop. An examination of roots for *Phialophora* spp. would have been revealing from 1966 to 1971 in view of more recent findings (11,14,15), since the undersown ryegrass treatment resulted in considerably less take-all during this period. This differential effect of undersown ryegrass disappeared after 1971.

The influence of higher nitrogen level is of interest because it reduced take-all consistently from 1967 to 1974 and increased it during the last 4 years of these sequences (Fig. 5). There was no effect of the higher nitrogen regime on disease during the peak in 1966, although Shipton (12) found that nitrogen reduced take-all in spring-sown wheat and barley only at the point of maximum incidence and severity.

Of the 8 years out of 14 when the effect of treatment on eyespot was not significant, 7 had less than 30% eyespot (average for year) and 6 had less than 20% infection. While this suggests less likelihood of significant treatment effects in low-infection seasons, the exception to this was the eyespot infection of less than 20% in 1965 when effect of treatment was significant. Whereas eyespot trends for all three cultural treatments follow the same basic pattern from year to year, the control deviated from the ryegrass and mustard treatments between 1968 and 1969 and between 1971 and 1972 (Fig. 4), thus contributing to the interaction between treatment and year. Of the three treatments, the mustard sown on stubble, which had the most eyespot in a number of seasons, gave the most uniform and vigorous crop over the years and was the only treatment free of the perennial weed couch (*Agropyron repens*). High levels of couch in the controls suppressed crop vigor in many seasons, which would have tended to reduce eyespot levels as would the suppressing effect of competition in undersown ryegrass plots. However, over the duration of the different sequences, the influence of the higher nitrogen level on eyespot was minimal.

SUMMARY

Spring barley was grown over an 18-year period on two sites 50 miles apart, and levels of take-all and eyespot caused respectively by *G. graminis* var. *tritici* and *P. herpotrichoides* were assessed each season. Annual disease levels for the 18 years are presented from replicated field plots where perennial weeds were controlled and standard cultural practices maintained. The influence of different agronomic treatments (undersown ryegrass, autumn-sown mustard on rotavated stubble, nitrogen level) on the same plots each year on disease incidence was also examined over a 14-year period on one site.

Eyespot seasonal fluctuations were similar on the two sites and were predominantly influenced by weather. The take-all peak occurred early in the sequence but did not occur in the same season at both centers, and it was greater on one site than on the other; this was followed by a series of minor peaks and troughs that tended to concur. Take-all level on the site that had the most pronounced disease peak also fluctuated most during the "decline" phase. Eyespot and take-all trends concurred to a degree suggesting the effects of seasonal variables common to both diseases. Effect of agronomic treatment over 14 years was not significant for either disease. Effect of season and of interaction between treatment and season was highly significant for both diseases. Over the 14 seasons, treatment differences were significant in 5 years for take-all and 6 for eyespot.

ACKNOWLEDGMENTS

I wish to thank A. Guinness, Son & Co. (Dublin) Ltd. for partial financial support for this project over 18 years; A. Hegarty, Statistics Department, An Foras Taluntais, for statistical analyses; and A. Shannon, C. Sheehy, and P. Drea for technical assistance.

LITERATURE CITED

1. Chambers, S. C. 1962. Root diseases in wheat on clover ley—factors under investigation. II. Relative susceptibility of wheat and barley. W. Aust. Dep. Agric. J. 3:521-522.
2. Cunningham, P. C. 1967. Disease incidence and yield reduction of wheat and barley following artificial soil infestation with *Ophiobolus graminis*. Proc. Br. Insectic. Fungic. Conf., 4th, 1:85-91.
3. Cunningham, P. C. 1983. Effects of different spring cereal sequences on soil-borne pathogen and grain yields. Ir. J. Agric. Res. 22:225-243.
4. Etheridge, J. 1967. Eyespot and take-all on Broadbalk. Rep. Rothamsted Exp. Stn. 1966:131.
5. Glynne, M. D. 1935. Incidence of take-all on wheat and barley on experimental plots at Woburn. Ann. Appl. Biol. 22:225-235.
6. Glynne, M. D. 1942. *Cercosporella herpotrichoides* Fron, causing eyespot of wheat in Great Britain. Ann. Appl. Biol. 24:254-264.
7. Glynne, M. D. 1951. Effects of cultural treatments on wheat and on the incidence of eyespot, lodging, take-all and weeds. Field experiments 1945-48. Ann. Appl. Biol. 38:665-688.
8. Glynne, M. D., and Moore, F. J. 1949. Effect of previous crops on the incidence of eyespot on winter wheat. Ann. Appl. Biol. 36:341-351.
9. Hornby, D., and Henden, D. R. 1983. A secondary outbreak of take-all. Rothamsted Rep. 1982 (Part 1):197-198.
10. Jensen, H. P., and Jorgensen, J. H. 1973. Reactions of five cereal species to the take-all fungus (*Gaeumannomyces graminis*) in the field. Phytopathol. Z. 78:193-203.
11. Scott, P. R. 1970. *Phialophora radicicola*, an avirulent parasite on wheat and grass roots. Trans. Br. Mycol. Soc. 55:163-167.
12. Shipton, P. J. 1972. Take-all in spring-sown cereals under continuous cultivation: Disease progress and decline in

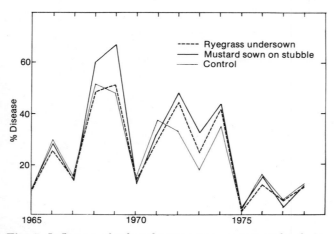

Fig. 5. Influence of cultural treatments on eyespot levels in spring barley over a 14-year period at Clonroche, County Wexford.

relation to crop succession and nitrogen. Ann. Appl. Biol. 71:33-46.

13. Slope, D. B., and Cox, J. 1964. Continuous wheat growing and the decline of take-all. Rep. Rothamsted Exp. Stn. 1963:108.

14. Slope, D. B., Salt, G. A., Broom, E. W., and Gutteridge, R. J. 1978. Occurrence of *Phialophora radicicola* var. *graminicola* and *Gaeumannomyces graminis* var. *tritici* on roots of wheat in field crops. Ann. Appl. Biol. 88:239-246.

15. Wong, P. T. W. 1981. Biological control by cross-protection. Pages 343-352 in: Biology and Control of Take-all. M. J. C. Asher and P. J. Shipton, eds. Academic Press, London.

Mycophagous Amoebas from Arable, Pasture, and Forest Soils

S. CHAKRABORTY, Department of Plant Pathology, Waite Agriculture Research Institute, Glen Osmond, South Australia, 5064, and K. M. OLD, Division of Forest Research, CSIRO, Canberra, Australian Capital Territory, 2600

The ability of some genera of soil amoebas to feed on fungal cells has been recognized for many years. However, with few exceptions (8), there were few systematic studies of this phenomenon until Old (11) and Anderson and Patrick (1) showed that amoebas caused perforation of fungal spores in soil.

Particular attention has been paid to the genera *Arachnula*, *Vampyrella*, and *Theratromyxa* belonging to the Proteomyxida in the Honigberg et al classification (10) of the protozoa (2,9,13,15). Other genera of interest include *Thecamoeba* (16), *Cashia* (17), and *Mayorella* (2).

This paper summarizes recent research that demonstrated mycophagy by *Gephyramoeba* spp., an unidentified leptomyxid closely resembling *Leptomyxa flabellata*, *Saccamoeba*, and *Mayorella* and confirms the mycophagy of *Thecamoeba granifera* subsp. *minor*. We also compare feeding mechanisms of these genera with those of *Arachnula* and *Theratromyxa*. Amoeba morphology and feeding are described in detail in Old (12), Old and Oros (15), Chakraborty and Old (3), and Chakraborty et al (4).

THE AMOEBAS AND THEIR FEEDING MECHANISMS

Table 1 lists the genera that we have tested for mycophagy with three plant-pathogenic fungi. *A. impatiens*, *Theratromyxa* sp., and *Mayorella* sp. have been isolated from many forest soils and several pasture and arable soils in southern Australia. Other species were obtained from a permanent pasture soil shown to suppress *Gaeumannomyces graminis* var. *tritici* (19).

Feeding trials were conducted in vitro using cloned amoebal cultures, fungal cells, and bacteria in water films a few millimeters in depth. Prolonged observation of cultures by phase contrast light microscopy was followed by scanning electron microscopy (SEM) of coverslips bearing fungal structures and amoebas.

Arachnula impatiens. The life cycle and feeding mechanisms of *A. impatiens* have been detailed elsewhere (14). Trophozoites partly or wholly engulf propagules and excise disk-shaped portions of cell walls. The amoeba then enters the cell lumen and ingests the protoplast. Cell debris extruded from spore lumina is further digested within digestive cysts. Several spores can be attacked simultaneously, and under favorable conditions a single trophozoite can grow to more than 1,000 μm in length.

Theratromyxa sp. Trophozoites of this genus are of a more determinate size than those of *A. impatiens* (about 100–150 μm in length). They move by advancing crescent-shaped fronts of filipodia followed by the rest of the cell. Fine filipodia make contact with the spore, and the trophozoite flows around it. Sometimes the amoeba moves on without further interaction with the fungus, but it may form a spheroidal mass around the spore and form a digestive cyst. Conidia of *Cochliobolus sativus* initially with one to five septa become aseptate, and the protoplast is completely disrupted. Plasmotomy occurs within the cyst, and up to five daughter trophozoites emerge from minute holes in the cyst wall. The only evidence for mycophagy for this amoeba has been obtained using *C. sativus* (2,15). From engulfment to evacuation of the cyst can take 2 days, compared with the 4–6 hr required by *A. impatiens* to lyse cells. Conidia lysed by *Theratromyxa* sp. bear many fine perforations approximately 0.5 μm in diameter. Usually a single conidium is attacked, but rarely two spores have been found in a single digestive cyst.

Gephyramoeba sp. This amoeba resembles *G. deliculata* Goodey (18) except that trophozoites are multinucleate instead of uninucleate. Trophozoites can attain a size of 80–750 × 50–130 μm in liquid culture by anastomoses. Mycophagy has been recorded for conidia

Table 1. Mycophagy of 11 soil amoebas on three plant-pathogenic fungi

Amoeba	Fungus[a]	Mycophagy	Perforation[b]
Proteomyxida			
Arachnula impatiens	Ggt	+	+
	Cs	+	+
	Pc	+	+
Theratromyxa sp.	Cs	+	+
Gephyramoeba sp.	Cs	+	+
	Pc	+	+
Unidentified leptomyxid	Ggt	+	+
	Cs	−	−
	Pc	+	+
Unidentified vampyrellid	Ggt	−	−
	Cs	−	−
	Pc	−	−
Amoebida			
Acanthamoeba sp.	Ggt	−	−
	Cs	−	−
	Pc	NT[c]	−
Mayorella spp.	Ggt	+	−
	Cs	+	−
Platyamoeba sp.	Ggt	−	−
	Cs	−	−
Saccamoeba sp.	Ggt	+	−
	Cs	+	+
Thecamoeba granifera subsp. *minor*	Cs	+	+
Thecamoeba spp.	Ggt	−	−
	Cs	−	−

[a]Ggt = *Gaeumannomyces graminis* var. *tritici*, Cs = *Cochliobolus sativus*, and Pc = *Phytophthora cinnamomi*.
[b]As observed by scanning electron microscopy.
[c]Not tested.

of *C. sativus* and chlamydospores of *Phytophthora cinnamomi*. Like *A. impatiens*, *Gephyramoeba* sp. can attack several fungal propagules simultaneously and penetrate spore walls locally within areas engulfed by the trophozoite. The trophozoite then flows in through the perforation and spore contents are lysed, either partially or completely. SEM observations showed perforations 1.0–3.0 μm in diameter, but the annular depressions characteristic of spores attacked by *A. impatiens* were absent.

Unidentified leptomyxid. The leptomyxid described by Chakraborty and Old (3) resembles *Leptomyxa flabellata* Goodey in most respects. The feeding mechanisms of this amoeba are remarkable in that they appear to depend on the morphology and dimensions of the host propagule. Table 1 shows that *P. cinnamomi* and hyaline hyphae of *G. graminis* var. *tritici* were parasitized, but the pigmented fungus *C. sativus* was unaffected. When feeding on intact thalli of either fungus, the amoeba surrounds segments of hyphae, which lose their protoplasts after 30–60 min. This rapid lysis destroys large numbers of fungal cells in a few days' incubation. SEM examination revealed many small perforations 0.5–1.5 μm in diameter in hyphal walls. Hyphal fragments, or separated chlamydospores of *P. cinnamomi* incubated with the fungus, are lysed in a different manner. The amoeba completely surrounds a propagule and forms a large digestive vacuole. Over a period of 16–17 hr, chlamydospores contained in these vacuoles are reduced to formless debris.

Saccamoeba **sp.** Members of this genus, together with *Mayorella* spp. and *Thecamoeba* spp., lack the large branched or reticulate trophozoites of the genera discussed above. Their taxonomy is better understood, and, as part of the order Amoebida, *Saccamoeba* spp. are more typical of soil amoebas isolated by conventional means from soils. Trophozoites of *Saccamoeba* are limax; i.e., they are more or less cylindric and move by advancing a single pseudopodium. When feeding on fungal hyphae, trophozoites form a spheroidal mass and physically bend and twist fungal cells to include them within the amoebal cell. Further details of the feeding mechanism need fine structural studies; however, after about 1 hr, segments of the hyphae are empty and the amoeba moves on. Hyphal fragments may be retained longer in the amoebal cell to be egested as cell debris. Mycophagy has been recorded for hyphae of *G. graminis* var. *tritici* and conidia of *C. sativus*.

Thecamoeba granifera **subsp.** *minor.* Feeding by this amoeba on *Fusarium oxysporum* Schlecht. emend. Snyder and Hansen was described by Pussard et al (17), and hyaline hyphae of *G. graminis* var. *tritici* were attacked in an identical fashion (4). Conidia of *C. sativus* were also perforated and lysed, and many conidia were broken in half as well as perforated by the amoeba.

Mayorella **spp.** Anderson and Patrick (2) suggested that members of *Mayorella* can be mycophagous as well as parasitic on other amoebae such as *V. lateritia*. We confirmed mycophagy by *Mayorella* on *G. graminis* var. *tritici* and *C. sativus*. In axenic culture with *C. sativus*, trophozoites moved over the substratum and accumulated as many as five conidia per cell. Conidia were often egested in an apparently viable condition, but when feeding occurred, large vacuoles and those described above for the unidentified leptomyxid

containing one or more conidia formed with the trophozoite. Conidia within the vacuole became aseptate and were finally egested by the trophozoite (4). So far, perforations have not been seen in conidia suspended in *Mayorella* cultures.

DISCUSSION

It is becoming increasingly clear that many soil amoebas are potentially mycophagous. Mycophagy of filamentous fungi is shared by at least 13 genera, allowing for possible synonymy between *A. impatiens* and *V. lateritia* (6). Mycophagous genera include *Arachnula*, *Vampyrella*, *Theratromyxa*, *Gephyramoeba*, *Leptomyxa*, *Thecamoeba*, *Cashia*, *Mayorella*, *Saccamoeba*, *Acanthamoeba* (8), *Hartmannella* (C. Palzer, *personal communication*), and the testate amoebas, *Geococcus* and *Arcella* (7). As can be seen in Table 1, mycophagy by one species does not imply a genus-wide property. *Thecamoeba* spp. did not feed on fungi utilized by *T. granifera* subsp. *minor*. Nevertheless, of nine amoebal species selected for feeding trials on the basis of their common isolation from a pasture soil suppressive to take-all disease of wheat, five were found to be mycophagous (4).

Some amoebas, e.g. *A. impatiens*, have a very wide range of prey including bacteria, flagellates, algae, and nematodes in addition to fungi (13), whereas *T. granifera* subsp. *minor* and *Cashia mycophaga* were reported to be obligate mycophages (16,17).

Besides the variations among mycophagous amoebas in feeding mechanisms, fungal propagules also vary in their susceptibility to perforation and lysis, both between and within fungal species. Old and Darbyshire (13) commented on the apparent inability of *A. impatiens* to perforate spores of several species. For example, ascospores of *Sordaria fimicola* (Roberge) Cesati and Notaris, conidia of *Aspergillus niger* van Tiegh, and zygospores of *Zygorrhynchus heterogamus* (*unpublished data*) were unaffected by the amoeba. This may be due to the presence of enzyme-resistant components in the spore walls and, in the case of the hydrophobic conidia of *Aspergillus niger* and *Penicillium frequentans* Westling (8), the inability of amoebas to engulf "dry" spores. In the work discussed here the unidentified leptomyxid perforated cell walls of *P. cinnamomi* and hyaline hyphae of *G. graminis* var. *tritici* but failed to damage *C. sativus* spores even though these were transported within the trophozoite for several hours (3).

From the limited evidence available it appears that many mycophagous genera are able to utilize a range of different fungal species and propagule types, and some are obligate fungal feeders. They are virtually worldwide in distribution, occurring in native and plantation forest ecosystems, pasture, and arable lands. As such they cannot fail to have some impact on soil fungus populations, and the ability of fungal propagules to resist or escape attack may be of significant ecological importance.

The growing body of knowledge of the occurrence, identity, and activity of mycophagous amoebas in soils supports the contention that the phenomenon is of more than academic interest. Chakraborty and Warcup (5) have proposed a possible role for these organisms in the suppression of take-all in a permanent pasture soil.

They also showed that soil populations of mycophagous amoebas can be markedly affected by cropping practices, and it is conceivable that the beneficial activities of these organisms could be increased. Further developments in this field will require studies of soil population dynamics, detailed investigations of feeding mechanisms and of the interactions between the amoebas and their prey, and the resolution of taxonomic relationships in the Proteomyxida.

LITERATURE CITED

1. Anderson, T. R., and Patrick, Z. A. 1978. Mycophagous amoeboid organisms from soil that perforate spores of *Thielaviopsis basicola* and *Cochliobolus sativus*. Phytopathology 68:1618-1626.
2. Anderson, T. R., and Patrick, Z. A. 1980. Soil vampyrellid amoebae that cause small perforations in conidia of *Cochliobolus sativus*. Soil Biol. Biochem. 12:159-167.
3. Chakraborty, S., and Old, K. M. 1982. Mycophagous soil amoeba: Interactions with three plant pathogenic fungi. Soil Biol. Biochem. 14:247-255.
4. Chakraborty, S., Old, K. M., and Warcup, J. 1983. Amoebae from a take-all suppressive soil which feed on *Gaeumannomyces graminis tritici* and other soil fungi. Soil Biol. Biochem. 15:17-24.
5. Chakraborty, S., and Warcup, J. 1983. Soil amoebae and saprophytic survival of *Gaeumannomyces graminis tritici* in a suppressive pasture soil. Soil Biol. Biochem. 15:181-185.
6. Dobell, C. 1913. Observations on the life history of Cienkowski's '*Arachnula*'. Arch. Protistenkd. 31:317-353.
7. Drechsler, C. 1936. A *Fusarium*-like species of *Dactylella* capturing and consuming testaceous rhizopods. J. Wash. Acad. Sci. 26:397-404.
8. Heal, O. W. 1963. Soil fungi as food for amoebae. Pages 289-297 in: Soil Organisms. J. Doeksen and J. van der Drift, eds. North-Holland, Amsterdam.
9. Homma, Y., Sitton, J. W., Cook, R. J., and Old, K. M. 1979. Perforation and destruction of pigmented hyphae of *Gaeumannomyces graminis* by vampyrellid amoebae from Pacific Northwest wheat field soils. Phytopathology 69:1118-1122.
10. Honigberg, B. M., Balamuth, W., Bovee, E. C., Corliss, J. O., Godjics, M., Hall, R. P., Kudo, R. R., Levine, N. D., Loeblich, A. R., Weiser, J., and Weinrich, D. M. 1964. A revised classification of the phylum Protozoa. J. Protozool. 11:7-20.
11. Old, K. M. 1977. Giant soil amoebae cause perforation and lysis of spores of *Cochliobolus sativus*. Trans. Br. Mycol. Soc. 68:277-281.
12. Old, K. M. 1978. Fine structure of perforation of *Cochliobolus sativus* conidia by giant amoebae. Soil Biol. Biochem. 10:509-516.
13. Old, K. M., and Darbyshire, J. F. 1978. Soil fungi as food for giant amoebae. Soil Biol. Biochem. 19:93-100.
14. Old, K. M., and Darbyshire, J. F. 1980. *Arachnula impatiens* Cienk., a mycophagous giant amoeba from soil. Protistologica 16:277-287.
15. Old, K. M., and Oros, J. M. 1980. Mycophagous amoebae in Australian forest soils. Soil Biol. Biochem. 12:169-175.
16. Pussard, M., Alabouvette, C., Lamaitre, I., and Pons, R. 1980. Une nouvelle amibe mycophage endogee *Cashia mycophaga* n. sp. (Hartmannellidae, Amoebida). Protistologica 16:433-451.
17. Pussard, M., Alabouvette, C., and Pons, R. 1979. Etude preliminaire d'une amibe mycophage *Thecamoeba granifera* ssp. *minor* (Thecamoebidae, Amoebida). Protistologica 15:139-149.
18. Pussard, M., and Pons, R. 1976. Etude des genres *Leptomyxa* et *Gephyramoeba* (Protozoa, Sarcodina). III. *Gephyramoeba delicatula* Goodey, 1915. Protistologica 12:351-383.
19. Wildermuth, G. B. 1980. Suppression of take-all by some Australian soils. Aust. J. Agric. Res. 31:251-258.

Northern Poor Root Syndrome of Sugarcane in Australia

B. J. CROFT, K. J. CHANDLER, R. C. MAGAREY, and C. C. RYAN, Bureau of Sugar Experiment Stations, Indooroopilly, Queensland, 4068, Australia

In the tropical high-rainfall regions of North Queensland, especially in the Innisfail and Babinda districts, a condition of sugarcane (*Saccharum* interspecific hybrid) known as northern poor root syndrome (NPRS) is associated with serious crop losses. Many primary roots in affected root systems develop a soft flaccid rot of the cortex. Both primary and secondary roots show lesions, general discoloration, and restricted development, and the finer root system is poorly developed. Plant symptoms are premature wilting, unthrifty growth, and a tendency for ratoon stools to tip from the ground, thus reducing the number of subsequent stools.

Soil treatments with methyl bromide, dazomet, and ethylene dibromide (EDB) greatly improve growth, indicating major biological components in the case of NPRS (Bureau of Sugar Experiment Stations, *unpublished*). *Pythium graminicola* Subra. (1,4,10) and *P. arrhenomanes* Drechs. (5–7,11) have caused root diseases of sugarcane in other countries. No study of soil fungi associated with sugarcane roots in Australia has been reported.

Very high populations of parasitic nematodes have been recorded at all sites showing NPRS (K. J. Chandler, *unpublished*). The most prevalent nematode is *Pratylenchus zeae* Graham, which has been described as a pathogen of sugarcane, alone or in association with *Pythium graminicola* (9). Treatments with nonvolatile nematicides, including aldicarb, reduce nematode populations but improve growth only slightly (3) and do not seem to reduce the incidence of root rot seen in NPRS.

Soils in NPRS-affected areas are highly leached, with low pH and low levels of calcium and magnesium (8). However, the role of these soil factors in NPRS is not known.

This paper reports the results of investigations into some soil fungi and nematodes.

MATERIALS AND METHODS

Fungal isolations. Roots showing a range of NPRS symptoms were washed for 2 hr in running tap water, cut into 5-mm sections, surface-disinfected for 1 min in 50% ethanol, washed twice in sterile distilled water, and dried on sterile filter paper. The dried root pieces were placed on potato agar plus 500 ppm penicillin, 50 ppm pimaricin, and 30 ppm polymyxin B sulfate (PA + 3P) and on potato agar plus 100 ppm PCNB, 50 ppm streptomycin, and 60 ppm rose bengal (PA + PSR). In the rotten primary roots, many spherical spores with large, blunt projections were observed under the microscope, and clumps of these spores were transferred to PA + 3P medium.

Pathogenicity tests of fungi. Inoculum was grown for 2–3 weeks on double-autoclaved sand plus potato-dextrose broth (6:1 w/v) at 28°C in the dark. Twenty grams of inoculum was thoroughly mixed with 1.46 kg (oven dry weight) of Mourilyan sand, which had been steam-pasteurized for 30 min at greater than 70°C. The soil was placed in terra cotta pots into which pregerminated sets of the cultivar Q90 were planted. Plants were maintained in a glasshouse at constant 28°C and 80–90% relative humidity. Uninoculated controls were included in each test. After 8 weeks, the dry weight of all tops and roots was determined. Before drying, root systems were rated for percentage rotten shoot roots and fine root growth relative to the uninoculated control, which was arbitrarily rated as 10.

Nematicide field trials. Five replicated field trials were established, including plots treated with EDB injected to the subsoil at 30-cm centers (133–190 kg a.i./ha), plots treated with fenamiphos surface-sprayed and lightly incorporated with scuffler tines (9.5–17.5 kg a.i./ha), and plots treated with combinations of these two chemicals. Plots were 40 m long, with six rows per plot. All treatments were applied to fallow ground before planting.

Soil and root samples were taken between 5 and 20 weeks after planting. Soil samples were collected with a 3.5-cm auger, with six subsamples per plot. Each subsample was taken close to plant bases. Roots from three separate plants were taken for nematode extraction. Extractions were made using a modified Baermann tray procedure (2).

RESULTS

Fungal isolations. Many fungi were isolated from NPRS-affected roots, including species of *Pythium, Curvularia, Trichoderma, Penicillium, Cunninghamella, Fusarium*, and many unidentified species. Two species were found to be pathogenic. Representative isolates of one species show characteristics in common with both *Pythium graminicola* and *P. arrhenomanes*. As suggested by Hendrix and Campbell (4), we regard these isolates as belonging to the species group *P. graminicola-P. arrhenomanes*. The other pathogenic species is an unidentified oomycete.

Pathogenicity tests. The results of the pathogenicity test shown in Table 1 illustrate the effect of two isolates each of *P. graminicola-P. arrhenomanes* and the unidentified oomycete on inoculated plants. This test also included one treatment where isolates of the two fungi were inoculated concurrently. Both isolates of *P.*

graminicola-P. arrhenomanes significantly reduced the dry weight of tops and roots and the fine root rating ($P < 0.01$). The root system was discolored with stubbing of the fine roots, and red-black lesions $1-4 \times 0.2-1$ mm were present on the primary and secondary roots.

The two isolates of the unidentified oomycete did not significantly reduce top growth but significantly reduced the dry weight of roots and the fine root rating and significantly increased the percentage of rotten shoot roots ($P < 0.01$). The rotten roots were packed with distinctive ornamented spores which were typical of those seen in rotten roots in the field. These fungi did not appear to directly attack the finer roots, and the reduction in fine root rating was attributed to the rotting of the primary roots.

The reduction in growth of tops and roots was no greater in plants inoculated with both species than in plants inoculated with *P. graminicola-P. arrhenomanes* alone. Significantly fewer roots were rotted in plants inoculated with both species than in plants inoculated with the unidentified oomycete alone. All fungi were reisolated from their respective treatments.

The unidentified oomycete has smooth-walled, hyaline mycelium $2.0-9.0$ μm (mostly $3.0-6.0$ μm) in diameter. Filamentous, inflated, sporangium-like structures are formed when cultures are immersed in water, but no zoospore or vesicle production has been observed. The fungus is homothallic, with oospores $27.9-46.5$ μm in diameter. The oospores are thick-walled ($2.5-5.5$ μm) and have large, blunt projections ($4.5-9.5$ μm) on their outer wall. These oospores are produced abundantly on cornmeal agar. Antheridia are both monoclinous and hypogynous.

Nematode populations. At all the experimental sites, the dominant nematode species seems to be *Pratylenchus zeae* Graham, with smaller and less consistent occurrences of *Helicotylenchus dihystera* (Cobb), *Meloidogyne javanica* (Treub) Chitwood, *Radopholus williamsi* Siddiqi, *Paratrichodorus minor* (Colbran), *Macroposthonia* sp., and *Xiphinema* sp. and isolated occurrences of *Radopholus similis* (Cobb), *Tylenchorhynchus martini* Fielding, and *Rotylenchulus* sp.

No nematodes could be detected in any of the EDB/fenamiphos combination treatments 8 weeks after planting, whereas moderate populations were present in control plots. In general, EDB gave the best nematode control deeper in the rhizosphere, whereas fenamiphos provided good control only in the surface layers. Combination treatments at the higher rates kept nematode populations below detectable levels for longer than 16 weeks.

Symptoms of soft, flaccid root rot were always as prevalent in the absence of nematodes as in the control plots. All nematicide treatments, especially those containing EDB, slightly improved fine root development and visibly improved stalk growth.

DISCUSSION

Two fungi isolated from NPRS-affected roots have been shown to be pathogenic to sugarcane in this investigation. *P. graminicola-P. arrhenomanes*, which has been reported as a pathogen of sugarcane in many countries, has now been reliably reported for the first time as a pathogen of sugarcane in Australia. In pot

Table 1. Effect of two isolates of *Pythium graminicola - P. arrhenomanes* and two isolates of an unidentified oomycete on shoot and root growth of young sugarcane plants in pots

Treatment	Dry weight of tops (g)[a]	Dry weight of roots (g)	Rotten shoot roots (%)	Fine root rating[b]
Control	4.01	3.65	1.2 (2.7)[c]	10.0
P. graminicola-P. arrhenomanes, isolate T10-1F	1.33	1.18	3.3 (4.4)	2.3
P. graminicola-P. arrhenomanes, isolate T22-1A	1.23	1.07	1.2 (3.5)	2.1
Unidentified oomycete, isolate T5-2A	3.84	1.77	62.4 (52.9)	4.3
Unidentified oomycete, isolate T17-1I	4.01	2.27	79.2 (64.0)	6.3
Isolate T10 + T17-1I	1.79	1.36	31.2 (33.5)	2.2
LSD: $P < 0.05$	0.71	0.74	12.8 (10.9)	1.7
$P < 0.01$	0.96	1.00	17.3 (14.7)	2.3

[a] All values are the mean of six replicates.
[b] Fine root rating was relative to the control, which was arbitrarily rated as 10.
[c] Values in parentheses are arcsin transformations of the percentage of rotten shoot roots.

tests, *P. graminicola-P. arrhenomanes* mainly attacks the fine roots. The extent to which this fungus occurs on NPRS-affected farms has not been determined because of the difficulty of diagnosing its presence through symptoms alone, and further studies will be necessary to elucidate its role in NPRS.

The other fungus is as yet unidentified but has some characteristics in common with the genus *Pythium*. The soft rot of the primary roots caused by this fungus occurs in all severely affected NPRS fields.

The number of saprophytic and parasitic nematode species found in NPRS-affected soil and the responses obtained with EDB and fenamiphos may suggest nematode involvement in NPRS. Although fungi and nematodes are suspected to be important factors in NPRS, their contributions to yield reduction have not been clearly demonstrated. Pathogenicity tests and nematode field trials suggest that the unidentified oomycete does not require the presence of nematodes to infect roots. *P. graminicola-P. arrhenomanes* severely debilitates sugarcane root systems in pot trials in the absence of nematodes.

Research is continuing in relation to both biological and agronomic aspects of the problem.

LITERATURE CITED

1. Adair, C. 1972. Ecological relationships of *Pythium* species in the rhizosphere of sugarcane and possible significance in yield decline. Hawaii. Plant. Rec. 58(17):213-239.
2. Chandler, K. J. 1978. Non-volatile nematicides: An initial assessment in North Queensland sugarcane fields. Proc. Aust. Soc. Sugar Cane Technol. 1978 Conf., pp. 85-91.
3. Chandler, K. J. 1980. Continued experiments with non-volatile nematicides in North Queensland sugarcane fields. Proc. Aust. Soc. Sugar Cane Technol. 1980 Conf., pp. 75-82.
4. Hendrix, F. F., and Campbell, W. A. 1973. *Pythium* as plant pathogens. Annu. Rev. Phytopathol. 11:77-98.
5. Lii-Jang, L. 1980. Pythium root rot of sugarcane in Puerto Rico. 1. Pathogenicity and identification. Puerto Rico J. Agric. 64(1):54-62.
6. Rands, R. D. 1961. Root rot. Pages 287-304 in: Sugar Cane

Diseases of the World. Vol. 1. J. P. Martin, E. V. Abbott, and C. G. Hughes, eds. Elsevier, Amsterdam.

7. Rands, R. D., and Dopp, E. 1938. Pythium root rot of sugarcane. U.S. Dep. Agric. Tech. Bull. 666.

8. Ridge, D. R., Hurney, A. P., and Haysom, M. B. C. 1980. Some aspects of calcium and magnesium nutrition in North Queensland. Proc. Aust. Soc. Sugar Cane Technol. 1980 Conf., pp. 55-61.

9. Santo, G. S., and Holtzmann, O. V. 1970. Interrelationships of *Pratylenchus zeae* and *Pythium graminicola* on sugarcane. Phytopathology 60:1537.

10. Srinivasan, K. V. 1968. The role of the rhizosphere microflora in the resistance of sugarcane in Pythium root rot. Proc. 13th Congr. ISSCT, pp. 1224-1236. Taiwan.

11. Watanabe, T. 1974. Fungi isolated from the underground parts of sugarcane in relation to the poor ratooning in Taiwan. 2. *Pythium* and *Phytiogeton*. Trans. Mycol. Soc. Jpn. 15:343-357.

Effects of Soil Insects on Populations and Germination of Fungal Propagules

E. A. CURL and R. T. GUDAUSKAS, Department of Botany, Plant Pathology, and Microbiology; J. D. HARPER, Department of Zoology-Entomology; and C. M. PETERSON, Department of Botany, Plant Pathology, and Microbiology, Auburn University, Auburn, AL 36849, U.S.A.

Interactions between components of the soil microflora and the microfauna (excluding nematodes) have been largely neglected in studies relating to the ecology of soilborne plant pathogens. The microarthropods, comprised principally of the Collembola (springtails) and the Acarina (mites), are the most abundant of the soil-inhabiting animals exclusive of nematodes and protozoa (7).

The order Collembola consists of minute animals, usually 1–3 mm long, in the subclass Apterygota. The collembolan suborder Arthropleona is well represented in soil below the surface layer and, like the microflora, is generally more abundant in close proximity to roots (11). Many species are predominantly mycophagous (9). Wiggins and Curl (10) suggested that the feeding activities of these insects might affect the competitive advantage of a pathogen at the root surface and influence disease incidence or severity. Subsequently, a suppressive effect of mycophagous Collembola on *Rhizoctonia solani* Kuehn and cotton-seedling disease was reported (5). An overview of the potential role of mycophagous insects in rhizosphere ecology relating to root disease has been published (6).

The present investigation was designed to obtain more specific information on the destructive feeding capacity of two collembolan species in relation to inoculum density of selected pathogenic fungi and the biological control agent *Trichoderma harzianum* Rifai.

MATERIALS AND METHODS

The collembolan species, *Proisotoma minuta* Tullberg (Isotomidae) (Fig. 1) and *Onychiurus encarpatus* Denis (Poduridae), used in these studies were extracted from soil close to the rhizosphere of various crop plants by the modified Tullgren extraction system described by Wiggins and Curl (10). The insects were further multiplied and maintained in deep petri dishes on a substrate of plaster of Paris and charcoal (9:1) with bakers' yeast as a food source. The two species were always used in mixtures, with *P. minuta* comprising about 90% of the population.

Effect on mycelia. An assessment was made of the capacity of the insects to restrict mycelial growth of six soil fungi on the surface of Czapek's agar within a 24-hr period. Mycelial disks (7 mm in diameter) from young potato-dextrose agar (PDA) cultures of *Rhizoctonia solani*, *Fusarium oxysporum* Schlect. f. sp. *vasinfectum* (Atk.) Snyder and Hansen, *Sclerotium rolfsii* Sacc., *Macrophomina phaseolina* (Tassi) Goid., *Verticillium*

dahliae Kleb., and *Trichoderma harzianum* were applied to the agar surface 12 mm from the petri dish edge. In dishes that were to receive insects, a disk of sterile PDA was also placed near the opposite edge. Approximately 200 insects per dish were added immediately to these dishes; other dishes were not infested. These cultures were prepared in triplicate.

After 24 hr, mycelial extension in the presence and absence of insects and the insect distribution pattern or "attitude" in relation to food source were determined by microscopic examination.

Insect attraction to and effect on spores. Before soil studies, tests were conducted to observe spore-insect relationships on a water-agar surface. Chlamydospores of *F. oxysporum* f. sp. *vasinfectum* were produced in a soil extract liquid medium and separated from the mycelium by sonification (sonicator cell disrupter, model W. 185F, Heat Systems Ultrasonic, Inc.) for 2 min at power control setting 4, followed by ultrafine screening. Conidia of *T. harzianum* were obtained by flooding PDA cultures with sterile water. Spores were washed on 0.45-μm Millipore filters and concentrations adjusted by hemacytometer counts to about 1×10^5 spores per milliliter for *Fusarium* and 2×10^6/ml for *Trichoderma*.

Four 13-mm-diameter circles were drawn equidistant on the bottoms of 15 petri dishes containing water agar. A single drop of spore suspension was placed within two of the circles on the agar surface of 10 dishes, and a drop of spore-wash water was placed within the other two circles. Approximately 150 Collembola were placed in the center of each of these dishes. For the remaining five dishes, all four circles received a spore drop but no insects were added. After 12 hr at 27°C, the number of Collembola per circle, the number of spores per microscopic field, and the percentage of germination of spores were recorded.

Spores in soil. Suspensions of *F. oxysporum* f. sp. *vasinfectum* chlamydospores and *T. harzianum* conidia were prepared as described above. One milliliter was applied to 25-mm Millipore membrane filters mounted over vacuum, leaving about 10^5 *Fusarium* spores or 10^6 *Trichoderma* spores on each filter. These filters were buried in either sterilized or nonsterilized sandy loam (80 g) adjusted to 50–60% moisture-holding capacity and contained in 100-ml jars with loosely applied screw caps. The soil used was collected from a fertile, experimental cotton (*Gossypium hirsutum* L.) field at Auburn, Alabama, and air-dried to exclude natural insect populations. The cropping history of the field has been described (11). Insects were introduced into some jars at approximately 200/jar (1,600/kg of soil); other jars of soil were kept without insects. After 10 hr at 27°C, the filters were gently removed and processed for microscopic examination according to the original method of Adams (1). Data recorded for *Fusarium* consisted of the number of chlamydospores in two microscopic fields on each of four filter disks per treatment and the percentage of germination of spores. For *Trichoderma*, data were taken from three microscopic fields on three filter disks. Following the experiment, insects were extracted from the soil to verify survival.

Sclerotia. Laboratory tests were performed to assess the direct effects of Collembola on sclerotia of three pathogenic fungi. Mature sclerotia of *S. rolfsii* were produced on PDA cultures, and microsclerotia of *V. dahliae* and *M. phaseolina* were obtained from 3-week-old cultures on Czapek's agar. Microsclerotia were separated from the agar and mycelia by chopping in a microblender with 35 ml of sterile water for 2 min followed by passage through a fine-mesh sieve and resuspension in water. The sclerotium concentration was adjusted to about 100 microsclerotia per milliliter.

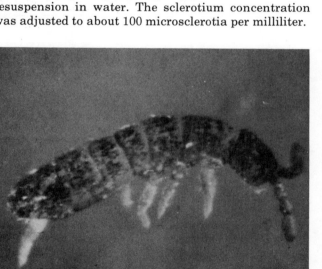

Fig. 1. *Proisotoma minuta,* a soil-dwelling, mycophagous insect of the order Collembola (×100).

Twenty sclerotia of *S. rolfsii* were placed in each of 12 petri dishes on the packed and smoothed surface of nonsterilized sandy loam (10% moisture) taken from a fertile cotton field. Approximately 200 Collembola were added to six of the dishes, and six were left noninfested. For the microsclerotial fungi, one drop of sclerotial suspension was applied to the surface of 2% water agar 15 mm from the dish edge, and a drop of sterile water was applied near the opposite edge. Collembola were added to the centers of five dishes, and five remained noninfested. After 16 hr of incubation in the dark at 27°C, the numbers of sclerotia and percentage of germination were recorded.

Finally, a test was conducted with microsclerotia of *M. phaseolina* in which 1 ml of a sclerotium suspension (100/ml) was applied to Millipore filters and buried in jars of nonsterilized soil in the manner described for spores. Collembola (200/jar) were added or not added and, after 24 hr, filters were recovered and the percentage of germination was recorded for sclerotia within five stereoscopic microscope fields for five disks of each treatment.

RESULTS

Effect on mycelia. Comparisons of infested and noninfested cultures clearly demonstrated the capacity of Collembola to suppress colony growth of certain fungi (Fig. 2). The distribution pattern of insects on the agar surface indicated their strong attraction to the fungal food source, although frequently no hyphal extension was visible to the unaided eye (Fig. 3). The insects were not attracted to sterile PDA disks. Apparently, visible hyphae on *R. solani* disks were consumed so rapidly that the insects dispersed widely in search of other food sources. In contrast, it is evident that the mycelia of *S. rolfsii* and *T. harzianum* repelled insects. However, insect activity accounted for a 75% reduction in germination of *S. rolfsii* sclerotia on the surface of nonsterilized soil and reductions of 65 and 50% for microsclerotia of *V. dahliae* and *M. phaseolina,* respectively, on an agar surface.

Results of a single preliminary test with microsclerotia of *M. phaseolina* on buried Millipore filters in nonsterilized soil showed very low germination overall,

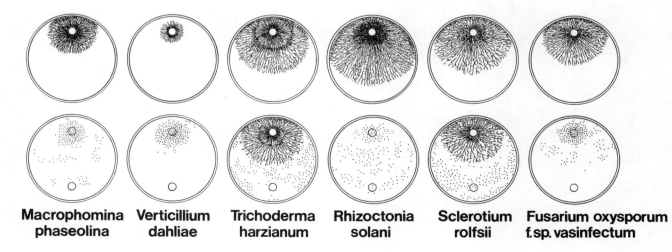

| Macrophomina phaseolina | Verticillium dahliae | Trichoderma harzianum | Rhizoctonia solani | Sclerotium rolfsii | Fusarium oxysporum f. sp. vasinfectum |

Fig. 2. Suppression of fungal growth (bottom) by Collembola on Czapek's agar, compared with colony growth in the absence of insects (top).

but they also showed a significant reduction when insects were present. The mean germination in nonsterilized soil without insects was 25.7%, compared with 14.3% in nonsterilized soil + insects. This appeared to result from grazing on the sclerotial germ tubes, preventing their normal development.

DISCUSSION

Most species of Collembola are regarded as saprophagous and microphagous, some being more specifically mycophagous and rarely known to injure plant roots. Their distribution and feeding activities are largely related to the availability of the microflora (4); therefore, populations are often larger in the root zone than distant from roots (11). That these very active insects can consume relatively large amounts of fungal material is evident from reports of mycelia and spores found in the gut contents (2,3,8). They have not, however, been given proper consideration as interacting components of the rhizosphere where activities of root-infecting fungi may be influenced.

The present study has established that two representative species of the soil-dwelling Collembola are attracted to mycelia and spores as food sources and can cause significant reductions in inoculum density. Although the mycophagous Collembola are generally believed to be unspecialized feeders, having no preferences for different fungi, their partial aversion to mycelia of *S. rolfsii* and *T. harzianum* indicates that exceptions occur.

The Collembola tested can reduce spore populations even in nonsterilized soil where other food sources are available (Table 1). Many ingested spores of *F. oxysporum* f. sp. *vasinfectum* and *T. harzianum* passed

Table 1. Effects of mycophagous Collembola on fungal spores in sterilized and nonsterilized soil

Treatment	*Fusarium* chlamydospores		*Trichoderma* conidia	
	Number of spores[w]	Germination[x] (%)	Number of spores[y]	Germination[x] (%)
Sterile soil				
Without insects	171.6 a[z]	74.6 a	120.9 a	35.5 a
With insects	76.4 b	47.2 b	20.8 b	29.6 a
Nonsterile soil				
Without insects	195.5 a	30.6 a	194.2 a	31.4 a
With insects	16.1 b	21.6 a	47.6 b	33.6 a

[w] Mean number of spores counted in two fields on each of four Millipore filter disks.
[x] Germination represents both noningested spores and spores ingested by insects and redeposited in fecal pellets on the filter disks.
[y] Mean number of spores counted in three fields on each of three Millipore filter disks.
[z] Pairs of means followed by the same letter do not differ from each other at the 0.01 level of probability according to the Student *t* test.

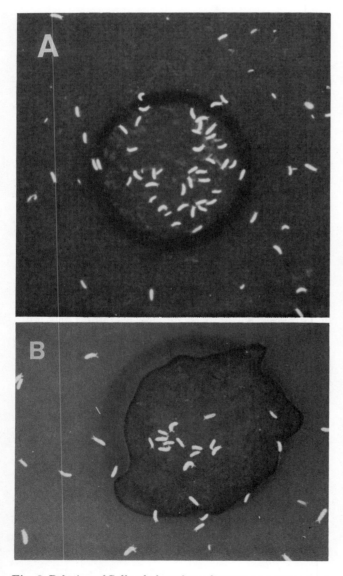

Fig. 3. Relation of Collembola to fungal spores on a water agar surface. (**A**) Attraction of insects to a drop of washed chlamydospores of *Fusarium oxysporum* f. sp. *vasinfectum* and (**B**) conidia of *Trichoderma harzianum*.

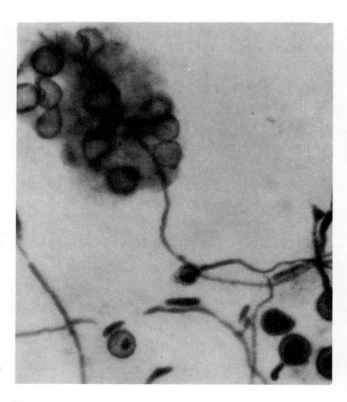

Fig. 4. Insect fecal pellet containing chlamydospores of *Fusarium oxysporum* f. sp. *vasinfectum* (×600).

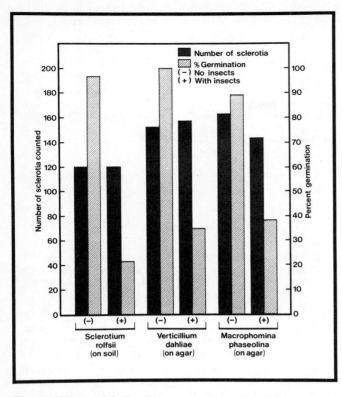

Fig. 5. Effects of Collembola on the sclerotia of three plant pathogenic fungi.

through the animal gut and were redeposited intact in fecal pellets (Fig. 4). The percentage of viability of these spores was not determined in this study; however, the combined germinability of noningested chlamydospores and those released from disrupted fecal pellets was reduced in sterilized soil.

The Collembola in this study apparently cannot readily destroy sclerotia; rather, they consume the hyphae of germinating sclerotia so efficiently that, presumably, they might render the propagules ineffective as viable inocula (Fig. 5). It is not understood why these insects feed on very young hyphae emerging from sclerotia of *S. rolfsii* and ingest conidia of *T. harzianum*, yet do not favor the mycelia of these fungi as food sources.

It seems evident that soil microarthropods, especially the Collembola, when present in sufficient numbers, may play an important role in soil-fungus ecology, conceivably altering the inoculum density and potential of plant pathogens and affecting the activities of natural biological control agents such as *Trichoderma* spp. Since these animals are attracted to fungal food sources of high density, their impact upon the quantitative and qualitative nature of the rhizosphere flora should be investigated further.

ACKNOWLEDGMENTS

We wish to express our appreciation to Susan Scott and Susan Ledbetter for technical assistance.

LITERATURE CITED

1. Adams, P. B. 1967. A buried membrane filter method for studying behavior of soil fungi. Phytopathology 57:602-603.
2. Anderson, J. M., and Healey, I. N. 1972. Seasonal and interspecific variation in major components of the gut contents of some woodland Collembola. J. Anim. Ecol. 41:359-368.
3. Christen, A. A. 1975. Some fungi associated with Collembola. Rev. Ecol. Biol. Sol. 12:723-728.
4. Christiansen, K. 1964. Bionomics of Collembola. Annu. Rev. Entomol. 9:147-178.
5. Curl, E. A. 1979. Effects of mycophagous Collembola on *Rhizoctonia solani* and cotton seedling disease. Pages 253-269 in: Soil-Borne Plant Pathogens. B. Schippers and W. Gams, eds. Academic Press, London.
6. Curl, E. A. 1982. The rhizosphere: Relation to pathogen behavior and root disease. Plant Dis. 66:624-630.
7. Kevan, D. K. M. 1965. The soil fauna—its nature and biology. Pages 33-51 in: Ecology of Soil-Borne Plant Pathogens. K. F. Baker and W. C. Snyder, eds. University of California Press, Berkeley.
8. Poole, T. B. 1959. Studies on the food of Collembola in a Douglas fir plantation. Proc. Zool. Soc. Lond. 132:71-82.
9. Wallwork, J. A. 1970. Ecology of Soil Animals. McGraw-Hill, London.
10. Wiggins, E. A., and Curl, E. A. 1979. Interactions of Collembola and microflora of cotton rhizosphere. Phytopathology 69:244-249.
11. Wiggins, E. A., Curl, E. A., and Harper, J. D. 1979. Effects of soil fertility and cotton rhizosphere on populations of Collembola. Pedobiologia 19:17-82.

A Technique to Compare Growth in Soil of *Gaeumannomyces graminis* var. *tritici* over a Range of Matric Potentials

O. F. GLENN, C. A. PARKER, and K. SIVASITHAMPARAM, School of Agriculture, Soil Science and Plant Nutrition, University of Western Australia, Nedlands, 6009

Take-all disease of wheat causes extensive loss of yield in many countries, including Australia (9). The fungal pathogen responsible, *Gaeumannomyces graminis* var. *tritici*, was described by Garrett (7) to be an ecologically obligate parasite. It became widely accepted that *G. graminis* var. *tritici* has poor competitive saprophytic ability (3,8,12).

Other observations have provided data that clarify the description of competitive saprophytic ability. Lucas (12) concluded that colonization of wheat straw by *G. graminis* var. *tritici* in the field is negligible but does happen. Brown and Hornby (2) reported that the mycelium of *G. graminis* var. *tritici* could grow up to 5 mm through soil in the absence of host plants. Parker (13) and Fang and Parker (5), from laboratory observations, noted that *G. graminis* var. *tritici* is capable of growing out into the soil and speculated that the fungus could colonize new substrates in the field. Infection of sterile wheat straws in the absence of host plants up to 90 mm from the original source of inoculum has been observed (11). Shipton (16) in a review of this area concluded that it is now clear that mycelium of *G. graminis* var. *tritici* has an unmistakable, although limited (3–5 mm), capacity for saprophytic growth in unsterile soil.

Because *G. graminis* var. *tritici* grows through soil in the sandwich system (11), it is possible to investigate its saprophytic growth in soil, an approach that Cunningham (4) pointed out was not hitherto possible. Using this technique we are investigating the effect on the saprophytic growth of *G. graminis* var. *tritici* of a range of physical and chemical variables in the soil environment. The data should be of value in predicting changes in inoculum potential in fields with different climatic and cropping histories.

Although the soil sandwich technique is very effective, it has limitations when used to make comparisons between soils held at the same percentage of water-holding capacity (14) because of differences in bulk density and texture. Matric potential is a better measure of the forces acting in soil to retain water, therefore making it less readily available to the fungus. It is therefore a more important measurement than water content (10).

An alternative technique has been developed based on an early method used to determine moisture characteristic curve (15). The technique enables the study of the growth of the fungus over a range of matric potential from field capacity to saturation in different soils of varying curves as well as investigating other aspects of the saprophytic growth of *G. graminis* var. *tritici*.

METHODS

The organism, *G. graminis* var. *tritici* isolate WUF 2, has been used previously (6,11). It was grown on a malt extract broth agar (17) with streptomycin added (MEBA + S) and incubated at 25°C. Inoculum was prepared by sterilizing 5.7-mm straw pieces, autoclaved at 15 lb for 15 min on two consecutive days. These were placed around the perimeter of a 7-day-old colony of WUF 2 on MEBA + S and incubated at 17° ± 1°C for a minimum of 21 days and used within 6 weeks. The soil used was a yellow sand plain soil from the Merredin area, which is described by Bettenay and Hingston (1) as a yellow earth of the Norpa series of the Ulva Association. It contains 15% clay in the surface (0–10 cm) layer, which increases to around 25% below 30 cm (W. Porter, *personal communication*). The soil was air-dried, then sieved through a 2.0-mm sieve. The pH (measured in 0.01M CaCl₂) was raised from 4.2 to 5.5 with calcium carbonate. Bulk density, after packing into the column, was calculated to be 1.57.

The columns. Columns were made from acrylic plastic tubing 100 cm long and with a 30-mm internal and 40-mm external diameter. They were split in half longitudinally, and 5-mm holes were drilled at 100-mm intervals through the center of one side. Semicircles of Membra Fil cellulose ester membrane filter 76 mm in diameter and 0.3-μm pore size (Nuclepore Corp., Pleasanton, CA 94566) were attached to the other side of the column so that their centers were directly opposite the 5-mm inoculation ports. The two halves of the column were taped securely together and the ports covered with waterproof tape. The base was closed with cotton gauze secured by a stainless steel hose clip. The water line and future sampling points were clearly marked; the column was then clamped upright, resting one edge on a small block in a beaker (Fig. 1).

While it was being filled with air-dry sand, the column was vibrated continuously until no further settling was observed. The column was lowered into a beaker filled with deionized water to the previously marked line, and deionized water was poured into the top of the column until the soil was saturated and water drained freely. The top of the column was covered to reduce evaporation from the surface. The water level was checked daily for a week and water removed if necessary, until drainage virtually stopped and the column was in equilibrium with the free water (D. J. McFarlane, *personal communication*).

Inoculation and incubation. Soil cores were removed with a cork borer through the 5-mm inoculation ports. The straw pieces were then placed in the center of the equilibrated soil column, the soil core was replaced, and the hole retaped (Fig. 1). The inoculated straw piece was thus at least 1 cm from the filter. The columns were incubated at room temperature (15–25°C) for 5 weeks.

Water content and fungal growth. The columns were opened by removing one of the longitudinal tapes in such a way as to expose the cellulose ester membranes. Ten-millimeter soil slices were removed at 50-mm intervals, weighed, dried, and reweighed. The data were used to plot the moisture characteristic curve (Fig. 2).

The filters were removed and examined at ×30 under a stereomicroscope, and the area covered by black runner hyphae of *G. graminis* var. *tritici* was recorded. These figures were converted to percentage total filter area for comparison (Fig. 3). In this first experiment, the only variable was the inoculum: two columns were inoculated with WUF 2-infected straw and one with sterile straw.

RESULTS

The results, which are most clearly expressed in the figures, show the ease with which growth data can be correlated with moisture content or matric potential. Over this range of water content, the maximum area of

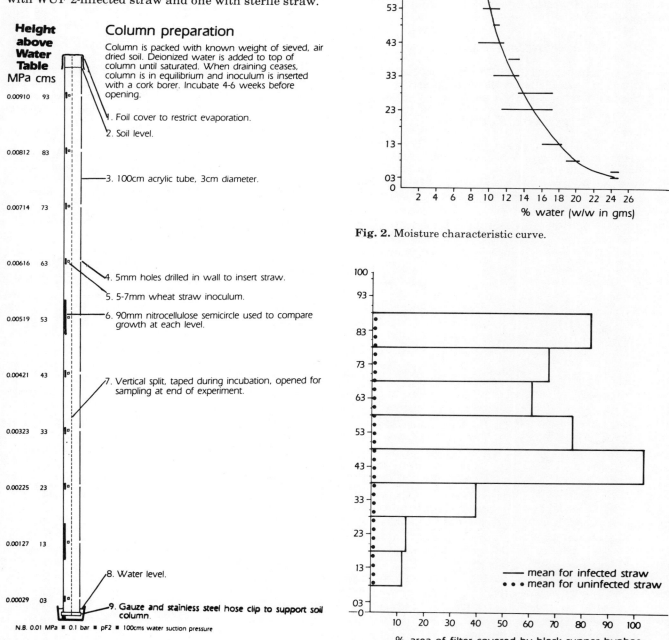

Fig. 1. Preparation of column for study of growth of *Gaeumannomyces graminis* var. *tritici*.

Fig. 2. Moisture characteristic curve.

Fig 3. Fungal growth.

filter covered by growth occurred at 0.0042 MPa when the water content was 11% w/w, or a 44% water-holding capacity.

DISCUSSION

The technique described is suitable for comparing the saprophytic growth of *G. graminis* var. *tritici* in a variety of soils over a range of matric potentials from 0.01 MPa down to almost zero. The fungus can grow at least 1 cm through soil, which supports the observations of Grose et al (11). We are now using the system to monitor the growth of *G. graminis* var. *tritici* when other parameters such as bulk density are varied. It could also be used to study the effect of cropping history on the capacity of the soil to support fungal growth; on amended soils, it could be used for nutritional studies. There is apparently no reason why this technique could not be applied to investigation of the behavior of other soil microbes.

ACKNOWLEDGMENT

We are grateful to the Rural Credits Development Fund of the Reserve Bank of Australia for its financial support, without which this work could not have been performed.

LITERATURE CITED

1. Bettenay, E., and Hingston, F. J. 1961. The Soils and Land Use of the Merredin Area, Western Australia. Soils Land Use Ser.41, CSIRO Div. Soils.
2. Brown, M. E., and Hornby, D. 1971. Behaviour of *Ophiobolus graminis* on slides buried in soil in the presence or absence of wheat seedlings. Trans. Br. Mycol. Soc. 56:95-103.
3. Butler, F. C. 1953. Saprophytic behaviour of some cereal root rot fungi. I. Saprophytic colonisation of wheat straw. Ann. Appl. Biol. 40:284-297.
4. Cunningham, P. C. 1975. Some consequences of cereal monoculture on *Gaeumannomyces graminis* (Sacc.) Arx & Olivier and the take-all disease. EPPO Bull. 5:297-317.
5. Fang, C. S., and Parker, C. A. 1975. The importance of inoculum in assessing the competitive ability of *Gaeumannomyces graminis* var. *tritici*. Proc. Aust. Spec. Conf. Soil Biol., 3d, pp. 15-16, CSIRO, Adelaide.
6. Fang, C. S., and Parker, C. A. 1981. An L-drying method for preservation of *Gaeumannomyces graminis* var. *tritici*. Trans. Br. Mycol. Soc. 77:103-106.
7. Garrett, S. D. 1956. Biology of Root-infecting Fungi. Cambridge University Press, Cambridge, pp. 93-162.
8. Garrett, S. D. 1970. Pathogenic Root-infecting Fungi. Cambridge University Press, Cambridge, p. 113.
9. Garrett, S. D. 1981. Introduction. Pages 1-11 in: Biology and Control of Take-All. M. J. C. Asher and P. J. Shipton, eds. Academic Press, London.
10. Griffin, D. M. 1970. Ecology of Soil Fungi. Chapman and Hall, London, pp. 76-82.
11. Grose, M., Parker, C. A., and Sivasithamparam, K. 1984. Growth of *Gaeumannomyces graminis* var. *tritici* in soil: Effects of temperature and water potential. Soil Biol. Biochem. 16:211-216.
12. Lucas, R. L. 1955. A comparative study of *Ophiobolus graminis* and *Fusarium culmorum* in saprophytic colonisation of wheat straw. Ann. Appl. Biol. 43:134-143.
13. Parker, C. A. 1974. The control of cereal root rot diseases in Western Australia. Aust. Plant Pathol. Soc. Newsl. 3(2):51.
14. Piper, C. S. 1947. Soil and Plant Analysis. Adelaide University Press, South Australia, p. 82.
15. Puri, A. N. 1949. Soils—Their Physics and Chemistry. Reinhold Publishing Corp., New York, pp. 346-357.
16. Shipton, P. J. 1981. Saprophytic survival between crops. Page 297 in: Biology and Control of Take-All. M. J. C. Asher and P. J. Shipton, eds. Academic Press, London.
17. Sivasithamparam, K., and Parker, C. A. 1978. Effect of infection of seminal and nodal roots by the take-all fungus on tiller numbers and shoot weight of wheat. Soil Biol. Biochem. 10:365-368.

Use of Aerial Photography for Assessing Soilborne Disease

J. R. HARRIS, CSIRO Division of Soils, Glen Osmond, South Australia, 5064

HISTORICAL BENCHMARKS

Aerial photography had its beginning in the 1850s with balloons as aerial platforms and cumbersome equipment carrying insensitive photographic emulsions as sensors. It has extended through this century from rotary and fixed-wing aircraft to rockets and spacecraft; from sensitive photographic emulsions in compact and precision cameras to electronic scanners, spectrophotometers, and scatterometers sensing the ultraviolet, visible, near infrared, and thermal infrared up to microwave radiations. There have been many historical benchmarks (13) in the application of remote sensing technology to resource analysis during the last 50 years. It has become an accepted practice in research and management techniques for agricultural, horticultural, forestry, and natural ecosystems.

Fungous disease. The first successful applications of aerial photography to plant pathology date back to 1927 (36,37,51) when oblique photography of cotton crops by panchromatic film from heights of 250–400 ft above terrain discriminated those plants killed by *Phymatotrichopsis omnivora* (Duggar) Henneb. Their sudden wilting opened the canopy to expose light-colored soil beneath. This sort of photography could not give early warning of infection before the onset of wilting, but it did disclose the spread of the disease from recognizable foci of infection.

An important advance in disease assessment came with the 1952 use of color-infrared (CIR) film for the aerial photography of nursery plots of cereals (12). Changes of reflectance occur when plants are stressed, droughty, polluted, diseased, or lose vigor from nutrient deficiency (4,7,12,14,17,20,24,54). Leaf rust, black stem rust, and barley yellow dwarf virus were detected in CIR aerial photographs, and the spread of the epiphytotics was observed (12). The virus disease barley yellow dwarf stood out more distinctly than the fungous foliar infections (11,12). From earlier studies of infected potato leaves by terrestrial photography (5), virus-infected plants were known to have high infrared reflectance.

Late blight of potato caused by *Phytophthora infestans* (Mont.) de Bary received attention in the 1960s in both the United Kingdom (9,10) and the United States (29,30). The sites of initial outbreak and the spread of infection across potato fields could be identified on aerial photographs. The first attempts to quantify yield loss in photographs by densitometry occurred in this study (30). In some potato fields, Verticillium wilt could also be detected (30,61).

Nematode disease. Some diseases could be associated with specific soil conditions or types. The "docking" disorder of sugar beet involving nematode attack by species of *Trichodorus* or *Longidorus*, and sometimes combined with such fungi as *Rhizoctonia solani* Kuehn, is an example (6,15). So is take-all of wheat caused by *Gaeumannomyces graminis* (Sacc.) von Arx & Olivier var. *tritici* Walker in U.K. wheat fields, where the disease becomes more severe in wetter sites (6,8) or where soil structure has deteriorated (62). The aerial photographs show clear associations of disease incidence with soil patterns.

Bacterial and viral disease. Epiphytotics of bacterial diseases of field beans have been studied by aerial photography: halo blight caused by *Pseudomonas syringae* van Hall (syn. *P. phaseolicola* Burk.) in the United Kingdom (8); and the common bacterial blight caused by *Xanthomonas phaseoli* (E. F. Sm.) Dows. and fuscous blight caused by *X. phaseoli* f. sp. *fuscans* (Burkh.) Starr & Burkh. in Canada (41,56–58,60). Bacteria spread in diffuse patterns across fields by raindrop scatter in contrast to discrete foci of soilborne and windborne infections.

A similar contrast occurred with aphid infestation of winter cereals in East Anglia (18). When barley yellow dwarf virus infection occurred in early sown crops and sooty molds (mostly *Cladosporium* spp.) developed, the patterns were distinctive and could be distinguished from the more diffuse patterns associated with late aphid feeding. Pattern analysis is an important guide to disease etiology.

Disease in forests. Aerial photography has found many applications in forest management, and literature reviews (19,35) cite numerous disease and pest problems. Attention is drawn here to only two examples because they are soilborne fungous diseases. *Fomes annosus* (Fr.) Cooke causes a root and heart rot of pines that spreads to adjacent trees. The disease can be contained if outbreaks are located early from aerial photographs (46). *Phytophthora cinnamomi* Rands is a destructive parasite of eucalypt forest in Australia. It first attacks understory shrubs, then spreads to the dominant overstory tree. Containment depends upon presumptive location by aerial survey of sites for which the infection can be confirmed by ground sampling (22). Aerial photography thus has a role in disease control.

Disease in field crops. In barely half a century, aerial photography has developed into a tool that can be used for studying patterns of disease in paddocks (8,18), for detecting outbreaks of disease in inaccessible or remote areas (22), for estimating disease incidence on a regional basis (28), and for crop loss assessment (2,3,10,58). Although soilborne diseases attracted early attention as subjects for aerial photography, they have been much neglected since. The advances in remote sensing techniques seem to have bypassed this field of plant pathology. Table 1 carries the quite short list of references to papers that report studies of soilborne diseases of field crops by aerial photography. Other soilborne diseases must have been photographed, but they do not feature in the literature.

TECHNIQUES FOR ACQUIRING AND ASSESSING IMAGERY

Image acquisition. There are accounts (12,30) of early methods of photographing from light aircraft. Advanced techniques for using large-format cameras are given in the Manual of Color Aerial Photography (50) and the Manual of Remote Sensing (44). In most cases where relatively simple aircraft operations with small-format (35 mm and 70 mm) cameras can be used, there are excellent sources of flight planning information and of camera techniques (1,16,21,31,61). Data relating to the exposure and processing of Kodak aerial films are provided in Kodak technical manuals (25–27).

Aircraft operations are of prime importance in plant disease studies. Current earth resource satellite systems such as Landsat 4 and Bhaskara and projected systems such as SPOT and MOS-1 (53) provide imagery at scales (representative fraction = 1:1,000,000 to 1:200,000) that are too small to be useful for estimates of disease within paddocks (17) where the information is needed at larger scales (R.F. = 1:20,000 to 1:2,000). Landsat data have been transformed to provide a green index number (GIN) that indicates vegetative drought stress over large geographic areas (52); this has a strong bearing upon crop yields.

Further, most of the important changes in reflectance that result from abnormal plant growth (7,14,17) occur in the near infrared and red wavelengths, which are detected efficiently and cheaply on CIR film. It is possible to obtain critical information from low-level obliques with hand-held 35-mm cameras (21,49,61); but large-format cameras are required for high-altitude flights, such as those of the corn blight watch (28) that studied an epiphytotic on a regional basis.

Thermal infrared and radar scanning systems have not been particularly successful in disease studies. They do provide information about soil moisture status from thermal inertia of soil and drought stress by changes in crop canopy temperatures that involve the "stress degree day" concept (32). In a comparative study in Canada (40), these methods were inferior to CIR photography.

Image analyses. Aerial photography can be interpreted from visual inspection of single scenes or stereoscopy of pairs, but the best approach to quantitative data comes from densitometric scanning (10,23,42,59). When the density of each picture point (pixel) that makes up the scene can be digitized, it is possible to carry out complicated electronic image analyses. Sets of adjacent pixels may be tested for homogeneity (within classes of like reflectance) and for statistical separability (between classes); image may be enhanced by stretching contrast and by sharpening edges; adding, determining the ratio, and recombining reflectance data to produce new false-color images; and providing outputs of thematic maps. Many of the needs of plant pathology are less demanding than this, but the advantages in quantifying the scene into data—even by visual ranking methods (48)—are considerable. Electronic scanning of CIR photographs (2,10,23, 24,40,59,60) improves accuracy of crop loss models (58). Computer techniques of determining the ratio of near infrared:red reflectance improve the sensitivity over the use of either band alone (54). There are photographic montage methods of density slicing that have used Agfacontour film (38,39).

Table 1. Published accounts of remote sensing of root diseases in field crops

Disease	Causal agents	Authors	Comment
Root rot of cotton	*Phymatotrichopsis omnivora* (Duggar) Henneb. syn *Phymatotrichum omnivorum* (Shear) Duggar	Taubenhaus et al (36,37,51) Pratt et al (42)	Earliest application using panchromatic film. Later study used CIR (42)
Verticillium wilt (early dying) of potato	*Verticillium albo-atrum* Reinke & Berth; *V. dahliae* Kleb.	Manzer and Cooper (29,30) Watson and Hoyle (61)	Yellowing shown best on color film to distinguish it from late blight on CIR
Take-all of wheat	*Gaeumannomyces graminis* (Sacc.) von Arx & Olivier var. *tritici* Walker	Brenchley and Dadd (9) Bell (6) Yarham (62)	Difficult to distinguish between cereal diseases on aerial photographs (6)
Black root of sugar beet	*Aphanomyces cochlioides* Drechs. associated with *Rhizoctonia solani* Kuehn and *Pythium aphanidermatum* (Eds.) Fitz.	Schneider and Safir (48)	Damping-off in acute stage; depressed top growth in mild
Fusarium root rot of pea	*Fusarium solani* (Mart.) Appel & Woll. f. sp. *pisi* (Jones) Snyder & Hansen; *F. oxysporum* Schlecht. f. sp. *pisi* (van Hall) Snyder & Hansen	Basu et al (2) Haglund and Jarmin (20)	Symptoms indistinguishable from drought. Yield loss estimation made from aerial photographs (2)
Lucerne (alfalfa) decline	Winterkill in absence of *Phytophthora megasperma* Drechs. Fungi found present were *Fusarium solani* (Mart.) Appel & Woll., *Pythium ultimum* Trow., *P. torulosum* Coker & Pratt, plus a range of lesion nematodes	Wallen et al (59) Basu et al (3)	Authors not prepared to accept minor pathogens as having more important role than winterkill
Lucerne root rot	*Phymatotrichopsis omnivora* (Duggar) Henneb.; *R. solani* Kuehn	Pratt et al (42)	Distinguishes between two diseases from pattern of patches and times of occurrence

In all cases, good-quality imagery is essential. In many of the papers (6,10,11,24,54) there are warnings that important differences show up only in imagery taken at particular times (42,58), or that the effects of root disease may be indistinguishable from other causes of stress, such as drought (2), winterkill (3,59), and salinity (4,44).

Verification of disease incidence by ground truth remains at the core of a sampling procedure that aerial photography complements but cannot supplant. Remote sensing is at best an extrapolation procedure that extends the information from reference sampling sites.

PROBLEMS PECULIAR TO ROOT DISEASES OF CEREALS

Patches in cereal crops. There is increasing appreciation (19,55) that a root disease of a cereal is more likely to be of complex etiology rather than caused by a single organism. Early studies of interactions within mixed infections (63,64) were overlooked until increasing evidence (33,34,43,45,47) substantiated the damage capabilities of the so-called minor pathogens that colonized roots in complex suites. The similarities and the contrasts between the fungi associated with foot rot and root rot of spring wheat in Australia (33) and winter wheat in Europe (45) provide examples of how the same ecological niche can be filled by different suites.

Persistence of patches. Sometimes unthrifty patches of crop growth recur at intervals. This may be associated with a soil type, a drainage pattern, a weed or pest infestation, or propagule carry-over of a pathogen. Many of these conditions are interrelated and thereby affect alternate rotation practices; e.g., stunting in a cereal crop may carry over into affected pasture plants and species composition (43). Persisting patches in serial aerial photographs usually suggest the intrusion of soil or environmental factors into disease etiology (6,8,55,62), and attempts should be directed toward defining the predisposing factors.

Conclusions. There are few, if any, precise assessments of crop losses in cereals from root disease. For the most part, root disease is less than disastrous; although subtle, it is widespread and significant in economic terms. The causal organisms coexist on the roots in complex suites that may be subject to synergisms and antagonisms not evident when the component fungi, bacteria, and nematodes are tested for pathogenicity one at a time. Diseases assignable to single causes are rare (19). Predisposing factors can be many and variable, but they exert considerable influence.

Plant pathologists have not responded adequately to this challenge by using aerial photography for crop loss assessment. There has not been a comprehensive review of remote sensing techniques for plant pathology for nearly 20 years (8). More use has been made of aerial photography for studying one disease, the bacterial blight of field beans in Canada, than can be amassed for the whole literature of cereal root diseases of the world.

LITERATURE CITED

1. Anonymous. 1978. Manual of Practice. Low altitude aerial surveillance for water resources control. State Water Resources Control Board, Division of Planning and Research, State of California, Sacramento.
2. Basu, P. K., Jackson, H. R., and Wallen, V. R. 1978. Estimation of pea yield loss from severe root rot and drought stress using aerial photographs and a loss conversion factor. Can. J. Plant Sci. 58:159-164.
3. Basu, P. K., Jackson, H. R., and Wallen, V. R. 1978. Alfalfa decline and its cause in mixed hay fields determined by aerial photography and ground survey. Can. J. Plant Sci. 58:1041-1048.
4. Bauer, M. E. 1975. The role of remote sensing in determining the distribution and yield of crops. Adv. Agron. 27:271-304.
5. Bawden, F. G. 1933. Infrared photography and plant virus diseases. Nature (London) 132:168.
6. Bell, T. S. 1972. Remote sensing for the identification of crops and crop diseases. Pages 153-166 in: Environmental Remote Sensing. E. C. Barrett and L. C. Curtis, eds. Bristol Symposium on Remote Sensing, University of Bristol, Oct. 1972.
7. Brach, E. J., and Mack, A. R. 1972. Difference in reflectance properties of diseased plants grown under different environments. Can. Symp. Remote Sensing, 1st, 103-107.
8. Brenchley, G. H. 1968. Aerial photography for the study of plant diseases. Annu. Rev. Phytopathol. 6:1-22.
9. Brenchley, G. H., and Dadd, C. V. 1962. Potato blight recording by aerial photography. N.A.A.S. Q. Rev. (London) 57:21-25.
10. Chiang, H. C., and Wallen, V. R. 1977. Detection and assessment of crop diseases and insect infestations by aerial photography. In: Crop Loss Assessment Methods. L. Chiarappa, ed. Section 3.1.1. FAO Manual, C.A.B., Farnham Royal, 2d suppl. 12 pp.
11. Clark, R. V., Galway, D. A., and Paliwal, T. C. 1981. Aerial infrared photography for disease detection in field plots of barley, oats and wheat. (Abstr.) Phytopathology 71:867.
12. Colwell, R. N. 1956. Determining the prevalence of certain cereal crop diseases by means of aerial photography. Hilgardia 26:223-286.
13. Colwell, R. N. 1979. Some historical benchmarks in the development of color aerial photography. Proc. Workshop Color Aerial Photogr. Plant Sci. Related Fields, 7th, 1-14. Am. Soc. Photogramm., Falls Church, VA.
14. DeCarolis, C., and Amodeo, P. 1980. Basic problems in the reflectance and emittance properties of vegetation. Pages 69-79 in: Remote Sensing Application in Agriculture and Hydrology. G. Fraysse, ed. A. A. Balkema, Rotterdam.
15. Dunning, R. A., and Cooke, D. A. 1967. Docking disorder. Br. Sugar Beet Rev. 36:23-29.
16. Fleming, J., and Dixon, R. G. 1981. Basic guide to small-format hand-held oblique aerial photography. Can. Center for Remote Sensing, Users' Manual 81-82, p. 63. Can. Dep. Energy Mines Resour., Ottawa.
17. Fritz, E. L., and Pennypacker, S. P. 1975. Attempts to use satellite data to detect vegetative damage and alteration caused by air and soil pollutants. Phytopathology 65:1056-1060.
18. Greaves, D. A., Hooper, A. J., and Walpole, B. J. 1983. Identification of barley yellow dwarf virus and cereal aphid infestations in winter wheat by aerial photography. Plant Pathol. 32:159-172.
19. Grogan, R. G. 1981. The science and art of plant-disease diagnosis. Annu. Rev. Phytopathol. 19:333-351.
20. Haglund, W. A., and Jarmin, M. L. 1978. Aerial photography for detection and identification of *Fusarium oxysporum-pisi* wilt of peas. Plant Dis. Rep. 62:570-575.
21. Harris, J. R., and Haney, T. G. 1973. Techniques of oblique aerial photography of agricultural field trials. CSIRO Div. Soils Tech. Pap. 40 pp.
22. Hogg, J. A., and Wester, G. 1975. Detection of die-back disease in the Brisbane Ranges by aerial photography. Aust. J. Bot. 23:775-781.
23. Jackson, H. R., and Wallen, V. R. 1975. Microdensitometer measurements of sequential aerial photographs of field beans infected with bacterial blight. Phytopathology

65:961-968.

24. Kannegieter, A. 1980. An experiment using multispectral photography for the detection and damage assessment of disease infection in winter-wheat: Agronomic considerations. ITC J. 1980–82:189-234.

25. Kodak. 1968. Applied Infrared Photography. Kodak Tech. Publ. M-28. Eastman Kodak Company, Rochester, NY.

26. Kodak. 1972. Kodak Aerial Films and Photographic Plates. Kodak Tech. Publ. M-61. Eastman Kodak Company, Rochester, NY.

27. Kodak. 1976. Kodak Data for Aerial Photography. Kodak Tech. Publ. M-29, 4th ed. Eastman Kodak Company, Rochester, NY.

28. MacDonald, R. B., Bauer, M. E., Allen, R. D., Clifton, J. W., Erikson, J. D., and Landgrebe, D. A. 1972. Results of the 1971 Corn Blight Watch Experiment. LARS Print 100272, Laboratory for Applications of Remote Sensing, Purdue University, Lafayette, IN.

29. Manzer, F. E., and Cooper, G. R. 1963. Infrared photography of potato late blight. (Abstr.) Phytopathology 53:350.

30. Manzer, F. E., and Cooper, G. R. 1967. Aerial photographic methods of potato disease detection. Bull. Maine Agric. Exp. Stn. 646. 14 pp.

31. McCown, R. L., Tolson, D. J., and Clay, H. J. 1973. Low cost aerial photography for agricultural research. J. Aust. Inst. Agric. Sci. 39:227-232.

32. Millard, J. P., Jackson, R. D., Goettelman, R. C., Reginato, R. J., and Idso, S. B. 1978. Crop water-stress assessment using an airborne thermal scanner. Photogramm. Eng. Remote Sensing 44:77-85.

33. Moen, R., and Harris, J. R. 1980. Fungi associated with the root-rot syndrome of cereal disease in soils of low fertility in South Australia. Working Pap., Int. Congr. Dryland Farming, 1st, Adelaide, 1:163-167.

34. Moen, R., and Harris, J. R. 1985. The Rhizoctonia disease complex of wheat. Pages 48-50 in: Ecology and Management of Soilborne Plant Pathogens. C. A. Parker, A. D. Rovira, K. J. Moore, P. T. W. Wong, and J. F. Kollmorgen, eds. American Phytopathological Society, St. Paul, MN.

35. Myers, B. J. 1974. The application of color aerial photography to forestry. For. Timber Bur. Canberra, Leafl. 124, 20 pp. Canberra Aust. Gov. Publ. Service.

36. Neblette, C. B. 1927. Aerial photography for study of plant diseases. Photo Era Mag. 58:346.

37. Neblette, C. B. 1928. Airplane photography for plant disease surveys. Photo Era Mag. 59:179.

38. Nielsen, U. 1972. Agfacontour film for interpretation. Photogramm. Eng. 38:1099-1105.

39. Nielsen, U. 1974. Photographic image enhancement. Can. For. Serv. Publ. 1345. Dep. Environ., Ottawa.

40. Paquin, R., and Ladouceur, G. 1980. Efficacité des images radar et infrarouge thermique, et de la photo colour infrarouge pour l'inventaire des cultures. Can. J. Plant Sci. 60:1077-1085.

41. Philpotts, L. E., and Wallen, V. R. 1969. IR color for crop disease identification. Photogramm. Eng. 35:1116-1125.

42. Pratt, R. M., Snow, G. F., and Carpenter, T. R. 1963. Recognition of Phymatotrichum root rot of cotton and alfalfa in an aerial survey. (Abstr.) Phytopathology 53:1141.

43. Price, R. D. 1970. Stunted patches and deadheads in Victorian cereal crops. Dep. Agric. Tech. Publ. 23, Victoria, Australia. 165 pp.

44. Reeves, R. G., Anson, A., and Landen, D. 1975. Manual of Remote Sensing. Am. Soc. Photogramm., Falls Church, VA.

45. Reinecke, P., Duben, J., and Fehrmann, H. 1979. Antagonism between fungi of the root rot complex of cereals. Pages 327-336 in: Soil-Borne Plant Pathogens. B. Schippers and W. Gams, eds. Academic Press, London.

46. Rishbeth, J. 1977. Air photography and plant disease. Pages 103-113 in: The Uses of Air Photography, 2nd ed. J. K. S. St. Joseph, ed. John Baker, London.

47. Salt, G. A. 1979. The increasing interest in "minor pathogens." Pages 289-312 in: Soil-Borne Plant Pathogens. B. Schippers and W. Gams, eds. Academic Press, London.

48. Schneider, C. L., and Safir, G. R. 1975. Infrared aerial photography estimation of yield potential in sugarbeets exposed to blackroot disease. Plant Dis. Rep. 59:627-631.

49. Seevers, P. M. 1979. Practical application and physiological relationship of color infrared photography relative to monitoring agricultural crops. Proc. Bienn. Workshop Color Aerial Photogr. Plant Sci. Related Fields, 7th, 201-208. Am. Soc. Photogramm., Falls Church, VA.

50. Smith, J. A., and Anson, A. 1968. Manual of Color Aerial Photography. Am. Soc. Photogramm., Falls Church, VA.

51. Taubenhaus, J. J., Ezekiel, W. N., and Neblette, C. B. 1929. Airplane photography in the study of cotton root rot. Phytopathology 19:1025-1029.

52. Thompson, D. R., and Wehmanen, O. A. 1979. Using Landsat digital data to detect moisture stress. Photogramm. Eng. Remote Sensing 45:201-207.

53. Vellupillai, D. 1983. International satellite directory. Flight data. Flight Int. 123 (3862):1311-1318, 1322-1324, 1329-1340.

54. Walburg, G., Bauer, M. E., Daughtry, C. S. T., and Housley, T. L. 1982. Effects of nitrogen nutrition on the growth, yield and reflectance characteristics of corn canopies. Agron. J. 74:677-683.

55. Wallace, H. R. 1978. The diagnosis of plant diseases of complex etiology. Annu. Rev. Phytopathol. 16:379-402.

56. Wallen, V. R., Galway, D., Jackson, H. R., and Philpotts, L. E. 1973. Aerial survey for bacterial blight, 1979. Can. Plant Dis. Surv. 53:96-98.

57. Wallen, V. R., and Jackson, H. R. 1971. Aerial photography as a survey technique for the assessment of bacterial blight of field beans. Can. Plant Dis. Surv. 51:163-169.

58. Wallen, V. R., and Jackson, H. R. 1975. Model for yield loss determination of bacterial blight of field beans utilizing aerial infrared photography combined with field plot studies. Phytopathology 65:942-948.

59. Wallen, V. R., Jackson, H. R., Basu, P. K., Baenziger, H., and Dixon, R. G. 1977. An electronically scanned aerial photographic technique to measure winter injury in alfalfa. Can. J. Plant Sci. 57:647-651.

60. Wallen, V. R., and Philpotts, L. E. 1971. Disease assessment with IR-color. Photogramm. Eng. 37:443-446.

61. Watson, R. D., and Hoyle, R. J., Jr. 1975. Remote sensing in Idaho. Idaho Agric. Exp. Stn. Misc. Ser. 19. 24 pp.

62. Yarham, D. J. 1981. Practical aspects of epidemiology and control. Pages 353-384 in: Biology and Control of Take-all. M. J. C. Asher and P. J. Shipton, eds. Academic Press, London.

63. Zogg, H. 1950. Über Mischenifektionen bei Fusskrankheiten des Getreides. Schweiz. Z. Allg. Pathol. Bakteriol. 13:574-579.

64. Zogg, H. 1952. Studien über die Pathogenität von Erregergemischen bei Getreidefusskrankheiten. Phytopathol. Z. 18:1-54.

Isolation and Characterization of Plasmid DNA in the Fungus *Rhizoctonia solani*

T. HASHIBA, National Institute of Agricultural Sciences, Ibaraki; M. HYAKUMACHI, Hokkaido University, Sapporo; Y. HOMMA, Shikoku National Agricultural Experiment Station, Kagawa; and M. YAMADA, National Institute of Agricultural Sciences, Ibaraki, Japan

Viruses or viruslike particles have been reported in more than 30 species of plant-pathogenic fungi (12). In several cases, the effects of the viruses on pathogenicity of the fungus have been studied (2,3,5,6,11). Several reports have been published on degenerative diseases of fungi, but among the plant-pathogenic fungi few species have been reported to exhibit degenerative-type symptoms. Castanho and Butler (2–4) reported that a degenerative decline in severely diseased isolate obtained from healthy isolate during routine transfers of *Rhizoctonia solani* Kuehn has been associated with an assortment of double-stranded RNA (dsRNA) segments of three different sizes; in one isolate, these were of molecular weight 2.2, 1.5, and 1.1×10^6, but so far no virus particles have been observed. Furthermore, certain isolates of *Endothia parasitica* contain dsRNA that is transmissible by hyphal anastomosis and can induce poor growth in culture and diminish pathogenicity (5,6).

In the present paper we report detection of a linear plasmid DNA molecule in three weakly pathogenic isolates of *R. solani* by biophysical and electron microscopic methods. Three pathogenic isolates contained no detectable plasmid DNA.

MATERIALS AND METHODS

The three weakly pathogenic isolates of anastomosis group 4 (AG 4) of *R. solani* used in this study were obtained from soil infested by *R. solani*, 1978 and 1980, as described previously (14,15). Pathogenicity of these isolates had been determined by Hyakumachi et al (16) and Homma et al (13). The weakly pathogenic isolates 1668 RI-1, 1271 RI-64, and 1272 RI-1 were isolated from the soil after inoculation of strongly pathogenic isolates 1668, 1271, and 1272 of *R. solani* into the soil repeatedly with radish replants. Cultures were maintained on potato-sucrose agar.

For mass mycelial propagation, isolates were grown in potato-sucrose liquid medium containing Polypeptone without shaking. Fifty petri dishes each containing 20 ml of Polypeptone liquid medium were inoculated with 1 ml of a culture homogenate of each isolate (previously grown 1–2 days in potato-sucrose liquid medium). The mycelium grown for 15 hr was harvested on a 150-μm steel sieve and washed with distilled water.

Protoplasts from *R. solani* were prepared using a modified method of Hashiba (8) and Hashiba and Yamada (9). All manipulations during the preparation of DNA from protoplasts were carried out at room

temperature unless otherwise indicated. Protoplasts were transferred to a 1.5-ml Eppendorf tube and centrifuged for 5 min. The supernatant was carefully removed with a fine-tip aspirator. Plasmid DNA was detected by a modified procedure of Birnboim and Doly (1).

The DNA samples were subjected to 1% agarose gel electrophoresis. A DNA monolayer for electron microscopy was prepared by the aqueous and formamide techniques as described by Kleinschmidt

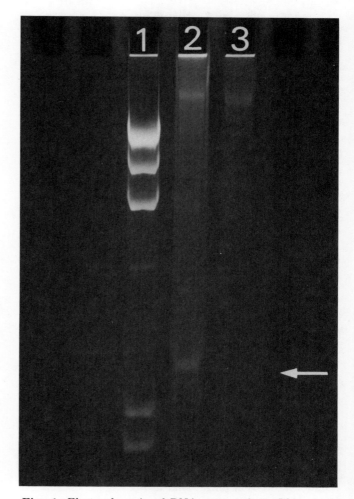

Fig. 1. Electrophoresis of DNA extract from *Rhizoctonia solani* (isolates 1271 and 1271 RI-64) in 1% agarose gels. Lane 1 contains the Hind III fragments of phage lambda DNA; lane 2, DNA extracted from isolate 1271 RI-64; lane 3, DNA extracted from isolate 1271. The arrow indicates plasmid DNA.

and Zahn (17). The transformation procedure was a modification of the protocol described by Hinnen et al (10) for *Saccharomyces cerevisiae*.

RESULTS

Detection of plasmid DNA. Alkaline lysates of protoplasts obtained from various isolates of *R. solani* were examined for the existence of plasmids by agarose gel electrophoresis. In investigating three weakly pathogenic isolates of *R. solani*—1668 RI-1, 1271 RI-64, and 1272 RI-1—a plasmidlike DNA band was found in addition to the top band corresponding to nuclear and mitochondrial DNAs (Fig. 1, lane 2). No plasmidlike DNA was detected among three strongly pathogenic isolates of *R. solani* 1668, 1271, and 1272 (Fig. 1, lane 3). All bands disappeared when treated with DNase (Fig. 2, lane 4) but were resistant to the RNase treatment (Fig. 2, lane 3). From these results, the DNA band found in addition to nuclear and mitochondrial DNA bands is interpreted to be plasmid DNA and designated pRS64. The plasmid band was separately isolated from a number of gels by an electrophoretic elution procedure for further study.

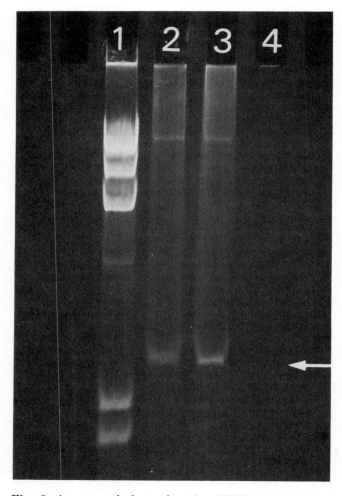

Fig. 2. Agarose gel electrophoresis of DNA isolated from isolate 1271 RI-64 of *Rhizoctonia solani*. Lane 1 contains the Hind III fragments of phage lambda DNA; lane 2, DNA without treatment; lane 3, DNA digested with 10 µg RNase (bovine pancreatic ribonuclease A) per milliliter; lane 4, DNA digested with 10 µg DNase (RNase-free deoxyribonuclease) per milliliter. The arrow indicates plasmid DNA.

Structure and size of plasmid DNA. In electron microscopy, the pRS64 plasmid fractions purified from gels were found to consist of homogeneous populations of linear DNA (Fig. 3). No circular DNA molecules as observed for the marker pBR322 were detected. The average size of these linear molecules was 0.77 µm corresponding to a molecular weight of 1.6×10^6 or 2.5 kilobase, assuming that 1 µm is equivalent to a molecular weight of 2.07×10^6.

The molecular weight of plasmid DNA was estimated on the basis of mobility by electrophoresis in comparison with linear DNA fragments obtained from

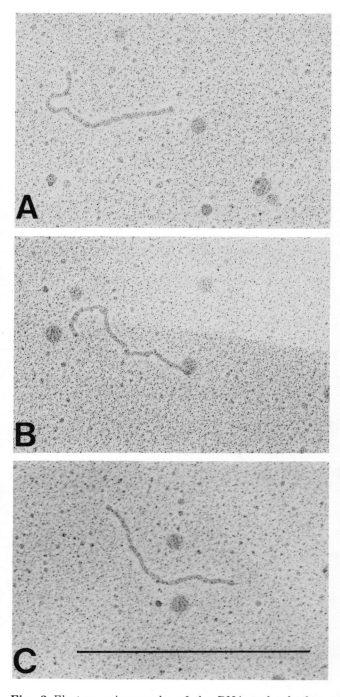

Fig. 3 Electron micrographs of the DNA molecule from *Rhizoctonia solani* by the formamide technique. The molecules are from **(A)** isolate 1271 RI-64, **(B)** isolate 1668 RI-1, and **(C)** isolate 1272 RI-1. Bar represents 1 µm.

Hind III digestion of λ DNA. As a result, the pRS64 DNA was shown to have molecular weight of 1.8×10^6. The value was in agreement with the size estimated by electron microscopy.

Transformation of protoplasts by plasmid DNA. A pathogenic isolate 1271 of *R. solani* was transformed by plasmid pRS64 DNA. Total DNA was extracted from the transformants and subjected to agarose gel electrophoresis. The original 1271 isolate contained no detectable plasmid DNA, but the isolated transformants (1271 RI-64-T) contained the pRS64 plasmid associated with weakly pathogenic isolate 1271 RI-64.

DISCUSSION

We searched for plasmids among three weakly pathogenic isolates of *R. solani* and found one novel DNA plasmid, pRS64 (molecular weight, $1.6–1.8 \times 10^6$). These molecules do not sediment with mitochondria, since they are resistant to RNase but sensitive to DNase. Furthermore, the 1271 isolate of *R. solani* lacking the DNA was transformed into weakly pathogenic strains with the DNA. From these results, the DNA is interpreted to be plasmid DNA and designated pRS64. This plasmid was characterized by a linear structure unlike the circular structure of bacterial plasmids and 2-μm DNA found in yeast. Weakly pathogenic cultures containing plasmids grew poorly and produced oxalic acid in medium (15,16). In *Zea mays*, both circular and linear DNA of comparable size have been reported (19). Linear DNA molecules have also been described in a few eukaryotes, e.g., the yeast *Kluyveromyces lactis* (7).

R. solani isolates 1271 RI-64, 1272 RI-1, and 1668 RI-1 were found to show decreased pathogenicity. All three pathogenic isolates of *R. solani* lacked the pRS64 plasmid. Pathogenic strains were transformed into weakly pathogenic strains with pRS64 plasmid. Thus, it seems highly probable that the weakly pathogenic character of *R. solani* isolates is under the control of the pRS64 plasmid. Martini et al (18) have reported that a single band of cccDNA was extracted from two different races of *R. solani* differing only in their host specificities. Each of the single bands migrated at consistently different rates on the gels. The fact that the two bands were different and paralleled the difference in phytopathogenicity of the two strains suggests a possible relationship between the presence of extrachromosomal DNA and phytopathogenicity.

ACKNOWLEDGMENT

We wish to thank T. Ishikawa, University of Tokyo, for helpful suggestions and encouragement throughout the experiments.

LITERATURE CITED

1. Birnboim, H. C., and Doly, J. 1979. A rapid alkaline extraction procedure for screening recombinant plasmid DNA. Nucl. Acid Res. 7:1513-1523.
2. Castanho, B., and Butler, E. E. 1978. Rhizoctonia decline: A degenerative disease of *Rhizoctonia solani*. Phytopathology 68:1505-1510.
3. Castanho, B., and Butler, E. E. 1978. Rhizoctonia decline: Studies on hypovirulence and potential use in biological control. Phytopathology 68:1511-1514.
4. Castanho, B., Butler, E. E., and Shepherd, R. J. 1978. The association of double-stranded RNA with Rhizoctonia decline. Phytopathology 68:1515-1519.
5. Dadds, J. A. 1980. Association of type 1 viral-like dsRNA with club-shaped particles in hypovirulent strains of *Endothia parasitica*. Virology 107:1-12.
6. Day, P. R., Dodds, J. A., Elliston, J. E., Jaynes, R. A., and Anagnostakis, S. L. 1977. Double-stranded RNA in *Endothia parasitica*. Phytopathology 67:1393-1396.
7. Gunge, N., Tamaru, A., Ozawa, F., and Sakaguchi, K. 1981. Isolation and characterization of linear deoxyribonucleic acid plasmids from *Kluyveromyces lactis* and the plasmid associated killer character. J. Bacteriol. 145:382-390.
8. Hashiba, T. 1982. Formation and fusion of protoplasts from *Rhizoctonia solani*. Jpn. J. Med. Mycol. 23:143-150.
9. Hashiba, T., and Yamada, M. 1982. Formation and purification of protoplasts from *Rhizoctonia solani*. Phytopathology 72: 849-853.
10. Hinnen, A., Hicks, J. B., and Fink, G. R. 1978. Transformation of yeast. Proc. Natl. Acad. Sci. USA 75:1929-1933.
11. Holling, M. 1978. Mycoviruses: Viruses that infect fungi. Adv. Virus Res. 22:3-53.
12. Holling, M. 1982. Mycoviruses and plant pathology. Plant. Dis. 66:1106-1112.
13. Homma, Y., Tamao, Y., and Katsube, T. 1983. Suppressive factors to Japanese radish damping-off in the soil reinoculated with *Rhizoctonia solani*. (Abstr.) Ann. Phytopathol. Soc. Jpn. 47:388.
14. Homma, Y., Yamashita, Y., Kuba, T., and Ishii, M. 1981. Difference among isolates in decline of Japanese radish damping-off in the soil reinoculated with *Rhizoctonia solani* Kühn. (Abstr.) Ann. Phytopathol. Soc. Jpn. 49:388.
15. Hyakumachi, M., Homma, Y., and Ui, T. 1983. Properties of diseased isolates of *Rhizoctonia solani* AG-4. (Abstr.) Ann. Phytopathol. Soc. Jpn. 49:120-121.
16. Hyakumachi, M., Homma, Y., and Ui, T. 1983. Properties of diseased isolates of *Rhizoctonia solani* AG-4. II. Cross protection with diseased isolates. (Abstr.) Ann. Phytopathol. Soc. Jpn. 49:374-375.
17. Kleinschmidt, A., and Zahn, R. K. 1959. Über deoxyribonucleinsäure-molekeln in protein-mischfilmen. Z. Naturforsch. 146:770.
18. Martini, G., Grimaldi, G., and Guardiola. J. 1978. Extrachromosomal DNA in phytopathogenic fungi. Pages 197-200 in: Genetic Engineering. H. W. Boyer and S. Nicosia, eds. Elsevier/North-Holland Biomedical Press.
19. Pring, D. R., Levings, C. S., III, Hu, W. W. L., and Timothy, D. H. 1977. Unique DNA associated with mitochondria in the "S"-type cytoplasm of male-sterile maize. Proc. Natl. Acad. Sci. USA 74:2904-2908.

Sharp Eyespot of Cereals and Rhizoctonia of Potato

T. W. HOLLINS, G. J. JELLIS, and P. R. SCOTT, Plant Breeding Institute, Cambridge, U.K.

Diseases caused by *Rhizoctonia* on cereals (sharp eyespot) and potato (black scurf and stem canker) were for many years attributed to strains of *Rhizoctonia solani* Kuehn (1,3). However, Boerema and Verhoeven (2) showed the cause of sharp eyespot to be a separate species that they named *R. cerealis* van der Hoeven, although they recognized that *R. solani* could also infect cereals.

Sharp eyespot has recently become more common in Britain (4) and appears equally common in winter wheat following potatoes and on wheat following wheat. This suggests that *R. solani* can cause sharp eyespot as shown in the United States (9) or that potatoes can be an alternative host for *R. cerealis*.

EXPERIMENT

Specificity of *Rhizoctonia* species. We isolated *Rhizoctonia* from cereals with sharp eyespot and from potatoes with black scurf and stem canker, collected mainly from eastern and southern England, and compared the isolates with known cultures of *R. cerealis* and *R. solani*. Only *R. cerealis* was obtained from

cereals (over 100 isolates) and only *R. solani* from potatoes (17 isolates).

Young plants of wheat and potato were inoculated with isolates of *R. cerealis* and *R. solani* in the glasshouse and growth rooms in six experiments over 4 years. *R. solani* infected potato substantially and not wheat, whereas *R. cerealis* infected wheat substantially and potato slightly (Fig. 1). Thus it seems that in Britain *R. cerealis* is the usual pathogen of cereals and *R. solani* the usual pathogen of potatoes. Our results also showed that potato can be an alternative host for *R. cerealis*, perhaps explaining in part the high frequency of sharp eyespot on wheat following potatoes. Furthermore, we found that an isolate from Germany, similar to *R. cerealis*, caused much disease on wheat and potato, confirming the findings of Hoffmann (6). However, we consider that in Britain inoculation with *R. cerealis* should provide an adequate test for resistance to sharp eyespot in cereals.

Effects of *R. cerealis* on cereal species and cultivars. Sharp eyespot was first recognized in Britain by Glynne and Ritchie (5), but it has been considered of minor importance (6). To assess the relative resistance of four cereal species, we inoculated wheat, barley, oats,

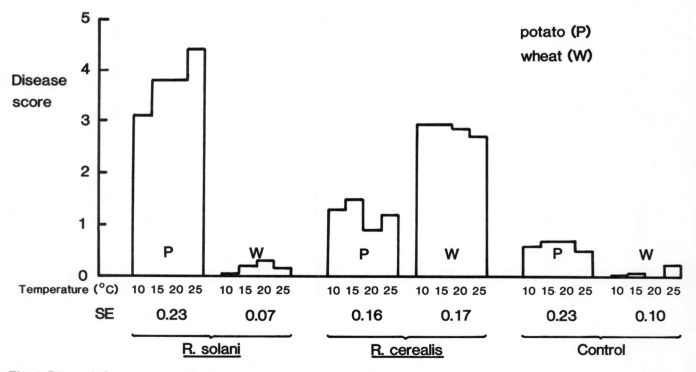

Fig. 1. Disease index on potato and wheat at four temperatures in growth rooms as caused by isolates of *Rhizoctonia solani* and *Rhizoctonia cerealis*. Mean of two experiments. Standard error 0.23 for *R. solani* on potato, 0.07 on wheat; 0.16 for *R. cerealis* on potato, 0.17 on wheat; control potato, 0.23 and 0.10.

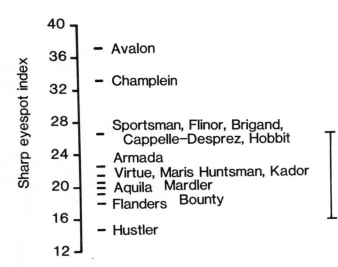

Inoculated field plots
(mean of 1980, 1981 and 1982)

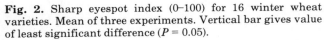

Fig. 2. Sharp eyespot index (0–100) for 16 winter wheat varieties. Mean of three experiments. Vertical bar gives value of least significant difference ($P = 0.05$).

and rye with *R. cerealis* in 3-m² plots at sowing in October. Inoculation reduced the number of plants per square meter in spring by 38, 50, 69, and 34% respectively (means of 1981 and 1982; six replicates each year). Surviving plants were able to compensate partially, however, except where disease was severe, so that by summer the number of fertile shoots per square meter was reduced by only 17, 22, 53, and 5% respectively. Yield reductions were more closely related to the amount of winter killing than to stem disease in summer; oats and barley were more affected than wheat and rye by winter killing (Table 1). Yield assessments on single shoots showed that even lesions girdling the stem and perhaps covering much of the first or second internode did little damage to yield. Severe lesions (those causing tissue softening) were too infrequent for reliable yield estimates.

R. cerealis is therefore capable of causing considerable damage to cereal crops during the winter by killing young plants, although such severe effects appear unusual in agriculture. It is probable that stem lesions on adult plants are the main source of potential damage.

For three consecutive years, plots of winter wheat measuring 0.1 m² were inoculated at sowing in October with a mixture of isolates of *R. cerealis*. Significant ($P < 0.05$) and consistent differences were found in the amount of sharp eyespot on commercial wheat cultivars in July (Fig. 2). Hustler and Flanders were always less diseased than Avalon and Champlein, although the ranking order of other cultivars was not always consistent.

Because variation in resistance to *R. cerealis* already exists among commercial winter wheat varieties, selection for resistance in a breeding program could eliminate more susceptible genotypes.

Table 1. Effect of inoculation of four cereals with *Rhizoctonia cerealis* in 3-m² plots[a]

| Cereal | Percentage reduction | | Disease index (0–100) |
	Fertile shoots (per m²)	Yield	
Wheat	17	5	20
Barley	21	14	18
Oats	53	46	7
Rye	5	5	12
LSD (5%)	...	...	87

[a] Means of 1981 and 1982 trials, each with six replicates.

DISCUSSION

There is little doubt that the causal fungus of sharp eyespot in cereals is *R. cerealis*. There is also good evidence to show that *R. solani* can infect cereal roots causing barley stunt disorder (7) and occasionally cereal stems (9), but much of the earlier work describing *R. solani* as a pathogen of cereals probably referred to *R. cerealis*.

We have demonstrated that *R. cerealis* can affect cereal crops by killing young plants during the winter as well as by causing stem lesions on adult plants. Such winter killing appears infrequent in British agriculture, and we have been unable to develop sufficiently severe stem lesions to reduce yield in trials at Cambridge. Although we have observed crops of winter wheat whose yield was probably reduced by sharp eyespot lesions, many of the stem lesions commonly seen during June and July on winter cereals probably cause minimal yield loss.

ACKNOWLEDGMENTS

We thank H. M. Crook, J. R. Paine, and N. C. Starling for much practical assistance.

LITERATURE CITED

1. Anonymous. 1967. Eyespot and sharp eyespot of wheat and barley. Minist. Agric. Fish. Advis. Leafl. 321. 7 pp.
2. Boerema, G. H., and Verhoeven, A. A. 1977. Check list for scientific names of common parasitic fungi. Series 2B. Fungi on field crops: Cereals and grasses. Neth. J. Plant Pathol. 83:165-204.
3. Brenchley, G. M., and Wilcox, H. J. 1979. Potato Diseases. Her Majesty's Stationery Office, London.
4. Clarkson, J. D. S., and Polley, R. W. 1981. Assessment of losses caused by stem base and root diseases of cereals. Proc. Br. Crop Prot. Conf.—Pests and Diseases, pp. 223-231.
5. Glynne, M. D., and Ritchie, W. M. 1943. Sharp eyespot of wheat caused by *Corticium* (*Rhizoctonia*) *solani*. Nature 152:160.
6. Hoffman, G. M. 1980. "Rhizoctonia"-ähnliche Pilze und *Rhizoctonia solani* Kühn (*Thanatephorous cucumeris* (Frank) Donk) an Getreide. Z. Pflanzenkr. Pflanzenschutz 87:317-327.
7. Murray, D. I. L. 1981. *Rhizoctonia solani* causing barley stunt disorder. Trans. Br. Mycol. Soc. 76:383-395.
8. Pitt, D. 1966. Studies on sharp eyespot disease of cereals. III. Effects of the disease on the wheat host and the incidence of disease in the field. Ann. Appl. Biol. 58:299-308.
9. Sterne, R. E., and Jones, J. P. 1978. Sharp eyespot of wheat in Arkansas caused by *Rhizoctonia solani*. Plant Dis. Rep. 62:56-60.

Saprophytic Survival of *Gaeumannomyces graminis* var. *tritici* in the Victorian Mallee, Australia

J. F. KOLLMORGEN and P. E. RIDGE, Victorian Crops Research Institute, Horsham, Victoria, 3400, and D. N. WALSGOTT, Mallee Research Station, Walpeup, Victoria, 3507, Australia

Take-all (*Gaeumannomyces graminis* (Sacc.) von Arx & Olivier var. *tritici* Walker) is a serious disease of wheat in the Mallee region of Victoria. The fungus persists between crops on volunteer cereals, grass weeds, or infected stubble. Depth of incorporation of stubble appears to be important. Chambers and Flentje (3) showed that *G. graminis* var. *tritici* survived better on segments of wheat straw buried at 2.5 cm than at 15.2 cm below the surface of a red-brown earth in South Australia. However, the effects of depth of burial on saprophytic survival have not been studied previously in Mallee soils.

Disk plows or tined scarifiers that incorporate stubble in the top 10 cm of soil 7–8 months before sowing are commonly used to cultivate fallows in the Mallee. Recently, however, there has been a trend toward tillage systems that retain stubble on the soil surface until just before sowing as a protection against erosion, and it is probable that this practice will affect the carry-over of *G. graminis* var. *tritici*.

We describe a series of experiments on the saprophytic survival of the fungus in wheat stubble both at the soil surface and at various depths in soil in the Mallee. The results are discussed in relation to the likely effects of changed tillage practices on the incidence of take-all. Some of the findings have previously been published (6).

MATERIALS AND METHODS

Investigations were initiated in 1981 and 1982 in a sandy-loam soil described by Chambers (2) at the Mallee Research Station, Walpeup, Victoria. Unless stated otherwise, experiments were of randomized-block design with four replications of each treatment.

Experiments with segments of wheat straw. In experiments 1, 2, and 3, segments of wheat straw, each 2.5 cm long with a node at one end, were sterilized by autoclaving at 100 kPa for 45 min on three successive occasions and then inoculated with cornmeal sand (1) cultures of *G. graminis* var. *tritici*. After incubation at 25°C for 4–5 weeks, straws were washed out of this medium and placed in fiberglass meshs bags measuring 16×20 cm (mesh size 1.5×1.5 mm). The bags were buried at depths of 0, 5, 10, or 15 cm, and about 75 cc of soil from the appropriate depth was added to each bag to ensure straw-to-soil contact. Experiments 1 and 2 were commenced on 14 July 1981 and experiment 3 on 26 August 1981. Cropping histories for the experimental sites were wheat-pasture-pasture (experiment 1), pasture-fallow in August 1980 (experiment 2), and oats-pasture-fallow in early August 1981 (experiment 3). At various times after burial (Table 1), straws were recovered and assessed for viable hyphae of *G. graminis* var. *tritici* by the wheat seedling test (5). Experiment 4 included five different types of straw (from cultivar Halberd) and four crossbreds, each with a different level of cellulose-digestible dry matter. Survival of *G. graminis* var. *tritici* on each straw type was measured at 0, 5, 10, and 15 cm at the site described for experiment 2. The experiment was initiated on 14 July 1981, and experimental procedures were as described previously except that the design was a randomized-block factorial (straw type × depth of burial).

Experiments with wheat crowns. For experiments 5–7, wheat crowns were grown in pots in the glasshouse and were infected by sowing seed between agar disks taken from actively growing cultures of *G. graminis* var. *tritici*. At maturity, the root system of each infected plant was washed and the crowns harvested by cutting plants 5 cm above and 5 cm below the base of the tillers. The crowns were inserted in plastic mesh bags 20×20 cm (mesh size 3×3 mm), and soil was added as in previous experiments. Subsequently (Tables 1 and 2),

Table 1. Percentage survival of *Gaeumannomyces graminis* var. *tritici* in wheat stubble at various depths in soil at the Mallee Research Station

| Depth (cm) | Segments of wheat straw[w] | | | | | | | | Crowns[x] | |
| | Experiment 1 | | Experiment 2 | | Experiment 3 | | Experiment 4[y] | | Experiment 5 | |
	10 wk	24 wk	10 wk	16 wk	8 wk	32 wk	10 wk	25 wk	7 wk	18 wk
0	58.5 a[z]	49.2 a	67.7 a	66.1 a	85.0 a	52.2 a	91.8 a	85.6 a	91.8 a	54.5 a
5	4.2 b	1.8 b	23.7 b	0.9 b	58.8 b	46.7 a	11.1 b	1.3 b	20.4 b	3.6 b
10	3.2 b	3.1 b	20.4 b	1.1 b	42.0 b	10.3 b	8.5 b	1.0 b	3.7 b	0.0
15	3.4 b	6.2 b	32.2 ab	3.3 b	40.7 b	6.0 b	11.4 b	2.7 b	3.7 b	0.0

[w] Based on 140 straws (experiments 1–3) and 72 straws (experiment 4).
[x] Based on 28 crowns (experiment 5).
[y] Straw type × survival interaction not significant ($P \leqslant 0.05$).
[z] Within each column, results with no letter in common are significantly different ($P \leqslant 0.05$) as determined by Duncan's multiple range test; means not followed by a letter are not included in the statistical analysis.

Table 2. Percentage survival of *Gaeumannomyces graminis* var. *tritici* in wheat crowns at the soil surface and at 10 cm after various cropping sequences at the Mallee Research Station[y]

Depth (cm)	Fallow-wheat-wheat	Fallow-wheat-rapeseed	Fallow-wheat-lupin
0	76.0 a[z]	66.1 a	55.4 a
10	23.9 b	28.8 b	12.4 b

[y]Based on 36 crowns; survival evaluated after 15 weeks.
[z]Within each column, results with no letter in common are significantly different ($P \leq 0.10$).

crowns were recovered and inserted in pots containing 300 cc of steam-treated soil and placed in the glasshouse. Two wheat seedlings were grown in each pot and, at tillering, their roots were examined for *G. graminis* var. *tritici* lesions. In experiment 5 (Table 1), crowns were positioned 0, 5, 10, and 15 cm below the soil surface on 11 September 1981 at the site used for experiment 3. In experiment 6 (Table 2), crowns were positioned on the soil surface or at a depth of 10 cm on 18 February 1982 in plots with previous cropping sequences of fallow-wheat-wheat, fallow-wheat-lupin, and fallow-wheat-rapeseed in 1979, 1980, and 1981, respectively. There were three replications of each cropping sequence and one bag at each depth in each plot. Experimental design was therefore a randomized-block factorial with three replications. Experiment 7 was begun on 1 April 1982 at a site with a cropping history of wheat-pasture-pasture. Bags containing crowns (five per bag) were placed on the soil surface and, at 2-month intervals over the following 18 months, crowns in four bags were bioassayed to detect viable hyphae of *G. graminis* var. *tritici*.

RESULTS

Survival of *G. graminis* var. *tritici* in straw segments and crowns was generally significantly greater on the soil surface than at 5, 10, or 15 cm (Table 1). There was no significant interaction ($P < 0.05$) between straw type and survival of *G. graminis* var. *tritici* (experiment 4). Thus, irrespective of cultivar or crossbred from which the straw was derived, survival on it was higher on the soil surface than at 5, 10, or 15 cm.

Table 2 shows that survival of *G. graminis* var. *tritici* in crowns was greater ($P < 0.10$) on the soil surface than at 10 cm after all three cropping sequences (experiment 6). In experiment 7, the bioassay showed that living hyphae of *G. graminis* var. *tritici* were present in 67% of infected crowns after 12 months (April 1983); but in the following 2 months there was a sharp decline in viability, and at 14 months (June 1983) no living hyphae were detected in any crowns.

DISCUSSION

The finding that burial of stubble colonized by *G. graminis* var. *tritici* at 5–15 cm caused a marked reduction in survival of the fungus is consistent with earlier observations by Chambers and Flentje (3). Garrett (5) demonstrated that warm and moist soil conditions stimulated a decline in viability of *G. graminis* var. *tritici*. However, in the Mallee, the soil surface is dry for prolonged periods, especially during the summer months, and moist conditions are usually

only encountered at depths below 5 cm. Although soil temperatures above 45°C occur in summer, these are unlikely to be lethal to *G. graminis* var. *tritici*, which has been shown to survive soil temperatures of 71°C (4). Thus, survival of *G. graminis* var. *tritici* might be expected to be higher on the soil surface than at depths below 5 cm. The rapid decline in viability of *G. graminis* var. *tritici* in crowns on the soil surface (experiment 7) after 12 months coincided with autumn rains that provided a warm, moist environment at the soil surface.

Survival of *G. graminis* var. *tritici* on artificially inoculated wheat straws has been shown to be poorer than that on naturally infected stubble (7). However, the purpose of the present study was to determine the relative effects of depth of burial rather than the absolute effects on survival, and there is no reason to suppose that artificially inoculated wheat straws would respond differentially to depth of burial. This is supported by the fact that results with crowns (experiments 5 and 6), which were not autoclaved, are in general agreement with those with artificially inoculated straws.

Although the data from this study suggest that deep plowing and soil inversion would be an effective control measure for take-all in the Mallee, this practice is inconsistent with the need to till soil economically and with minimal risk of erosion. It is clear, however, that tillage systems that leave infected stubble on the soil surface until just before seeding will favor carry-over of *G. graminis* var. *tritici*. In these systems, presowing cultivations and seeding itself incorporate at least 60–70% of the stubble into the upper 10 cm of the soil profile where it is ideally positioned to provide a source of inoculum to infect developing seedlings. The effect of reduced tillage systems would be of particular concern in short rotation sequences involving successive crops of susceptible cereals where there is limited time for decomposition of infected stubble.

ACKNOWLEDGMENTS

Funds for this investigation were provided by the Wheat Industry Research Committee of Victoria and the Commonwealth Wheat Industry Research Council.

LITERATURE CITED

1. Butler, F. C. 1953. Saprophytic behaviour of some cereal root rot fungi. III. Saprophytic survival in wheat straw buried in soil. Ann. Appl. Biol. 40:305-311.
2. Chambers, S. C. 1970. Pathogenic variation in *Ophiobolus graminis*. Aust. J. Biol. Sci. 23:1099-1103.
3. Chambers, S. C., and Flentje, N. T. 1968. Saprophytic survival of *Ophiobolus graminis* on various hosts. Aust. J. Biol. Sci. 21:1153-1161.
4. Fellows, H. 1941. Effect of certain environmental conditions on the prevalence of *Ophiobolus graminis* in soil. J. Agric. Res. 63:715-726.
5. Garrett, S. D. 1938. Soil conditions and the take-all disease of wheat. III. Decomposition of the resting mycelium of *Ophiobolus graminis* in infected wheat stubble buried in the soil. Ann. Appl. Biol. 25:742-766.
6. Kollmorgen, J. F., and Walsgott, D. N. 1984. Saprophytic survival of *Gaeumannomyces graminis* var. *tritici* at various depths in soil. Trans. Br. Mycol. Soc. 82:346-348.
7. MacNish, G. C., and Dodman, R. L. 1973. Survival of *Gaeumannomyces graminis* var. *tritici* in the field. Aust. J. Biol. Sci. 26:1309-1317.

The Changing Nature of Stalk Rot of Maize Caused by *Gibberella zeae*

THOR KOMMEDAHL and CAROL E. WINDELS, Department of Plant Pathology, University of Minnesota, St. Paul 55108, U.S.A.

Wherever maize (*Zea mays*) has been grown, stalk rot has proved to be a chronic disease that causes an appreciable loss in yield (3). The effect of stalk infection is either to interfere with the accumulation or storage of nutrients in the grain or to weaken stalks so that plants break and fall and ears cannot be harvested.

Three stalk rot pathogens have been reported to be important over a wide geographic area, under a wide variety of conditions, and over a long period of time: *Diplodia zeae*, *Gibberella zeae* (*Fusarium roseum* 'Graminearum'), and *F. moniliforme*. Because *D. zeae* was thought early to be the principal stalk rot pathogen, it received considerable attention in research (3). Later studies dealt with *Fusarium* species.

Other pathogenic species are reported to be important in specific locations or under restricted environmental conditions in certain seasons, and they are described in several reviews (3,4,6). In addition, a dozen or so fungi can cause stalk rot when maize plants are inoculated with them, or if plants are injured either by insects, hail, or tillage equipment. Also, such organisms frequently occur in stalk rot tissue as secondary organisms that may function as part of a complex to exacerbate damage from the primary pathogen.

Historically, *F. roseum* 'Graminearum' has been the most widespread stalk rot fungus consistently associated with stalk rot in the Corn Belt of the United States (6) as well as in other parts of the world (1,2,12). Early workers (3,4) consistently reported the occurrence of this fungus in rotted stalks and found perithecia of *G. zeae* on overwintering debris lying on the soil surface. Maize breeders used this fungus to inoculate maize in testing for stalk rot resistance in new inbreds or hybrids. Often *D. zeae* was found also and, in fact, was once reputed to be of greater importance than *G. zeae* in some parts of the Corn Belt. However, in recent years, *D. zeae* has seldom been isolated. This low incidence of *D. zeae* has been attributed to the success in breeding hybrids resistant to stalk rot that have reduced natural populations of this fungus, since maize is the only host. *Fusarium* species, on the other hand, have a wide range of hosts among the cereals and grasses.

In recent decades, more attention has been given to *F. moniliforme*, probably because it has been isolated so consistently from stalks over a wide geographic area where maize is grown. It impressed many maize researchers with its almost universal occurrence in grains (6,10) and stalks (3,4,6,14,15). Moreover, it was reported in air samples (16), rain splash (16), in or on insects (17,21), in hail-damaged stalks (11), and in downed ears from preceding crops of maize (21). Some workers have described how *F. moniliforme* washes into sheaths during a rain storm after anthesis when sheath collars tend to loosen around stalks. Rain or dew as well

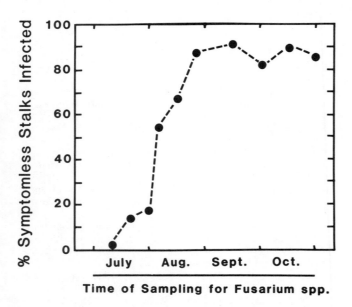

Fig. 1. Incidence of *Fusarium* species isolated from stalks of symptomless plants of maize at 2-week intervals over a 2-year period.

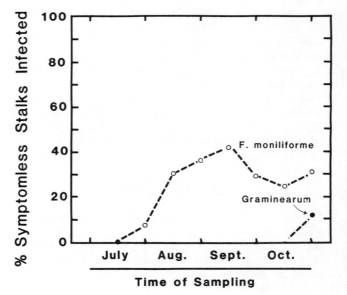

Fig. 2. Incidence of *Fusarium moniliforme* and *F. roseum* 'Graminearum' in stalks of symptomless plants of maize at 2-week intervals over a 2-year period.

as pollen and spores are carried into these open sheaths to provide a substrate and create a habitat for germinating spores of *F. moniliforme.* Infection then occurs at the nodes, and such infected plants are subject to breakage at these weakened nodes.

Several surveys of farmers' fields as well as a systematic schedule for isolating *Fusarium* species from stalks indicated that in Minnesota, *F. moniliforme* was a nearly ubiquitous parasite in stalks from about anthesis to maturity (5,6), which confirmed in part the observations that had been made much earlier by Koehler (4) and others (3). The incidence of *Fusarium* species present in symptomless plants is shown in Fig. 1, in which five *Fusarium* species are combined; it also includes several cultivars or populations of *F. roseum.* Surveys made in commercial fields in Minnesota over about a decade revealed that plants were almost universally infected with *F. moniliforme* without any symptoms of stalk rot (5,7–9,18,20). Not only *F. moniliforme,* but *F. tricinctum, F. oxysporum, F. solani,* as well as *F. roseum* 'Equiseti,' 'Acuminatum,' and occasionally 'Graminearum' were isolated (6).

Only after the plants start to ripen, or turn from green to a brownish tinge, did the incidence of stalks infected with *F. roseum* 'Graminearum' start to occur (Fig. 2). In another study (*unpublished*), the incidence of Graminearum-infected stalks increased from 40% in green (or turning to brown) stalks to 80% in brown and disintegrating stalks, when concurrent isolations were made in October–November from hundreds of plants. When stalks that had stood all winter in the fields were sampled at random in spring, nearly 60% were infected with Graminearum.

FUSARIUM ROSEUM 'GRAMINEARUM'

Graminearum can be isolated from roots of maize seedlings shortly after seed germination, especially when maize is grown on fields with a history of continuous maize culture and in which stalk fragments are buried by conventional tillage into the root zone of maize plants. If conditions favor disease development, seedlings may succumb to seedling blight. However, if environmental conditions shift to favor seedling development, Graminearum appears to be arrested in its growth and is confined to roots while the plant begins its "grand period of growth." When anthesis takes place, several *Fusarium* species, especially *F. moniliforme,* colonize stalks although no apparent damage or symptoms are seen. When environmental conditions, biotic or abiotic, change to create stress, the quiescent Graminearum resumes growth in roots and begins the invasion of the basal portions of stalks. Depending on the environment, rotting can occur rapidly. If stress is imposed in August when grains are denting and plants are still green, symptoms resembling wilt or early frost are visible, followed by browning and typical stalk rot symptoms. If stress occurs toward the end of the season, stalk rot symptoms are not visible until plants are maturing and becoming moribund. This leads to lodging and stalk breakage and loss of ears from inability of harvesting machinery to strip ears from stalks. As shown in Fig. 2, Graminearum appears late in the season and by November at least 80% of the disintegrated stalks and 60% of the stalks collected at random in spring are infected with Graminearum.

Seventy percent of these isolates proved to be pathogenic to maize when inoculated with them.

The source of inoculum of Graminearum is likely to be basal stem fragments of infected stalks (13,18). Of 900 such fragments left in the field on the ground for 1 year (18 fields), 30% yielded Graminearum. Soil free from debris rarely yielded Graminearum (19). When seedlings or mature plants were dug from fields of continuously cropped maize, roots were frequently found to have invaded old stalk tissues.

RELATIVE IMPORTANCE OF *FUSARIUM* SPECIES IN STALK ROT

The earlier workers who claimed that Graminearum was the predominant stalk rot pathogen, except in areas where *D. zeae* was not common, were probably right. Later workers who concluded that *F. moniliforme* was the major stalk rot pathogen were overly impressed with the almost universal presence of this fungus on all plant parts from anthesis to maturity. They assumed that stalks infected with *F. moniliforme* had been rotted by that pathogen.

We think that this is not so. *F. moniliforme* is probably a common parasite of maize that causes damage probably only under extraordinary circumstances. Normally, plants are not affected adversely by *F. moniliforme* as they grow to maturity. As plants mature their resistance mechanisms are no longer functional, and Graminearum is likely to overwhelm *F. moniliforme* and rapidly destroy stalk tissues so that plants are weakened and are blown down by wind. *F. moniliforme* and other species may play a role in hastening the decay of maize tissues and initiate the appropriate ecological sequence that ends with *G. zeae* as the climax species.

LITERATURE CITED

1. Burgess, L. W., Dodman, R. L., Pont, W., and Mayers, P. 1981. Fusarium diseases of wheat, maize and grain sorghum in eastern Australia. Pages 64-76 in: *Fusarium*: Diseases, Biology, and Taxonomy. P. E. Nelson, T. A. Toussoun, and R. J. Cook, eds. Pennsylvania State University Press, University Park. 457 pp.
2. Cassini, R. 1981. Fusarium diseases of cereals in Western Europe. Pages 56-63 in: *Fusarium*: Diseases, Biology, and Taxonomy. P. E. Nelson, T. A. Toussoun, and R. J. Cook. eds. Pennsylvania State University Press, University Park. 457 pp.
3. Christensen, J. J., and Wilcoxson, R. D. 1966. Stalk rot of corn. Monogr. 3. American Phytopathological Society, St. Paul, MN. 59 pp.
4. Koehler, B. 1960. Cornstalk rots in Illinois. Ill. Agric. Exp. Stn. Bull. 658. 90 pp.
5. Kommedahl, T., and Windels, C. E. 1977. Fusarium stalk rot and common smut in cornfields of southern Minnesota in 1976. Plant Dis. Rep. 61:259-261.
6. Kommedahl, T., and Windels, C. E. 1981. Root-, stalk-, and ear-infecting *Fusarium* species on corn in the USA. Pages 94-103 in: *Fusarium*: Diseases, Biology, and Taxonomy. P. E. Nelson, T. A. Toussoun, and R. J. Cook, eds. Pennsylvania State University Press, University Park. 457 pp.
7. Kommedahl, T., Windels, C. E., and Johnson, H. G. 1974. Corn stalk rot survey methods and results in Minnesota in 1973. Plant Dis. Rep. 58:363-366.
8. Kommedahl, T., Windels, C. E., and Stucker, R. E. 1979. Occurrence of *Fusarium* species in roots and stalks of symptomless corn plants during the growing season.

Phytopathology 69:961-966.

9. Kommedahl, T., Windels, C. E., and Wiley, H. B. 1978. *Fusarium*-infected stalks and other diseases of corn in Minnesota in 1977. Plant Dis. Rep. 69:692-694.

10. Kucharek, T. A., and Kommedahl, T. 1966. Kernel infection and corn stalk rot caused by *Fusarium moniliforme*. Phytopathology 56:983-984.

11. Littlefield, L. J. 1964. Effects of hail damage on yield and stalk rot infection in corn. Plant Dis. Rep. 48:169.

12. Maric, A. 1981. Fusarium diseases of wheat and corn in eastern Europe and the Soviet Union. Pages 77-93 in: *Fusarium*: Diseases, Biology, and Taxonomy. P. E. Nelson, T. A. Toussoun, and R. J. Cook, eds. Pennsylvania State University Press, University Park. 457 pp.

13. Nyvall, R. F. 1970. Chlamydospores of *Fusarium roseum* 'Graminearum' as survival structures. Phytopathology 60:1175-1177.

14. Nyvall, R. F., and Kommedahl, T. 1968. Individual thickened hyphae as survival structures of *Fusarium moniliforme* in corn. Phytopathology 58:1704-1707.

15. Nyvall, R. F., and Kommedahl, T. 1970. Saprophytism and survival of *Fusarium moniliforme* in corn stalks. Phytopathology 60:1233-1235.

16 Ooka, J. J., and Kommedahl, T. 1977. Wind and rain dispersal of *Fusarium moniliforme* in corn fields. Phytopathology 67:1023-1026.

17. Palmer, L. T., and Kommedahl, T. 1969. Root-infecting *Fusarium* species in relation to rootworm infestation in corn. Phytopathology 59:1613-1617.

18. Warren, H. L., and Kommedahl, T. 1973. Prevalence and pathogenicity to corn of *Fusarium* species from corn roots, rhizosphere, residues, and soil. Phytopathology 63:1288-1290.

19. Windels, C. E., and Kommedahl, T. 1974. Population differences in indigenous *Fusarium* species by corn culture of prairie soil. Am. J. Bot. 61:141-145.

20. Windels, C. E., and Kommedahl, T. 1976. *Fusarium* species in roots and stalks of corn in Minnesota in 1974 and 1975. (Abstr.) Proc. Am. Phytopathol. Soc. 3:292.

21. Windels, C. E., Windels, M. B., and Kommedahl, T. 1976. Association of *Fusarium* species with picnic beetles on corn ears. Phytopathology 66:328-331.

Collar Rot of Passion Fruit Possibly Caused by *Nectria haematococca* in Taiwan

Y. S. LIN and H. J. CHANG, Department of Plant Pathology, TARI, Wufeng, Taichung, Taiwan, Republic of China

The commercial plantings of purple passion fruit (*Passiflora edulis* Sims) in Taiwan were heavily damaged by a disease inciting collar rot and eventual girdling. The rotted collar zone usually swelled and split. Abundant perithecia of *Nectria haematococca* Berk. and Br. (syn. *Hypomyces solani* (Berk. and Br.) Snyder and Hansen) were found on the surface (Fig. 1). Cultures of this fungus with an asexual stage of homothallic *Fusarium solani* (Mart.) Sacc. were consistently isolated. P. J. Ann (*personal communication*) had previously obtained *Phytophthora parasitica* Dastur from rhizosphere soil of passion fruit, never from diseased tissue, and proved it to be pathogenic in greenhouse tests. Leu and Lee (5) found that *P. parasitica* from diseased leaves and fruits could attack leaf and fruit only, not the root. We have conducted pathogenicity tests of *N. haematococca* and *P. parasitica* on purple and yellow passion fruit in greenhouse and field. Resistant plants of yellow passion fruit were selected and used as root stocks in passion fruit cultivation in Taiwan. This paper reports the possible cause and the control of this disease.

MATERIALS AND METHODS

Isolation of possible pathogens. Pieces of diseased tissue with red perithecia were placed on the inner side of the lids of petri dishes overnight. The ascospores were ejected onto water agar by the next morning. Single ascospores were isolated. Diseased stem or root tissues without perithecia were also placed on pentachloro-nitrobenzene (PCNB) medium (9), BNPRA+HMI medium (3-hydroxy-5-methylisoxazole incorporated into potato-dextrose agar containing benomyl, nystatin, PCNB, rifampicin, and ampicillin) (6), and water agar (2%) for isolation of *Fusarium* spp., *Phytophthora* spp., and others. In other isolation attempts, soil samples were collected from the rhizosphere of diseased plants and put in sterilized water (300 g/700 ml) in a beaker. Pieces of passion fruit leaves were then floated on the water surface and incubated at 24°C for several days. All fungi that grew on the cut leaves were isolated. Cultures of pathogenic and saprophytic *F. solani* were also obtained from roots or basal stems of peas, beans, cucumber, coffee, and trifoliate orange for comparison purposes.

Inoculations in greenhouse. Several lines of yellow passion fruit (*P. edulis* f. *flavicarpa*) obtained from the Fengshan Tropical Horticultural Experimental Station and of purple passion fruit were used in inoculation experiments conducted in the greenhouse at temperatures of 20–30°C. The yellow passion fruit plants used were still in segregation, as evident from plant appearance. A virgin soil (loam) was infested with a spore suspension of *F. solani* (isolate FSPF, Table 1) and incubated in the greenhouse for 10 days. The infested soil had an inoculum density of 10^5 propagules per gram of soil when assayed on PCNB medium. Into this soil, seeds of yellow or purple passion fruit were sown. Another form of inoculum was sterilized wheat grains colonized by isolate FSPF in flasks. The grains were then dried, ground, and blended with virgin soil at 0.1, 0.5, or 2% (w/w). Checks were made with virgin soil amended with 0.1, 0.5, or 2% (w/w) sterilized wheat grains (Table 2). Lines of yellow passion fruit were preliminarily screened for resistance to isolate FSPF. The test seedlings were grown in pots. On inoculation, their roots were wounded with a scalpel cutting into the soil, and a spore suspension was poured into the pots. The collar zones of surviving plants (usually 2 months old) from the preliminary screening were further wounded and subjected to a second inoculation by adding to the pot 10 ml of ascospore or conidium suspension (10^5 spores per milliliter) of isolate FSPF, or 10 ml of zoospore suspension (10^4 spores per milliliter) of *P. parasitica*, or attaching to the wound a disk of mycelium (1 cm in diameter) of FAPF. Cross inoculations with *F. solani* from passion fruit (isolate FSPF) and *F. solani* from other sources (Table 1) were also made to passion fruit and other plants including peas, beans, cucumber, and trifoliate orange by the wounding inoculation method in greenhouse. The results for passion fruit are given in Table 3.

Inoculation in the field. A field experiment was conducted at the Taiwan Agricultural Research Institute. The seedlings of yellow and purple passion fruit were grown in pots for 2 months, then transplanted to the field. After 2 more months, the field-grown plants of passion fruit were inoculated either with a zoospore suspension (10^4 spores per milliliter) of *P. parasitica* or with mycelium disks of isolate FSPF applied to the wounded or unwounded collar zone. Table 4 presents the survival data of plants wounded at the collar zone. Twenty-four plants were used per treatment. The trial, which was set up in April with a completely randomized design, included three replicates. The dead plants were counted and subjected to an isolation of the causal organism 1 year after inoculation.

Contribution 1120 from Taiwan Agricultural Research Institute.

RESULTS

Isolation of possible pathogens. *F. solani* was consistently isolated from the field specimens of diseased plants of passion fruit. An isolate from passion fruit and other isolates from peas, beans, cucumber, coffee, trifoliate orange, and sugarcane soil were subjected to single-spore isolation on potato-dextrose agar. All cultures were incubated at 20–25°C under a 12-hr light period. An *F. solani* isolate supplied by Dr. Shirley N. Smith of the University of California at Berkeley was subjected to the same isolation procedure. A comparison revealed that only the isolate from passion fruit (FSPF) and that originating from the United States (FSUSA) were homothallic cultures. These isolates produced perithecia identifiable as *N. haematococca* (1).

P. parasitica was occasionally obtained from rhizosphere soil but rarely from diseased tissues of passion fruit plants, which usually developed a rot extending from the collar zone up to the stem that caused vines to die (Fig. 2).

Inoculation in greenhouse. Pathogenicity tests conducted by infesting soils with wheat grain inoculum or chlamydospores of isolate FSPF yielded damping-off symptoms different from those observed in the field (Table 2). However, isolate FSPF was pathogenic to purple passion fruit by wound inoculation using spore suspension or mycelium disk as inoculum (Table 3). Close examination revealed that the plants wilted as a consequence of stem girdling from collar rot. The fungus could not attack unwounded plants, and yellow passion fruit was resistant (Table 3). Most inoculated plants of yellow passion fruit recovered within 2 weeks when spore suspension was used as inoculum. However, a few plants died a year after inoculation with mycelial disks.

Table 1. Sources of fungi tested

Designation	Fungus	Perithecium[a]	Source
FSUSA	*Fusarium solani*	+	Supplied by S. N. Smith (U.S.A.)
FSPF	*F. solani*	+	Diseased tissue of passion fruit
	F. solani	+	Perithecia on diseased tissue of passion fruit
FSP	*F. solani* f. sp. *pisi*	−	Root of peas
FSB	*F. solani*	−	Root of beans
FSC	*F. solani*	−	Root of cucumber
FSCO	*F. solani*	−	Root of coffee
FSTO	*F. solani*	−	Perithecia on diseased basal stem of trifoliate orange
FSS	*F. solani*	−	Soil of sugarcane field
PP	*Phytophthora parasitica*		Rhizosphere soil of passion fruit

[a]Perithecia produced on single-spored cultures.

Table 2. Number of passion fruit plants surviving in virgin soil infested with wheat grain inoculum or chlamydospores of *Fusarium solani* (FSPF) from passion fruit[a]

Variety	Wheat inoculum/soil (w/w)				Chlamydospores (10^5 propagules/g of soil)
	Check	0.1%	0.5%	2.0%	
Purple	30	21	21	6	18
Yellow	30	30	24	6	20

[a]Thirty plants were used per treatment. Dead plants showed symptoms of damping-off only.

Fig. 1. Collar rot of passion fruit with perithecia on the surface (arrow) in field.

Table 3. Degree of infection of yellow and purple passion fruit plants inoculated 382 days previously in the greenhouse with different fungi

		Mycelium disk						Spore suspension					
		Yellow variety			Purple variety			Yellow variety			Purple variety		
Treatment	Fungus[a]	0[b]	1	2	0	1	2	0	1	2	0	1	2
Wounded	F	10	11	9	3	5	22	27	3	0	11	1	18
	P	21	0	9[c]	1	0	29	30	0	0	3	0	27
	P + F	13	10	7	0	0	30	21	6	3	0	1	29
	Check	30	0	0	29	0	1[c]	30	0	0	30	0	0
Not wounded	F	30	0	0	30	0	0	30	0	0	27	0	3
	P	...	...	...	...	...	...	30	0	0	4	0	26
	P + F	...	...	...	...	...	...	29	0	1	1	0	29
	Check	...	...	...	...	...	...	30	0	0	30	0	0

[a]F = homothallic *Fusarium solani* (FSPF); P = *Phytophthora parasitica*.
[b]Disease rating: 0 = healthy, 1 = lesions, 2 = wilted.
[c]Killed by homothallic *Fusarium solani*.

P. parasitica was pathogenic to both wounded and unwounded purple passion fruit. The inoculated plants developed dead vines within 2 weeks. Yellow passion fruit, however, showed some degree of resistance. Inoculation with *P. parasitica* and isolate FSPF to passion fruit resulted in a synergistic effect and caused wilt symptoms within a week (Table 3).

The U.S. isolate of *F. solani* (FSUSA), thought to be a saprophyte, had a similar cultural appearance on potato-dextrose agar and pathogenicity to passion fruit to our isolate (FSPF). However, neither isolate was pathogenic on peas, beans, cucumber, or trifoliate orange. Moreover, pathogenic and nonpathogenic isolates of *F. solani* from peas, beans, cucumber, coffee, trifoliate orange, and soil from a sugarcane field failed to attack passion fruit after wound inoculation.

Inoculation in the field. The results (Table 4) were similar to the greenhouse inoculation tests in that the purple passion fruit was susceptible and the yellow one was resistant. A few check plants of purple passion fruit died. Inspection of the roots of the diseased plants revealed a typical collar rot. Homothallic *F. solani* was also reisolated from such diseased plants.

DISCUSSION

Brun (reviewed in Emechebe and Mukiibi [4]) isolated *Hypomyces* sp. from passion fruit with a canker symptom. Simmonds (12) stated that a basal rot was caused by an unidentified organism when plants were wounded. This disease was different from Fusarium wilt. Young (16) found *F. sambucinum* attacking the wounded basal stem and causing a collar rot. In 1976, Emechebe and Mukiibi (4) provided further evidence that *N. haematococca*, the ascogenous state of homothallic *F. solani*, could attack wounded basal stems. However, Reichle et al (11) and Snyder and Hansen (13) believed that most homothallic *F. solani* were saprophytic. Our results, based on both greenhouse and field trials, leave little doubt that the two homothallic *F. solani* isolates FSPF and FSUSA cause collar rot of passion fruit. Plants showed damping-off symptoms only when inoculated with wheat grain inoculum or chlamydospores of isolate FSPF (Table 2). On the contrary, when plants were inoculated with isolated FSPF and FSUSA by the wounding inoculation method (Tables 3 and 4), the diseased plants showed collar rot symptoms similar to those found in the natural field. The isolates attacked passion fruit plants through wounds, purple plants usually being susceptible while some lines of yellow passion fruit were resistant. However, neither isolate was pathogenic to peas, beans, cucumber, or trifoliate orange. Moreover, *F. solani* from peas, beans, cucumber, coffee, trifoliate orange, and soil of sugarcane field failed to attack passion fruit by wound inoculation. It is of interest that *N. haematococca*, which occurs widely in the tropics, could damage citrus in Taiwan (3). Our results with isolates FSPF and FSTO from perithecia on diseased passion fruit and trifoliate orange, respectively, indicate their differences in pathogenicity and sexual reproduction (Table 1). Further studies are needed.

Several previous reports (12,14,15) stated that *P. parasitica* caused collar rot of passion fruit. Isolates of *P. parasitica* were mostly from young diseased tissue. We also obtained this fungus from inoculated plants in greenhouse and field but we were unable to isolate the same fungus from natural diseased specimens. Only a few isolates of *P. parasitica* were occasionally recovered from rhizosphere soil of passion fruit plants having dead vines in the field. P. J. Ann (*personal communication*) also did not obtain this fungus from diseased plants affected with collar rot . Leu and Lee (5) reported that *P. parasitica* could not attack roots of purple passion fruit. However, they failed to describe the inoculation method in detail. Our isolate of *P. parasitica* had very strong virulence on purple passion fruit, causing dead vines in inoculation tests. Yellow passion fruit was resistant. Field observations revealed that most diseased plants of purple and yellow passion fruit had collar rot, but not dead vines. We believe that *P.*

Fig. 2. Dead vine of passion fruit in field.

Table 4. Average number of passion fruit plants surviving when inoculated with *Fusarium solani* (FSPF) and *Phytophthora parasitica* by wound method in field[a]

| Inoculum[b] | Variety | | Fungus reisolated | |
	Yellow	Purple	F	P
F	23	11	+	−
P	24	4	+	+
P + F	20	3	+	+
Check	24	19	+	−

[a]Completely randomized design with three replicates. Twenty-four plants were used per treatment.
[b]F = homothallic *Fusarium solani* (FSPF); P = *Phytophthora parasitica*.

parasitica may be a pathogen of passion fruit, though not an important one, under field conditions of Taiwan.

Yellow passion fruit has been shown to be resistant to *F. oxysporum* f. sp. *passiflorae* (10), *P. parasitica* (2,7), and an undiagnosed wilt disease in Hawaii (8). Emechebe and Mukiibi (4) proved the pathogenicity of *N. haematococca* on purple passion fruit but did not test the fungus for its reaction on yellow passion fruit. Our results, therefore, provide strong evidence that some lines of yellow passion fruit are resistant to this fungus. During 1981–1982, the resistant plants were selected and used as resistant root stocks for an 800-ha passion fruit plantation in Taiwan, indicating that this disease is well controlled.

ACKNOWLEDGMENT

This research was supported in part by the Council of Agricultural Planning and Development.

LITERATURE CITED

1. Booth, C. 1971. The Genus *Fusarium*. Commonw. Mycol. Inst., Kew, Surrey, England. 237 pp.
2. Brodrick, H. T., Milne, D. L., Wood, R., and Mulder, J. 1976. Control of Phytophthora stem-rot of granadillas in South Africa. Citrus Subtrop. Fruit J. 508:15-17.
3. Cheng, C. C. 1955. A new citrus disease caused by a *Nectria* sp. Bull. Entome-Phytopathol. Soc. (Chung-hsing Univ., Taiwan) 6(2):14-16. (Abstract in Chinese)
4. Emechebe, A. M., and Mukiibi, J. 1976. Nectria collar and root rot of passion fruit in Uganda. Plant Dis. Rep. 60:227-231.
5. Leu, L. S., and Lee, C. C. 1976. Phytophthora blight of *Passiflora edulis* Sims. Plant Prot. Bull. (Taiwan, R.O.C.) 18:286-292 (In Chinese with English summary).
6. Masago, H., Yoshikawa, M., Fukada, M., and Nakanishi, N. 1977. Selective inhibition of *Pythium* spp. on a medium for direct isolation of *Phytophthora* spp. from soils and plants. Phytopathology 67:425-428.
7. Milne, D. L., De Villiers, E. A., Logie, J. M., Bredell, G. S., Barnard, C. J., and Kuhne, F. A. 1977. Growing grafted granadillas. Citrus Subtrop. Fruit J. 524:16-18.
8. Nakasone, H. Y., Hirano, R., and Ito, P. 1967. Preliminary observations on the inheritance of several factors in the passion fruit (*Passiflora edulis* L. and forma *flavicarpa*). Hawaii Agric. Exp. Stn. Tech. Prog. Rep. 161.
9. Nash, S. M., and Snyder, W. C. 1962. Quantitative estimation by plate counts of propagules of the bean root rot *Fusarium* in field soils. Phytopathology 52:567-572.
10. Purss, G. S. 1958. Studies of the resistance of species of *Passiflora* to Fusarium wilt (*F. oxysporum* f. sp. *passiflorae*). Queensl. J. Agric. Sci. 15:95-99.
11. Reichle, R. E., Snyder, W. C., and Matuo, T. 1964. *Hypomyces* stage of *Fusarium solani* f. *pisi*. Nature 203:664-665.
12. Simmonds, J. H. 1936. Passion vine diseases. Queensl. Agric. J. 45:322-330.
13. Snyder, W. C., and Hansen, H. N. 1965. Species concept, genetics, and pathogenicity in *Hypomyces solani*. Phytopathology 44:338-342.
14. Turner, G. J. 1974. Phytophthora wilt and crown rot of *Passiflora edulis*. Trans. Br. Mycol. Soc. 62:59-63.
15. Van den Boom, T., and Huller, I. M. 1970. Phytophthora stem rot of passion fruit, *Passiflora edulis*, in South Africa. Phytophylactica 2:71-74.
16. Young, B. R. 1970. Root rot of passion fruit vine (*Passiflora edulis* Sims) in the Auckland area. N.Z. J. Agric. Res. 13:119-125.

Survival of *Phytophthora cinnamomi* in Eucalyptus Roots Buried in Forest Soils

ALEX MACKAY and GRETNA WESTE, Botany School, University of Melbourne, Parkville, Victoria, 3052, Australia

There is considerable evidence that *Phytophthora cinnamomi* Rands, a root and collar rot pathogen with a wide host range, may persist between hosts as resistant chlamydospores, either within root tissue or independently in host-free soil. Previous experiments have demonstrated but not clearly defined certain conditions that may control the survival of the fungus in soil and root debris. Mircetich and Zentmyer (2) recorded saprophytic growth of the fungus in nonsterile soils and survival up to 6 years in the absence of a host, provided the sandy loam remained moist. Reeves (4) and Shea et al (5) have demonstrated saprophytic growth and survival in nonsterile soils. Chlamydospores were shown to survive and even to increase in numbers in various nonsterile, host-free soils and gravels if these contained organic material and sufficient moisture (8). Maximum numbers of chlamydospores were produced at a matric potential of −5 bars. Infected seedling roots buried in a range of media contained viable chlamydospores after 30 days (6). Soil moisture potential affected percentage survival of the pathogen in root fragments buried for 56 days (3). Experiments have thus demonstrated that *P. cinnamomi* survives frequently as chlamydospores in buried roots and that such survival may depend on the maintenance of a suitable matric water potential. More experimental work is needed to define conditions limiting survival and to relate them to chlamydospore production and survival, both within host-free nonsterile soil and within living and dead roots. This paper reports studies of the survival of two isolates of *P. cinnamomi* in host roots of two species of *Eucalyptus* seedling roots buried in different conducive and suppressive forest soils and the effect of soil matric potential on such survival.

MATERIALS AND METHODS

Three-month-old seedlings of *Eucalyptus maculata* Hook. (field resistant) and *E. sieberi* L.A.S. Johnson (susceptible) were inoculated with axenically prepared zoospore suspensions of *P. cinnamomi* placed on the root tips. The zoospores were taken from two isolates of *P. cinnamomi* collected from different parts of Victoria, Australia. The inoculated roots were excised and buried in plastic boxes in four different types of soil, three conducive and one suppressive (1). The soils were treated in a pressure plate apparatus so that each soil had a matric potential of −1/3, −5, or −10 bars. The boxes were weighed, sealed inside plastic bags, and incubated at 21°C. After periods of 10, 100, and 200 days, the boxes

were reweighed and the roots were recovered and cut into segments. Portions of each segment were surface-sterilized and plated on $P_{10}VP$ agar. The percentage of root colonized by the pathogen was calculated, and the number of chlamydospores that formed as a result of zoospore inoculation was counted in teased-out root tissue with the aid of a microscope and hemocytometer.

RESULTS

Incubation for 10 days. *P. cinnamomi* was reisolated from all the inoculated roots; no chlamydospores were observed after 10 days of burial, however, and mycelium emerged from the cut ends and sides of the root segments. There was greater fungal colonization of *E. sieberi* than of *E. maculata* roots (Fig. 1). In the latter species, the fungus was confined to a localized lesion at the site of inoculation, whereas in *E. sieberi* the pathogen had extended along the main root and into the first lateral branch. Statistical analysis of the results after 10 days of burial showed that the difference in percentage of roots colonized between the two eucalypt species was significant but that there was no significant difference between pathogen isolates. The amount of water lost from the soil samples was less than 0.3%, indicating that the soil matric potentials remained relatively constant during the 10-day period. This was probably an important factor in the survival of the mycelium.

Incubation for 100 days. *P. cinnamomi* was reisolated from all the inoculated roots and grew from chlamydospores. The maximum water lost by the soils in 100 days was less than 1% and occurred in the wetter soils ($\Psi_m = -1/3$ bar). Therefore soil matric potentials remained relatively constant.

Because there were no significant differences between the two isolates of the pathogen with respect to chlamydospore numbers or to the percentage of root infected, these results were pooled. For the complete range of water potentials, differences between the two *Eucalyptus* species were significant with reference to the percentage of root colonized but not with respect to chlamydospore numbers. There was thus no difference between susceptible and field-resistant species with regard to viability of the pathogen.

Incubation for 200 days. The pathogen was not isolated from any roots. No chlamydospores were viable. Despite all precautions there had been considerable loss of moisture from all soils, and final matric potentials were less than −10 bars.

DISCUSSION

After 100 days, significant differences were recorded between soil types, soil matric water potentials, and *Eucalyptus* species. The two isolates of *P. cinnamomi* behaved similarly. The interaction of soil type and matric potential on root colonization was highly significant, and this was largely due to the difference between colonization of suppressive and conducive soils at −1/3 bar. The interaction of *Eucalyptus* species and soil matric water potential on root colonization was also highly significant, declining with decreasing water potential. In other words, lack of moisture overrode the difference in susceptibility, and dry roots were not heavily colonized. The variability in results at 100 days was only half that at 10 days.

Numbers of viable chlamydospores. The results

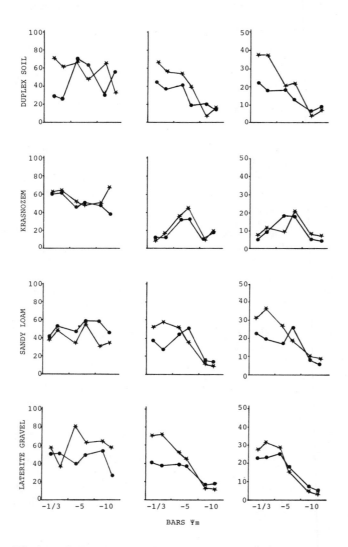

Fig. 1. Colonization of excised *Eucalyptus* roots by *Phytophthora cinnamomi* 10 days (column 1) and 100 days (column 2) after burial in four different soils; numbers of viable chlamydospores within excised roots 100 days after burial (column 3). The y-axis is soil matric water potential. Roots tested were *E. sieberi* (susceptible; starred line) and *E. maculata* (field resistant; circled line). Because there was no significant difference between the two isolates of *P. cinnamomi* tested, results were pooled. The difference in results was highly significant ($P < 0.01$) for the two *Eucalyptus* species.

obtained from chlamydospore counts were similar to and supported those from root colonization. There was a high correlation between the two sets of results ($R = 0.87$). Both soil type and soil matric water potential were highly significant factors influencing chlamydospore numbers. The interaction between soil type and soil water potential on chlamydospore numbers was highly significant, and the interaction between *Eucalyptus* species and soil water potential was significant. The lack of survival after 200 days was probably due to drying of the soils.

Soil matric water potential influenced all results: colonization of roots, chlamydospore numbers, viability of the fungus, and soil suppression (Fig. 1).

For the three conducive soils, the proportion of roots infected decreased from 50 to 20% with reducing water potential, and there were no significant differences between the soils. However, the suppressive soil inhibited isolations from roots buried at the matric water potential of −1/3 bar; at water potentials of −5 and −10 bars, the inhibition ceased, and pathogen survival decreased with reducing water potential as with the conducive soils. Maximum amount of root colonized and maximum number of chlamydospores therefore occurred at −1/3 bar in the three conducive soils but at −5 bars in the suppressive soil.

Survival of *P. cinnamomi* was greater in roots of the susceptible *E. sieberi* than in roots of the field-resistant *E. maculata* in that greater numbers of chlamydospores germinated from *E. sieberi* than from *E. maculata* in all conducive soils at −1/3 bar, but differences decreased in drier soils. Chlamydospores were present, but they occurred in low numbers in roots buried in the suppressive soil at −1/3 bar.

Chlamydospore numbers, the percentage of root colonized, and the survival of the pathogen were maximum in the susceptible species buried in conducive soils at −1/3 bar. However, suppressive soil had no inhibitory action at −5 or −10 bars matric water potential. Suppression is associated with high population densities of soil microorganisms (8). Major factors influencing chlamydospore survival are therefore probably microbial numbers and soil matric potentials. There are serious implications for forest managers in the 100% survival of the pathogen in roots of both susceptible and field-resistant hosts.

LITERATURE CITED

1. Marks, G. C., and Smith, I. W. 1983. Development of *Phytophthora cinnamomi* infection in roots of *Eucalyptus* species growing in a soil that suppresses Phytophthora root diseases. Aust. J. Bot. 31:239-245.
2. Mircetich, S. M., and Zentmyer, G. A. 1966. Production of oospores and chlamydospores of *Phytophthora cinnamomi* in roots and soil. Phytopathology 56:1076-1078.
3. Old, K. M., Oros, J. M., and Malafant, K. W. 1984. Survival of *Phytophthora cinnamomi* in root fragments in Australian forest soils. Trans. Br. Mycol. Soc. In press.
4. Reeves, R. 1975. Behaviour of *Phytophthora cinnamomi* Rands in different soils and water regimes. Soil Biol. Biochem. 7:19-24.
5. Shea, S. R., Gillen, K. J., and Leppard, W. I. 1980. Seasonal variation in population levels of *Phytophthora cinnamomi* Rands in soil in diseased, freely drained *Eucalyptus marginata* Sm. sites on the northern jarrah forest of

southwestern Australia. Prot. Ecol. 2:135-156.

6. Weste, G. 1983. *Phytophthora cinnamomi*: The dynamics of chlamydospore formation and survival. Phytopathol. Z. 106:163-176.

7. Weste, G., and Vithanage, K. 1977. Microbial populations of forest soils. Aust. J. Bot. 25:377-383.

8. Weste, G., and Vithanage, K. 1979. Survival of chlamydospores of *Phytophthora cinnamomi* in several non-sterile, host-free forest soils and gravels at different soil water potentials. Aust. J. Bot 27:1-9.

The Rhizoctonia Disease Complex of Wheat

RITA MOEN and J. R. HARRIS, CSIRO Division of Soils, Glen Osmond, South Australia, 5064

The Rhizoctonia disease complex of wheat and barley is associated with many endemic fungi, bacteria, viruses, and nematodes (4,8,9,11). Synergistic effects are commonplace (1), and because many of the associated organisms are categorized as minor pathogens (2) they are often overlooked as sources of damage to roots. *Rhizoctonia solani* Kuehn infections have been investigated in tests where the fungus-damaged roots are in an otherwise axenic environment; or, where the associated organisms have been tested for pathogenicity singly, they have proved to be less damaging than the isolates of *R. solani*.

MATERIALS AND METHODS

The Rhizoctonia disease complex of wheat was studied in pot experiments involving a moderately fertile, calcareous red-brown earth soil of principal profile form (10) Dr2.23 from Colley Hill, Eyre Peninsula. Wheat seeds cultivar Warigal were surface-sterilized with 1.5% sodium hypochlorite (15 min), washed three times in sterile water, and sown in small pots of perlite containing inoculum of *R. solani* that had been inoculated from active growth on potato-dextrose agar (PDA) plates. After 3 weeks, three seedlings were transplanted into duplicate pots containing 500 g of the Colley Hill soil in which the moisture was maintained at 17% w/w water-holding capacity. Parallel treatments compared natural soil to soil sterilized by autoclaving, and fertilized to unfertilized soil. The persistence of five strains of *R. solani* in infected tissue was investigated by sacrificing a set of plants each week beginning 19 days after transplanting. Root segments were plated onto PDA and a PDA amended with 50 ppm streptomycin sulfate, 50 ppm neomycin sulfate, and 250 ppm chloramphenicol (3). The growth rate of the wheat plants was measured by root weight, herbage weight, and leaf area (planimetrically).

RESULTS

By the fourth harvest (40 days), only one strain of *R. solani* was recoverable from wheat roots. This strain had persisted in both the natural soil and the sterilized soil; but after the second harvest (26 days), the four other strains were isolated only from sporadic lesions.

Despite the infections that had been established prior to transplanting, the plants grew comparatively well with few outward signs of disease. Plants in the natural soil outyielded those in the sterilized soil until after the fourth harvest. Total dry weight data for this experiment are presented in Table 1; however, the main effect of *R. solani* attack was found to be upon root weight, as shown in Table 2. The growth of roots in natural soil was always less than that in sterilized soil,

and it was reduced most severely for the most persistent strain of *R. solani* (H-5). Table 2 presents the percentage root growth for each strain compared with its uninoculated control.

Associated fungal infections. The number of fungi associated with wheat roots throughout the experiment continued to increase with time, the diversity of species being greatest in the natural soil and least in the sterilized soil. At the first harvest, nine species were identified; by the fourth harvest, this had increased to 14. Two species of *Fusarium, F. equiseti* (Corda) Sacc. and *F. avenaceum* (Corda ex Fr.) Sacc. dominated; less frequent were *Cochliobolus sativus* (Ito & Kuribay) Drechs. ex Dastur, *Trichoderma viride* Pers. ex Gray, and *Alternaria alternata* (Fr.) Keissler; appearing later were *Macrophomina phaseolina* (Tassi) Goid., *Microdochium bolleyi* (Sprague) de Hoog & Herman., *Periconia macrospinosa* Lefeb. & Johnson, and species of *Penicillium* and *Aspergillus*.

In the sterilized soil, most of the fungi were seedborne, as was demonstrated by plating out surface-sterilized seed. The isolates were dominated by *Alternaria alternata*, with lesser numbers of *Microdochium bolleyi* and, later, *Aspergillus* sp. (*niger* group) and *Penicillium* spp. The five inoculated strains of *R. solani* caused no significant differences in plant weight between strains and uninoculated controls, a situation different from that in natural soil.

The earliest stages of a *Rhizoctonia* infection may be mildly stimulating to the growth of the wheat plant. All inoculated strains of *R. solani* in the sterilized soil increased herbage weights and leaf areas, but not root weights, over uninoculated controls in the early harvests. This was never the case in the natural soil, so it must be assumed that secondary infections by associated microorganisms had early deleterious effects upon plant growth (Table 1).

When we tested the response of wheat to an inoculated *Rhizoctonia* infection in four sandy soils of low fertility in a previous experiment, the amount of damage to the plant varied considerably from soil to soil. This was an expression of the differences in microfloras and possibly of some soil factors.

Associated bacterial infections. When root segments were plated onto agar media, bacteria were isolated from both clean and rotting roots. This was most apparent in samples from the sterilized soil where fungal outgrowths were fewer, but it was a consistent feature from the natural soil also. When clean root tips were plated, it was obvious that bacteria colonized root surfaces faster than fungi. Replating to purify by selecting discrete colonies yielded three classes of bacteria: Bacillaceae (*Bacillus*), Enterobacteriaceae (*Erwinia*), and a miscellaneous group.

A range of these bacteria was screened for in vitro antagonism against four pathogenic fungi: *F. graminearum* Schwabe, *Gaeumannomyces graminis* (Sacc.) von Arx & Olivier var. *tritici* Walker, *Cochliobolus sativus*, and *R. solani*. Some that exhibited large zones of inhibition were tested for their ability to protect wheat from infection by *C. sativus* and *R. solani*. Two strains of a fluorescent *Pseudomonas* sp. that had been selected for antagonism to *G. graminis* var. *tritici* (14) were included in the trial.

Wheat (cultivar Warigal) was surface-sterilized, pregerminated on sterile filter paper, and transferred into tubes of a sterilized, coarse (Waikerie) sand moistened with plant nutrient solution. The two pathogenic fungi were grown on PDA and provided inocula of 8-mm plugs placed 4 cm below the seed. The test bacteria were washed from PDA plates with sterile water, and the suspension was added to the sand above the emerging seedling. Each test bacterium was used in combination with each pathogenic fungus and an uninoculated (fungus-free) control.

None of the bacteria had any effect upon the growth of wheat in terms of measurements of root length (13). However, in the presence of *C. sativus*, all of the bacteria except one strain of *P. fluorescens* increased the severity of disease, as was evident from decreases in root length. The fluorescent pseudomonad produced a mild stimulus of root growth, indicating that some protection was provided.

Rhizoctonia damage was more severe when any one of the bacteria was present, and the reduction in root growth of wheat was marked.

Influence of nutrition. The etiology of the Rhizoctonia disease complex may influence plant nutrition. An extension of the same pot experiment with wheat inoculated by *Rhizoctonia*, using natural and sterilized Colley Hill soil, was used to provide an additional comparison between unfertilized soil and soil fertilized by NPK between tillering and heading.

Up to the stage of tillering, the growth of wheat in the sterilized soil lagged behind that in natural soil. After tillering, growth in sterilized soil increased dry matter production tenfold by time of heading; in natural soil the increase was fivefold, or half that of the growth rate in sterilized soil (Table 1). We attributed this to the effects of the secondary infection, which occurred strongly in the natural soil but was mild in the sterilized soil (4).

When fertilizer was applied, the overall response was greater in natural soil than in sterilized soil. There were no significant differences between the inoculated *Rhizoctonia* strains and the uninoculated controls in the sterilized soil. In the natural soil, the most virulent strains of *R. solani* did show treatment differences, particularly by the depression of root weights shown in Table 2. In Table 1, the low slope and low intercept values of the regression equations for the first four harvests indicate the strains of *R. solani* that have been most aggressive. When the production of roots had been reduced by secondary infections, the plant was less able to respond to NPK fertilization and the deficit widened. This lack of response was most marked for strain H-5 in natural soil, where root growth was only 34% of that of controls.

DISCUSSION

The failure to persist in both soil series indicated that *R. solani* is a primary pathogen of juvenile tissues and is unable to maintain aggressive pathogenicity as tissues mature. We conclude that damage to the wheat root system is not only associated with a short-lived *Rhizoctonia* infection, but that it continues to be sustained by the associated minor pathogens. When these minor pathogens were excluded by sterilizing the test soil, the plants recovered rapidly and damage was negligible.

Disease expression also increased with bacterial involvement. Far from there being any evidence that antagonism endowed protection against fungal attack, the opposite held: disease expression was exacerbated as the bacteria participated actively in the rotting process. Perhaps this has been one of the reasons why previous attempts (6,7) to use bacterial antagonists, particularly *Bacillus subtilis* (Ehrenb.) Cohn, against *R. solani* infections have failed, although an alternative explanation was suggested at the time.

The ability of a bacterium to produce in vitro antagonism against a fungus is a characteristic that correlates poorly with protection in the infection court on the root surface (12). In the case of *Rhizoctonia* attack, where the fungus is able to induce necrosis in advance of hyphal penetration (5), the presence of bacteria increases the severity of rotting. This suggests that they invade necrotic cells in advance of hyphal colonization.

Table 1. Weight of wheat plants infected with *Rhizoctonia solani* and transplanted into sterilized and natural Colley Hill soil[a]

Soil treatment *R. solani* strain	First four harvests (to tillering)					Final harvest (heading)	
	19 days	26 days	33 days	40 days	Linear regression	Unfertilized	Fertilized[b]
Sterilized							
R-1	187	290	445	889	$y = 32.3x - 500$	6,742	8,273
R-2	200	292	478	811	$y = 28.8x - 406$	6,703	8,441
S-4	165	271	418	851	$y = 31.5x - 503$	6,653	8,245
H-5	213	223	377	863	$y = 30.0x - 467$	6,547	8,165
R-7	201	247	425	820	$y = 29.1x - 433$	6,715	8,051
Nil control	207	262	402	861	$y = 30.0x - 453$	6,900	7,657
Natural							
R-1	198	290	435	945	$y = 34.1x - 539$	3,598	5,037
R-2	179	276	470	928	$y = 34.9x - 565$	3,356	4,426
S-4	230	283	475	825	$y = 28.2x - 380$	3,562	4,841
H-5	189	256	415	547	$y = 17.6x - 168$	2,437	2,773
R-7	202	277	465	904	$y = 32.8x - 504$	3,266	4,580
Nil control	238	333	490	945	$y = 32.6x - 459$	3,569	5,123

[a] Given as mean total dry weight, milligrams per pot.
[b] NPK fertilizer was applied after tillering.

Table 2. Mean root growth as a percentage of the uninoculated controls for *Rhizoctonia solani* inoculated onto wheat

Soil treatment *R. solani* strain	First four harvests (to tillering)				Final harvest (heading)	
	19 days	26 days	33 days	40 days	Unfertilized	Fertilized[a]
Sterilized						
R-1	85	95	106	105	92	106
R-2	88	104	122	96	85	115
S-4	69	83	99	108	109	111
H-5	97	80	91	108	87	100
R-7	100	87	100	101	99	106
Natural						
R-1	72	83	78	97	109	107
R-2	66	78	89	90	90	68
S-4	84	78	89	71	97	85
H-5	68	75	72	44	44	34
R-7	75	84	89	88	94	79

[a]NPK fertilizer was applied after tillering.

The range of fungi causing secondary infections continued to change during the growth of the wheat plant. Inoculated *R. solani* infections had been supplanted completely by the tillering stage. The early *Fusarium-Cochliobolus* association gradually gave way to a community dominated by *Microdochium bolleyi* and a *Coprinus*-like basidiomycete. The range of associated fungi continued to widen to include *Drechslera biseptata* (Sacc. & Roum.) Rich. & Fraser, *Acremonium kiliense* Grütz, *F. merismoides* Corda, *Epicoccum purpurescens* Ehrenb. ex Schlect, *Phoma chrysanthemicola* Hollós, *Embellisia chlamydospora* (Hoes, Burehl & Shaw) Simmons, and unidentified species of *Aspergillus* (*restrictus* group), *Penicillium*, and *Sporotrichum*.

The severity of the short-lived *Rhizoctonia* attack influenced the severity of the infections that followed. But the major damage in the disease syndrome came from the minor pathogens of the common root rot suite that continued their attack throughout the growing season. Data in Table 2 show that plants with the greatest secondary infection had the least ability to recover when nutrition was improved. The complex must be regarded as a whole and might be more accurately described as *Rhizoctonia*-induced common root rot.

LITERATURE CITED

1. Bateman, D. F. 1970. Pathogenesis and disease. Pages 161-171 in: *Rhizoctonia solani*: Biology and Pathology. J. R. Parmeter, Jr., ed. University of California Press, Berkeley.
2. Colhoun, J. 1979. Minor pathogens and complex root diseases: Concluding remarks. Pages 425-430 in: Soil-Borne Plant Pathogens. B. Schippers and W. Gams, eds. Academic Press, London.
3. Harris, J. R., and Moen, R. 1983. Replacement of *Rhizoctonia solani* on wheat seedlings by a succession of root-rot fungi. Trans. Br. Mycol. Soc. In press.
4. Harris, J. R., and Moen, R. 1983. Secondary infection and fertilizer response of wheat initially infected with *Rhizoctonia solani*. Trans. Br. Mycol. Soc. In press.
5. Kerr, A. 1956. Some interactions between plant roots and pathogenic soil fungi. Aust. J. Biol. Sci. 9:45-52.
6. Merriman, P. R., Price, R. D., and Baker, K. F. 1974. The effect of inoculation of seed with antagonists of *Rhizoctonia solani* on the growth of wheat. Aust. J. Agric. Res. 25:213-218.
7. Merriman, P. R., Price, R. D., Kollmorgen, J. F., Piggott, T., and Ridge, E. H. 1974. Effect of seed inoculation with *Bacillus subtilis* and *Streptomyces griseus* on the growth of cereals and carrots. Aust. J. Agric. Res. 25:219-226.
8. Murray, D. I. L. 1981. *Rhizoctonia solani* causing barley stunt disorder. Trans. Br. Mycol. Soc. 76:383-395.
9. Murray, D. I. L., and Gadd, G. M. 1981. Preliminary studies on *Microdochium bolleyi* with special reference to colonization of barley. Trans. Br. Mycol. Soc. 76:397-403.
10. Northcote, K. H. 1979. A Factual Key for the Recognition of Australian Soils, 4th ed. Rellim Technical Publications, Adelaide. 124 pp.
11. Price, R. D. 1970. Stunted patches and deadheads in Victorian cereal crops. Victoria Dep. Agric. Tech. Publ. 23.
12. Sivasithamparam, K., and Parker, C. A. 1978. Effects of certain isolates of bacteria and actinomycetes on *Gaeumannomyces graminis* var. *tritici* and take-all of wheat. Aust. J. Bot. 26:773-782.
13. Tennant, D. 1975. A test of a modified line intersect method of estimating root length. J. Ecol. 63:995-1001.
14. Weller, D. M., and Cook, R. J. 1983. Suppression of take-all of wheat by seed treatments with fluorescent pseudomonads. Phytopathology 73:463-469.

Population and Survival of Sclerotia of *Rhizoctonia solani* in Soil

TAKASHI NAIKI, Faculty of Agriculture, Gifu University, Gifu 501-11, Japan

Rhizoctonia solani Kuehn is a common soilborne pathogen that has many plant hosts, forms sclerotia in soil, and survives for a long period in the absence of a host, either as sclerotia or thick-walled brown hyphae in plant debris (3,8,10). The number of survival units and their distribution in soil is particularly important in the epidemiology of the disease caused by this fungus.

This paper presents information on the population and survival of sclerotia of *R. solani* in sugar beet (*Beta vulgaris* L.) fields where root rot occurred.

MATERIALS AND METHODS

Soil samples were collected from between mature sugar beet plants in different fields of Hokkaido during the harvest season. Samples of rhizosphere soil (RS) were obtained by collecting the soil that adhered to beet roots after they were removed from fields. Nonrhizosphere soil (NRS) was sampled at depths of 0–5, 6–10, 11–15, and 16–20 cm and at distances of 0–5, 6–10, 11–15, and 16–20 cm from the root crown, using a stainless steel core sampler (5 cm in diameter and 5 cm long). RS and NRS samples were collected from three or four random locations in each field and stored at 4° C until assessed.

The number of sclerotia in soil was determined by a sieving-flotation method with 2% H_2O_2 solution (15). This method allows recovery of sclerotia of *R. solani* from soil without loss of viability, which was determined by plating washed sclerotia (two to three washes in sterile water) on acidified water agar. The percentage of germination was recorded as the average of three counts of about 40–50 sclerotia per count.

RESULTS

Anastomosis groups. *R. solani* was isolated from four different sugar beet fields. All of the isolates were classified into three anastomosis groups (AG) designated by Ogoshi (14). Among 126 sclerotial isolates of *R. solani*, 120 (95.2%) belonged to AG 2 type 2 (AG 2-2), 4 to AG 1, and 2 to AG 5. These results indicate that most of the sclerotia of *R. solani* in fields sampled were formed by AG 2-2, which is the root rot fungus of sugar beet.

Density and distribution of sclerotia. The vertical and horizontal distribution of sclerotia in soil around the roots of the most severely diseased plants (completely decayed) is shown in Fig. 1.

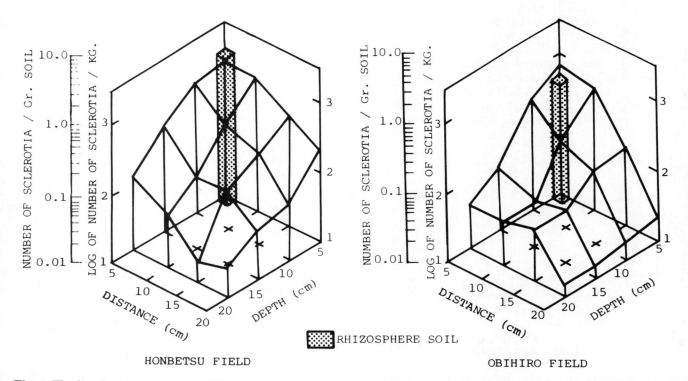

Fig. 1. The distribution of sclerotia of *Rhizoctonia solani* in rhizosphere and nonrhizosphere soils around a heavily infected sugar beet root in two continuous cropping fields at harvest.

More sclerotia were found in Honbetsu field (sandy loam, alluvial soil) than in Obihiro field (sandy loam, volcanic ash soil) in both RS and NRS. In both fields, about 80% of the sclerotia recovered were in RS and in NRS in the shallowest zone (0–10 cm depth) closest to the diseased root (0–10 cm). The number of sclerotia decreased with sampling depth and distance away from an infected root. A few sclerotia were still recovered from the deepest sampling depth and at the greatest distance from an infected root. Although there was a positive correlation between sclerotial number and root rot severity in both RS and NRS, the RS always contained more sclerotia than NRS regardless of disease severity and the field. Some sclerotia occurred even around healthy roots: 0.04–0.12 sclerotia per gram of dry soil in RS and 0.03 in NRS, respectively. By comparison, 1.43–2.5 and 0.83–1.01 sclerotia per gram of dry soil were found in RS and in NRS, respectively, from the most heavily infected root.

The relationship between soil particle size and sclerotial number in sugar beet fields was examined by sieving soil from around heavily infected plants through mesh measuring 0.25, 0.5, 1.0, 2.0, and 3.36 mm. The number of sclerotia obtained in the residual soil particles in each fraction was counted. Most sclerotia were associated with the large soil particles (≥3.36 mm). None was recovered from the finest fraction (<0.25 mm), and only a few sclerotia were retained on the 0.25-mm mesh sieve.

Size and viability of sclerotia. Most of the sclerotia (94.0%) recovered from sugar beet fields ranged between 0.5 and 3.36 mm in diameter, with the greatest number measuring 0.5 and 2.0 mm in diameter. A few very large sclerotia (≥3.36 mm in diameter) were also obtained.

Viability of sclerotia recovered from RS around the most heavily infected plants was high (>70%) (Fig. 2). However, germinability of sclerotia from NRS varied significantly in fields unless sclerotia were collected from around heavily infected plants. Germinability decreased gradually with increase in depth of sampling and distance from the infected root. Large sclerotia were more viable than small sclerotia. Germinability of sclerotia of AG 2-2 has been observed to decline more rapidly than that of other AGs (13). However, 270 days after incubation in natural soil, approximately 25% of sclerotia of this AG still contained a few intact cells with

cytoplasm randomly distributed within the sclerotia, even though they failed to germinate on water agar. Sclerotia produced in soil by AG 2-2 were brown to dark brown and rather soft with an indefinite shape compared with those of other AGs. They were constructed of very loosely arranged brown cells without any well-defined zone in internal morphology. With a longer incubation period, most of the sclerotial cells appeared devoid of cytoplasm and the peripheral cells collapsed to form a crustlike layer. The occurrence of bacterialike organisms within the sclerotial tissues and microbial breakdown were frequent as observed in short-lived sclerotia of *R. solani* (12).

DISCUSSION

Most of the sclerotia produced by *R. solani* in sugar beet fields are formed by AG 2-2, which is responsible for root rot of the plant. In RS and NRS from the most heavily infected roots of sugar beet, 1.43–2.5 and 0.83–1.01 sclerotia per gram of dry soil were detected, respectively. A few sclerotia were found even in soil around healthy sugar beet. RS always contained the most sclerotia, which agrees with the findings of Herzog (5). Most sclerotia occurred on the taproot of diseased plants and in the soil adjacent to the diseased root. Over 80% of sclerotia occurred in the top 10 cm of soil and within 10 cm from the diseased roots. Although the sclerotial number in soil increased significantly with root rot severity, it declined progressively with soil depth and increasing distance from diseased roots. Therefore, sclerotia of *R. solani* in soil are not uniform in their distribution. Similar results were obtained by Leach and Devey (9) for the distribution of *Sclerotium rolfsii* in soil.

Butler and Bracker (4) reported that the size of sclerotia of *R. solani* varies considerably among isolates. When the sclerotial size of AG 2-2 was compared under different cultural conditions, more big sclerotia were produced in pure culture than in soil. More big sclerotia were also recovered from naturally infested soil than from artificially infested soil (13), but sclerotial morphology was identical regardless of cultural conditions. In this study, most sclerotia produced in sugar beet fields were associated with the largest soil particles. The size of sclerotia ranged from 0.5 to more than 3.365 mm in diameter, with most ranging between 0.5 and 2.0 mm in diameter.

Sclerotial germination varied significantly in fields unless the sclerotia were collected from around heavily infected plants. However, large sclerotia from more heavily infected roots were always the most viable. There was also a progressive decline in sclerotial vigor and viability (based on germinability on agar) with increasing distance from an infected root and with increasing sampling depth. The variation in vigor and viability of sclerotia probably relates to distance from the food base, as reported by Blair (2). Mycoparasitic and other antagonistic effects of soil microorganisms on sclerotial germination and survival of *R. solani* in soil have been reported (1,6,8,12). There was a marked increase in the size of microbial populations around and on the surface of sclerotia in soil with the incubation period (11). The variation in sclerotial germination may also be associated with age of the sclerotia under the condition of antagonistic effects of soil microorganisms.

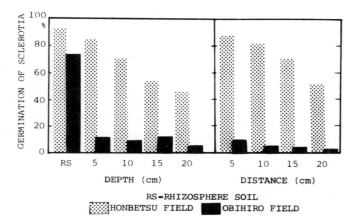

Fig. 2. The germinability of sclerotia of *Rhizoctonia solani* obtained from rhizosphere and nonrhizosphere soils of various depths and distances from a heavily infected sugar beet root.

Failure of the sclerotia to germinate on the agar medium does not necessarily mean that the sclerotia are dead. Contact with host plant roots has been reported to induce vigorous germination of sclerotia having only a few intact cells with cytoplasm (10,13). In fact, tests of this type indicate that over half of the sclerotia recovered from sugar beet fields were viable. Recently, Hyakumachi et al (7,8) showed a high positive correlation between the number of overwintered viable sclerotia, not plant debris, and the incidence of both damping-off and root rot of sugar beet. In addition, experience shows that the first outbreak of root rot of sugar beet in continuously cropped fields always occurs where the disease was most severe the previous year. Therefore, sclerotia were thought to be important as a primary source of inoculum in sugar beet fields.

ACKNOWLEDGMENTS

This research was conducted under the direction of Professor Tadao Ui at Hokkaido University. I am greatly indebted to him for his valuable suggestions during the experiments. I am also grateful to Dr. R. James Cook, U.S. Department of Agriculture, Washington State University, for critical reading of the manuscript.

LITERATURE CITED

1. Aluko, M. O., and Hering, T. F. 1972. The mechanisms associated with the antagonistic relationship between *Corticium solani* and *Gliocladium virens*. Trans. Br. Mycol. Soc. 55:173-179.
2. Blair, I. D. 1943. Behaviour of the fungus *Rhizoctonia solani* Kühn in the soil. Ann. Appl. Biol. 30:118-127.
3. Boosalis, M. G., and Scharen, A. L. 1959. Methods for microscopic detection of *Aphanomyces euteiches* and *Rhizoctonia solani* and for isolation of *Rhizoctonia solani* associated with plant debris. Phytopathology 49:192-198.
4. Butler, E. E., and Bracker, C. E. 1970. Morphology and cytology of *Rhizoctonia solani*. Pages 32-51 in: *Rhizoctonia solani*: Biology and Pathology. J. R. Parmeter, Jr., ed. University of California Press, Berkeley.
5. Herzog, W. 1961. Das Uberdauern und der Saprophytismus des Wurzeltoters *Rhizoctonia solani* Kühn im Boden. Phytopathol. Z. 40:379-415.
6. Herzog, W., and Wartenberg, H. 1960. Der Konservierungseffekt antibiotisher microorganismen und den Skerotien von *Rhizoctonia solani* Kühn. Ber. Dtsch. Bot. Ges. 73:346-348.
7. Hyakumachi, M., and Ui, T. 1982. The role of the overwintered plant debris and sclerotia as inoculum in the field following sugar beet root rot. Ann. Phytopathol. Soc. Jpn. 48:628-633.
8. Hyakumachi, M., Yamamoto, Y., and Ui, T. 1983. Survival and pathogenicity of sclerotia produced by sugar beet root rot fungus (*Rhizoctonia solani* AG-2 type-2) in the areas with different degrees of disease incidence. Ann. Phytopathol. Soc. Jpn. 49:18-21.
9. Leach, L. D., and Devey, A. E. 1938. Determining the sclerotial population of *Sclerotium rolfsii* by soil analysis and predicting losses of sugar beet on the basis of these analyses. J. Agric. Res. 56:619-632.
10. Naiki, T., and Ui, T. 1969. On the survival of the sclerotia of *Rhizoctonia solani* Kühn in soil. Mem. Fac. Agric. Hokkaido Univ. 6:430-436.
11. Naiki, T., and Ui, T. 1972. The microorganisms associated with the sclerotia of *Rhizoctonia solani* Kühn in soil and their effects on the viability of the pathogen. Mem. Fac. Agric. Hokkaido Univ. 8:252-265.
12. Naiki, T., and Ui, T. 1975. Ultrastructure of sclerotia of *Rhizoctonia solani* Kühn invaded and decayed by soil microorganisms. Soil Biol. Biochem. 7:301-304.
13. Naiki, T., and Ui, T. 1978. Ecological and morphological characteristics of the sclerotia of *Rhizoctonia solani* Kühn produced in soil. Soil Biol. Biochem. 10:471-478.
14. Ogoshi, A. 1976. Studies on the grouping of *Rhizoctonia solani* Kühn with hyphal anastomosis and on the perfect stage of groups. Bull. Natl. Inst. Agric. Sci., Ser. C 30:1-63.
15. Ui, T., Naiki, T., and Akimoto, M. 1976. A sieving-flotation technique using hydrogen peroxide solution for determination of sclerotial population of *Rhizoctonia solani* Kühn in soil. Ann. Phytopathol. Soc. Jpn. 42:46-48.

Rhizoctonia in South Australian Wheat Fields

S. M. NEATE, Department of Plant Pathology, Waite Agricultural Research Institute, Glen Osmond, South Australia, 5064

In South Australia, the first report of *Rhizoctonia solani* Kuehn causing a "bare patch" disease of cereals was by Samuel in 1928 (12). A description of the disease was later given by Samuel and Garrett (13). Stunting of diseased plants becomes apparent about 3 weeks after sowing, and stunted plants form distinct circular to irregular patches that are 30 cm to 200 m in diameter (5). Stunting is caused by rotting of seminal roots (13).

Three pathogenic strains of *R. solani* have been reported from cereal fields in South Australia. Samuel and Garrett (13), Kerr (10), Flentje (7), and De Beer (6) found a "root strain" that was pathogenic to roots of wheat and other plants and that they suggested caused bare patch. Kerr (10) and De Beer (6) found a "stem strain" that attacked stems and roots and caused preemergent damping-off of wheat, and Kerr (10) found a "crucifer strain" that was pathogenic to crucifers but not wheat. "Nonpathogenic" strains of *R. solani* have also been found in South Australian wheat fields (6,7,10). All strains are present both within and outside patches in the field, but populations of the root strain are higher within patches (6,10). Representatives of all the pathogenic and nonpathogenic strains have been fruited, and all formed the teleomorph *Thanatephorus cucumeris* (Frank) Donk (6,7,9,10).

Rhizoctonia anamorphs of *Ceratobasidium cornigerum* (Bourd.) Rogers, *Waitea circinata* Warcup and Talbot, and *Iodophanus carneus* (Pers. per Pers.) Korf have also been shown to be present in wheat fields in South Australia (18–21), but little is known about their pathogenicity. Warcup and Talbot (20) reported *C. cornigerum* to be nonpathogenic to wheat, but Burpee et al (3) have shown that *C. cornigerum* and other binucleate *Rhizoctonia* strains may be pathogenic to some plants.

This study examines the identity and pathogenicity of *Rhizoctonia* spp. in South Australian wheat fields.

MATERIALS AND METHODS

Localities. The sites investigated were Avon in the mid north and Coonalpyn in the upper southeast of South Australia. The soil at Avon is sand over limestone grading to calcareous clay loam; at Coonalpyn the soil is sand over limestone. Avon was in the wheat phase of a wheat-grassy pasture rotation, and Coonalpyn was in the wheat phase of a wheat-clover (*Trifolium subterraneum* L.) pasture rotation.

Isolates. *Rhizoctonia* was isolated from diseased wheat and weeds both within and outside of "bare patches" by plating lesioned roots without surface sterilization onto weak Czapek-Dox plus yeast agar (NDY/6) (17) containing 150 ppm of streptomycin sulfate. Isolation from soil was by a modification of the method of Weinhold (22). About 500 g of soil taken to a depth of 12 cm was washed through a sieve of 250-μm mesh. An aliquot of the particles retained on the sieve was added to cooled NDY/6 agar containing 150 ppm of streptomycin sulfate and 100 ppm of tetracycline hydrochloride. Root and debris plates were incubated at 15–20°C. Hyphal tips of *Rhizoctonia*-like fungi were removed after 12–18 hr.

The number of nuclei in cells of vegetative hyphae of isolates was determined by staining with 1% aniline blue in 50% glycerine slightly acidified with HCl (16). The isolates were divided into anastomosis groups (AG) by placing hyphae of a known anastomosis group opposite the isolate to be tested on a thin layer of distilled water agar on a microscope slide and incubating for 24–48 hr (14). Teleomorphs were produced mainly by fruiting of isolates by the soil-on-agar method (15,20), using NDY agar and Urrbrae clay as a "casing soil." *Waitea* fruited on cornmeal agar after several weeks in diffuse light at 15–20°C (18).

Pathogenicity tests. Pathogenicity tests were carried out in a soil mixture comprising 50% UC mix (1) and 50% sterilized coarse river sand. A modification of De Beer's method (6) was used. About 1 g of suction-dried mycelium from an isolate grown for 18 days in liquid *Rhizoctonia* medium (6) was washed with distilled water, macerated in further distilled water, and poured

Table 1. Number of isolates in each group of *Rhizoctonia* from Coonalpyn and Avon

Group[a]	Coonalpyn	Avon
A	28	98
B	77	7
C	8	0
D	3	0
E	0	1

[a] A = dark brown with age, B = straw-colored with age, C = orange to dark brown sclerotia, D = pink to orange isolates, E = others.

Table 2. Origin of isolates in different groups of *Rhizoctonia* from Coonalpyn and Avon

Origin	Group A	B	C	D
Lucerne	+	+	−	−
Lupin	+	+	−	−
Clover	+	+	−	−
Ryegrass[a]	+	+	−	−
Wheat	+	+	−	−
Soil				
Organic matter	+	+	−	+
Sclerotia	−	−	+	−

[a] *Lolium rigidum* Gaud.

onto soil in a 220-ml plastic cup. A 1.5-cm layer of soil was added, the seeds placed on this, and a further 1.5-cm layer of soil added.

The test plants were *Lupinus angustifolius* L. 'Unicrop' (lupin), *Medicago sativa* L. 'Hunter River' (lucerne), *Raphanus sativus* L. 'Long White Icicle' (radish), *Triticum aestivum* L. 'Halberd' (wheat), *Lactuca sativa* L. 'Great Lakes' (lettuce), and *Lycopersicon esculentum* Mill. 'Rutgers' (tomato). These plants represent plant families that Kerr (10) and Flentje and Saksena (8) used to determine the pathogenicity of strains of *R. solani*. The plants were grown at 15–24° C in a glasshouse, with pots watered to field capacity every 4 days.

Plants were visually rated for disease on a scale of 1–7 where 1 was one or two superficial lesions on roots of less than 2 mm in diameter and 7 was severe damage to the taproot or destruction of most of the seminal roots.

RESULTS

Rhizoctonia isolates from plants or soil were grouped on morphological features into the following groups: A) isolates becoming dark brown with age, B) isolates becoming straw-colored with age, C) isolates with hyaline mycelium and orange to dark brown sclerotia, D) pink to orange isolates, and E) other isolates. All isolates from groups A–D were examined for nuclear number and used in fruiting experiments, and selected group A isolates were anastomosis tested. Selected isolates from groups A–D from Coonalpyn were used in pathogenicity experiments.

Table 1 records the numbers of isolates in each group obtained from Coonalpyn and Avon. Isolates from

Table 3. Number of isolates of *Rhizoctonia* from Coonalpyn and Avon with binucleate or multinucleate hyphal cells

Group	Nuclei[a]	Avon	Coonalpyn
A	M	90	17
	B	0	0
B	M	0	1
	B	7	37
C	M	...	2
	B	...	0
D	M	...	0

[a] M = multinucleate, B = binucleate.

Table 4. Number of isolates of *Rhizoctonia* from Coonalpyn and Avon producing their teleomorph

Group	Locality	Thanate-phorus cucumeris	Cerato-basidium cornigerum	Iodophanus carneus	Waitea circinata
A	Avon	68	0	0	0
	Coonalpyn	11	0	0	0
B	Avon	0	0	0	0
	Coonalpyn	0	24	0	0
C	Coonalpyn	0	0	0	1
D	Coonalpyn	0	0	3	0

Table 5. Pathogenicity of *Rhizoctonia* isolates from different sources

Group	Identity	Origin	Nuclei per cell[a]	Lucerne	Lupin	Lettuce	Wheat	Radish	Tomato
A	*Thanatephorus cucumeris*	Lucerne	M	5.0[b]	3.3	3.0	1.5	1.0	3.0
	T. cucumeris	Clover	M	5.0	4.0	3.0	3.0	1.0	3.0
	T. cucumeris	Wheat	...[c]	5.0	6.0	2.8	6.0	6.0	7.0
	R. solani	Lupin	M	4.3	3.3	5.0	3.0	3.0	5.0
	R. solani	Ryegrass	M	5.0	2.5	2.5	2.5	2.0	3.2
	R. solani	Wheat	M	2.4	3.0	2.0	2.9	2.6	5.0
	R. solani	Organic matter	M	6.5	4.0	0	3.0	6.0	4.3
	R. solani	Organic matter	M	2.5	4.5	1.2	6.5	0.4	0.8
B	*Ceratobasidium cornigerum*	Lucerne	...	0.5	0	0	0	0	0
	C. cornigerum	Lucerne	...	2.5	0.3	0	0	0	2.0
	C. cornigerum	Lucerne	B	0	2.0	0	0	0	0
	C. cornigerum	Lupin	...	7.0	4.2	1.0	0	2.6	2.0
	C. cornigerum	Clover	B	1.0	0.3	1.5	0	0.5	1.1
	C. cornigerum	Clover	B	3.0	0.3	2.0	0	1.6	2.0
	C. cornigerum	Ryegrass	B	5.6	2.5	0	0	0.5	0
	C. cornigerum	Wheat	B	0	2.0	1.5	0	0	0
	C. cornigerum	Organic matter	B	...	2.2	0.7	0	1.8	0
	Rhizoctonia sp.	Wheat	B	5.0	0.8	0	0.2	0	1.0
	Rhizoctonia sp.	Wheat	B	0	0	...	0	0	0
C	*Waitea circinata*	Sclerotium	M	5.0	5.0	0	2.0	0	0.5
	Rhizoctonia sp.	Sclerotium	...	0.8	6.0	0	2.0	0	0
	Rhizoctonia sp.	Sclerotium	...	0	5.0	0	2.0	0	0.8
D	*Iodophanus carneus*	Organic matter	...	0	0	0	0	0	0
	I. carneus	Organic matter	...	0	0.3	...	0	0	0

[a] M = multinucleate, B = binucleate.
[b] Scale of 0–7.
[c] Not determined.

groups A and B were obtained more frequently than other groups at both sites, but they differed in their frequency at each site. The reason for the difference in proportion of isolates in groups A and B at the two sites is not known.

Groups A and B were isolated from a large number of plant species and organic matter in soil, but groups C and D were only isolated from soil (Table 2). Isolates from groups A, C, and D were multinucleate; with the exception of one isolate from Avon, isolates in group B were binucleate (Table 3).

Not all isolates fruited, but those in group A that fruited produced the teleomorph *T. cucumeris*, in group B the teleomorph *C. cornigerum*, in group C the teleomorph *W. circinata*, and in group D the teleomorph *I. carneus* (Table 4).

All *R. solani* isolates tested for pathogenicity were tested against isolates from AG 1, 2, 3, and 4. None anastomosed with any of these anastomosis groups. All the isolates self-anastomosed and anastomosed with each other.

Table 5 records the pathogenicity of selected isolates of *Rhizoctonia*. Most group A, B, and C isolates produced lesions on several of the test plants, but only one group D isolate produced a slight lesion on lettuce.

DISCUSSION

Grouping the isolates morphologically was, with one exception, a reliable means of separating the species at Coonalpyn and Avon (Tables 2–4), even though group A isolates (*R. solani*) varied in color and some isolates took several weeks to develop their brown pigment.

R. solani isolates in AG 2-1 and AG 4 have been found in South Australian wheat fields, but most of the wheat-field isolates belong to an undescribed anastomosis group (Neate, *unpublished*). All of the *R. solani* isolates tested here belonged to this group.

Although the anastomosis groups (CAG) (3,4) of the binucleate isolates were not determined, none of the group B isolates resembled CAG 3, 4, or 5 because members of these groups are similar in cultural morphology to *R. solani* (4). *R. cerealis* van der Hoeven, which causes the sharp eyespot disease of cereals in Europe (2) and whose teleomorph is an undescribed *Ceratobasidium* (D. I. L. Murray, *personal communication*), and *R. solani* AG 3, which causes a stunt disorder of barley and oats in Britain (11), were not found in this study.

Most *R. solani* isolates were pathogenic to all the hosts tested, but *C. cornigerum* was pathogenic to fewer hosts and was less pathogenic than *R. solani*. Although *C. cornigerum* was isolated from wheat, it was not pathogenic to that species. Some *C. cornigerum* isolates, however, were severely pathogenic to lucerne and lupin. *W. circinata* was pathogenic to wheat and the legumes, but *I. carneus* is probably nonpathogenic to the plants tested.

The data show that *Rhizoctonia* species other than *R. solani* are more common in some wheat fields in South Australia than has previously been recognized. *C. cornigerum*, *W. circinata*, and *I. carneus* have not been reported from wheat fields in North America, Europe, or other states of Australia.

LITERATURE CITED

1. Baker, K. F. 1957. The U.C. system for producing healthy container grown plants. Calif. Agric. Exp. Stn. Man. 23.
2. Boerema, G. H., and Verhoeven, A. A. 1977. Check list for scientific names of common parasitic fungi. Series 26: Fungi on the field crops: Cereals and grasses. Neth. J. Plant Pathol. 83:165-204.
3. Burpee, L. L., Sanders, P. L., Cole, H., Jr., and Sherwood, R. T. 1980. Pathogenicity of *Ceratobasidium cornigerum* and related fungi representing five anastomosis groups. Phytopathology 70:843-846.
4. Burpee, L. L., Sanders, P. L., Cole, H., Jr., and Sherwood, R. T. 1980. Anastomosis groups among isolates of *Ceratobasidium cornigerum* and related fungi. Mycologia 72:689-701.
5. Butler, F. C. 1961. Root and foot rot diseases of wheat. Sci. Bull. Dep. Agric. N.S.W. 77 pp.
6. De Beer, J. F. 1965. Studies on the ecology of *Rhizoctonia solani* Kühn. Ph.D. thesis, University of Adelaide, South Australia.
7. Flentje, N. T. 1956. Studies on *Pellicularia filamentosa* (Pat.) Rogers. I. Formation of the perfect stage. Trans. Br. Mycol. Soc. 39:343-356.
8. Flentje, N. T., and Saksena, H. K. 1957. Studies on *Pellicularia filamentosa* (Pat.) Rogers. II. Occurrence and distribution of pathogenic strains. Trans. Br. Mycol. Soc. 40:95-108.
9. Flentje, N. T., and Stretton, H. M. 1964. Mechanisms of variation in *Thanatephorus cucumeris* and *T. praticolus*. Aust. J. Biol. Sci. 17:684-704.
10. Kerr, A. 1955. Studies on the parasitic and saprophytic activities of *Pellicularia filamentosa* (Pat.) Rogers and *Sclerotinia homeocarpa* Bennett. Ph.D. thesis, University of Adelaide, South Australia.
11. Murray, D. I. L. 1981. *Rhizoctonia solani* causing barley stunt disorder. Trans. Br. Mycol. Soc. 76:383-395.
12. Samuel, G. 1928. Two "stunting" diseases of wheat and oats. J. Dep. Agric. South Aust. 32:40-43.
13. Samuel, G., and Garrett, S. D. 1932. *Rhizoctonia solani* on cereals in South Australia. Phytopathology 22:827-836.
14. Sanders, P. L., Burpee, L. L., and Cole, H., Jr. 1978. Preliminary studies on binucleate turfgrass pathogens that resemble *Rhizoctonia solani*. Phytopathology 68:145-148.
15. Stretton, H. M., McKenzie, A. R., Baker, K. F., and Flentje, N. T. 1964. Formation of the basidial stage of some isolates of *Rhizoctonia*. Phytopathology 54:1093-1095.
16. Tu, C. C., and Kimbrough, J. W. 1973. A rapid staining technique for *Rhizoctonia solani* and related fungi. Mycologia 65:941-944.
17. Warcup, J. H. 1955. Isolation of fungi from hyphae present in soil. Nature (London) 175:953-954.
18. Warcup, J. H., and Talbot, P. H. B. 1962. Ecology and identity of mycelia isolated from soil. Trans. Br. Mycol. Soc. 45:495-518.
19. Warcup, J. H., and Talbot, P. H. B. 1963. Ecology and identity of mycelia isolated from soil. II. Trans. Br. Mycol. Soc. 46:465-472.
20. Warcup, J. H., and Talbot, P. H. B. 1965. Ecology and identity of mycelia isolated from soil. III. Trans. Br. Mycol. Soc. 48:249-259.
21. Warcup, J. H., and Talbot, P. H. B. 1966. Perfect states of some *Rhizoctonias*. Trans. Br. Mycol. Soc. 49:427-435.
22. Weinhold, A. R. 1977. Population of *Rhizoctonia solani* in agricultural soils determined by a screening procedure. Phytopathology 67:566-569.

Anastomosis Groups of *Rhizoctonia solani* and Binucleate *Rhizoctonia*

AKIRA OGOSHI and TADAO UI, Department of Plant Pathology, Faculty of Agriculture, Hokkaido University, Kita-ku, Sapporo 060, Japan

RHIZOCTONIA SOLANI

Studies in Japan. The first report of *Rhizoctonia solani* Kuehn in Japan, published in 1902, concerned the damping-off of eggplant. In 1906, *Hypochnus sasakii* was reported as the pathogen on camphor tree. It was later reported that this fungus is identical with *Sclerotium irregulare*, which was reported as the sheath blight fungus of rice plant. In Japan, the sheath blight caused by *R. solani* is one of the most serious diseases of rice, next to rice blast caused by *Pyricularia oryzae*. It had been thought that *H. sasakii* and *R. solani* were different species.

In 1932, Matsumoto et al (4) reported that *H. sasakii* and *R. solani* are able to be divided by hyphal anastomosis. The isolates of *H. sasakii* fused with each other, but there was no fusion between *H. sasakii* and *R. solani* isolates. In 1966, Watanabe and Matsuda (10) divided *R. solani* isolates in Japan into seven types in terms of their morphology, pathology, physiology, and other characters (Table 1). In 1976, five anastomosis groups (AG) were reported (5). Later reports added AG 6, AG bridging isolates (AG BI) (3), and AG 7 (2).

Groups of *R. solani*. Anastomosis is useful when considering groupings of *R. solani*, but the behavior of the fungus cannot be completely understood in those terms. For example, AG 1 causes not only sheath blight on rice plants but also web or leaf blight on many plants. However, AG 1 is divided into at least two subgroups or types—Sasakii and web blight—based on pathogenicity. Usually the Sasakii type does not cause web blight and the web blight type does not cause sheath blight; they also differ in their cultural appearance. We recently found AG 1 (IC) in Japan from sugar beets and buckwheats, a type that coincides with AG 1 type 3 as described by Sherwood (9).

R. solani can be divided into 12 groups by means of a combination of hyphal fusion, pathology, and morphology (Table 1). The isolate of IA of AG 1 fuses very well not only with other isolates of IA but also with isolates of IB and IC. In contrast with AG 1, the fusion between AG 2-1 and AG 2-2 is less frequent, although the rate of fusion among isolates differs. Isolates of IIIB and IV fuse very well, as though they were the same group.

Most interesting are the AG BI isolates, which fuse rarely with AG 2-1, AG 3, and AG 6, but frequently with AG 2-2. This behavior is significant from the viewpoint of the phylogeny of *R. solani*.

Geographic distribution of *R. solani*. Distribution of *R. solani* isolates and groups in Japan was determined according to reports by Ogoshi and by Watanabe and Matsuda (5,10). The group and origin of isolates used in their reports were investigated. Geographic latitude from Kyushu to Hokkaido covers from 30 to 45°, i.e., from the temperate to the subarctic zones. All groups are found all over Japan.

On the basis of published reports, we investigated the distribution of anastomosis groups all over the world. We can see in general that AG 1 through AG 4 have been reported in many countries and AG 5 in some countries. AG 6, AG 7, and AG BI have not been reported outside Japan; because these groups have only recently been found, however, we think they may occur in other countries. AG 6 and AG BI have been isolated only from uncultivated lands. Conversely, AG 5 and AG 7 have been found in cultivated lands. The reason is not known.

Host plants and distribution of *R. solani*. We investigated the groups and origins of *R. solani* isolates as described in the literature. AG 1 has been isolated mainly from the Leguminosae and the Graminae; isolates from the former family may be AG 1 (IB) and from the latter may be AG 1 (IA). Several AG 2-1 types have been isolated from the Cruciferae, but AG 2-2 was isolated from the Chenopodiaceae (IV)—mainly from sugar beets—or from the Graminae (IIIB). Most AG 3 types have been isolated from the Solanaceae, AG 4 from the Chenopodiaceae, Leguminosae, and Solanaceae, and AG 5 from the Leguminosae and from soils. From this evidence, we see why AG 2 was called "var. *brassicae*" or "Cruciferen Gruppe" and AG 3 was called "Kartoffel Gruppe."

Anastomosis groups of *R. solani* are found all over the world, but their distribution depends on the crops that are cultivated there. This conclusion is supported by the evidence that one field is occupied by a specific

Table 1. Groups of *Rhizoctonia solani* and their diseases

Group	Type[a]	Disease
AG 1 (IA)	Sasakii	Sheath blight of rice plant
AG 1 (IB)	Web blight	Web blight and leaf blight
AG 1 (IC)		Not known[b]
AG 2-1	Winter crops	Damping-off of crucifers
AG 2-2 (IIIB)	Rush	Sheath blight of mat rush
AG 2-2 (IV)	Root rot	Root rot of sugar beet
AG 3	Potato	Black scurf of potato
AG 4	Praticola	Damping-off of various plants
AG 5		Not known
AG 6		Not known
AG 7		Not known
AG BI		Not known

[a] Watanabe and Matsuda (10).
[b] Distinct disease has not been reported.

anastomosis group that is pathogenic to the crop growing in the field (8).

BINUCLEATE *RHIZOCTONIA*

Anastomosis groups. The Japanese plant pathologist Sawada studied and described many diseases and pathogens in Taiwan, including *Hypochnus cinnamomi* on the camphor tree, *H. setariae* on foxtail millet, and *Sclerotium oryzae-sativae* on rice plants; the last was also reported in Japan. All these fungi are binucleate.

There are many binucleate *Rhizoctonia* species, and we cannot easily distinguish or identify one species from another. We thus used hyphal fusion to group species other than *R. solani* and found that the occurrence of hyphal fusion among binucleate *Rhizoctonia* isolates was similar to that in *R. solani*. We differentiated 15 anastomosis groups in binucleate *Rhizoctonia* (7), calling them AG A, AG B, and so on. It is now thought that there may be more groups.

Hyphal anastomosis between Japanese and North American isolates. Almost at the same time, Burpee et al (1) reported anastomosis groups in North American binucleate *Rhizoctonia*, calling them CAG 1, CAG 2, and so on. It then became important to clarify the relation between both groups, and we compared them with the cooperation of Burpee and Adams.

The results showed that AG A corresponds to CAG 2, AG D to CAG 1, and AG F to CAG 4; however, AG E corresponded to CAG 3 and CAG 6 (6). CAG 5 and CAG 7 did not fuse with any Japanese isolates, so these two groups may be independent. Further studies on other isolates of CAG 3 and CAG 6 should be carried out.

Distribution in Japan and diseases caused by binucleate *Rhizoctonia*. Anastomosis groups of binucleate *Rhizoctonia* are found all over Japan, from north to south, as well as of *R. solani*. The distribution shows no characteristic features.

The important diseases caused by binucleate *Rhizoctonia* in Japan are strawberry root rot (by AG A), gray sclerotium disease of rice plant (by AG Ba), brown sclerotium disease of rice plant (by AG Bb), and foot rot of cereals (by AG D). Many other diseases caused by various anastomosis groups have been reported, with some diseases caused by several groups.

Although there are many groups in the binucleate *Rhizoctonia*, only some are pathogenic and cause distinct diseases, whereas others are weak parasites. Still other groups may be saprophytic or symbiotic in nature.

LITERATURE CITED

1. Burpee, L. L., Sanders, P. L., Cole, H. Jr., and Sherwood, R. T. 1980. Anastomosis groups among isolates of *Ceratobasidium cornigerum* and related fungi. Mycologia 72:689-701.
2. Homma, Y., Yamashita, Y., and Ishii, M. 1983. A new anastomosis group (AG-7) of *Rhizoctonia solani* Kühn from Japanese radish fields. Ann. Phytopathol. Soc. Jpn. 44:189-190.
3. Kuninaga, S., Yokosawa, R., and Ogoshi, A. 1978. Anastomosis grouping of *Rhizoctonia solani* Kühn isolated from noncultivated soils. Ann. Phytopathol. Soc. Jpn. 44:591-598.
4. Matsumoto, T., Yamamoto, W., and Hirane, S. 1932. Physiology and parasitology of the *Rhizoctonia* generally referred to as *Hypochnus sasakii* Shirai. J. Soc. Trop. Agric. 4:370-388.
5. Ogoshi, A. 1976. Studies on the grouping of *Rhizoctonia solani* Kühn with hyphal anastomosis and on the perfect stages of groups. Bull. Natl. Inst. Agric. Sci. Ser. C. 30:1-63.
6. Ogoshi, A., Oniki, M., Araki, T., and Ui, T. 1983. Studies on the anastomosis groups of binucleate *Rhizoctonia* and their perfect states. J. Fac. Agric. Hokkaido Univ. 61:244-260.
7. Ogoshi, A., Oniki, M., Sakai, R., and Ui, T. 1979. Anastomosis grouping among isolates of binucleate *Rhizoctonia*. Trans. Mycol. Soc. Jpn. 20:33-39.
8. Ogoshi, A., and Ui, T. 1983. Diversity of clones within an anastomosis group of *Rhizoctonia solani* Kühn in a field. Ann. Phytopathol. Soc. Jpn. 49:239-245.
9. Sherwood, R. T. 1969. Morphology and physiology in four anastomosis groups of *Thanatephorus cucumeris*. Phytopathology 59:1924-1929.
10. Watanabe, B., and Matsuda, A. 1966. Studies on the grouping of *Rhizoctonia solani* Kühn pathogenic to upland crops. Appointed Experiment (Plant Diseases and Insect Pests) 7:1-131.

A Study of Pepper Wilt in Northern Iraq

AWATIF A. RAHIM and FAYADH M. SHARIF, Department of Biology, College of Science, University of Salahuddin, Arbil, Iraq

Wilt is a worldwide disease of pepper that has been reported from the United States, Italy, Greece (7), Yugoslavia (2), Turkey (3), and Iran (4). In Tamim Govarnarate of Iraq, it was reported to be caused by *Phytophthora capsici* Leonian (6). Our preliminary investigation of the disease in northern Iraq showed that wilt is also caused by two species of *Fusarium*. Interest in interactions between different root fungi upon disease (10) has been revived, as there is evidence for an interaction between *P. capsici* and *Rhizoctonia solani* Kuehn on pepper (5).

This study examined the distribution, etiology, and host range of the causal agents; varietal resistance of available pepper cultivars; and possible interactions between the causal agents in disease expression.

MATERIALS AND METHODS

Seedlings of *Capsicum frutescens* 'California Wonder' grown in plastic pots containing sterilized mixed loam in the greenhouse were used throughout the experiments. In each experiment, 50 seedlings were used.

Preparation and use of isolates. Pathogens were isolated by conventional procedures (8) from roots or basal stems of diseased pepper plants collected from different fields in the area. Pure cultures of isolated fungi were sent to specialized institutions for confirmation of our identification. Inoculum of the fungus was prepared by mixing a young culture (7 days old) of the fungus grown at 25°C on potato-dextrose agar (PDA) with 50 ml of distilled water using a blender; the mixture was then filtered through cheesecloth. The filtrate, which contains spores and mycelial fragments, was used as inoculum, 20 ml being added to each pot 15 days after planting. Isolated fungi were then tested for their pathogenicity on pepper plants. Koch's postulates were fulfilled for each fungus. Plants were considered diseased when they showed wilting, which normally ended with their death.

The colony growth of fungi on four different media was compared for determination of growth rate. Two of the media—PDA and cornmeal agar—were natural, and the other two—Richard's medium and Czapek's agar—were synthetic. Sterilized media distributed in petri dishes were each inoculated in the center with an agar disk (5 mm in diameter) of young growth of the fungus and incubated at 25°C. The diameters of the colonies were measured every 24 hr. Six replicates were used.

Sets of six replicates of PDA plates were inoculated as above, then incubated at 15, 20, 25, 30, and 35°C to determine the temperature range for growth.

Inoculation of seedlings. Groups of pepper plants at three developmental stages—seedling stage (two to four leaves), 1 month old, and flowering stage—were inoculated with each fungus separately. Percentage of diseased plants was recorded. In tests of varietal resistance, seedlings of five available pepper cultivars were used: *Capsicum frutescens* 'California Wonder' and *C. annuum* 'Yellow Wonder,' 'Sivrija,' 'Belkapija,' and 'Mousili.'

Seedlings of plants belonging to Solanaceae or Cucurbitaceae were also tested for their susceptibility to infection. These plants included *Solanum melongena* 'Black Beauty,' *Cucumis sativus* 'Beta Alpha,' *Cucurbita maxima* 'Ascandrani,' and *Citrullus vulgaris* 'Charleston Gray.'

Four groups of pepper seedlings were inoculated each with a mixed inoculum of the fungi in the following combinations: group 1, *P. capsici* + *F. solani*; group 2, *P. capsici* + *F. oxysporum*; group 3, *F. solani* + *F.*

Table 1. Percentage of disease in pepper plants inoculated with *Fusarium solani, F. oxysporum,* and *Phytophthora capsici*

Incubation period (days)	Percentage of diseased plants		
	F. solani	*F. oxysporum*	*P. capsici*
4	20	0	20
8	40	25	50
12	60	45	90
16	80	60	100
20	90	85	100

Table 2. Effect of age of pepper plants on susceptibility to infection with *Fusarium solani, F. oxysporum,* and *Phytophthora capsici*

Incubation period (days)	Age[a]	Percentage of diseased plants		
		F. solani	*F. oxysporum*	*P. capsici*
4	A	5	20	10
	B	30	15	15
	C	0	0	0
8	A	75	100	100
	B	75	75	95
	C	50	40	80
12	A	80	100	100
	B	80	95	100
	C	80	65	85
16	A	100	100	100
	B	100	100	100
	C	85	85	90
20	A	100	100	100
	B	100	100	100
	C	95	85	90

[a] A = seedlings (two to four leaves), B = 1 month old, and C = flowering stage.

oxysporum; and group 4, *P. capsici* + *F. oxysporum* + *F. solani*. After percentages of diseased plants were recorded, pathogens were isolated and identified.

Interaction between pathogens on culture media. Mycelial disks (5 mm in diameter) of the pathogens taken from young cultures were inoculated on fresh PDA plates. In each plate, two disks of two different fungi were separated by a proper distance in the following combinations: *F. solani* + *P. capsici*; *F. oxysporum* + *P. capsici*; and *F. solani* + *F. oxysporum*. Inoculated plates were incubated at 25°C, and the diameter of each colony was measured every 24 hr. As a control, similar plates were inoculated each with two mycelial disks of the same fungus. Six replicates were used.

RESULTS

Three species of fungi were isolated from diseased pepper plants: *P. capsici* from fields of Sulaimaniyah Govarnarate and *F. solani* Mart. and *F. oxysporum* var. *redolens* Wollenw. in Tamim Govarnarate. Identifica-

tion was confirmed by Centraalbureau voor Schimmelcultures (The Netherlands) and Commonwealth Mycological Institute (England). All three pathogens preferred PDA medium to cornmeal agar, Czapek's agar, and Richard's agar, although the other media also supported growth. The growth curves of the fungi on PDA showed that the optimum temperature range for growth of *P. capsici* is around 25°C, that of *F. solani* is 20–27°C, and that of *F. oxysporum* is around 25–30°C.

All three fungi were shown to be powerful pathogens of pepper (Table 1). Pepper plants were susceptible to infection by the three fungi at all developmental stages tested, although susceptibility tended to decrease with increasing age of the plants (Table 2). Although all the cultivars tested were susceptible to infection by these fungi, the local variety Mousili showed some resistance against *P. capsici*; the percentage of diseased plants of this cultivar did not exceed 70% 20 days after inoculation (Table 3).

Of the plants species tested, all were susceptible except watermelon, which showed a good resistance to

Table 3. Percentage of disease in pepper varieties inoculated with *Fusarium solani, F. oxysporum,* and *Phytophthora capsici*

Incubation period (days)	Fungus	Percentage of disease in variety				
		Mousili	Sivrija	Belkapija	Yellow Wonder	Calif. Wonder
4	*F. solani*	5	35	10	20	30
	F. oxysporum	0	30	10	30	15
	P. capsici	0	35	20	20	10
8	*F. solani*	75	100	60	50	75
	F. oxysporum	80	100	65	50	60
	P. capsici	55	100	55	45	35
12	*F. solani*	75	100	100	100	100
	F. oxysporum	85	100	85	80	95
	P. capsici	55	100	85	85	75
16	*F. solani*	80	100	100	100	100
	F. oxysporum	85	100	90	100	100
	P. capsici	60	100	95	100	85
20	*F. solani*	80	100	100	100	100
	F. oxysporum	85	100	100	100	100
	P. capsici	70	100	100	100	100

Table 4. Percentage of disease in plants belonging to Solanaceae and Cucurbitaceae when inoculated with *Fusarium solani, F. oxysporum,* and *Phytophthora capsici*

Incubation period (days)	Fungus	Percentage of disease				
		Eggplant	Tomato	Watermelon	Cucurbit	Cucumber
4	*F. solani*	35	20	0	0	25
	F. oxysporum	30	15	0	0	30
	P. capsici	15	10	0	15	20
8	*F. solani*	55	60	30	65	45
	F. oxysporum	40	50	0	25	80
	P. capsici	25	30	0	60	50
12	*F. solani*	70	90	45	90	50
	F. oxysporum	45	70	0	60	100
	P. capsici	40	50	25	65	90
16	*F. solani*	80	95	50	95	100
	F. oxysporum	75	80	0	85	100
	P. capsici	50	75	35	90	100
20	*F. solani*	100	100	70	100	100
	F. oxysporum	75	90	0	90	100
	P. capsici	55	75	55	100	100

Table 5. Effect of interaction between *Fusarium solani, F. oxysporum,* and *Phytophthora capsici* on percentage of diseased pepper plants

Incubation period (days)	Percentage of diseased plants inoculated with						
	F. solani	*F. oxysporum*	*P. capsici*	*F. solani* + *F. oxysporum*	*F. solani* + *P. capsici*	*F. oxysporum* + *P. capsici*	*F. solani* + *F. oxysporum* + *P. capsici*
4	21	15	20	30	20	10	10
8	40	25	50	55	30	15	25
12	62	45	91	55	50	50	50
16	80	62	100	65	55	50	70
20	90	86	100	75	65	50	75

Table 6. Growth rate of the fungi *Fusarium solani, F. oxysporum,* and *Phytophthora capsici* cultured in different combinations on potato-dextrose agar at 25°C

Hours	Colony diameter (mm)					
	P. capsici + *P. capsici*	*F. solani* + *F. solani*	*F. oxysporum* + *F. oxysporum*	*P. capsici* + *F. solani*	*P. capsici* + *F. oxysporum*	*F. solani* + *F. oxysporum*
24	14	12	6	12	16	9
	15	11	7	11	7	8
48	38	22	19	37	39	21
	38	20	21	25	19	19
72	54	32	29	54	53	30
	54	30	30	32	29	29

F. oxysporum and a lesser degree to *P. capsici*; percentage of infection with these fungi was 20 and 75, respectively (Table 4).

The fungi exhibited an inhibitory interaction in their pathogenicity on pepper plants in mixed inoculations when compared with individual inoculations (Table 5). This inhibition was more evident in plants inoculated with *F. oxysporum* + *P. capsici* and *F. solani* + *P. capsici*, where percentages of diseased plants did not exceed 50 and 65, respectively. *F. oxysporum* was isolated from plants with mixed inoculum of the other two fungi, whereas *F. solani* was isolated from those inoculated with mixed inoculum of *F. solani* + *P. capsici*. In all cases, only one fungus of the inoculated mixture was isolated from the inoculated plants.

No interaction was observed between the three fungi on culture media (Table 6). The growth rate of each fungus was not affected by the presence of another fungus on the same plate, and no inhibition zone was formed.

DISCUSSION

This study found the causal agents of pepper wilt in northern Iraq to be *P. capsici, F. solani,* and *F. oxysporum* var. *redolens. P. capsici* had already been reported as a pathogen (6), whereas the other two are new for Iraq on pepper. Because the disease is widespread in the region, it needs extensive study especially in terms of control.

Disease symptoms were in accordance with those described (9), except we did not observe symptoms on the leaves. The incubation period of the disease ranged between 4 and 16 days, which also agrees with previous reports (8). The relative increase in plant resistance to

infection with age is in part the result of the increase in xylem elements in the tissues (1).

Mixed inoculation of pepper plants with the fungi in different combinations reduced their pathogenicity, especially in the case of *P. capsici* + *F. oxysporum* and *P. capsici* + *F. solani*. This inhibition was less evident in the other combinations. Similar results were observed between *F. solani* f. *pisi* and *F. oxysporum* f. *pisi* on pea plants and *F. oxysporum* var. *redolens* and *F. moniliforme* on asparagus (9). Interference was also observed between *P. capsici* and virulent or avirulent isolates of *R. solani* (5).

Absence of interaction between fungi on culture media suggests that the infected plant or the plant-pathogen complex is involved in altering disease development. This point needs further study as it relates to our understanding of the disease in the field.

LITERATURE CITED

1. Agrios, G. N. 1978. Plant Pathology, 2nd ed. Academic Press, New York. 700 pp.
2. Aleksic, Z., Aleksic, D., and Sutic, D. 1974. Some results of investigation of *P. capsici* on pepper. Zast Bilja 25 (128/129): 229-240; in Biol. Abstr. 51309 (9) 60 1975.
3. Cinar, A., and Bicici, M. 1977. Control of *P. capsici* on red pepper. J. Turk. Phytopathol. 6(3):119-124.
4. Ershad, D., and Hille, M. 1975. Study of pepper root rot in Iran. Iran. J. Plant Pathol. 11:1-2; in Biol. Abstr. 62914 (11) 62 1976.
5. Kubaisi, A. F. 1981. Epidemiology and control of *Phytophthora capsici* and *Rhizoctonia solani*, the causal agents of root and stem rot of pepper. M.Sc. thesis, College of Science, University of Baghdad, Iraq.
6. Mustafa, F. H. 1974. A list of the common plant diseases in Iraq. Ministry of Agriculture and Agrarian Reform.

Directorate General of Plant Protection, Bull. 74. 28 pp.

7. Tomkins, C. M., and Tucker, C. M. 1941. Root rot of pepper and pumpkin caused by *Phytophthora capsici*. J. Agric. Res. 63:417-426.

8. Tuite, J. 1969. Plant Pathological Methods for Fungi and Bacteria. Burgess Publishing Company, Minneapolis. 239 pp.

9. Weber, G. F. 1932. Blight of peppers in Florida caused by *Phytophthora capsici*. Phytopathology 22:775-780.

10. Wood, R. K. S. 1967. Physiological Plant Pathology. Blackwell Scientific Publication, Oxford and Edinburgh. 570 pp.

Rhizoctonia on Small-Grain Cereals in Great Britain

M. J. RICHARDSON, Department of Agriculture and Fisheries for Scotland, Agricultural Scientific Services, East Craigs, Edinburgh, EH12 8NJ, and R. J. COOK, Agricultural Development and Advisory Service, Cambridge, CB2 2DR, U.K.

This review is largely concerned with sharp eyespot on winter wheat, which is the most widespread disease of cereals caused by *Rhizoctonia* in Britain. We have little information relating to other cereals, although infection is sometimes severe on barley and there are reports of moderate infection on maize and oats. Recent information has been obtained during surveys of Scottish cereal crops from 1970 to 1976 (15) and from surveys and investigational work by colleagues in the Agricultural Development and Advisory Service of England and Wales (ADAS), particularly King (10) and Clarkson and Cook (2).

THE FUNGI

Until about 5 years ago, most workers identified the cause of sharp eyespot as *R. solani* Kuehn. It was apparent, however, that isolates that caused sharp eyespot were different from most other isolates of *R. solani* (Table 1). It is now generally accepted that the sharp eyespot fungus is distinct from *R. solani* and it was described as a new species, *R. cerealis* van der Hoeven, by Boerema and Verhoeven (1). Because the fungus is morphologically, taxonomically, and pathologically distinct from *R. solani*, early references linking *R. solani* with sharp eyespot can be taken as referring to *R. cerealis* with confidence. The teleomorph is *Ceratobasidium cereale* Murray & Burpee (12).

R. solani is a ubiquitous pathogen occurring on a wide range of hosts, but in Britain it is rarely isolated from cereals. When it does occur, it is associated with stunting as a result of root damage. Dillon Weston and Garrett (5) observed partial failures of crops on seven farms in Norfolk, England. Mostly barley, the crops included patches measuring up to 0.1 ha containing backward, stunted plants with infected roots. The disease, known as barley stunt, is widely distributed in eastern and central England (Gladders, *unpublished*). It is most commonly recorded on loamy sand soils and on either winter barley or spring barley. Infection is recorded in crops following grass as well as long runs of cereals. Murray (11) reported similar symptoms in five Scottish barley crops. Identification of *R. solani* as the cause of the barley stunt disorder was confirmed when the isolates produced the teleomorph, *Thanatephorus cucumeris* (Frank) Donk.

OCCURRENCE AND ECONOMIC IMPORTANCE

Sharp eyespot is widely distributed throughout Great Britain. Results from ADAS winter wheat disease surveys (King, *unpublished*) show that it is more common in the west than in the east (Table 2), and Scottish surveys show that it is most common on winter wheat but that it also occurs frequently on barley and oats (Table 3). An examination of the incidence of sharp eyespot in winter wheat crops, obtained from various surveys over a 20-year period (Table 4), shows that there has been little change in incidence. The relative importance of sharp eyespot as a cause of loss has been underestimated, and it now appears to be of greater importance as a result of the control of other plant base pathogens. Analysis of data from the ADAS wheat surveys also indicates that previous cropping has some effect on disease incidence (Table 5), although the differences are not significant.

The reason for the increased incidence of the disease following potatoes is not understood, particularly as reports in the literature suggest that *R. cerealis* does not infect potatoes and that *R. solani* rarely causes severe sharp eyespot lesions on wheat. Crops following potatoes are unlikely to be sown before mid-October, so there is not likely to be any interaction between sowing date and previous crop. Severe infection has been noted by the second author in several crops of first winter wheat following another arable break crop—oilseed rape; the reasons for this association are also unclear.

Sharp eyespot has not always been considered as a potentially serious cause of loss. Glynne and Ritchie (8) reported it as frequent but only at levels of less than 1%; consequently, they considered it more a complication in diagnosing true eyespot (*Pseudocercosporella herpotrichoides* (Fron) Deighton) than a cause of serious loss. Doling and Batts (6) thought it unlikely that the disease materially affected yield because the lesions were generally superficial. Pitt (14) reported that in 1964, a year of comparatively high disease incidence, there were few crops with levels of infection high enough to affect

Table 1. Differences between *Rhizoctonia cerealis* and *R. solani*

Character	*R. cerealis*	*R. solani*
Teleomorph	*Ceratobasidium cereale*	*Thanatephorus cucumeris*
Anamorph hyphae	Narrow, 5.2 (2.5–8) μm in diam; binucleate	Broad, 8.9 (5.5–12) μm in diam; 4–16 nucleate
Sclerotia	None or few; pale in color	Frequent, coarse
Growth rate	Slow	Fast
Cereal disease	Sharp eyespot	Barley stunt
Host range	Widespread on Gramineae	Rare on barley in UK; frequent on wide range of non-Gramineae hosts

yield. He considered the effects on establishment to be more important, and he believed that in severely attacked crops lodging would be more important than yield reduction from infected tillers. Croxall et al (3), however, recognized it as the most frequent plant base disease in surveys of winter wheat in the West Midlands of England in 1962–1963 and noted that the crop with the highest level of infection (80%) possibly suffered a loss of 25%. Crops with average levels of infection (10–25%) were considered to be losing about 10% of field potential, taking no account of loss at the establishment stage.

Subsequent assessments of losses caused by sharp eyespot in Great Britain have confirmed the earlier observations of Croxall et al (3). Richardson et al (16) found that diseased tillers at GS 75 yielded 8% less than unaffected tillers. No attempt was made to discriminate between different levels of damage, and without any indication of the mean level of tiller infection in the national crop, no absolute estimate of loss could be made. Clarkson and Cook (2) reported more detailed assessments and established that within the three infection categories of slight, moderate, and severe there was little difference between loss caused by slight and moderate infections; losses from severely infected tillers, however, were much higher (Table 6). Information on disease incidence is also available from surveys of tiller diseases made by ADAS during the period 1975–1982 (King, *unpublished*). Using data from these surveys, yield loss owing to sharp eyespot during the period 1975–1982 has been estimated to range from 0.04 to 0.87%, with a mean of 0.38%. The loss caused by sharp eyespot in British wheat crops in 1980 and 1982, when severe infection was widespread, can be estimated at about 80,000 t (0.8%), which at 1982/3 prices of £118 per tonne would be valued at £9.4 million. Gilligan (7) has noted, however, that the sampling method used—collection of single tillers "at random" during a journey through a crop—results in overestimation of the amount of infection present by a factor of about 1.6, although the reason for this is not clear. A revised estimate, therefore, taking sampling bias into account, would be in the order of £5.9 million per annum from a crop of £1,170 million total value. Even on the annual average losses of 0.38%, total loss would be about £3 million.

CONTROL

Hollins and Scott (9) demonstrated that there is a range of resistance to *R. cerealis* available in winter wheat cultivars grown in Britain. There were significant and consistent differences for some cultivars

Table 2. Mean relative infection with sharp eyespot disease in England and Wales, 1975–1982[a]

| Area | Severity | | |
	Slight	Moderate	Severe
North	113	140	140
East	81	71*[b]	60*
South	103	101	94
Wales and west	158*	176*	258*

[a] Corrected to account for annual variation.
[b] Asterisk indicates significant difference from national mean (= 100).

Table 3. *Rhizoctonia cerealis* infection levels in Scotland

| Crop | Number sampled | Years | Percentage infection | | Percentage crops infected | |
			GS 11–13[a]	GS 75	GS 11–13	GS 75
Winter wheat	340	1971–1974	<1	5–7	5–40	45–65
Spring barley	471	1971–1973	<0.5	3	1–5	30
Spring oats	452	1974–1976	<0.5	2	0–4	5–30

[a] Growth stage (18).

Table 4. *Rhizoctonia cerealis* infection levels in winter wheat crops in Great Britain, 1962–1982

Year	Percentage fields infected	Mean percentage stems infected	Number of fields surveyed
1962	68	24	21[a]
1963	100	17	54[a]
1963	84	25	23[b]
1964	94	31	72[b]
1965	64	20	76[b]
1972	64	7	104[c]
1973	46	5	102[c]
1974	65	7	96[c]
1975	34	3	333[d]
1976	57	8	315[d]
1977	60	9	348[d]
1978	51	7	310[d]
1979	52	6	301[d]
1980	81	18	306[d]
1981	69	15	300[d]
1982	60	12	320[d]

[a] Croxall et al (3); lesions on tiller samples.
[b] Pitt (14); lesions on tiller samples.
[c] Richardson (*unpublished*, from Scottish surveys); isolation from adult plant bases.
[d] King (*unpublished*, ADAS wheat disease surveys); lesions on tiller samples.

Table 5. Effect of previous crop on relative severity of sharp eyespot (national mean = 100)

| Previous crop | Severity | | |
	Slight	Moderate	Severe
Wheat	96	92	99
Other cereal	97	100	106
Grass	114	105	112
Potato	116	130	114
Pulse	99	103	99

Table 6. Mean percentage loss of grain yield components according to severity of sharp eyespot[a]

| Yield component | Severity | | |
	Slight	Moderate	Severe
Grain number per ear	2.4	0.9	19.6
1,000-grain weight	0.4	4.2	10.8
Grain yield per ear	2.8	5.4	26.4

[a] Weighted according to incidence of each severity category in 18 fields (2).

over three seasons, but with variation if the whole range of cultivars tested was examined. The significant difference was about 11 on a scale of 0 (resistant) to 100 (susceptible) (18). Hustler, Flanders, Bounty, and Aquila were most resistant, with values of 15–20, and Champlein and Avalon were most susceptible, with values of 30–40.

Sharp eyespot in winter wheat is reduced by late sowing, but the benefit of such a practice must be set against the penalties of later sowing. Clarkson (*unpublished*) recorded 53% severe sharp eyespot in part of a crop sown on 27 September and 19% in the part sown on 9 November. In a more extensive analysis of data from unsprayed crops surveyed from 1975 to 1981, there was a regular and significant reduction in moderate and severe sharp eyespot as sowing date progressed from September (140% of national mean infection) to early December (40% of national mean infection) (2). It may be that there is a link between these observations and those of Pitt (13), who found that infection was favored by cool, dry conditions and was most severe when it occurred in the earlier stages of growth. With reduction in evapotranspiration as autumn progresses, one can expect British soils to become wetter, thus not favoring infection of late-sown crops although temperatures also fall during this period.

No such effects of sowing date have been noted for winter barley, although it should be noted that the disease is now more commonly seen on winter barley than in the past and that crops are now generally sown much earlier.

Attempts have been made to find chemicals active against sharp eyespot. Benomyl and carbendazim have proved to be useful components of selective media for isolating *R. cerealis*. Cook (*unpublished*) found that various fungicides—iprodione, thiophanate methyl, oxycarboxin, carbendazim + maneb, tolclofos methyl, and prochloraz—did not significantly reduce the amount of infection (with 50% infected tillers in the untreated crop) or increase yield, although iprodione and tolclofos methyl consistently reduced the infection index by comparison with the untreated control. Davies and Price (4) found, however, that tolclofos methyl (at higher rates than used by the second author) and iprodione reduced severe infection, although with no yield benefit. Results of spray timing experiments (Cook, *unpublished*) with tolclofos methyl and iprodione suggest that more than one application of fungicide is needed to reduce infection. Van der Hoeven and Bollen (17) found that infection of rye was increased following sprays with benomyl to control *P. herpotrichoides*. The fungus is more sensitive in vitro to benomyl but not markedly so, with an ED_{50} of 3.2 ppm, compared with ED_{50} of 0.5–1.5 ppm for *Fusarium nivale* (Fr.) Ces. and *F. culmorum* (W. G. Sm.) Sacc. There is no differential sensitivity in isolates from sprayed and unsprayed fields, so they concluded that the increase in sharp eyespot was caused by adverse effects of benomyl on microbial antagonists to *R. cerealis* in the soil.

The fungicides presently available and used to control other cereal pathogens are thus unlikely to cause any significant reduction in sharp eyespot levels; by reducing competition from other pathogens and antagonists, they may well be responsible for the increased awareness of sharp eyespot. Work on control of the disease with fungicides is, however, continuing.

ACKNOWLEDGMENTS

We are grateful to J. King, J. Clarkson, and P. Gladders for permission to quote unpublished data.

LITERATURE CITED

1. Boerema, G. H., and Verhoeven, A. A. 1977. Check-list for scientific names of common parasitic fungi. Series 2b. Fungi on field crops: Cereals and grasses. Neth. J. Plant Pathol. 83:165-204.
2. Clarkson, J. D. S, and Cook, R. J. 1983. Effect of sharp eyespot (*Rhizoctonia cerealis*) on yield loss in winter wheat, and of some agronomic factors on disease incidence. Plant Pathol. 32:421-428.
3. Croxall, H. E., Dale, W. T., and Knight, B. C. 1964. The incidence of soilborne diseases in the West Midlands 1959–63. Proc. Br. Insecticides and Fungicide Conf. 1963:39-44.
4. Davies, W. P., and Price, K. R. 1983. Sensitivity of sharp eyespot of wheat and *Rhizoctonia cerealis* to fungicides. Ann. Appl. Biol. Suppl., Tests of Agrochemicals and Cultivars 4:54-55.
5. Dillon Weston, W. A. R., and Garrett, S. D. 1943. *Rhizoctonia solani* associated with a root rot of cereals in Norfolk. Ann. Appl. Biol. 30:79.
6. Doling, D. A., and Batts, C. C. V. 1960. Effect of previous cropping on eyespot and take-all in four varieties of winter wheat. Plant Pathol. 9:115-118.
7. Gilligan, C. A. 1982. Size and shape of sampling units for estimating incidence of sharp eyespot, *Rhizoctonia cerealis*, in plots of wheat. J. Agric. Sci. Cambridge 79:461-464.
8. Glynne, M. D., and Ritchie, W. M. 1943. Sharp eyespot of wheat caused by *Corticium* (*Rhizoctonia*) *solani*. Nature (London) 152:161.
9. Hollins, T. W., and Scott, P. R. 1983. Resistance of wheat cultivars to sharp eyespot caused by *Rhizoctonia cerealis*. Ann. Appl. Biol. Suppl., Tests of Agrochemicals and Cultivars 4:126-127.
10. King, J. 1977. Surveys of diseases of winter wheat in England and Wales, 1970–75. Plant Pathol. 26:8-20.
11. Murray, D. I. L. 1981. *Rhizoctonia solani* causing barley stunt disorder. Trans. Br. Mycol. Soc. 76:383-391.
12. Murray, D. I. L., and Burpee, L. L. 1984. *Ceratobasidium cereale* sp. nov., the teleomorph of *Rhizoctonia cerealis*. Trans. Br. Mycol. Soc. 82:170-172.
13. Pitt, D. 1964. Studies on sharp eyespot disease of cereals. I. Disease symptoms and pathogenicity of isolates of *Rhizoctonia solani* Kühn and the influence of soil factors and temperature on disease development. Ann. Appl. Biol. 54:77-89.
14. Pitt, D. 1966. Studies on sharp eyespot disease of cereals. III. Effects of the disease on the wheat host and the incidence of disease in the field. Ann. Appl. Biol. 58:299-308.
15. Richardson, M. J. 1975. Cereals—Why only 35 cwt per acre? ADAS Q. Rev. 16:152-163.
16. Richardson, M. J., Whittle, A. M., and Jacks, M. 1976. Yield-loss relationships in cereals. Plant Pathol. 25:21-30.
17. Van der Hoeven, E. P., and Bollen, G. J. 1980. Effect of benomyl on soil fungi associated with rye. 1. Effect on the incidence of sharp eyespot caused by *Rhizoctonia cerealis*. Neth. J. Plant Pathol. 86:163-180.
18. Zadoks, J. C., Chang, T. T., and Konzak, C. F. 1974. A decimal code for the growth stages of cereals. Weed Res. 14:415-421.

Fungal Invasion of Clover and Grass Roots in New Zealand Pasture Soils

R. A. SKIPP, G. C. M. LATCH, and M. J. CHRISTENSEN, Plant Diseases Division, DSIR, Palmerston North, New Zealand

Most of the 8 million hectares of sown pasture in New Zealand has been developed during the last 100 years. Indigenous forests have been cleared, burned, and replaced with pastures of introduced species, mainly ryegrass (*Lolium perenne* L.) and white clover (*Trifolium repens* L.). Although the general soil mycoflora of such pastures has been studied (9,15), specific information about fungi invading the roots of pasture plants has only recently been published (11–14). This paper reviews accounts of fungal invasion of white clover roots (11–14) and reports comparative data for ryegrass roots.

MATERIALS AND METHODS

Turf samples (150 × 150 mm; 100 mm deep) were taken from grazed ryegrass-white clover pastures at sites throughout New Zealand. Nodal roots from white clover stolons in turfs from 48 sites were investigated during 1980–1981 (13); adventitious roots of ryegrass from 20 sites were investigated during 1980–1983. Recent additional studies were made on samples of roots and soil from other pastures. Previous reports have described methods for examination of roots from turf samples by light microscopy (11), plate culture isolation of fungi from segments of surface-sterilized root (13), testing the ability of fungal isolates to invade living seedling root tissue (12), and studying root invasion of seedlings grown in sieved pasture soil (11). For isolation data presented in parentheses below, S = the percentage of sites from which a fungus was recorded and I = the percentage of segments from which fungus was isolated.

RESULTS AND DISCUSSION

Light microscopy of roots from pasture soil. Young roots of ryegrass and clover were white, becoming uniformly brown with age (14). The epidermis and cortex of thickened root bases of white clover nodal roots were often compressed or shed to reveal the white stele. Ryegrass root cortical tissues were usually intact, but root hairs were often broken.

Clover roots from most sites were extensively invaded by ramifying hyphae of mycorrhizal fungi (13). Cleared roots from pastures (14) and from seedlings grown in pasture soil (11) commonly had more than 30% of 1-mm segments of their length infected by aseptate mycorrhizal mycelium. Septate mycelium (i.e., the majority of identified isolates—see below) was less obvious in healthy young roots, often limited to a few cells (11,14). However, in the cortex of older roots and in roots damaged during emergence of clover cyst nematode (*Heterodera trifolii* Goffart)(11), septate hyphae were as abundant as mycorrhizae. The stele was rarely penetrated by mycelium even in root bases where cortical tissues packed with fungal hyphae had become detached. Oospores were occasionally seen in roots of seedlings (11) and field-grown clover (13,14), as were cystosori of *Polymyxa graminis* Ledingham and sporangia of *Olpidium* sp.

Cleared ryegrass roots from seven of the 20 sample sites were examined for fungal invasion (100 1-mm segments of root length per sample). Mycorrhizae were present in 37% of segments. Two types of mycorrhizal invasion were observed: a broader endogonaceous type that showed direct penetration and infection via root hairs, and a fine endophyte. Fungi with dark septate mycelium occupied 51% of root segments; hyaline septate mycelium, 22% of segments. Old brown roots were extensively invaded by fungi (see also 16). Groups of pigmented fungal cells characteristic of *Phialophora radicicola* Cain var. *graminicola* Deacon (2,5) were found in 16% of root segments. Oospores and resting spores of *Olpidium* sp. and *Lagenocystis* sp. (5,10) were seen occasionally.

Fungi isolated only from white clover. *Bimuria novae-zelandiae* D. Hawksw., Chea & Sheridan (S 90%, I 16.2%) was the most widely distributed and most frequently isolated of all fungal species from white clover roots (13). This recently described ascomycete (4) has not been reported outside New Zealand. The fungus was isolated from apparently healthy seedlings growing in pasture soil (11). Isolates invaded seedling roots growing in inoculated soil and caused root necrosis (12).

Phoma chrysanthemicola Hollós (S 83%, I 4.9%) and a *Chrysosporium* sp. (S 46%, I 3.4%) were also encountered frequently. Both fungi invaded seedling roots from a pasture soil (11) where the *Chrysosporium* sp. formed a major part of the seedling rhizoplane mycoflora. Necrotic root lesions developed on seedlings grown in soil inoculated with either fungus (12). A *Colletotrichum* sp. (S 31%, I 1.8%) was common in samples from North Island hill country.

Thielaviopsis basicola (Berk. & Br.) Ferraris was rarely isolated from, or seen in, roots from turf samples, but conidia were seen on necrotic roots of seedlings grown in some North Island pasture soils. Isolates caused extensive necrosis of seedling roots (12).

Fungi isolated only from ryegrass. *Phialophora radicicola* (S 95%, I 18.7%) was the most abundant species isolated from ryegrass roots. The fungus grew from more than 30% of segments from four of the sites.

Tests with 20 isolates confirmed that the fungus could invade seedling root tissues. Pigmented cells of the fungus (2) were produced within the root cortex, but no necrotic lesions formed. *Gaeumannomyces graminis* (Sacc.) von Arx & Olivier (S 30%, I 0.4%), although found infrequently, was more invasive than *P. radicicola*, causing necrotic lesions that extended into the stele (see also 2).

Of two seedborne endophytic fungi known to occur in ryegrass, *Acremonium* sp. and *Gliocladium* sp. (6), only the *Gliocladium* sp. was isolated from ryegrass roots. It was encountered infrequently in the survey (S 10%, I 0.3%) but was readily isolated from plants from infected seed (6).

Fungi isolated from both white clover and ryegrass. *Codinaea fertilis* Hughes & Kendrick was commonly isolated from roots of white clover (S 50%, I 7.7%) and ryegrass (S 85%, I 9.1%). The greatest proportion of clover (48%) and ryegrass (31%) root segments found infected by *C. fertilis* was from a Northland site; the fungus was not found in the south and east of the South Island. *C. fertilis* is reported to be a warm-temperature, root-rotting pathogen of white clover (1,8), but it has not previously been considered as a root invader of ryegrass.

Fusarium oxysporum Schlecht. was the most commonly isolated species of *Fusarium* from white clover (S 88%, I 6.2%) and ryegrass (S 85%, I 7.0%). It was a major component of the rhizoplane flora of seedling clover roots (11). Mechanical damage, such as that caused by clover cyst nematode (11) or during secondary thickening, allows rapid invasion by *F. oxysporum* and other *Fusarium* spp. (7). *F. oxysporum* isolates penetrated and formed lesions on clover seedling roots (12). Invasion of ryegrass roots in inoculated soil was halted by the formation of lignitubers. *F. culmorum* (W. G. Smith) Sacc. and *F. avenaceum* (Corda ex Fr.) Sacc. were occasionally isolated from roots, and isolates caused lesions on inoculated clover and ryegrass, respectively.

Sterile dark mycelial forms were abundant on roots of white clover (S 100%, I 11.1%) and ryegrass (S 100%, I 22.6%). Sterile hyaline forms were also common on white clover (S 96%, I 7.8%) and ryegrass (S 100%, I 8.7%) roots. Sterile fungi were found to invade roots of clover (12) and ryegrass seedlings; although able to cause lesions, they showed differing degrees of aggressiveness.

Basidiomycete isolates with clamp connections were more frequently obtained from ryegrass roots (S 70%, I 4.0%) than clover roots (S 54%, I 0.9%). Cultural characteristics of the isolates suggested that the two plants were host to different basidiomycete species. All six isolates from clover tested invaded inoculated clover seedling roots, sometimes with extensive necrosis (12); 27 isolates from ryegrass invaded seedling roots similarly. "*Rhizoctonias*" were also isolated from roots: *Ceratobasidium cornigerum* (Bourd.) Rogers from white clover (S 67%, I 1.5%) and *Thanatephorus cucumeris* (Frank) Donk from ryegrass (S 25%, I 0.4%). *T. cucumeris* was consistently isolated from roots of damped-off clover seedlings grown in soil from several pasture sites.

Species of *Cylindrocarpon* (white clover: S 67%, I 2.3%; ryegrass: S 70%, I 1.4%), *Acremonium* (white clover: S 79%, I 1.8%; ryegrass: S 80%, I 1.4%), and *Gliocladium* (white clover: S 88%, I 2.4%; ryegrass: S 50%, I 1.1%) were common in roots. Isolates from white clover caused necrotic lesions on clover seedling roots (12). Isolates of *Periconia* spp. (white clover: S 27%, I 0.3%; ryegrass: S 75%, I 3.7%) and *Trichoderma* spp. (white clover: S 48%, I 1.0%; ryegrass: S 60%, I 3.1%) rarely penetrated seedling roots but caused stunting and necrosis, presumably by production of toxin. Other fungi found in ryegrass and clover roots included species of *Diheterospora*, *Penicillium*, *Phoma*, *Phomopsis*, and *Rhinocladiella*.

Oomycetes were rarely isolated from clover ryegrass roots. Damping-off of white clover seedlings was rarely observed, but lucern (*Medicago sativa* L.) sown into soil from several of the pasture sites suffered severe damage. *Pythium irregulare* Buis. was isolated from damped-off seedlings.

Significance of root-invading fungi. Roots of ryegrass and white clover from each pasture site contained many (sometimes more than 15) of the above fungi in their internal mycofloras. Many common rhizosphere fungi (9,15) were absent from root tissues, and the predominant root-invading fungi of ryegrass were absent from clover and vice versa. Senescence and damage probably aided colonization (e.g., by *Fusarium* spp.), but most isolates could invade young roots, often causing necrosis. Aggressive invaders such as *T. basicola* and *G. graminis* were rarely encountered, and most roots from pastures were free from "disease." The less aggressive fungi should probably be regarded as "minor pathogens" (10) that take refuge in roots from the competitive soil environment (12).

Colonization of root tissues by relatively innocuous fungi may provide benefits for both host and fungus. Endomycorrhizae aid nutrient uptake and help protect roots from fungal and nematode attack (3). "Acremonium endophytes" from ryegrass and tall fescue (*Festuca arundinacea* Schreb.) give protection from insects, and a "Phialophora endophyte" from tall fescue roots inhibits growth of other fungi in culture (6, *unpublished*). *Phialophora radicicola* infection protects wheat and grasses from *G. graminis* (2). Thus, the diverse internal mycofloras of pasture plants may provide a buffer against serious damage from pathogens. Problems of declining pasture productivity experienced in many areas (1,5,7,8) may result from reduced diversity in the root mycoflora, which favors attack by damaging pathogens.

LITERATURE CITED

1. Campbell, C. L. 1980. Root rot of ladino clover induced by *Codinaea fertilis*. Plant Dis. 64:959-960.
2. Deacon, J. W. 1974. Further studies on *Phialophora radicicola* and *Gaeumannomyces graminis* on roots and stem bases of grasses and cereals. Trans. Br. Mycol. Soc. 63:307-327.
3. Dehne, H. W. 1982. Interaction between vesicular-arbuscular mycorrhizal fungi and plant pathogens. Phytopathology 72:1115-1119.
4. Hawksworths, D. L., Chea, G. Y., and Sheridan, J. E. 1979. *Bimuria novae-zelandiae* gen. et sp. nov., a remarkable ascomycete isolated from a New Zealand barley field. N.Z. J. Bot. 17:267-273.
5. Labruère, R. E. 1979. Resowing problems of old pastures. Pages 313-326 in: Soil-borne Plant Pathogens. B. Schippers and W. Gams, eds. Academic Press, New York.
6. Latch, G. C. M. 1983. Incidence and control of ryegrass endophytes in New Zealand. Proc. Forage and Turfgrass Endophyte Workshop, Oregon, pp. 29-34.
7. Leath, K. T., Lukezic, F. L., Crittenden, H. W., Elliott, E. S., Halisky, P. M., Howard, F. L., and Ostazeski, S. A. 1971.

The Fusarium root rot complex of selected forage legumes in the northeast. Penn. Agric. Exp. Stn. Bull. 777. 64 pp.

8. Menzies, S. A. 1973. Root rot of clover caused by *Codinaea fertilis*. N.Z. J. Agric. Res. 16:239-245.

9. Ruscoe, Q. W. 1973. Changes in the mycofloras of pasture soils after long term irrigation. N.Z. J. Sci. 16:9-20.

10. Salt, G. A. 1979. The increasing interest in "minor pathogens." Pages 289-312 in: Soil-borne Plant Pathogens. B. Schippers and W. Gams, eds. Academic Press, New York.

11. Skipp, R. A., and Christensen, M. J. 1981. Invasion of white clover roots by fungi and other soil micro-organisms. I. Surface colonisation and invasion of roots growing in sieved pasture soil in the glasshouse. N.Z. J. Agric. Res. 24:235-241.

12. Skipp, R. A., and Christensen, M. J. 1982. Invasion of white clover roots by fungi and other soil micro-organisms. III.

The capacity of fungi isolated from white clover roots to invade seedling root tissue. N.Z. J. Agric. Res. 25:97-101.

13. Skipp, R. A., and Christensen, M. J. 1983. Invasion of white clover roots by fungi and other soil micro-organisms. IV. Survey of root-invading fungi and nematodes. N.Z. J. Agric. Res. 26:151-155.

14. Skipp, R. A., Christensen, M. J., and Caradus, J. R. 1982. Invasion of white clover roots by fungi and other soil micro-organisms. II. Invasion of roots in grazed pastures. N.Z. J. Agric. Res. 25:87-95.

15. Thornton, R. H. 1965. Studies of fungi in pasture soils. I. Fungi associated with live roots. N.Z. J. Agric. Res. 8:417-449.

16. Waid, J. S. 1957. Distribution of fungi within the decomposing tissue of ryegrass roots. Trans. Br. Mycol. Soc. 40:391-406.

Pathogenic *Rhizoctonia* and Orchids

J. H. WARCUP, Waite Agricultural Research Institute, The University of Adelaide, Glen Osmond, South Australia, 5064

In nature, members of the Orchidaceae are closely associated with fungal mycelia. Orchids typically pass through a saprophytic seedling (protocorm) stage or, more rarely, they remain saprophytic throughout their life. In most adult plants, whether saprophytes or not, the absorbing organs are infected by similar fungi that are mainly intracellular, forming coils of hyphae within cortical cells. Later these coils are usually "digested" by the orchid. A considerable number of orchid fungi have been obtained in culture; most, especially from green orchids, are members of the form genus *Rhizoctonia*.

Many species of *Rhizoctonia*, including *R. solani*, are well-known pathogens of germinating seed, immature roots, leaves, or fruit of various plants. Yet there is little information on whether pathogenic species of *Rhizoctonia* are a different group from orchid endophytes, though Downie (4) and Williamson and Hadley (13) have shown that some pathogenic strains of *R. solani* may also be orchid symbionts. This paper explores whether other pathogen species of *Rhizoctonia* may also be orchid symbionts.

MATERIALS AND METHODS

Symbiotic seed germination tests were done as detailed previously (8,9), except that the weak oatmeal medium (10) was used in most tests. Water agar was also used for germination experiments with some strains of *R. solani*. The orchids used as test species were *Microtis unifolia* (G. Forst.) Reichb., *Prasophyllum regium* R. Rogers, and *Orchis morio* L. All are terrestrial orchids, the first two from Australia and the third from Europe. Previous work had shown that all three would associate with a wide range of *Rhizoctonia* species from orchids.

The isolates of *Rhizoctonia* came from several sources, the Centraalbureau voor Schimmelcultures, Baarn, Netherlands (CBS); the Waite Institute collection (F); and Dr. D. I. L. Murray, Kelmscott, Western Australia. Some of the strains used are detailed in Table 1.

RESULTS AND DISCUSSION

Representative results of the symbiotic germination tests are given in Table 1. If several isolates of a species or anastomosis group gave similar results, only one or two isolates are recorded.

In all, eight isolates of *R. cerealis* van der Hoeven, the cause of root decay and sharp eyespot disease of wheat or rye or yellow patch of turfgrass, were tested as orchid symbionts. The isolates were of diverse origin—U.S.A., Scotland, Germany, and Japan. Although different

Table 1. Host and origin of species of *Rhizoctonia* tested as orchid symbionts

Species	Culture	Host and origin	Protocorm development[a] *Microtis unifolia*	*Prasophyllum regium*	*Orchis morio*
R. cerealis	CBS 236.77 T[b]	Rye, Netherlands	+	+	−
	BN1	Agrostis, U.S.A.	+	+	+
R. fragariae	CBS 335.62 T	Strawberry, Canada	+	(+)	−
	BN4	Unknown	+	+	+
R. solani					
AG 1[c]	F165	*Picea*, Canada	−	?	?
AG 2-1	F48	Soil, S. Australia	+	+	+
AG 2-1	TG1	Unknown, Japan	+	+	−
AG 2-2	C127	Unknown, Japan	(+)	+	+
AG 3	F151	Soil, U.S.A.	−	+	(+)
AG 3	F86	Potato, S. Australia	+	+ P	−
AG 4	F207	Pine, Canada	+	+	+
AG 4	F42	Lettuce, England	−	+	+
Root strain	F91	Wheat, S. Australia	−	(+) P	(+) P
R. zeae	CBS 384.34 T	Maize, U.S.A.	0	0	0
R. oryzae	CBS 273.38	Rice, U.S.A.	0	0	0
	(CBS) 541	Rice, U.S.A.	0	0	0
Rhizoctonia sp. (*Waitea circinata*)	F220	Soil, S. Australia	0	0	0
Rhizoctonia sp. (*Corticium koleroga*)	CBS 154.35	Coffee, unknown	0	0	0

[a]+ = Protocorm development to form shoot; (+) = partial development of protocorm; P = some or all protocorms parasitized; 0 = no symbiosis; − = no test.
[b]T type culture.
[c]Anastomosis group.

isolates varied in their ability to cause protocorm development of the test orchids, all gave complete germination of at least one orchid species. *R. cerealis* is in anastomosis group 1 (CAG 1) of *Ceratobasidium* (2). Members of CAG 1 attack gramineous hosts but appear nonpathogenic or weakly pathogenic on species of other families (3). *R. cerealis* is an undescribed species of *Ceratobasidium* (Murray, *personal communication*).

The type culture of *R. fragariae* Husain and McKeen from strawberry (*Fragaria ananassa* Hort.) and an isolate of *Ceratobasidium cornigerum* (Bourd.) Rogers (CAG 2) with a probable *R. fragariae* anamorph were tested as orchid symbionts. Both were found to stimulate protocorm development. This is not surprising because many other isolates of *C. cornigerum* are known to be orchid symbionts (10). Isolates of *C. cornigerum* pathogenic to lucerne (*Medicago sativa* L.) and lupin (*Lupinus albus* L.) are known to be capable of acting as orchid symbionts (11).

R. solani Kuehn is now regarded not as a simple species but a collection of noninterbreeding populations (1). These populations correspond to the anastomosis groups (AG1–5), though Parmeter et al (5) comment that not all isolates can be placed in these groups. Hence it is of interest to see whether isolates from different anastomosis groups may act as orchid symbionts.

When Williamson and Hadley (13) examined pathogenic isolates of *R. solani* for symbiotic activity with orchids, they used four isolates from South Australia: F16 from a wheat root; F48 from soil but pathogenic to crucifers; F82 from soil and nonpathogenic to all hosts tested; and F87 from soil but parasitic on stems of many hosts. They found that isolates F48 and F82 gave the most stable relationships with the orchids tested, whereas F87 showed a trend to postcompatibility parasitism and F16 infrequently became symbiotic and was parasitic in most infections. These results are generally substantiated here.

Isolate F48 (AG 2-1) generally gave good protocorm development, whereas F91—the "root strain" from wheat and similar to F16—gave few protocorms, and on the weak oatmeal agar these were usually parasitized. However, on water agar at 15°C a few protocorms developed to produce shoots. Isolate F87 belongs to AG 4 (Neate, *personal communication*) and is similar to strains F42 and F207 used successfully here. The potato isolate (AG 3) successfully germinated seed of *M. unifolia*, although it showed a tendency to parasitism on *P. regium*. Other AG 4 isolates, not recorded in Table 1, also showed a tendency to parasitize protocorms of various orchids unless grown on very weak media. The one AG 1 isolate used here appeared to stimulate protocorm development slightly in one test, but it was completely nonsymbiotic in another. Hence further work is needed to see whether members of this anastomosis group are orchid symbionts.

Although some members of each of anastomosis groups 1–4 have produced teleomorphs of the collective species *Thanatephorus cucumeris* (Frank) Donk (7), Anderson (1) remarked that all strains of AG 4 are of the "*praticola*" type and represent a good biological species. He suggested that the name *Thanatephorus praticola* (Frank) Kotila be used for this group.

R. zeae Voorhees isolated from a dry rot of ears of corn (*Zea mays* L.). did not associate with seed of any of the test orchids. *R. oryzae* Ryker and Gooch, the cause of

sheath spot of rice (*Oryza sativa* L.), also did not associate with the test orchids.

Waitea circinata Warcup and Talbot has a *Rhizoctonia* anamorph (12). The species may also be mildly pathogenic to roots of lucerne, lupin, and wheat (Neate, *personal communication*). Experiments with the test orchid and with other orchids using four isolates of the fungus from soil in South Australia gave no indication that the fungus is an orchid symbiont. *R. oryzae* and *W. circinata* have some features in common: they have pink to salmon to brown, irregular sclerotia; they grow well at temperatures of about 30°C; and they seem not to be orchid symbionts. One may wonder if they are related.

Corticium koleroga (Cooke) Höhnel (*Koleroga noxia* Donk) was considered by Talbot (6) to be a *Ceratobasidium,* but he considered that the teleomorphs of the various web blights needed reinvestigation before the species could be satisfactorily classified. The isolate from coffee (*Coffea* sp.) used here was not symbiotic with the test orchids. However, because many species of *Ceratobasidium* are orchid symbionts (11), further examination of this fungus and other web blights is desirable. Perhaps the test orchids themselves were inappropriate, as web blight fungi would seem more likely to associate with epiphytic than terrestrial orchids; but more appropriate seed was not available when these tests were made.

LITERATURE CITED

1. Anderson, N. A. 1982. The genetics and pathology of *Rhizoctonia solani.* Annu. Rev. Phytopathol. 20:329-347.
2. Burpee, L. L., Sanders, P. L., and Cole, H. 1980. Anastomosis groups among isolates of *Ceratobasidium cornigerum* and related fungi. Mycologia 72:689-701.
3. Burpee, L. L., Sanders, P. L., Cole, H., Jr., and Sherwood, R. T. 1980. Pathogenicity of *Ceratobasidium cornigerum* and related fungi representing five anastomosis groups. Phytopathology 70:843-846.
4. Downie, D. G. 1959. The mycorrhiza of *Orchis purpurella.* Trans. Bot. Soc. Edinburgh 38:16-29.
5. Parmeter, J. R., Sherwood, R. T., and Platt, W. D. 1969. Anastomosis grouping among isolates of *Thanatephorus cucumeris.* Phytopathology 59:1270-1278.
6. Talbot, P. H. B. 1965. Studies of "*Pellicularia*" and associated genera of Hymenomycetes. Persoonia 3:371-406.
7. Talbot, P. H. B. 1970. Taxonomy and nomenclature of the perfect stage in *Rhizoctonia solani.* Pages 20-31 in: *Rhizoctonia solani*: Biology and Pathology. J. R. Parmeter, Jr., ed. University of California Press, Berkeley.
8. Warcup, J. H. 1973. Symbiotic germination of some Australian terrestrial orchids. New Phytol. 72:387-392.
9. Warcup, J. H. 1975. Factors affecting symbiotic germination of orchid seed. Pages 87-104 in: Endomycorrhizas. F. E. Sanders, B. Mosse, and P. H. Tinker, eds. Academic Press, London.
10. Warcup, J. H. 1981. The mycorrhizal relationships of Australian orchids. New Phytol. 87:371-381.
11. Warcup, J. H. 1981. Orchid mycorrhizal fungi. Pages 51-63 in: Proceedings of the Orchid Symposium, Sydney, Australia, 1981. L. Lawler and R. D. Kerr, eds. Harbour Press, Sydney.
12. Warcup, J. H., and Talbot, P. H. B. 1962. Ecology and identity of mycelia isolated from soil. Trans. Br. Mycol. Soc. 45:495-518.
13. Williamson, B., and Hadley, G. 1970. Penetration and infection of orchid protocorms by *Thanatephorus cucumeris* and other *Rhizoctonia* isolates. Phytopathology 60:1092-1096.

Origin and Distribution of *Phytophthora cinnamomi*

G. A. ZENTMYER, Department of Plant Pathology, University of California, Riverside 92521, U.S.A.

The origin of the widespread pathogen *Phytophthora cinnamomi* Rands is a matter of much speculation and conjecture. New records of its occurrence continue to appear. This paper covers both of these aspects, with emphasis on samples taken from native hosts in the western United States and in Latin America.

In his report of stripe canker of cinnamon in Sumatra, Rands (3) commented in 1922 that "the causal fungus is believed to have come into the plantings along with its natural host which is indigenous to this region." There are indications that this speculation in regard to Sumatra may have merit, with some data pointing to possible centers of origin in a large region extending from northeastern Australia into the New Guinea-Celebes-Malaysian region (5) and into other parts of eastern Asia. Ko et al (2) have presented evidence for the inclusion of Taiwan in the area of origin of *P. cinnamomi*. In their studies, both A1 and A2 types of the pathogen were isolated from native hosts in Taiwan.

Both mating types of *P. cinnamomi* have also been found in eastern Australia, and Broadbent and Baker (1) have pointed out that trees and shrubs in vulnerable situations for disease development in this region often show some resistance to *P. cinnamomi*. The situation in New Guinea is somewhat obscure, as only the A1 type has been isolated from the native Nothofagus forest (4).

There has also been research on the possible origin of *P. cinnamomi* in the Americas. In addition, there has been some recent investigation in South Africa, highlighted by S. von Broembsen's (*personal communication*) discovery of the A1 mating type of *P. cinnamomi* in native vegetation and as zoospores in streams draining remote mountainous areas of the Cape Province in that country.

Our research on the origin of *P. cinnamomi* has involved hundreds of soil and root samples collected over the past 35 years from native forests and other native vegetation in Central and South America, Mexico, the Caribbean islands, and the West Coast of the United States (California, Oregon, and Washington). The samples in Latin America have been collected primarily in the course of the search for indigenous avocado trees (*Persea americana* Mill.) and other species of *Persea* that might have resistance to avocado root rot caused by *P. cinnamomi*.

MATERIALS AND METHODS

Soil and root samples were collected from moist areas if possible, where the native hosts were growing. Roots from these samples were cultured on selective media (usually $P_{10}VP$); small pieces of feeder roots (1–2 mm in diameter and approximately 1 cm long) were selected with necrotic lesions if possible. In the case of recent samples taken in the United States, soil and root samples were also trapped in the laboratory or greenhouse, by placing susceptible *Persea indica* seedlings in a soil-water suspension in plastic cups (7).

In the samples collected in field trips in Latin America, prepoured plates with selective media were used; plates were examined in from 3 to 5 days in various laboratories in the countries involved (experiment station or university facilities). Also in these countries, root samples were taken from diseased trees in many areas where the avocado has been brought into cultivation, to determine whether the trees were infected with *P. cinnamomi*.

In Latin America the native hosts sampled included *Persea alba*, *P. americana*, *P. americana* var. *gigantea* (*P. gigantea*), *P. americana* var. *nubigena* (*P. nubigena*), *P. caerulea*, *P. donnell-smithii*, *P. lingue*, *P. schiedeana*, *P. steyermarkii*, *P. subcordata*, *P. vesticula* (*P. popenoei*), and several species in the genera *Aiouea*, *Beilschmedia*, *Cinchona*, *Nectandra*, *Ocotea*, and *Phoebe*. Collections were made in the following countries: Argentina, Brazil, Chile, Colombia, Costa Rica, Cuba, Ecuador, El Salvador, Guatemala, Haiti, Honduras, Mexico, Peru, Puerto Rico, St. Croix, Trinidad, and Venezuela.

In the United States, samples were collected in the past 4 years in California, Oregon, and Washington. Native hosts sampled in the forests on the West Coast of these states included *Cupressus macrocarpa*, *Pinus ponderosa*, *Pinus radiata*, *Pseudotsuga menziesii*, and *Sequoia sempervirens*. Efforts were made to collect samples primarily from moist areas where *P. cinnamomi*, if present, would have the best possibility of survival and chances would be better for isolation. In the 1950s and 1960s, over 300 samples were collected from native vegetation in southern California, primarily from the low chaparral growth that includes many different low-growing shrubs; included were the genera *Adenostoma*, *Arctostaphylos*, *Ceanothus*, *Phofinia*, and *Rhamnus* and a few trees such as *Platanus racemosa* in creek bottoms. Samples were taken above and some distance from cultivated areas to minimize the opportunity for infestation of the area with *P. cinnamomi* from avocado groves or other possible sources of inoculum.

RESULTS

Latin America. Approximately 400 samples have been collected from the various countries noted in Latin America over the past 33 years. Many of these are from avocado trees under cultivation, and *P. cinnamomi* was readily recovered from about 25% of these trees; most showed positive symptoms of Phytophthora root rot.

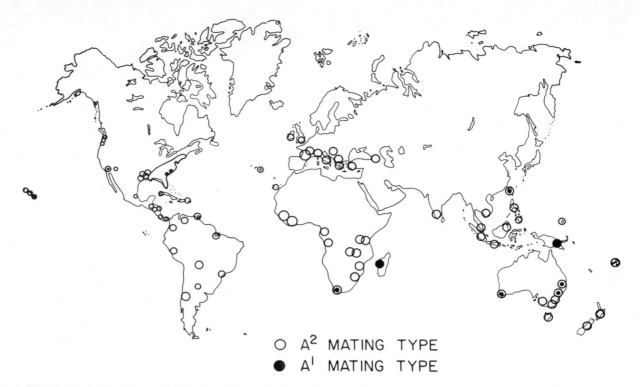

O   A² MATING TYPE
●   A¹ MATING TYPE

Fig. 1. Distribution of the two mating types of *Phytophthora cinnamomi*.

In addition to the results from the Latin American samples summarized in 1977 (6), more samples were taken between 1978 and 1983. Results have been similar in all of these samples: in no case was *P. cinnamomi* recovered from any native trees of *Persea* or any of the other genera noted above growing in undisturbed areas (rain forest, tropical swamp, montane forest, cloud forest, etc.) remote from cultivation. Many of the samples were taken from areas of very high rainfall, in swamps, and in soils containing considerable clay. In no case were any native trees found with symptoms of Phytophthora root rot. In several instances, *P. cinnamomi* was recovered from native species of *Persea*; but in each case cultivated crops were nearby, and the pathogen could have been introduced into the area by humans. This is also true of samples from native *Cinchona* in Costa Rica; *P. cinnamomi* was recovered from trunk cankers on native *Cinchona* trees that were, however, in an area not far from an old cinchona plantation where *P. cinnamomi* was also isolated from old surviving trees and sprouts.

Trapping of some of these Latin American samples with sensitive plants or other methods should be done, however, to increase the possibility of detecting the pathogen where extremely small amounts of inoculum might be present.

United States. The same results were obtained from samples collected from forest trees primarily along the coast of California, Oregon, and Washington. Most of these samples were taken between 1978 and 1982. In no case was *P. cinnamomi* isolated from the forest trees noted above; some of these were also growing in very wet areas, and in coastal areas where temperatures would not be sufficiently low in winter to kill *P. cinnamomi*. The pathogen was readily isolated from avocado trees in cultivation in southern California and from many ornamental plants in California. These isolates have all

been A2 mating type, except for a few A1 isolates from camellia.

Distribution. *P. cinnamomi* has been described from over 70 countries on nearly 1,000 hosts, which are predominantly woody plants. Figure 1 records the distribution of the two mating types of the fungus, based on material provided by the Commonwealth Mycological Institute, our cultures, and records in the literature. Obviously the A2 mating type is the common one worldwide, with the A1 very restricted in terms of hosts and geographic distribution. To date the A1 type has been described from only six countries: Australia, Malagasy Republic (Madagascar), Papua New Guinea, South Africa, Taiwan, and the United States. According to our records, the A1 type has been identified from the following genera: *Aotus, Banksia, Camellia, Eucalyptus, Macadamia, Nothofagus, Persea, Thuja,* and *Vitis.*

LITERATURE CITED

1. Broadbent, P., and Baker, K. F. 1974. Behaviour of *Phytophthora cinnamomi* in soils suppressive and conducive to root rot. Aust. J. Agric. Res. 25:121-137.
2. Ko, W. H., Chang, H. S., and Su, H. J. 1978. Isolates of *Phytophthora cinnamomi* from Taiwan as evidence for an Asian origin of the species. Trans. Br. Mycol. Soc. 71:496-499.
3. Rands, R. D. 1922. Streepkanker van kaneel, veroorzaakt door *Phytophthora cinnamomi* n. sp. (Stripe canker of cinnamon caused by *Phytophthora cinnamomi* n. sp.). Meded. Inst. Plantenziekt. 54:41.
4. Shaw, D. E., Cartledge, E. G., and Stamps, D. J. 1972. First records of *Phytophthora cinnamomi* in Papua New Guinea. Papua New Guinea Agric. J. 23:46-48.
5. Shepherd, C. J. 1975. *Phytophthora cinnamomi*—an ancient immigrant to Australia. Search 6:484-490.
6. Zentmyer, G. A. 1977. Origin of *Phytophthora cinnamomi*: Evidence that it is not an indigenous fungus in the Americas. Phytopathology 67:1373-1377.
7. Zentmyer, G. A., and Ohr, H. D. 1978. Avocado root rot. Univ. Calif. Div. Agric. Sci. Leafl. 2440. 15 pp.

Part II
Colonization of Roots and Rhizosphere

The Biology of the Rhizosphere

R. C. FOSTER, CSIRO Division of Soils, Glen Osmond, South Australia, 5064

An understanding of the precise nature of the root-soil interface is essential to studies of plant nutrition, root physiology, and plant root pathology. First, roots secrete large amounts of gel, and because the presence of microorganisms may double the loss of organics from roots (20), these major losses of soil organic matter represent a serious drain on the carbon resources of the plant. Second, gel accounts for 70–90% of the ion exchange capacity of the root and hence determines the ionic composition of the plant (12) and its competitive abilities in a mixed species association (18). Wheat cultivars that secrete copious gel are less susceptible to aluminum toxicity at low pH (17), and root gels and the microbial carbohydrates derived from them are important in maintaining soil crumb stability. The origin, nature, extent, and subsequent fate of gel is therefore of great importance. Third, microorganisms that colonize the root surface affect root parameters that control the efficiency of the root for nutrient uptake. Thus, infected roots may have more and wider cortical cell layers, fewer and shorter root hairs, longer roots, and increased lateral frequency when compared with sterile controls (see 20 for review). Some rhizosphere microorganisms absorb nutrients moving to the root surface; where concentrations are limiting, they may induce nutrient deficiency in the host plant. Conversely, materials secreted by rhizosphere fungi and bacteria may release nutrients (Ca, K, P) from soil components and make them available to the root (20). Recent work suggests that in nitrogen-deficient soils, nitrogen is fixed by bacteria in normal (nonnodule) roots (19) and that all economically important crops (except members of the Chenopodiaceae and Brassicaceae) have mycorrhizae that normally increase phosphate uptake by the host (see 20 for review).

The rhizosphere also contains populations of algae as high as $10^{3.5}$/g of soil; amoebas, 180×10^3/g of soil; actinomycetes, 46×10^6/g of soil; flagellates, 50×10^2/g of soil (8); fungi, $10^{4.5}$/g of soil; mites; and Collembolas. Many of these are important in nutrient cycling and soil formation processes (3), and some have been implicated in biological control of soilborne root diseases of economic importance (20).

In the oligotrophic conditions in the bulk soil, 60% of the bacteria are <0.3 μm in diameter (2) and hence near the limits of convenient resolution in soil sections by light microscopy. Although the root-soil interface has been intensely studied by microbiologists ever since Hiltner introduced the term "rhizosphere" in 1904, detailed consideration of the interrelationships between the soil particles, rhizosphere, microorganisms, and the root surface has had to await the development of methods for preparing ultrathin sections of soils and for preserving and staining unstable rhizosphere organics (e.g., root surface gels, microbial slimes, enzymes) in situ in natural soil fabrics (7,9,11).

CONDITIONS IN THE OUTER RHIZOSPHERE

Chemical analyses of materials extracted from soils have provided qualitative and quantitative data on the total amount of soil organics, but the precise location of the different classes of organics in the soil fabric is largely unknown. Work on the ultrastructure and histochemistry of the bulk soil and the outer rhizosphere has thus concentrated on the location of organics and microorganisms, particularly carbohydrates and their availability to symbionts and pathogens as they move to the root surface (9,11).

Transmission electron micrographs (TEM) of ultrathin sections (Figs. 1–8) show that in the outer rhizosphere metabolizable materials capable of supporting microbial cells are infrequent and widely and randomly distributed (Fig. 1). Histochemical tests suggest the following findings: most soil organics are low in carbohydrates and rich in polyphenols and/or metal ions so that they are chemically protected against microbial breakdown; many soil bacteria are physically separated by several microns from nutrient deposits; most polysaccharide deposits are enclosed in pores in clay fabrics <1 μm wide, too narrow for the entry of microorganisms and therefore unavailable to them (i.e., physically protected; Fig. 2). This may explain why merely grinding a soil doubles the respiration rate. Many of the bacteria <0.3 μm in diameter contain little or no storage reserves such as PHB, glycogen, and polyphosphate (10), suggesting size reduction due to starvation (4). This is consistent with the view that most soil bacteria have a much reduced metabolism, with generation times of 6–12 months, and are adapted to starvation conditions. Thus *Arthrobacter*, common in the outer rhizosphere and bulk soil (20), shows 50% survival after 80 days without added nutrient. Some gram-negative species are enclosed in sheaths of clay minerals that provide protection from adverse conditions; certainly capsule materials protect bacteria from the chloroform treatment of the Jenkinson biomass method (Figs. 3 and 4). Even after the lysis of the bacteria, their capsule materials persist and bind soil components into stable crumbs (9).

DIFFERENTIATION OF THE ROOT SURFACE

Despite the fact that root cap cells are coated with carbohydrate-rich gel, the root cap is usually devoid of microorganisms because cap cells are continually renewed from within and are sloughed off into the soil. A new, microbe-free surface is constantly presented to the

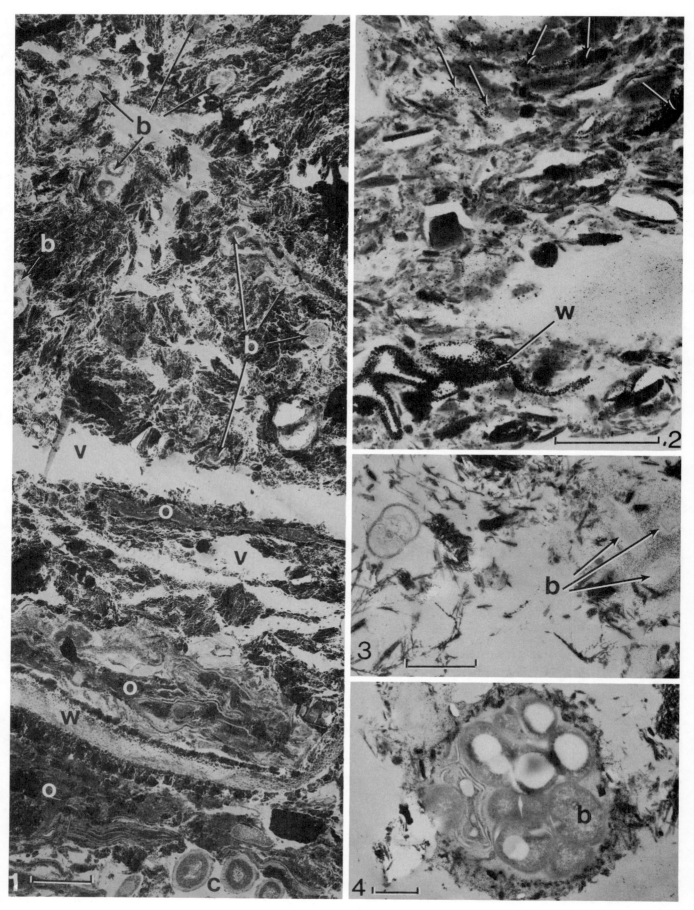

soil fabric, and bacteria contacted by the tip require time to develop into colonies.

The differentiating protoepidermal cells inside the root cap secrete a thickened outer tangential wall that is largely mucilaginous (7,15). When they emerge from beneath the cap in the extension zone, they extend 10–30 times their original length (1) so that the framework of cellulose microfibrils is reoriented from a transverse to an axial direction, and new wall material is continually exposed to the soil. Hence, like the root cap, the extension zone is also relatively microbe free. However, there is evidence that some pathogens are specifically attracted to the carbohydrates exposed during extension (14) or to the exudates that escape in large amounts from the root in this region (20).

When longitudinal extension is complete, the epidermal cells have an inner multilamellate secondary wall through which the root hairs emerge (15). Root hairs secrete gel at their tip, and the accumulation of gel, together with the loss of exudates from the root, results in a proliferation of microorganisms in the root hair zone. Other bacteria, some of which may fix nitrogen or secrete hormones that influence root morphology, proliferate in the intercellular spaces of the cortex apparently without eliciting a host defense response.

After a few days, root hairs and epidermal cells become suberized, and in tree species unsuberized roots constitute only 0.5% of the root length. Following suberization, in cereals the epidermal and outer cortical cells begin to autolyse, so that in a 10-day-old root segment fewer than 25% of cortical cells have a nucleus (13). Release of lysates and absence of cytoplasmic defense mechanisms mean that the surface cells are colonized by microorganisms (7,10,11). The cuticle is broken down by the incursion of soil minerals (7) and soil microorganisms (11), and the primary wall gels are released into the nearby soil and embed it (Fig. 5) to a distance of about 50–70 μm. This results in a further increase in the numbers of microorganisms in the rhizosphere, reaching a maximum at anthesis. Bacteria and fungi begin to remove the root surface gel and the primary wall of the epidermal cell, eventually exposing the microfibrils of the secondary wall. These contact the soil minerals to provide a means of direct transfer from the mineral surface into the root-free space.

Thus in contrast to the outer rhizosphere and bulk soil, the inner rhizosphere is energy rich. The root releases up to 30% of the photosynthate reaching it as readily metabolizable materials, so the rhizosphere supports a population of microorganisms that is 50–100 times larger than that in the bulk soil. This microflora is more diverse than that in the outer rhizosphere and is dominated by gram-negative bacteria, whereas *Bacillus* and *Arthrobacter* species are common in the outer rhizosphere. The individual bacteria are larger than those in the bulk soil, where 63% are <0.3 μm in diameter

and only 6% are >0.5 μm (2); in the rhizosphere, only 20% are <0.3 μm in diameter and more than 30% are >0.51 μm. Moreover, many are filled with storage materials such as PHB, glycogen granules, and polyphosphate (10), perhaps indicating a high C/N ratio in the exudates.

Microorganisms are not uniformly distributed on the root surface; most occur in discrete colonies (20), and their locations suggest that colonies arise from randomly disposed, infected soil organic fragments fortuitously contacted by the root as it penetrates the soil. Organisms especially proliferate in the grooves between the epidermal cells (Fig. 8), the site of root exudation (20) and gel accumulation. Some have an unusual morphology and there may be a unique rhizoplane microflora (10).

Estimates of microbial populations in the rhizosphere vary from 10/g of soil (using conventional plating techniques; 20) to 10/cc (by direct counting from electron micrographs; 8,10,16). Several reasons account for the discrepancies between the methods: routine plating techniques may detect only 0.1–1% of the organisms present; most light microscopy has been done on whole washed roots rather than on sections of the root/rhizosphere complex, so that only those microorganisms embedded in the root surface gel are counted; and TEM detects all microorganisms whether they are viable or not. Histochemical tests showed that only 25% of rhizoplane bacteria of old roots were secreting acid phosphatase (9), whereas all the cells on young roots were positive for succinic dehydrogenase (Figs. 6 and 7). Direct observations (8,10,16) as well as theoretical calculations suggest that most rhizosphere microorganisms occur within 10–50 μm of the rhizoplane (Table 1).

Various defense mechanisms have been reported when pathogenic or nonpathogenic bacteria infect aerial plant tissues. Thus nonpathogenic pseudomonads injected into the mesophyll of leaves became enclosed in fibrillar or granular materials or by a fine, electron dense pellicle secreted by the host (6), but similar

Table 1. Number of bacteria at different distances from the root surface as determined by transmission electron micrography

Pine[a]		Eucalypt[b]		Clover[c]	
Distance (μm)	Population (cells × 10^{10})	Distance (μm)	Population (cells × 10^{10})	Distance (μm)	Population (cells × 10^{10})
0–4	0	0–10[d]	120–160	0–1	12.0
4–8	14.4	0–10[e]	40–72	0–5	9.6
8–12	22.2	>10[d]	2.0	5–10	4.1
12–16	11.1			10–15	3.4
16–20	1.4			15–20	1.3

[a] Data from reference 9; ectotrophic mycorrhizae.
[b] Data from reference 16.
[c] Data from reference 10; endotrophic mycorrhizae.
[d] Ectotrophic mycorrhizae.
[e] Nonmycorrhizal.

Figs. 1–4. Transmission electron micrographs of ultrathin sections of soil fabrics from the outer rhizosphere. **(1)** A small colony of bacteria (c) is associated with soil organic matter (o). Other bacteria (b) scattered at random through the soil fabric are small compared with those in the rhizosphere and lack storage materials such as polyhydroxybutyrate and polyphosphate; v = microvoid, w = cell wall remnants. **(2)** After treatment with periodate-silver methenamine to reveal 2,4,diglycol groups, cell wall remnants are revealed as well as small deposits of carbohydrate (arrows) that are scattered throughout the soil fabric. **(3)** When soil fabrics are treated overnight with CCl₄ (Jenkinson's biomass method) and then incubated for 14 days, most bacteria are lysed and only their capsule materials remain (arrows). Cells protected by clays **(4)** or thick slime deposits are not lysed and proliferate. (Figure 1 reprinted from "Ultramicromorphology of some South Australian soils" by R. C. Foster, in *Modification of Soil Structure*, edited by W. W. Emerson, R. D. Bond, and A. R. Dexter, © 1978, John Wiley & Sons, Ltd. Reprinted by permission of John Wiley & Sons, Ltd.)

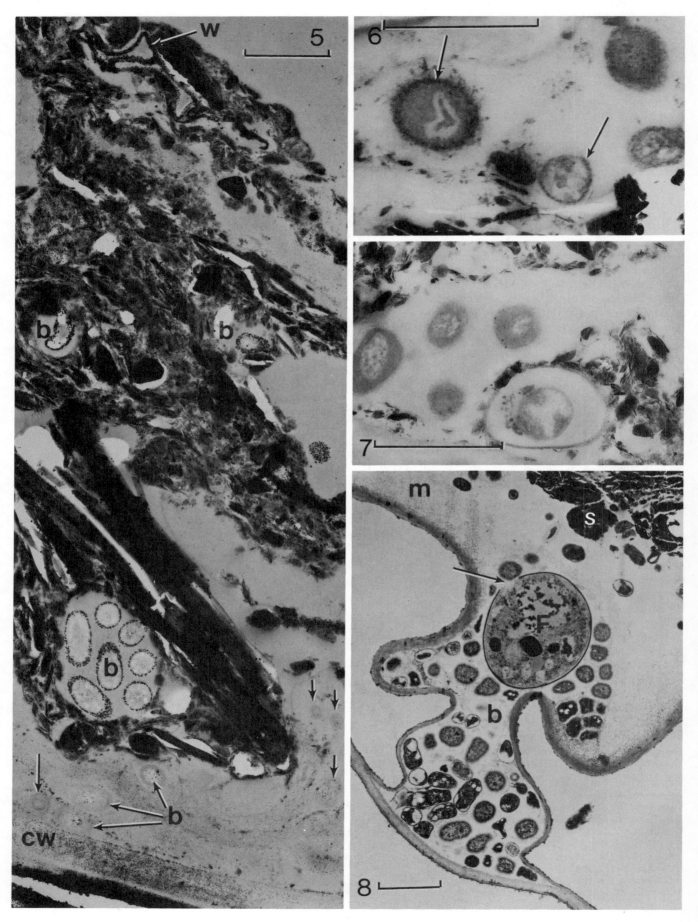

Figs. 5-8. Transmission electron micrographs of ultrathin sections of soil fabrics from the inner rhizosphere. **(5)** In the inner rhizosphere, most bacteria lie close to the rhizoplane (arrows); but small colonies and isolated bacteria (b) are found throughout the rhizosphere. The root surface mucilage does not stain for 2,3-diglycol groups, although the walls of some bacteria and old cellular remnants (w) are stained; cw = cell wall. Incubation in media specific for succinic dehydrogenase gives a positive reaction at the bacterial surface with **(6)** full medium and **(7)** substrate-free control. **(8)** Evidences for biological control of soilborne root disease. When wheat roots infected with *Gaeumannomyces graminis* are grown in suppressive soils, the cytoplasm of many hyphae (F) becomes disorganized and the cell wall lysed by nearby rhizosphere bacteria (arrow); m = mucilage. (Figure 5 reprinted, by permission, from ref. 7; Figure 8 reprinted, by permission, from ref. 11.)

reactions have not been described in response to normal rhizosphere microorganisms. However, deposits of electron dense material have been associated with large bacterial populations within the root (11).

INTERACTIONS BETWEEN RHIZOSPHERE MICROORGANISMS

In the root hair zone and beyond, microbial cover is only about 7–15% (20). Except at anthesis, when some overcrowding of colonies may be evident (10), there is little competition for space at the root surface, though there may be competition for energy-rich sites.

It would be surprising if an environment so rich in species and containing such dense populations of microorganisms were not exploited by predators, and the rhizosphere contains predatory bacteria such as *Bdellovibrio* and bacteriophage that attack various species of bacteria (10). Flagellates and amoebas that browse upon the bacteria and assist in recycling N and P locked up in dead cells (3) are sometimes encountered in rhizospheres (5). In the presence of the root pathogen *Gaeumannomyces*, large populations of bacteria accumulate in the root surface gel and destroy it, and some bacteria also become attached to the hyphae and destroy them, thus effecting a degree of biological control of the disease (Fig. 8).

CONCLUSIONS

Traditional chemical and microbiological analyses of components extracted from soils tell us which microorganisms and biochemicals are present in the soil but are unable to indicate where they were located with respect to each other in the intact soil fabric. Despite the technical difficulties encountered in preparing ultrathin sections of soils, ultrastructural histochemical techniques are the only ones readily available that provide information on the identity of microorganisms, their organic substrates, and their location simultaneously.

ACKNOWLEDGMENTS

I thank Y. K. McEwan for excellent technical assistance in preparing specimens for electron microscopy, T. W. Cock for preparing the plates, and A. D. Rovira for supplying the root in Fig. 8.

LITERATURE CITED

1. Avers, C. J. 1957. An analysis of differences in growth rate of trichoblast and hairless cells in the root epidermis of *Phleum pratense*. Am. J. Bot. 44:686-690.
2. Balkwill, D. L., Labeda, D. P., and Casida, L. E., Jr. 1975. Simplified procedures for releasing and concentrating microorganisms from soil for transmission electron microscopy viewing as thin sectioned and frozen etched preparations. Can. J. Bot. 21:252-262.
3. Clarholm, M. 1982. Protozoan grazing of bacteria in soil—Impact and importance. Microb. Ecol. 7:343-350.
4. Crozat, Y., Cleyet-Marel, J. C., Giraud, J. J., and Obabon, M. 1982. Survival rates of *Rhizobium japonicum* populations introduced into different soils. Soil Biol. Biochem. 14:401-405.
5. Darbyshire, J. F., and Greaves, M. P. 1967. Protozoa and bacteria in the rhizosphere of *Sinapis alba* L., *Trifolium repens* L. and *Lolium perenne* L. Can. J. Microbiol. 13:1057-1068.
6. Fett, W. F., and Jones, S. B. 1982. Role of bacterial immobilization in race-specific resistance of soybean to *Pseudomonas syringae* pv. *glycinea*. Phytopathology 72:488-492.
7. Foster, R. C. 1982. The ultrastructure and histochemistry of the rhizosphere. New Phytol. 89:263-273.
8. Foster, R. C., and Marks, G. C. 1967. Observations on the mycorrhizas of forest trees. II. The rhizosphere of *Pinus radiata* D. Don. Aust. J. Biol. Sci. 19:1027-1038.
9. Foster, R. C., and Martin, J. K. 1978. In situ analysis of soil components of biological origin. Pages 75-111 in: Soil Biochemistry. Vol. 5. E. A. Paul and J. N. Ladd, eds. Marcel Dekker, New York. 480 pp.
10. Foster, R. C., and Rovira, A. D. 1978. The ultrastructure of the rhizosphere of *Trifolium subterraneum* L. Pages 278-290 in: Microbial Ecology. M. W. Loutit and J. A. R. Miles, eds. Springer-Verlag, New York. 452 pp.
11. Foster, R. C., Rovira, A. D., and Cock, T. W. 1983. Ultrastructure of the Root-Soil Interface. American Phytopathological Society, St. Paul, MN. 157 pp.
12. Haynes, R. J. 1980. Ion exchange properties of roots and ionic interactions within the root apoplasm: Their role in ion accumulation by plants. Bot. Rev. 46:75-99.
13. Henry, C. M., and Deacon, J. W. 1981. Natural, non-pathogenic death of the cortex of wheat and barley seminal roots, as evidenced by nuclear staining with acridine orange. Plant Soil 60:255-274.
14. Hinch, J. M., and Clarke, A. E. 1980. Adhesion of fungal zoospores to root surfaces is mediated by carbohydrate determinants of the root slime. Physiol. Plant Pathol. 16:303-307.
15. Leech, J. H., Mollenhauer, H. H., and Whaley, W. G. 1963. Ultrastructural changes in the root apex. Symp. Soc. Exp. Biol. 17:74-84.
16. Malaczjuk, N. 1979. The microflora of unsuberized roots of *Eucalyptus callophylla* R. Br. and *Eucalyptus marginata* Don ex Sm. seedlings grown in soils suppressive to *Phytophthora cinnamomi* Rands. II. Mycorrhizal roots and associated microflora. Aust. J. Bot. 27:255-272.
17. Mugwira, L. M., and Elgawhary, S. M. 1979. Aluminum accumulation and tolerance of triticale and wheat in relation to root cation exchange capacity. Soil Sci. Soc. Am. Proc. 43:736-740.
18. Nordstrom, L. O. 1982. Variability between species K/Ca ratios of successional plants. Plant Soil 65:137-139.
19. Pohlmann, A. A., and McColl, J. G. 1982. Nitrogen fixation in the rhizosphere and rhizoplane of barley. Plant Soil 69:341-352.
20. Rovira, A. D., Bowen, G. D., and Foster, R. C. 1983. Significance of rhizosphere microflora and mycorrhizas in plant nutrition. Pages 61-93 in: Encyclopedia of Plant Physiology. New Ser. Vol. 15. A. Lauchli and R. L. Bieleski, eds. Springer-Verlag, Berlin.

Mode of Colonization of Roots by *Verticillium* and *Fusarium*

J. S. GERIK and O. C. HUISMAN, Department of Plant Pathology, University of California, Berkeley 94720, U.S.A.

Fungal colonization of roots has been studied only to a limited extent in the past because of the relative difficulty involved in such investigations. Most of these studies have involved young plants growing in small pots in the greenhouse. Although such a system allows for a better control of environmental parameters, it is not conducive for normal root growth over an extended period of time because of the small soil volume in the system. A field system represents a more suitable environment in which plant roots are able to grow and to encounter fungal propagules in a more normal situation.

Nonpathogenic strains of *Fusarium oxysporum* have been of particular interest in many studies of fungal colonization of roots because of the relative abundance of these fungi on the surface of roots of many plant species growing in a wide variety of soils. It has been reported to be the dominant fungal species of the roots of both dwarf bean and barley growing in the same soil in England (6). Even when present in low inoculum densities, *F. oxysporum* was able to compete and colonize better than many other soil fungi that may be present at inoculum densities many times greater (5).

Verticillium dahliae has been shown to colonize the roots of a wide range of plant species, including both those immune and susceptible to systemic infection (1,3). Evans and Gleeson (1), in their examination of young plants grown in the greenhouse in naturally infected soil, found that root colony densities of *V. dahliae* were directly related to inoculum density. Colonies were small, randomly scattered along the root surface, and probably confined to the rhizoplane because they were sensitive to brief mercury treatment. In our work with field-grown cotton roots we obtained similar results. We also found that colony densities of *V. dahliae* on roots were constant throughout the growing season and were similar for young roots with attached apexes and older roots. The latter observation suggests that nearly all fungal colonization occurs early in the life of roots, i.e., near the tip (2).

This study was undertaken to determine the dynamics of colonization of cotton roots by *F. oxysporum* and *V. dahliae* in a field in the San Joaquin valley of California.

MATERIALS AND METHODS

Roots of cotton plants were obtained throughout the growing season from a field infested with *V. dahliae* at the West Side Field Station in the San Joaquin valley of California. All soil material was washed from the roots; root segments with root tips were separated from the others, the latter being labeled bulk roots. All roots were washed three times in a sterile sodium hexametaphosphate (1%) and Tergitol NP-10 (0.1%) solution. The roots were plated on pectate-cellophane extract media that restrict fungal colony expansion and were incubated at 22°C for 10 days. The emerging fungal colonies were viewed under a dissecting microscope and identified. The distance from the center of the colonies to the apex of the root was determined with the aid of an ocular micrometer and recorded.

Fungal colony morphology was studied by use of specific stains. Using a soluble protein fraction as antigen, antiserum specific to the fungal genera was produced in rabbits. Root tissue was incubated in the serum, and the antibodies were localized by use of an anti-rabbit IgG alkaline phosphatase conjugate. When incubated in a substrate solution, an insoluble enzymatic product precipitated along the hyphal surface allowing for visible observation of the hyphae in the root tissue.

The antiserum produced against *V. dahliae* was highly specific to *Verticillium* spp. A weak cross-reaction with *F. oxysporum* was fully removed by cross-adsorption of the serum with this fungus. The antisera prepared against *F. oxysporum* did not distinguish between the two strains encountered by us in cotton roots (see below). It appears to be specific for *F. oxysporum* and did not react with *V. dahliae*.

RESULTS

F. oxysporum was the most abundant fungal species observed emerging from plated root segments, accounting for over 50% of all fungal colonies. Other fungal species occurred in frequencies of 20% or less. Two strains of *F. oxysporum* were distinguishable on the basis of colony morphology: one was a sporodochial type, and the other a sclerotial type. Both occurred in about equal frequencies on roots.

V. dahliae was found to colonize the root tissue near the apex only sparingly (Fig. 1). The highest frequencies of colonization occurred at 1 cm or more from the root tip, except early in the growing season when slightly higher frequencies were observed closer to the root apex. Colony frequencies in the bulk root samples (representing older or more distal root tissue) were similar to or slightly higher than those observed at 2 cm from the tip. In general, the colony frequency increased with distance from the root tip until a point of maximum colonization was reached at 1–2 cm from the tip. The considerable variance (C.V. = 27% for *V. dahliae* and 30% for *F. oxysporum*) of the data was related to the decreasing frequency with which long roots with attached apexes were isolated from soil (average length of segments was

5 mm). The most abundantly represented tissue (that near the apex) had low colony densities, whereas the distal tissue, especially that over 8 mm, consisted of increasingly smaller sample size.

Both *F. oxysporum* types were able to colonize root tissue more quickly than *V. dahliae*. Maximal colony frequencies were reached within 5 mm from the root apex (Fig. 1). The colonization pattern of the two *F. oxysporum* strains exhibited one important difference. Although the colony frequency for the sporodochial strain was maintained on older root tissue at the maximum values achieved near the apex, those for the sclerotial strain declined dramatically to low levels in the older tissue (Fig. 1).

The prepared antisera permitted the specific staining of *V. dahliae* and *F. oxysporum* colonies in root tissue. The colony density on roots observed with the *V. dahliae* antiserum proved to be the same as that determined by the bioassay involving the plating of root segments. The stained colonies of *V. dahliae* in cotton roots exhibited several interesting properties (Fig. 2). Most hyphae were located just outside the endodermis. Hyphal densities were significantly lower on the root surface and in the outer cortex. The colony appearance was consistent with growth of hyphae from the root surface toward the stele, followed by more extensive growth on the inner cortex parallel to and immediately adjacent to the stele. Also present were numerous single and isolated cortical cells that stained heavily. These cells appeared to be packed with numerous hyphae when viewed with higher magnification (Fig. 2B). Hyphae were also absent from areas of lateral root emergence (Fig. 2C).

Colonies stained with the *F. oxysporum* antisera were noticeably different from those stained with the *V. dahliae* antiserum. Most hyphae were confined to the root surface and the outer cortex. There was not noticeable abundance of *F. oxysporum* hyphae in the inner cortex. In many cases, hyphae growing parallel to the anticlinal wall of the cortical cells were observed. It appears that *F. oxysporum* did not readily invade the cortex of intact root tissue. The isolated, heavily stained cortical cells observed with *V. dahliae* were not present in root tissue stained with the *F. oxysporum* antisera.

DISCUSSION

The data support our hypothesis that colonization of roots by fungi occurs primarily near the root tip (2). Maximal colony densities are attained within 1–2 cm from the tip and increase little, if any, thereafter. Fungal propagules are probably responding to the high levels, compared with other root tissue, of exudates from the zone of elongation. The combination of higher microbial metabolism and lower exudation rates on older root tissue probably minimizes new colonization. That fungi differ in the speed with which they can respond to and colonize roots is suggested by our differential results with *V. dahliae* and *F. oxysporum*.

The *F. oxysporum* strains can apparently react more quickly to root stimuli and colonize the root closer to the tip than *V. dahliae*. The ability to react quickly to host stimuli may be the underlying factor responsible for the success *F. oxysporum* enjoys as a root colonizer. Parkinson and Pearson reported that even when inoculum of another root-colonizing fungus (a sterile dark form) was 50 times greater than that of *F. oxysporum*, the latter was still the predominant root colonizer (5). The dramatic drop in colony frequency exhibited by the sclerotial strain of *F. oxysporum* is an interesting phenomenon. Such a decline has been reported earlier for other root-colonizing fungi (6). At least two plausible explanations can be presented for this colonization pattern. The first is based on a poor competitive ability of this strain. The rapid response to root exudates characteristic of *F. oxysporum* would permit early contact with a passing root tip under low competitive conditions. However, as more competitive but more slowly responding fungi colonize the root tissue, the strain may not be able to maintain its initial foothold on the tissue. The second explanation is based on pathogenic ability. The presence of this fungus on the roots could lead to significant root tip mortality. Such behavior would also lead to greatly reduced colony frequency on living roots. The latter explanation is supported by data showing a positive correlation between root death and colony frequency of this fungus (G. L. Andersen, *personal communication*) and a negative correlation between the colony frequency of this fungus and other root-colonizing fungi (Gerik and Huisman, *unpublished*).

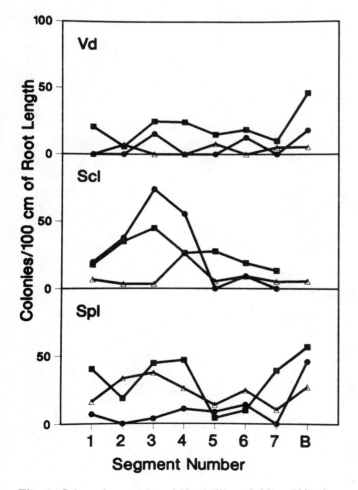

Fig. 1. Colony frequencies of *Verticillium dahliae* (Vd), the sclerotial isolate of *Fusarium oxysporum* (ScI), and the sporodochial isolate of *F. oxysporum* (SpI) on 1-mm intervals (1–5), an interval 5–10 mm (6), and an interval 10–16 mm (7) from the root tip and on bulk roots (B) on 5 April (squares), 11 June (circles), and 23 June (triangles).

The observations made with the specific stains show that *V. dahliae* can establish itself throughout the cortex of the root. The ability of the fungus to penetrate and colonize individual cells is evident from the frequent occurrence of heavily stained and mycelial packed cortical cells. The mechanism by which this selective colonization occurs and its role in the host-parasite interaction is not known. Much more extensive work on this topic is needed. One of us (Huisman) has established that cortical root colonies of *V. dahliae* are in several thousand fold excess of that needed for systemic infection of the average cotton plant. Thus the presence of an established cortical colony of this pathogen is not enough of a foothold for gaining access to the stelar tissue.

Areas of lateral root formation are zones of extensive cell disruption and presumably nutrient release, and they are thus possible entry points into the vascular system for *V. dahliae*. The staining assay reveals an apparent absence of hyphae in these areas of the root. Our observation suggests that it would be highly unlikely for points of lateral root emergence to be avenues for infection into the vascular system for *V. dahliae*. Mace et al (4) have reported high concentrations of gossypollike compounds at sites of lateral rot emergence in cotton roots. We observed that in fresh roots these areas developed increased levels of pigmentation in the presence of our substrate solution. This suggests the presence of phenolics, because the assay depends on the coupling of a diazo dye with a

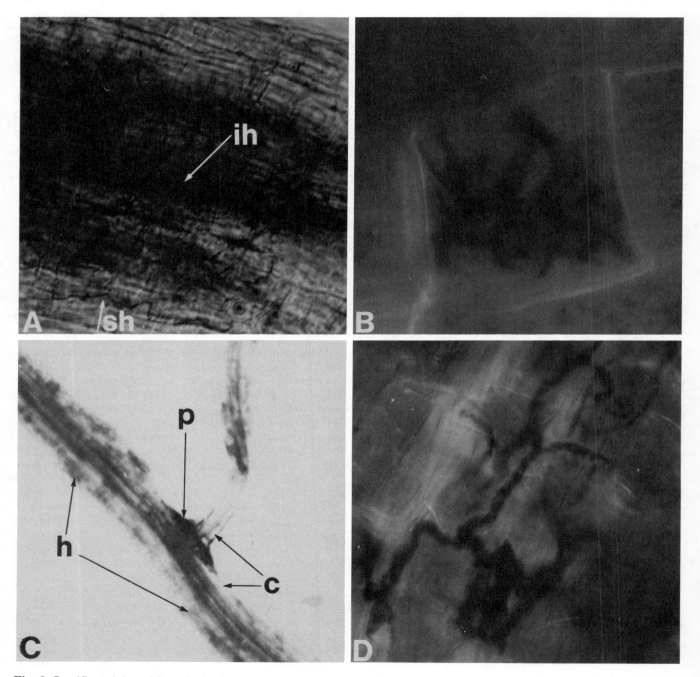

Fig. 2. Specific staining of fungal colonies on root tissue. (**A**) *Verticillium dahliae* colony showing hyphae on the surface (sh) and inner (ih) cortex. (**B**) Individual cortical cell heavily colonized by *V. dahliae*. (**C**) Area of lateral root emergence showing stained hyphae of *V. dahliae* (h), heavily pigmented area (p), and area clear of hyphae (c). (**D**) Hyphae of *Fusarium oxysporum* on root surface.

phenolic moiety. Perhaps the presence of such compounds inhibits fungal activity at these sites.

Our observations on *F. oxysporum* made with the staining technique suggest that the fungus is mostly an ephiphytic colonizer. Most of the stained colonies we observed were probably of the sporodochial type because the sclerotial type had much lower frequencies on the bulk or older roots. If the sclerotial types are causing increased root mortality, their colonization strategy may be different from that of the sporodochial type. The production of more specific sera to distinguish between these two isolates will help answer these questions.

The application of immunoenzymatic techniques has allowed for new insight into the study of the mode of fungal colonization of plant roots. Fungal colonies in root tissue stained with this method were easily observed under the dissecting microscope. The method is an improvement over fluorescent antibody techniques by eliminating the need for ultraviolet illumination. The refinement of this technique by the production of more specific sera will be of even more use in the future.

LITERATURE CITED

1. Evans, G., and Gleeson, A. C. 1973. Observations on the origin and nature of *Verticillium dahliae* colonizing plant roots. Aust. J. Biol. Sci. 26:151-161.
2. Huisman, O. C. 1982. Interactions of root growth dynamics to epidemiology of root-invading fungi. Annu. Rev. Phytopathol. 20:303-327.
3. Lacy, M. L., and Horner, C. E. 1966. Behavior of *Verticillium dahliae* in the rhizosphere and on roots of plants susceptible, resistant, and immune to wilt. Phytopathology 56:427-430.
4. Mace, M. E., Bell, A. A., and Stipanovic, R. D. 1974. Histochemistry and isolation of gossypol and related terpenoids in roots of cotton seedlings. Phytopathology 64:1297-1302.
5. Parkinson, D., and Pearson, R. 1967. Studies on fungi in the root region. VII. Competitive ability of sterile dark fungi. Plant Soil 27:120-130.
6. Parkinson, D., Taylor, G. S., and Pearson, R. 1963. Studies on fungi in the root region. I. The development of fungi on young roots. Plant Soil 19:332-349.

Dynamics of Root Colonization by the Take-all Fungus

CHRISTOPHER A. GILLIGAN, Department of Applied Biology, University of Cambridge, Cambridge, CB2 3DX, U.K.

The dynamics of root colonization of a parasite may be taken to mean the spread of the parasite on roots over time and space. The dynamics of root colonization by the take-all fungus, *Gaeumannomyces graminis* (Sacc.) von Arx & Olivier var. *tritici* Walker, cover a range of phenomena from details of the invasion and infection of individual cells to the progress of infection in an entire field. I propose to consider the dynamics of colonization as they relate to epidemiologic mechanisms at the macroscopic level of colonization along roots. To concentrate on cellular aspects would yield little information about the epidemiology of the fungus, whereas description of disease progress over a population of plants is insufficiently sensitive to assess the epidemiologic significance of components of the infection cycle such as the initiation and progress of infection (15).

Colonization of cereal roots by *G. graminis* var. *tritici* is initiated by hyphae growing out from inoculum close to the root in soil. Natural inoculum comprises fragments of previously colonized straw and root tissue. When the nutrient status of inoculum is low, there may be a preliminary feeding stage involving colonization of root hairs before that of the main root axis (1). Longitudinal colonization of the axis is by thickened, melanized runner hyphae that run along the surface of the root (surface runner hyphae) or that run between cell layers in the cortex (cortical runner hyphae) (11). Thin, hyaline hyphae emerge from runner hyphae and penetrate cortical cells and ultimately the stele. As progressive invasion of the stele occurs, xylem vessels become blocked and regions of dark discoloration, sometimes referred to as lesions, occur in the stele.

Host pathogen and environmental variables influence the rates of colonization of the stele, cortex, and surface of the root. Before discussing these, however, I propose to discuss the experimental methodology and statistical analysis of studies on root colonization.

METHODOLOGY AND STATISTICS

Many experimenters have preferred to use artificial inoculum when challenging individual roots because of the variability in natural inoculum. Some have used agar disks colonized by *G. graminis* var. *tritici* and placed immediately beneath a germinating seed, after Garrett (5); layers of fragmented inoculum, derived from fungal cultures on maize meal in sand (11); or so-called inoculum units such as colonized agar blocks (3) or colonized, whole cereal grain (8). Inoculum units have the advantage of ease of handling. They permit careful placement of inoculum at preselected sites on the infection court, especially when used in conjunction with growth tubes or boxes in which some roots grow against the surface of the container and are readily accessible (3,8,11). Layers of diffuse inoculum with variable inoculum size approximate more closely to natural inoculum. The density of inoculum, however, is usually much greater, and the microbial colonization of the inoculum fragments is likely to be atypical of that found in nature. Experimentation is inevitably a compromise, and the choice between these methods of inoculation clearly depends upon the nature of the hypotheses being tested.

In many experiments, a plug of colonized agar has been placed directly beneath seeds (3,5,6,11,21); usually, however, it would be better to use inoculum layers or units. Placing inoculum, richly supplied with nutrients, immediately beneath a germinating seed is intensely artificial and may impose a confusing bias on results. For example, the asymmetric pattern of stelar discoloration in which discoloration normally occurs only above sites of inoculation was not apparent until roots were inoculated away from the seed (11). Similarly, results from experiments involving inoculation immediately beneath seeds led Garrett (6) to propose the hypothesis that resistance of the host declined after stelar disruption occurred and hence that the rate of colonization up the root would be slower than that down the root. Observation of colonization above and below sites of inoculation distant from the seed led to rejection of this hypothesis (11).

Two variables are frequently used to monitor colonization of roots by runner hyphae above and below sites of inoculation. The variables are hyphal density, usually at fixed distances (24), and farthest extent of hyphal growth from the site of inoculation. Care should be taken to identify a priori hypotheses before the experiment is begun, thus allowing efficient use to be made of the analysis of variance. By the very nature of the treatments that are selected for an experiment, the experimenter is identifying (albeit implicitly) null hypotheses that are to be tested. These, and in most cases these alone, should be tested. Multiple comparison tests to test all possible pairs of means, such as Duncan's multiple range tests, should not be used where a priori hypotheses exist (14). Inappropriate use of the multiple comparison procedure weakens the sensitivity of tests of the important hypotheses.

Particular statistical problems relate to the analysis of sequential observations on the same experimental unit, such as a root. The observations of colonization at successive times on the same root are unlikely to be independent (17), a criterion required for fulfillment of the assumption of the analysis of variance. A simple way to overcome this difficulty is to construct orthogonal polynomial contrasts for each root. Suppose four successive observations were made on each root. There are three degrees of freedom, and the experimenter may be interested in the following three variables: extent of colonization at the first time, say t days after inoculation; the linear (average) rate of growth between t_1 and t_4; and the quadratic rate of growth between t_1 and t_4. The polynomial contrasts are constructed by

$$\sum_{i=1}^{4} k_i \, T_i,$$

where T_i is the extent of colonization at time i and k_i is the corresponding coefficient for the contrast. They may be obtained from tables if time intervals are equal or by calculation (16). Thus four observations per root are reduced to one for each contrast. The quadratic contrast assesses the quadratic deviation from linearity between times t_1 and t_4, i.e., the degree of "kink" in the relationship. Separate analyses of variance are then carried out for each variate with the factor for time removed from the analysis (8).

HOST FACTORS
AFFECTING COLONIZATION

Although some hypotheses are emerging (see below), there is still much to be learned about the effects of the host on the dynamics of colonization by the take-all fungus.

Natural infection from inoculum in soil is usually initiated at or slightly behind the root tip. Initially the root grows faster than do the runner hyphae, and tissue is available for colonization below as well as above the site of inoculation. Dark discoloration of the stele lags behind extension of runner hyphae. The discoloration is discontinuous (8). Units of inoculum may cause one or many dark lesions. Dark discoloration occurs only above the site of inoculation (8,11), except in the case of colonization of roots by weakly pathogenic isolates of *G. graminis* var. *tritici*, when stelar discoloration may be seen below sites of inoculation (3). Even so, Deacon and Henry (3) have shown that regions of discoloration expand proximally, and they suggest that distal lesions arise from the reinvasion of the xylem tissues by the fungus.

I have reported an asymmetric pattern of root colonization by runner hyphae for wheat seedlings growing in nutrient-amended sand (7–9). After an initial period of equal growth up and down the root, distal growth slowed relative to proximal growth. Initiation of the asymmetric pattern of growth was correlated on seminal roots of wheat with the onset of stelar blockage (8) and failed to occur when stelar blockage was retarded by growing plants at $10°C$ rather than $19°C$ (8). Others have not recorded an asymmetric pattern (3,4,11). These authors used different forms of inoculum but also a soil:sand mixture rather than sand as a medium for growth of the host. More work is needed to reconcile these results.

Several properties of host roots have been shown to affect the rate of colonization by *G. graminis* var. *tritici* (3,4,7–9). These include the origin and age of the root. I have recorded daily rates of colonization above

inoculation sites on lateral (derived from seminal), seminal, and adventitious roots of wheat of 1.4, 2.0, and 3.5 mm, respectively, during the first 9 days after inoculation at $19°C$ (7). Subsequently the daily rates of proximal extension of runner hyphae were similar on seminal and adventitious roots; the rate of colonization of lateral roots remained slow (7). Deacon and Henry (4) recorded more rapid extension of runner hyphae on old than on young portions of seminal roots of barley. Although the age of wheat roots did not affect the rate of extension of runner hyphae of a weakly pathogenic isolate of *G. graminis* var. *tritici* (see 3, Table 1), there was evidence for reduced symptoms of disease on older when compared with younger roots of wheat (3,4), the reverse of the case for barley (4).

The internal host mechanisms influencing the rate of colonization of roots are complex. Three closely related phenomena have been proposed, either implicitly or explicitly, in the literature to explain experimental results. These are nutrient supply to infected roots (7–9), natural death of root cortical cells with age (3,4,13), and microbial colonization of the rhizosphere and cortical cells (18,19). Further work is needed to elucidate and quantify the relationships between these phenomena.

PATHOGEN FACTORS

In addition to the genetic constitution of the pathogen population, one factor—the nutrient status of initial inoculum—stands out in importance in affecting the dynamics of root colonization by *G. graminis* var. *tritici*. In this context, the initial inoculum is to be considered in isolation from inoculum density. The latter influences the number of infections and interaction between infections.

Two effects of the initial inoculum can be distinguished. The first is quantal: either infection occurs or it does not. (In practice the response is unlikely to be unequivocally quantal. Some subjective criterion such as presence or absence of stelar discoloration must be used to decide whether or not infection has occurred.) The second effect is quantitative: given that infection is initiated, what further contribution does the initial inoculum make?

Wilkinson and Cook (25) analyzed the quantal effect of size of inoculum at the level of the population in two soils. They found that the critical size for inoculum to cause stelar discoloration of wheat roots was greater in a soil suppressive to *G. graminis* var. *tritici* than in a conducive soil. Using much larger inoculum (whole grains of millet, precolonized by *G. graminis* var. *tritici*), I investigated the quantitative effect of inoculum on the dynamics of colonization by varying the duration and dose of inoculum at a single site (9). It was evident that the effect of inoculum on rates of colonization of roots by surface runner hyphae persists for some time after the runner hyphae have become established on the root. For example, removal of the inoculum unit from contact with a root after 3 days reduced the rates of distal and proximal extension of surface runner hyphae relative to that for continuous presence of inoculum for 18 days (9).

The probability of obtaining more than one infection on the same root increases with increasing density and size of inoculum. Interaction between infections on the same root is determined as much by host as by pathogen. So far, we have little knowledge of the quantitative effects of interactions between infections. I have demonstrated, albeit with relatively large, artificial units of inoculum, that proximal infections on seminal roots of wheat slowed the rate of colonization by runner hyphae arising from a second distal inoculation (9). The effect was, of necessity, confounded with increasing age of the host plant between the two inoculations. The presence of distal second infections did not alter the course of proximal infections (9).

ENVIRONMENTAL FACTORS

Both abiotic and biotic environmental factors have been shown to affect the dynamics of colonization of the take-all fungus (20–23). Foremost among the biotic factors is the occurrence of fluorescent pseudomonads that have been implicated in specific antagonism to the fungus (2). Closely associated with these antagonists are the effects of the form of nitrogen on the expression of the disease, such that NH_4^+-N inhibits the disease whereas NO_3^--N favors the development of disease (12). How is this related to the dynamics of colonization? Smiley and Cook (23) proposed that the effect of form of N was mediated by pH since uptake of NH_4^+-N resulted in a fall in pH of the rhizosphere soil. The rate of growth of runner hyphae on wheat roots declined with decreasing rhizosphere pH (23). Below pH 5.0 the effect was direct, but between pH 5.0 and 6.6 the effect was attributed to the mediation of antagonistic microorganisms (23).

Subsequent work by Smiley (20) showed that the form of N did not alter the countable populations of microorganisms in the rhizosphere or rhizoplane but altered the numbers of bacteria and streptomycetes that inhibited the growth of the pathogen in vitro. Later work revealed, however, that antagonism to G. graminis var. tritici by Pseudomonas species appears to depend critically upon environmental conditions in the rhizosphere (22). Hence the expression of antagonism varies within and between soils. Smiley (22) suggests that antagonism in vivo inhibits the rate of colonization rather than the quantal response of presence or absence of infection.

Differences in the numbers and types of microorganisms in and on adventitious and seminal roots have been reported by Sivasithamparam et al (18,19). The differences may account in part for the more rapid rate of colonization of seminal than adventitious roots. Adventitious roots support significantly fewer microorganisms that show in vitro antagonism to G. graminis var. tritici than do seminal roots (19).

SYNTHESIS

The disparate information presented above on dynamics of colonization of roots by G. graminis var. tritici is essentially descriptive and sometimes analytical. How may we integrate the data so as to synthesize our knowledge of the system and thereby test the consistency of the hypotheses in the literature? Much could be attempted by argument; but, in part, on account of the complexity of the system, I favor the combined use of experimentation and mathematical modeling, especially computer simulation (10). Perhaps our first task should be to simulate the dynamics of growth of host roots.

ACKNOWLEDGMENT

Provision of funds by the Royal Society and the Fourth International Congress of Plant Pathology to attend the Congress is gratefully acknowledged.

LITERATURE CITED

1. Brown, M. E., and Hornby, D. 1971. Behavior of *Ophiobolus graminis* on slides buried in soil in presence or absence of wheat seedlings. Trans. Br. Mycol. Soc. 56:95-103.
2. Cook, R. J., and Rovira, A. D. 1976. The role of bacteria in the biological control of *Gaeumannomyces graminis* by suppressive soils. Soil Biol. Biochem. 8:269-273.
3. Deacon, J. W., and Henry, C. M. 1978. Studies on virulence of the take-all fungus, *Gaeumannomyces graminis*, with reference to methodology. Ann. Appl. Biol. 89:401-409.
4. Deacon, J. W., and Henry, C. M. 1980. Age of wheat and barley roots and infection by *Gaeumannomyces graminis* var. *tritici*. Soil Biol. Biochem. 12:113-118.
5. Garrett, S. D. 1936. Soil conditions and the take-all disease of wheat. Ann. Appl. Biol. 23:667-699.
6. Garrett, S. D. 1970. Pathogenic Root-Infecting Fungi. Cambridge University Press, London. 294 pp.
7. Gilligan, C. A. 1980. Colonization of lateral, seminal and adventitious roots of wheat by the take-all fungus, *Gaeumannomyces graminis* var. *tritici*. J. Agric. Sci. 94:325-329.
8. Gilligan, C. A. 1980. Dynamics of root colonization by the take-all fungus, *Gaeumannomyces graminis*. Soil Biol. Biochem. 12:507-512.
9. Gilligan, C. A. 1980. Inoculum potential of *Gaeumannomyces graminis* and disease potential of wheat roots. Trans. Br. Mycol. Soc. 75:419-424.
10. Gilligan, C. A. 1983. Mathematical modeling of soilborne plant pathogens. Annu. Rev. Phytopathol. 21:45-64.
11. Holden, J. 1976. Infection of wheat seminal roots by varieties of *Phialophora radicicola* and *Gaeumannomyces graminis*. Soil Biol. Biochem. 8:109-119.
12. Huber, D. M., Painter, C. G., McKay, H. C., and Peterson, D. L. 1968. Effect of nitrogen fertilization on take-all of winter wheat. Phytopathology 58:1470-1472.
13. Lewis, S. J., and Deacon, J. W. 1983. Effects of shading and powdery mildew infection on senescence of the root cortex and coleoptile of wheat and barley seedlings, and implications for root- and foot-rot fungi. Plant Soil 69:401-411.
14. Madden, L. V., Knoke, J. K., and Louie, R. 1982. Considerations for the use of multiple comparison procedures in phytopathological investigations. Phytopathology 72:1015-1017.
15. Prew, R. D. 1980. Studies on the spread of *Gaeumannomyces graminis* var. *tritici* in wheat. I. Autonomous spread. Ann. Appl. Biol. 94:391-396.
16. Ridgman, W. J. 1974. Experimentation in Biology: An Introduction to Design and Analysis. Blackie, London. 233 pp.
17. Rowell, J. G., and Walters, D. E. 1976. Analysing data with repeated observations on each experimental unit. J. Agric. Sci. Cambridge 87:423-432.
18. Sivasithamparam, K., Parker, C. A., and Edwards, C. S. 1979. Rhizosphere microorganisms of seminal and nodal roots of wheat grown in pots. Soil Biol. Biochem. 11:155-160.
19. Sivasithamparam, K., Parker, C. A., and Edwards, C. S. 1979. Bacterial antagonists to the take-all fungus and fluorescent pseudomonads in the rhizosphere of wheat. Soil Biol. Biochem. 11:161-165.
20. Smiley, R. W. 1978. Antagonists of *Gaeumannomyces graminis* from the rhizoplane of wheat in soils fertilized with ammonium- or nitrate-nitrogen. Soil Biol. Biochem. 10:169-174.
21. Smiley, R. W. 1978. Colonisation of wheat roots by *Gaeumannomyces graminis* inhibited by specific soils, microorganisms and ammonium nitrogen. Soil Biol. Biochem. 10:175-179.

22. Smiley, R. W. 1979. Wheat rhizosphere pseudomonads as antagonists of *Gaeumannomyces graminis*. Soil Biol. Biochem. 11:371-376.

23. Smiley, R. W., and Cook, R. J. 1973. Relationship between take-all of wheat and rhizosphere pH in soils fertilized with ammonium vs. nitrate-nitrogen. Phytopathology 63:882-890.

24. Wildermuth, G. B., and Rovira, A. D. 1977. Hyphal density as a measure of suppression of *Gaeumannomyces graminis* var. *tritici* on wheat roots. Soil Biol. Biochem. 9:203-205.

25. Wilkinson, H. T., and Cook, R. J. 1981. The effect of size and concentration of inoculum on the infection of wheat by *Gaeumannomyces graminis* var. *tritici* in different soils. (Abstr.) Phytopathology 71:265.

A Mathematical Model of Vesicular-Arbuscular Mycorrhizal Infection in Roots of *Trifolium subterraneum*

S. E. SMITH, Department of Agricultural Biochemistry, Waite Agricultural Research Institute, University of Adelaide, Glen Osmond, South Australia, 5064, and N. A. WALKER, Biophysics Laboratory, School of Biological Sciences, University of Sydney, New South Wales, 2006, Australia

Root systems of plants become infected by a wide range of different fungi that vary considerably in host range, nutrition, mode of survival in the absence of hosts, and in their effect upon the growth of the hosts. It has been common practice to distinguish between mycorrhizal fungi, which frequently (but not invariably) have a beneficial effect on the growth of host plants, and pathogenic root-infecting fungi, which cause disease. This artificial division reflects the interests of plant pathologists and plant physiologists concerned with the growth and yield of crop plants; it ignores the many similarities between mycorrhizal and other root-infecting fungi and the gradations and overlaps that exist between the two groups. This point was clearly made by Garrett (2), who wrote: "Although detailed discussion of mycorrhizal fungi may appear somewhat irrelevant to study of root disease fungi, yet those interested in the general biology and evolution of root infecting fungi can ill afford to neglect the symbiotic fungi."

Investigation of the ecology and epidemiology of any root-infecting fungus, whether mutualistic or parasitic, involves studies of many aspects of its biology. These include host range; reproduction and survival; penetration of the root tissues and the rate at which initial infection occurs; distribution of infection in the root systems and growth of the fungi in or on the roots; and hyphal colonization of the soil. To this list can be added the effect of root growth on the progress of infection and the influence of infection upon plant growth. Interest in these aspects of the biology of vesicular-arbuscular mycorrhizae is increasing because the fungi have potential importance in agriculture and because mycorrhizal infection is ubiquitous in natural ecosystems. An understanding of the symbiosis is thus a prerequisite to the understanding of nutrient cycling (3,5). Mathematical modeling provides a useful tool in the analysis of the interrelated events involved in mycorrhizal formation, just as it does in the analysis of disease epidemics (1,4,6,8,9,11,12).

In this paper we consider two processes involved in the formation of vesicular-arbuscular mycorrhizae: (*A*) the rate at which roots become infected from propagules in the soil and (*B*) the rate of fungal growth in the root cortex from individual entry points. These processes cannot be measured directly on living mycorrhizal root systems. Data are therefore collected from sequential, destructive harvests of replicate plants, followed by staining to reveal presence of the fungi. Application of a mathematical model is needed to calculate values of *A* and *B*.

We also consider the contribution of these processes and of rate of growth of the roots to the proportion of the root length that becomes infected in soils containing different densities of mycorrhizal propagules. We chose to vary the density of propagules, expecting this to result in variation in the rate of formation of entry points but not in the rate of growth of infection units. Parallels between models of mycorrhizal infection and of initiation and spread of root disease are briefly discussed.

MATERIALS AND METHODS

The methods have been published previously (10–12) and are given only briefly here.

The model of infection. The infection of young, uninfected plant roots growing in soil containing randomly distributed propagules of mycorrhizal fungi can be described by the equations:

$$dU/dt = A (L - L^*) \qquad (1)$$

$$dL^*/dt = B U (1 - L^*/L) \qquad (2)$$

where U is the number of entry points per plant, A is the frequency of infections in unit length of root in unit time, L is the length of root per plant, L^* is the length of infected root per plant, and B is the rate of spread of the fungus within the root cortex from a single entry point (11). Application of the model requires data for U, L, and L^* for each plant. Calculation of A and B thus yields information on the infection process and the behavior of the fungi that cannot be obtained from determinations of the proportion of root length infected.

We assume that new infections are formed only on uninfected regions of the root, and therefore the term $(L - L^*)$ is included in equation 1. The term $(1 - L^*/L)$ in equation 2 is introduced to allow for interpenetration of originally distinct infection units as hyphae grow within the cortex and come to overlap. A and B are mean values for the whole root system, or for main and lateral roots when data for these are collected separately; different regions of the root may be differently susceptible to infection (1,11).

Individual plants vary with respect to L, and this may produce correlated variation in U and L^*. The equations can be recast in terms of l, the fraction of the root length infected ($l = L^*/L$), and u, the frequency of infection units ($u = U/L$), as follows:

$$du/dt = A (1 - l) - g u \qquad (3)$$

and

$$dl/dt = B u (1 - l) - g l \qquad (4)$$

where g is the specific rate of growth in length of the root $[g = (dL/dt)/L]$.

Plants. Plants of *Trifolium subterraneum* were grown in mixtures of soil and steamed sand. Different proportions of nonsterile soil provided a range of densities of mycorrhizal propagules of several different fungi. The "undiluted" soil had a propagule density in the range of 1.8 to 4.0/g, determined by the probable numbers method under the same environmental conditions as the experiments. Results are expressed on the basis of the proportion of soil in the mixtures, from 0.05 to 1.0. Growth conditions were 14-hr day at 350 μE $m^{-2} s^{-1}$; day/night temperatures of 24°/15°C; six plants per pot containing 450 g of soil mixture at 12–15% (w/w) moisture content. Roots were cleared and stained with trypan blue. Observations on usually six to eight plants per treatment were made using a binocular dissecting microscope. Numbers of entry points (U) were counted directly, and root length (L) and infected root length (L^*) were determined by a grid intersect method.

RESULTS AND DISCUSSION

There was a linear relationship between A and propagule density in soil at the first harvest (Fig. 1). At this stage, it seems likely that entry points are formed only from propagules present in the soil at the beginning of the experiment (primary infections). We calculate that by 13 days, about one-fourth to one-half of these propagules have given rise to entry points (assuming one entry point per propagule). Primary infections presumably continue to occur, but a more complex relationship between A and propagule density develops

at the longer times (Fig. 2), possibly the result of ectotrophic spread of hyphae and the formation of secondary entry points. We have no way of distinguishing between primary and secondary infections. The data in Figs. 1 and 2 show a difference in the values of A in main and lateral roots. Differences in percentage of infection between main and lateral roots have been observed previously (7), but the reasons for the differences have not been determined. It is possible that if roots are more susceptible to infection immediately behind the apex (1,5,11), this could give rise to greater average values of A in the branching lateral root system than in the main root with a single root apex. Comparison of values of A in root systems with different branching patterns would be of interest.

There was some variation in the values of B in plants grown at different propagule densities (Fig. 3), but no clear indication of differences in rate of growth of infection units in main and lateral roots. As expected, B was independent of the propagule density in soil. Median values of B obtained from plants grown in all propagule densities declined during the experiment, with again no difference between main and lateral roots (Fig. 4). The decline may be the result of changes in the age structure of the population of infection units. In main roots, the mean length of the infection units (calculated from values of U and L^*) becomes constant after about 20 days (Fig. 5) corresponding with a fall in B (Fig. 4). Most of the infection units have apparently ceased to grow, and new infections are being formed at a very low rate. In lateral roots the mean length of the infection unit continued to increase throughout the experiment, but at a decreasing rate. Little is known about the physiologic changes that occur during the life of an infection unit, nor what proportion of an infection unit is physiologically active. Information on the rate of growth of the infection units as a function of age could provide a starting point for investigation of these phenomena.

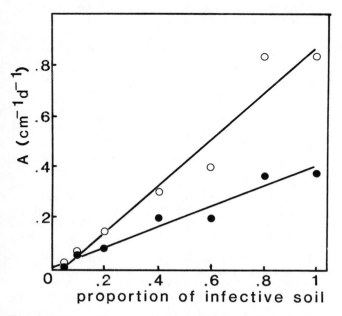

Fig. 1. Values of the rate of infection (A cm^{-1}d^{-1}) for the first harvest period on main roots 5–11 days (filled circles) and lateral roots 8–11 days (open circles) of *Trifolium subterraneum* as a function of the proportion of infective soil in soil/sand mixtures. The slopes and regression coefficients for main roots are 0.395 and 0.977; for lateral roots, 0.911 and 0.973.

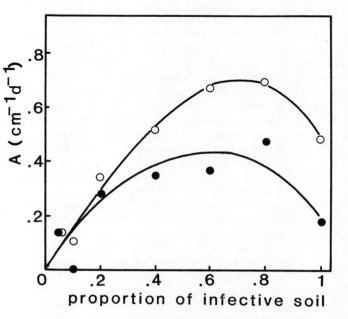

Fig. 2. Values of the rate of infection (A cm^{-1}d^{-1}) for the period 13–16 days on main (filled circles) and lateral (open circles) roots of *Trifolium subterraneum*.

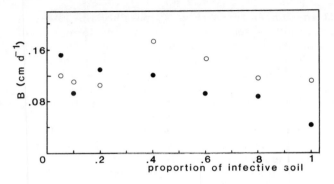

Fig. 3. Values of the rate of mycelial growth (B cm d^{-1}) in main (filled circles) and lateral (open circles) roots of *Trifolium subterraneum* for the period 5–16 days as a function of the proportion of infective soil in soil/sand mixtures.

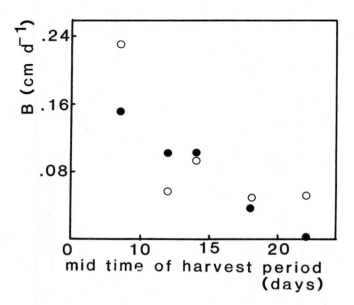

Fig. 4. Median values of the rate of mycelial growth (B cm d^{-1}) as a function of time in main (filled circles) and lateral (open circles) roots of *Trifolium subterraneum.*

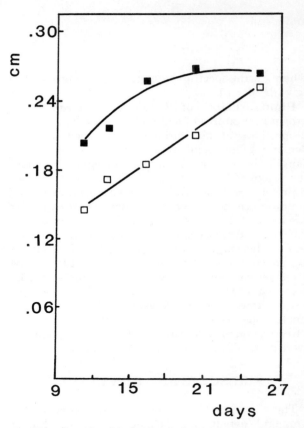

Fig. 5. Mean length of infection units (cm) in main (filled squares) and lateral (open squares) roots of *Trifolium subterraneum* as a function of time.

Future work should include studies with individual fungal species (rather than mixed inoculum) infecting a range of host plants. We also need more data on the time courses of A and of B and a study of the reasons behind these time courses. If such studies were correlated with measurements of mycorrhizal effects on phosphate uptake, metabolism, and growth of plants, they could be most useful in assessing the reasons for variations in effectiveness of different host/fungus combinations (8,9).

Modeling techniques have been applied more extensively to epidemiology of foliar pathogens than to epidemiology of root-infecting fungi. In a recent review, Huisman (6) emphasized the scarcity of data for root diseases that are susceptible to modeling in terms of the dynamics of root growth, rate of fungal infection, and rate of colonization of tissues. Such data are becoming available for vesicular-arbuscular mycorrhizae (1,8,9,11,12), and it is possible that methods developed for the study of symbiosis may be useful in work on plant diseases.

Conversely, some of the models already developed by plant pathologists can be applied to mycorrhizal systems. For example, it is possible to calculate the radius of interaction between roots and mycorrhizal propagules using Gilligan's model (4) for *Gaeumannomyces graminis*. The radius reached a maximum value of about 9 mm after 21 days (Walker and Smith, *unpublished*), which agrees very well with a value of 8 mm (at 30 days) obtained by direct measurement by Sanders and Sheikh (9). Such results confirm the possibility of applying similar principles to the modeling of disease and symbiosis.

ACKNOWLEDGMENTS

We would like to thank the Australian Research Grants Scheme and the Utah Foundation for financial support.

LITERATURE CITED

1. Buwalda, J. G., Ross, G. J. S., Stribley, D. P., and Tinker, P. B. 1982. The development of endomycorrhizal root systems. IV. The mathematical analysis of effects of phosphorus on the spread of vesicular-arbuscular mycorrhizal infection in root systems. New Phytol. 92:391-399.
2. Garrett, S. D. 1956. Biology of Root Infecting Fungi. Cambridge University Press, Cambridge. 293 pp.
3. Gianinazzi-Pearson, V., and Diem, H. G. 1982. Endomycorrhizae in the tropics. Pages 209-251 in: Microbiology of Tropical Soils and Plant Productivity. Y. R. Dommergues and H. G. Diem, eds. Martinus Nijhoff/Dr. W. Junk, The Hague. 328 pp.
4. Gilligan, C. A. 1979. Modeling rhizosphere infection. Phytopathology 69:782-784.
5. Harley, J. L., and Smith, S. E. 1983. Mycorrhizal Symbiosis. Academic Press, London. 496 pp.

6. Huisman, O. C. 1982. Interrelations of root growth dynamics to epidemiology of root-invading fungi. Annu. Rev. Phytopathol. 20:303-327.

7. Mosse, B. 1975. A microbiologist's view of root anatomy. Pages 39-66 in: Soil Microbiology. N. Walker, ed. Butterworths, London. 262 pp.

8. Sanders, F. E., Buwalda, J. G., and Tinker, P. B. 1983. A note on modelling methods for studies of ectomycorrhizal systems. Plant Soil 71:507-512.

9. Sanders, F. E., and Sheikh, N. A. 1983. The development of vesicular-arbuscular mycorrhiza in plant root systems. Plant Soil 71:223-246.

10. Smith, F. A., and Smith, S. E. 1981. Mycorrhizal infection and growth of *Trifolium subterraneum*: Comparison of natural and artificial inocula. New Phytol. 88:311-325.

11. Smith, S. E., and Walker, N. A. 1981. A quantitative study of mycorrhizal infection in *Trifolium*: Separate determination of rates of infection and of mycelial growth. New Phytol. 89:225-240.

12. Walker, N. A., and Smith, S. E. 1984. The quantitative study of mycorrhizal infection. II. The relation of rate of infection and speed of fungal growth to propagule density, the mean length of the infection unit and the limiting value of the fraction of the root infected. New Phytol. 96:55-69.

Rhizoplane Mycoflora of *Gahnia radula* and *Isopogon ceratophyllus* in Soils Infested and Free from *Phytophthora cinnamomi*

H. Y. YIP and G. WESTE, Botany School, University of Melbourne, Parkville, Victoria, 3052, Australia

Phytophthora cinnamomi Rands was first recorded in Victoria from the dry sclerophyll forests of the Brisbane Ranges in 1969 (15). Disease caused by this pathogen subsequently caused destruction of 75% of the rich understory in the areas invaded (24). The resulting ecological changes have been studied since that date and are still being recorded. The death of so many understory shrubs leads to a large increase in bare ground, which is followed in turn by a rapid increase in numbers of sedges, grass, and rush. Laboratory experiments have demonstrated that *P. cinnamomi* is a nonspecific pathogen and that the zoospores are attracted to the young roots of every species examined to date (10). The zoospores encyst on the roots and germinate, and the germ tube penetrates all roots. Resistance only develops after penetration and infection. Recent experiments have shown that zoospores even in low numbers infect the roots of sedge, grass, and rush and may cause a small necrotic lesion or necrotic root tip. These plants show no secondary symptoms and continue growth, although the pathogen remains alive in the root tissue and may disperse from it (13).

Soil microbial studies have demonstrated that soils with relatively high population densities of soil microorganisms exert significant control of *P. cinnamomi*, for example in forests of *Eucalyptus regnans* (mountain ash) on krasnozem. Soil from the Brisbane Ranges had a particularly low population density, the differences being highly significant for actinomycetes and bacteria and significant for fungi. Small microbial populations favor safe dispersal of zoospores through wet soil and uninhibited mycelial growth from infected roots (26,27).

The continued growth and increase in population density of the sedge, *Gahnia radula* (R. Br.) Benth. (Cyperaceae), through infested soil was compared with the 100% mortality of a totally susceptible species, such as the cone bush *Isopogon ceratophyllus* R. Br. (Proteaceae). We decided to compare the rhizoplane mycoflora of these two species before and after invasion by *P. cinnamomi*.

MATERIALS AND METHODS

The Brisbane Ranges are located 80 km southwest of Melbourne. Their vegetation, soils, and climate have been described (26).

Eight midseasonal samples of *G. radula* and *I. ceratophyllus* from sites infested and free from *P. cinnamomi* were taken during 1979 and 1980. *I. ceratopyllus* plants with obvious symptoms of dieback from infested sites were sampled from infection margins where the disease was recent; those behind the margins were either dead or dying. Plants were removed with a block of soil measuring about 15 cm^3 and containing intact root systems. The presence of the pathogen in all sites was regularly checked by baiting methods (5) and subsequent isolation onto antibiotic agar (8).

Roots were sampled as follows: *G. radula*—Old roots, 1.5–2 mm in diameter, first-order roots, cortex intact, dark brown; young roots, 0.5–0.75 mm in diameter, third-order roots, cortex and apex intact, white. *I. ceratophyllus*—Old roots, about 0.5 mm in diameter, from taproot, cortex intact, reddish brown; young roots, about 0.5 mm in diameter, from taproots, cortex and apex intact, white.

On each sampling date, twelve 20-mm lengths of both young and old roots were sampled from each plant species growing in soil either infested or free from *P. cinnamomi*. Root samples were serially washed (9), and each root length was then cut into 2-mm segments to provide 120 segments for each sample. Segments were plated half on 2% malt agar and the remainder on Czapek-yeast agar. The fungi isolated were identified and recorded.

The frequency of different fungi in a root sample was recorded as the number of root segments colonized by each fungus. Wilcoxson's two-sample test (19), a nonparametric method, was used for comparing the frequency of rhizoplane fungi on seasonal samples of young or old roots from noninfested and infested soils.

Soil tests. The organic matter content in soils of the sampling sites was estimated by the method of Walkely and Black (23). In each sampling site, soil samples were taken at five different depths from three random spots. The samples from each depth of a sampling site were bulked, thoroughly mixed, passed through a 2-mm sieve, and air dried. From each bulked depth sample, 0.5 g of soil in triplicate was digested in concentrated sulfuric acid in the presence of potassium dichromate. The digest was then titrated against N ferrous sulfate and the organic matter calculated as percentage of organic carbon. To determine soil pH, three samples of soil collected at random to a depth of 6 cm were obtained from each sampling site and thoroughly mixed to form a bulk sample. From each bulk sample, 10 g of soil in triplicate was mixed with 50 ml of deionized water, and the reaction of the mixture was measured with a pH meter.

RESULTS

Rhizoplane mycoflora of *G. radula* in noninfested soil. The species composition of fungi isolated regularly from young and old roots was similar (Table 1). *Penicillium spinulosum* strain A, *Aspergillus parvulus*, and *P. decumbens* may be considered as regularly occurring rhizoplane dominants because they were isolated frequently from both young and old roots in more seasons than other fungi. The total number of fungal isolates obtained from seasonal samples of old roots was always higher than from young roots.

Rhizoplane mycoflora of *I. ceratophyllus* in noninfested soil. The species composition of fungi regularly isolated from young roots was similar to that from old roots (Table 2). *Mortierella nana* and DS 52

Table 1. Frequency of fungi isolated from *Gahnia radula* rhizoplane in soil free from *Phytophthora cinnamomi* (no. of isolates)

Species	Old roots								Young roots							
	1979				1980				1979				1980			
	Sum.	Aut.	Win.	Spr.	Sum.	Aut.	Win.	Spr.	Sum.	Aut.	Win.	Spr.	Sum.	Aut.	Win.	Spr.
Aspergillus parvulus	10	2	16	4	66	43	24	44	11	···	5	2	20	18	13	17
Mortierella isabellina	···	···	2	1	3	11	1	4	···	···	···	···	4	2	5	3
M. nana	4	···	15	10	16	3	17	2	9	1	3	10	4	7	4	6
M. parvispora	13	6	29	20	5	14	16	4	5	···	···	4	3	···	1	5
M. pulchella	5	2	15	6	7	10	13	6	3	1	2	1	2	1	10	3
M. ramanniana var. *anguli-spora*	···	6	9	···	6	13	4	18	···	1	2	···	2	1	4	6
Mucor sp.	44	1	···	···	1	1	···	4	1	1	···	···	···	···	1	2
Oidiodendron spp.	···	···	···	11	···	2	···	···	6	5	2	7	4	10	6	6
Penicillium citreonigrum	1	3	7	···	27	5	3	19	1	···	1	6	11	4	13	22
P. decumbens	···	···	···	2	26	36	7	22	···	···	···	···	16	28	22	9
P. glabrum	···	1	2	2	4	1	19	8	···	3	···	3	12	2	18	4
P. raistrickii	···	1	5	···	2	1	7	8	···	4	1	···	7	1	11	9
P. restrictum series	3	···	1	3	10	1	16	3	2	1	7	1	6	4	8	7
P. spinulosum strain A	69	72	50	48	54	54	62	32	28	25	11	19	29	26	23	35
P. spinulosum strain B	3	···	1	1	4	3	3	7	1	7	1	···	···	2	···	1
P. spinulosum strain C	1	···	1	1	20	2	3	3	3	3	···	5	2	3	···	···
P. thomii	1	···	1	1	6	4	7	3	···	···	4	···	3	2	8	1
Trichoderma viride	13	15	12	5	11	27	19	31	4	6	43	1	7	10	2	12
Trichoderma sp.	30	70	···	···	1	···	···	6	···	2	20	···	···	9	1	3
DS 2	3	1	25	38	11	3	3	···	2	9	10	···	1	1	2	···
DS 8	···	···	···	1	4	···	4	1	4	19	···	4	5	9	2	2
DS 52 (dark sterile fungus)	···	···	1	···	···	···	···	1	4	13	3	14	1	18	5	4
Other infrequent species	6	16	49	15	38	55	33	36	31	23	35	33	46	29	27	36
Percentage of root segment colonized	100	97	100	95	100	100	100	100	82	96	99	68	100	100	99	98
Total isolates	208	196	241	169	320	290	261	259	115	124	150	110	185	187	186	193

Table 2. Frequency of fungi isolated from *Isopogon ceratophyllus* rhizoplane in soil free from *Phytophthora cinnamomi* (no. of isolates)

Species	Old roots								Young roots							
	1979				1980				1979				1980			
	Sum.	Aut.	Win.	Spr.	Sum.	Aut.	Win.	Spr.	Sum.	Aut.	Win.	Spr.	Sum.	Aut.	Win.	Spr.
Acremonium diversisporum	···	1	4	1	···	1	···	5	10	2	12	9	5	2	7	5
Aspergillus parvulus	···	···	···	2	6	20	2	6	2	···	2	···	12	6	11	4
Mortierella marburgensis	13	2	···	7	8	1	4	1	1	···	1	···	9	2	3	···
M. nana	16	7	13	24	32	32	34	19	4	9	10	19	32	30	42	16
Oidiodendron spp.	···	1	30	3	2	4	5	7	7	4	14	4	13	11	14	7
Penicillium citreonigrum	···	5	3	1	···	3	8	···	···	2	3	···	···	12	7	···
P. glabrum	17	5	···	1	1	1	···	1	···	···	···	3	···	3	2	···
P. raistrickii	11	···	1	2	···	1	5	···	3	···	4	···	···	4	2	···
P. restrictum series	···	3	5	8	7	6	9	4	14	···	3	3	6	11	12	11
P. simplicissimum	19	4	2	5	2	7	5	···	···	9	···	2	18	···	2	···
P. spinulosum strain A	5	···	···	2	3	3	5	9	8	···	···	5	5	5	···	11
P. spinulosum strain B	30	4	···	1	3	9	16	5	3	2	···	1	10	3	3	1
P. spinulosum strain C	20	···	···	1	17	23	4	5	7	2	···	5	7	26	8	6
P. thomii	11	3	1	3	3	4	1	2	7	2	···	···	···	3	2	···
Penicillium sp. 1	···	17	4	···	23	22	4	11	···	3	···	···	6	5	2	8
Trichoderma viride	14	···	···	···	···	···	1	1	2	1	···	···	2	1	1	2
DS 6	1	4	1	14	2	1	1	5	···	8	···	1	···	2	···	···
DS 8	···	···	···	···	4	7	3	20	···	···	···	···	···	···	···	1
DS 21	···	···	···	13	···	···	···	···	···	6	···	9	···	···	···	···
DS 39	···	···	1	···	···	···	···	···	···	19	1	···	···	···	···	3
DS 42	···	···	15	1	1	7	···	···	···	9	···	16	···	···	8	4
DS 52	2	5	39	26	7	3	11	24	3	18	33	36	3	20	10	21
Other species	46	60	45	42	52	32	50	52	45	23	18	25	33	20	20	23
Percentage of root segment colonized	100	87	97	99	98	100	97	97	95	90	74	87	97	98	89	85
Total isolates	205	121	164	157	172	187	168	177	116	119	101	138	161	166	156	133

were isolated in high frequency from both young and old roots in several seasons and are therefore regarded as the regularly occurring fungal dominants on the rhizoplane of *I. ceratophyllus*. The total number of fungal isolates obtained from old roots in most seasons was slightly higher than that from young roots.

Rhizoplane mycoflora in soil infested by *P. cinnamomi*. Comparison of these results (Tables 3 and 4) with those from noninfested soil (Tables 1 and 2) shows a basic similarity in species composition of regularly isolated fungi irrespective of the pathogen. The total number of isolates obtained from old roots of both *G. radula* and *I. ceratophyllus* in infested soil (Tables 3 and 4) in most seasons was higher than from young roots.

Changes in frequency of rhizoplane fungi in soil infested by *P. cinnamomi*. Species of fungi whose seasonal frequency was significantly different in the presence of the pathogen are presented for both *G. radula* and *I. ceratophyllus* (Tables 5 and 6). To assess the overall effect of the presence of the pathogen on the fungi, an index of percentage sensitivity (IS) has been provided for each fungus listed. The index was calculated for infested soils from the expression IS = (number of samples with significantly different fungal frequencies)/(total number of samples [or 16]) × 100.

The frequency of 16 species of fungi on *G. radula* rhizoplane was changed in the presence of *P. cinnamomi* (Table 5). The two fungal dominants *Penicillium spinulosum* strain A and *P. decumbens* with IS values of 43.7 and 31.2 respectively were highly sensitive either to the presence of the pathogen or to associated changes. The frequency of both species on roots in infested soil was significantly reduced in four seasons. The remaining fungal dominant, *A. parvulus*, was least affected by infested soil; its frequency on roots

in infested soil was significantly reduced only in one season. *M. nana* and *M. parvispora* with IS values of 43.7 and 31.2 were also highly sensitive to changes associated with pathogen invasion. On infested roots, the frequency of *M. nana* was significantly increased in six seasons; conversely, the frequency of *M. parvispora* was significantly reduced in five seasons.

Trichoderma sp. on *G. radula* roots (Table 5) has a high IS value of 25, but the effect of infested soil on the frequency of this species and of *Penicillium restrictum* series varied from a significant reduction to a significant increase in different seasons. The frequency of certain fungi on *G. radula* roots in infested soil, such as *Cylindrocarpon destructans*, was significantly affected only in single seasons. Inconsistent and nonrecurrent changes in frequency may or may not be associated with infested soil.

The presence of the pathogen on the rhizoplane mycoflora of *G. radula* was characterized by frequent significant reduction in the frequency of *Penicillium spinulosum* strain A, *P. decumbens*, and *M. parvispora* and by frequent significant increase in the frequency of *M. nana*.

The frequency of 15 species of fungi isolated from *I. ceratophyllus* roots in infested soil was significantly changed (Table 6). The IS values of all these fungi were low, ranging from 6.2 to 18.7. The frequency of 11 fungal species on infected roots was significantly altered only in single seasons. These include *M. nana*, which is one of the two fungal dominants of *I. ceratophyllus* roots. DS 52, the other fungal dominant, has an IS value of 18.7; but the effect of infested soil on its frequency was inconsistent. *Penicillium* sp. 1 and *P. spinulosum* strain C were significantly reduced in two and three seasons, respectively.

By contrast, the rhizoplane mycoflora of *I. cerato-*

Table 3. Frequency of fungi isolated from *Gahnia radula* rhizoplane in soil infested by *Phytophthora cinnamomi* (no. of isolates)

| Species | Old roots | | | | | | | | Young roots | | | | | | | |
| | 1979 | | | | 1980 | | | | 1979 | | | | 1980 | | | |
	Sum.	Aut.	Win.	Spr.	Sum.	Aut.	Win.	Spr.	Sum.	Aut.	Win.	Spr.	Sum.	Aut.	Win.	Spr.
Aspergillus parvulus	1	4	5	11	44	33	26	17	5	2	3	1	25	21	11	6
Cylindrocarpon destructans	31	⋯	⋯	⋯	⋯	⋯	⋯	3	5	⋯	⋯	⋯	⋯	⋯	⋯	⋯
Mortierella nana	6	12	23	29	28	35	21	50	23	3	15	11	38	19	13	51
M. parvispora	1	13	3	4	2	1	⋯	3	⋯	⋯	⋯	⋯	1	1	6	⋯
M. pulchella	5	5	8	8	13	5	9	6	⋯	2	5	5	⋯	8	10	2
Mucor sp.	10	1	2	4	⋯	20	8	3	2	⋯	8	3	⋯	⋯	1	2
Oidiodendron spp.	⋯	⋯	⋯	⋯	⋯	3	⋯	⋯	2	2	3	1	1	8	2	5
Penicillium citreonigrum	⋯	⋯	4	3	5	2	⋯	1	⋯	2	1	2	2	4	4	1
P. decumbens	⋯	1	⋯	2	4	5	7	3	⋯	3	8	3	15	5	2	1
P. glabrum	⋯	5	2	4	2	3	3	1	⋯	⋯	4	4	1	5	1	4
P. raistrickii	2	3	6	⋯	⋯	1	5	3	1	2	3	2	1	7	8	3
P. restrictum series	2	2	1	6	25	5	4	22	⋯	2	3	4	3	2	7	15
P. spinulosum strain A	⋯	39	37	37	23	42	60	32	⋯	2	8	13	15	22	32	9
P. spinulosum strain B	1	⋯	5	9	1	6	7	1	⋯	21	20	1	3	7	13	5
P. spinulosum strain C	4	1	4	4	13	⋯	1	2	1	⋯	1	1	2	11	3	6
P. thomii	2	⋯	5	7	6	6	5	3	1	1	1	⋯	2	3	10	3
Sesquicillium candelabrum	17	1	1	1	1	6	7	4	2	⋯	7	21	⋯	⋯	3	2
Trichoderma viride	21	46	3	14	5	22	17	7	6	1	4	10	9	3	7	3
Trichoderma sp.	⋯	2	2	6	1	14	20	3	⋯	1	⋯	1	⋯	⋯	6	⋯
DS 2	⋯	3	36	44	46	12	1	14	2	10	23	2	32	⋯	8	2
DS 8	⋯	⋯	1	2	⋯	⋯	⋯	⋯	⋯	⋯	⋯	15	⋯	10	7	6
DS 52	1	2	2	⋯	⋯	1	⋯	⋯	2	5	14	3	4	7	1	3
Other species	48	20	43	16	21	76	47	82	31	30	42	46	17	42	5	60
Percentage of root segment colonized	99	99	97	100	100	100	100	100	62	64	100	91	100	98	99	96
Total isolates	152	160	193	211	253	311	248	260	83	89	173	149	160	185	206	189

phyllus lacked fungal species showing significant changes in frequency in infested soil.

DISCUSSION

A total of 141 species of sporing fungi, including three species new to science and 64 different isolates of sterile fungi, have been cultured from the rhizoplane of *G. radula* and *I. ceratophyllus* during the 2-year period. The predominant genus was *Penicillium*, with a large number of species. The abundance of Penicillia may be related to a surface soil pH of 4.7–5.8 (11,24). The low number of *Aspergillus* species was probably associated with a characteristic preference for warm tropical soils (7,24). *Mortierella* with 14 species isolated was the codominant genus. Dark, sterile fungi were common, and they have also been reported as early colonizers of root cortex of rye grass (22).

Invasion of the dry sclerophyll *Eucalyptus* forest by *P. cinnamomi* resulted in death and disease of 75% of the understory within a period from 6 months to 3 years (26). During the first year after invasion, the soil of the extending disease front contained a high pathogen population, dead and dying plant material, and a 40-fold increase in number of saprophytic microorganisms (27,28). Two years later, the dense understory cover was replaced by bare ground, which increased from 30 to 90%; the soil microbial population had declined to one-fifth of its previous population. Soil fungi were reduced by a factor of 10. Susceptible shrubs such as *I. ceratophyllus* disappeared and field-resistant sedges such as *G. radula* gradually increased, providing a new and different understory. *I. ceratophyllus* and *G. radula* differed in many ways, including rhizoplane substrate, such as exudate, senescing cells, and colonizing surface. *I. ceratophyllus* formed a large root stock and a mass of fine proteoid roots that explored the surface litter in damp seasons. *G. radula* grew from an underground rhizome with clusters of swollen, hairy, dauciform roots in surface soil. When healthy, both provided extensive colonizing surfaces. Both proteoid roots and dauciform clusters tended to disappear in the presence of *P. cinnamomi*, but their rhizoplane mycoflora was not investigated.

Rhizoplane fungi colonize from soil populations and hence are involved with interactions such as nutrient competition and antagonisms. The soils sampled had a low organic content (organic carbon ranged from 1.2 to 5.9%) and a small microbial population, 10^5 microorganisms per gram of oven-dry soil compared with 10^7 per gram of garden soil (28). Hence competition and antagonisms were less than for richer soils.

Rhizoplane mycoflora was examined to show changes in species diversity and frequency resulting from seasonal influence, root age, host species, and presence of *P. cinnamomi*, with particular reference to susceptibility or resistance of host.

Seasonal variations in species composition and frequency of rhizoplane fungi were not consistent. Previous reports (27,28) demonstrated a highly significant variation, measured by the dilution plate method, as number of colonies per gram of oven-dry root on the same host rhizoplanes. Maximum numbers of colonies were recorded in winter and fewer in summer.

Table 4. Frequency of fungi isolated from *Isopogon ceratophyllus* rhizoplane in soil infested by *Phytophthora cinnamomi* (no. of isolates)

Species	Old roots 1979 Sum.	Aut.	Win.	Spr.	1980 Sum.	Aut.	Win.	Spr.	Young roots 1979 Sum.	Aut.	Win.	Spr.	1980 Sum.	Aut.	Win.	Spr.
Acremonium diversisporum	···	1	4	2	1	2	3	5	1	6	6	1	2	8	5	7
Aspergillus parvulus	2	3	4	6	1	10	5	1	7	···	···	3	6	4	4	···
Mortierella marburgensis	24	···	2	···	2	···	···	···	3	1	5	1	···	1	3	3
M. nana	8	14	13	13	28	34	29	8	14	8	10	13	22	56	19	27
M. pulchella	5	3	···	2	8	4	···	···	4	···	···	1	2	1	1	2
Mucor sp.	14	···	···	···	···	···	···	···	···	···	···	···	···	···	···	···
Oidiodendron spp.	···	1	11	3	7	7	6	25	1	5	4	21	8	12	13	10
Penicillium citreonigrum	···	5	2	3	···	···	4	3	···	4	···	···	···	···	1	7
P. glabrum	4	···	···	5	···	1	···	7	4	3	···	···	···	2	9	2
P. janczewskii	9	···	3	1	···	···	1	···	15	···	1	1	···	1	8	1
P. raistrickii	2	···	2	2	1	1	2	4	4	···	1	···	1	···	4	···
P. restrictum series	5	8	2	4	7	10	12	1	1	2	3	5	13	8	4	18
P. simplicissimum	2	2	1	4	7	···	1	4	2	8	2	···	8	···	8	5
P. spinulosum strain A	35	···	11	1	3	9	1	···	27	···	1	1	3	1	9	4
P. spinulosum strain B	3	7	7	3	2	2	1	2	8	···	10	···	5	···	···	5
P. spinulosum strain C	1	4	4	7	7	5	6	···	7	2	2	4	5	10	14	5
P. thomii	6	2	2	5	3	5	1	11	11	1	···	···	2	1	4	6
Penicillium sp. 1	···	1	3	8	1	16	···	1	···	4	6	2	···	1	3	2
Trichoderma viride	31	···	1	2	···	···	···	···	4	···	20	···	8	···	···	···
Trichoderma sp.	12	···	···	···	···	···	10	···	···	···	···	···	···	···	2	···
DS 6	···	3	21	17	3	2	1	1	···	···	···	2	6	2	1	3
DS 8	···	···	···	···	···	···	8	···	···	···	···	9	···	1	5	···
DS 10	···	···	10	4	1	9	3	5	···	···	6	···	1	2	9	4
DS 21	···	4	···	5	5	···	···	2	···	···	···	5	···	···	···	···
DS 39	···	···	4	···	1	5	···	7	···	2	4	1	···	···	1	···
DS 42	···	3	10	···	10	···	3	5	···	1	5	1	5	2	5	4
DS 50	···	1	···	···	···	···	···	···	···	26	···	···	···	···	···	···
DS 52	2	11	8	15	37	15	11	24	2	17	24	24	28	29	9	13
Other species	58	24	41	42	30	37	44	46	19	33	17	47	41	26	29	38
Percentage of root segment colonized	98	66	94	89	99	99	93	97	94	85	89	88	98	99	100	88
Total isolates	223	97	166	154	165	174	152	162	134	123	127	142	166	168	168	166

Higher numbers of fungi were isolated from old roots than from young roots, probably associated with an increase in exudates and senescing tissues (3,20). Old roots were more likely to sustain damage and therefore to increase exudation (6,18).

Rhizoplane fungi of the two plants differed in both species of fungal dominants and in the frequencies of regularly isolated fungi irrespective of the pathogen (Table 7). Evidently the roots were selective for rhizoplane fungi (12,15–17). Earlier work showed a highly significant difference in the number of colonies isolated from the different rhizoplanes (27,28).

G. radula and *I. ceratophyllus* differed in their reaction to *P. cinnamomi*. *I. ceratophyllus* was susceptible and died rapidly after infection, so that fungal components of the living rhizoplane could only be examined shortly after infection and before any disease-associated changes in vegetation had occurred. *G. radula* was a rare component of the healthy forest, but numbers increased in infested soil. *P. cinnamomi* infection was confined to a small necrotic zone behind the root tip, in which the fungus remained viable and reproduced (12,14).

The percentage of root colonized by fungi was approximately equal for both *Gahnia* and *Isopogon*. The total number of fungi isolated from roots ranged

Table 5. Significant differences associated with *Phytophthora cinnamomi* in the frequency of fungi isolated from *Gahnia radula* rhizoplane[a]

Species	IS[b]	Age of roots	1979 Summer	Autumn	Winter	Spring	1980 Summer	Autumn	Winter	Spring
Penicillium spinulosum strain A	43.7	Old	−***	−**			−*			
		Young	−***	−**			−**			−*
P. decumbens	31.2	Old						−**		−**
		Young					−*	−**	−**	
Mortierella parvispora	31.2	Old	−**		−*	−*		−*	−**	
P. citreonigrum	18.7	Old					−**			−*
		Young					−*		−**	
P. glabrum	18.7	Old							−*	
		Young					−*		−**	
Trichoderma viride	18.7	Old								−**
		Young			−*					−*
Mortierella nana	43.7	Old		+*		+*		+**		+***
		Young			+**		+**			+***
DS 2	18.7	Old					+**			+*
		Young					+***			
P. spinulosum strain B	12.5	Young			+*				+*	
Sesquicillium candelabrum	12.5	Old	+***							
		Young				+***				
Trichoderma sp.	25.0	Old	−***	−**				+**	+**	
P. restrictum series	12.5	Old							−*	+**
Cylindrocarpon destructans	6.2	Old	+***							
Mucor sp.	6.2	Old	+**							
Aspergillus parvulus	12.5	Old								−*
		Young								−***
Oidiodendron spp.	6.2	Old				−*				

[a] Significance determined by Wilcoxon's two-sample test: * = P≤0.05, ** = P≤0.01, and *** = P≤0.001; + = frequency significantly higher on roots in infested soil, and − = frequency significantly higher on roots in noninfested soil.
[b] Index of sensitivity.

Table 6. Significant differences associated with *Phytophthora cinnamomi* in the frequency of fungi isolated from *Isopogon ceratophyllus* rhizoplane[a]

Species	IS[b]	Age of roots	1979 Summer	Autumn	Winter	Spring	1980 Summer	Autumn	Winter	Spring
Penicillium sp. 1	12.5	Old		−**			−**			
P. spinulosum strain C	18.7	Old	−***				−**	−*		
DS 52	18.7	Old			−**		+*			
		Young					+*			
P. spinulosum strain B	18.7	Old	−*						−*	
		Young			+*					
Mortierella nana	6.2	Young						+*		
P. janczewskii	6.2	Young	+**							
P. spinulosum strain A	6.2	Old	+*							
Trichoderma viride	6.2	Young			+*					
Mucor sp.	6.2	Old	+*							
Oidiodendron spp.	6.2	Old								+*
DS 10	6.2	Young			+*					
P. simplicissimum	6.2	Old	−*							
Aspergillus parvulus	6.2	Young								−**
P. citreonigrum	6.2	Young					−**			
DS 8	6.2	Old								−*

[a] Significance determined by Wilcoxon's two-sample test: * = P≤0.05, ** = P≤0.01, and *** = P≤0.001; + = frequency significantly higher on roots in infested soil, and − = frequency significantly higher on roots in noninfested soil.
[b] Index of sensitivity.

Table 7. Total frequency of certain fungi regularly isolated from *Gahnia radula* and *Isopogon ceratophyllus* roots in noninfested soil[a] (no. of isolates)

Species	Fungi isolated from	
	G. radula	*I. ceratophyllus*
Acremonium diversisporum	1	64
Mortierella isabellina	36	6
M. marburgensis	11	52
M. parvispora	125	6
M. ramanniana var. *angulispora*	69	6
Mucor sp.	56	0
Penicillium simplicissimum	5	40
Penicillium sp. 1	0	105
Trichoderma viride	218	23
Trichoderma sp.	141	0
DS 2	109	5
DS 6	3	40
DS 42	9	61

[a] Total frequency for eight seasons from young and old roots combined.

from 196 to 320 for *Gahnia* roots, however, and only from 121 to 205 for *Isopogon* roots; the range reflects seasonal variation. There are thus fewer species of fungi on the latter (Tables 1 and 2). The *Gahnia* root system also has a much larger colonizing surface, with its many small adventitious roots, than the *Isopogon* root system, with its massive root stock and relatively few fine roots.

Trichoderma species, which are noted antagonists of *P. cinnamomi* (4), were common components of the *Gahnia* rhizoplane; but they were very rare or absent from the *Isopogon* rhizoplane in uninfested soils (Table 7). *Trichoderma* species were much rarer on both root systems in infested soils (Tables 3 and 4). Significant changes in the rhizoplane fungi occurred in both root systems in infested soils, some fungi increasing and others declining (Table 5). These changes may be related to the increased soil organic matter resulting from understory mortalities.

Present investigations have demonstrated that certain fungi, such as *Penicillium spinulosum* strain A and *P. decumbens* from the *Gahnia* rhizoplane, altered significantly in seasonal frequency after infection.

Factors that tended to change the rhizoplane mycoflora in the presence of *P. cinnamomi* were increase in necrotic tissue as substrate; a resultant increase in competitive saprophytes such as molds and bacteria; a reduction in living roots with the decline in understory vegetation; and continued survival and zoospore production by the pathogen in the small necrotic zones of *G. radula* roots.

Bowen (2) has shown that the concentration and availability of minerals influence root exudate composition. Changes in *Gahnia* and *Isopogon* mineral composition with infection by *P. cinnamomi* have also been recorded (25). Changes in rhizoplane mycoflora would therefore be expected in infested soil. Changes in vegetation lead to changes in soil organic content and in soil mycoflora (1,21). The major changes in understory associated with invasion by *P. cinnamomi* influence the rhizoplane fungi of *Gahnia* but not of *Isopogon*, which dies rapidly after infection.

A monocotyledon root system characterized by a large number of small roots and by their continual renewal is probably better adapted to resist a root pathogen than a perennial shrub with a massive root stock and relatively few small roots. Renewal of the root stock has never been observed. There is also the fundamental difference in

susceptibility between the two plant species, which we have yet to characterize. In addition, the larger component of antagonistic fungi in the *Gahnia* rhizoplane may play an important role in resistance to *P. cinnamomi*.

ACKNOWLEDGMENT

We gratefully acknowledge the assistance of Dr. H. J. Swart, School of Botany, University of Melbourne.

LITERATURE CITED

1. Badura, L., and Badurowa, M. 1964. Some observations on the mycoflora in the litter and soil of the beech forest in Lubsza region. Acta Soc. Bot. Pol. 33:507-527.
2. Bowen, G. D. 1969. Nutrient status effects on loss of amides and amino acids from pine roots. Plant Soil 30:139-142.
3. Bowen, G. D., and Rovira, A. D. 1976. Microbial colonization of plant roots. Annu. Rev. Phytopathol. 14:121-144.
4. Brasier, C. M. 1975. Stimulation of sex organ formation of *Phytophthora* by antagonistic species of *Trichoderma*. II. Ecological implications. New Phytol. 74:195-198.
5. Chee, K. H., and Newhook, F. J. 1965. Improved methods for use in studies on *Phytophthora cinnamomi* and other *Phytophthora* species. N.Z. J. Agric. Res. 3:88-95.
6. Dix, J. N. 1967. Mycostasis and root exudation: Factors influencing the colonization of bean roots by fungi. Trans. Br. Mycol. Soc. 50:23-31.
7. Domsch, K. H., and Gam. W. 1980. Compendium of Soil Fungi. Academic Press, London.
8. Eckert, J. W., and Tsao, P. H. 1962. A selective antibiotic medium for isolation of *Phytophthora* and *Pythium* from plant roots. Phytopathology 52:771-777.
9. Harley, J. L., and Waid, J. S. 1955. A method of studying active mycelia on roots and other surfaces in soil. Trans. Br. Mycol. Soc. 38:104-118.
10. Hinch, J., and Weste, G. 1979. Behaviour of *Phytophthora cinnamomi* zoospores on roots of Australian forest species. Aust. J. Bot. 27:679-691.
11. Jensen, H. L. 1931. The fungus flora of the soil. Soil Sci. 31:123-158.
12. Kovacs, M. F., Jr. 1971. Identification of aliphatic and aromatic acids in root and seed exudates of pea, cotton and barley. Plant Soil 34:34-41.
13. Phillips, D., and Weste, G. 1984. Field resistance in three native monocotyledon species which colonize indigenous sclerophyll forest after invasion by *Phytophthora cinnamomi*. Aust. J. Bot. 32:339-352.
14. Phillips, D., and Weste, G. 1985. *Phytophthora cinnamomi*: A study of resistance in three native monocotyledons that invade diseased Victorian forests. Pages 177-179 in: Ecology and Management of Soilborne Plant Pathogens. C. A. Parker, A. D. Rovira, K. J. Moore, P. T. W. Wong, and J. F. Kollmorgen, eds. American Phytopathological Society, St. Paul, MN.
15. Podger, F. D., and Ashton, D. H. 1970. *Phytophthora cinnamomi* in dying vegetation in the Brisbane Ranges, Victoria. Aust. For. Res. 4:33-36.
16. Rovira, A. D. 1965. Plant root exudates and their influence upon soil micro-organisms. Pages 170-185 in: K. F. Baker and W. C. Snyder, eds. Ecology of Soil-borne Plant Pathogens. University of California Press, Berkeley.
17. Rovira, A. D., Newman, E. I., Bowen, H. J., and Campbell, R. 1974. Quantitative assessment of the rhizoplane microflora by direct microscopy. Soil Biol. Biochem. 6:211-216.
18. Schroth, M. N., and Teakle, D. S. 1963. Influence of virus and fungus lesions on plant exudation and chlamydospore germination of *Fusarium solani* f. *phaseoli*. Phytopathology 53:610-612.
19. Sokal, R. R., and Rohlf, F. J. 1969. Biometry: The Principles and Practice of Statistics in Biological Research. W. H.

Freeman, San Francisco.

20. Starkey, R. L. 1938. Some influence of the development of higher plants upon the micro-organisms in soil. VI. Microscopic examination of the rhizoplane. Soil Sci. 45:207-249.

21. Thornton, R. H. 1960. Growth of fungi in some forest and grassland soils. Pages 84-91 in: The Ecology of Soil Fungi. D. Parkinson and J. S. Waid, eds. Liverpool University Press, Liverpool.

22. Waid, J. S. 1957. Distribution of fungi within decomposing tissues of rye-grass roots. Trans. Br. Mycol. Soc. 40:391-406.

23. Walkely, A., and Black, J. A. 1934. An examination of the degtjaroff method for determining soil organic matter and a proposed modification of the chromic acid titration method. Soil Sci. 37:29-38.

24. Warcup, J. H. 1951. The ecology of soil fungi. Trans. Br. Mycol. Soc. 34:376-399.

25. Weste, G., and Chaudhri, M. A. 1982. Changes in the mineral composition of plants infected by *Phytophthora cinnamomi*. Phytopathol. Z. 105:131-141.

26. Weste, G., and Taylor, P. 1971. The invasion of native forest by *Phytophthora cinnamomi*. I. Brisbane Ranges, Victoria. Aust. J. Bot. 19:281-294.

27. Weste, G., and Vithanage, K. 1977. Microbial populations of forest soils. Aust. J. Bot. 25:153-167.

28. Weste, G., and Vithanage, K. 1978. Effect of *Phytophthora cinnamomi* on microbial populations associated with the roots of forest flora. Aust. J. Bot. 26:153-167.

Part III
Biological Control

Soils Suppressive to Fusarium Wilt: Mechanisms and Management of Suppressiveness

C. ALABOUVETTE, Y. COUTEAUDIER, and J. LOUVET, I.N.R.A., 17 Rue Sully 21034 Dijon CEDEX, France

As long ago as 1892, Atkinson (9), describing Fusarium wilt of cotton, indicated that disease severity varied according to soil type. Since that time, the phenomenon of suppressiveness of soil to Fusarium wilts has been extensively studied and suppressive soils have been identified in many areas of the world. Various books and review articles deal with the results of much of the early work in this field (10,17,28). During the last 10 years and especially since the last International Congress of Phytopathology in 1978 (3), this work has expanded considerably; the original data now available have been interpreted in several ways. It is thus timely to review the different hypotheses and to draw attention to their points of agreement.

In parallel with studies of the mechanisms of soil suppressiveness, work has been devoted to applying what has been learned to biocontrol of diseases. There, too, approaches have been highly varied. However, definite progress has been made and it seems reasonable to expect a fruitful conclusion to several lines of research now underway.

CHARACTERISTICS OF SUPPRESSIVE SOILS

Despite the diversity of examples studied, several characteristics are common to all soils suppressive to Fusarium wilts. Most suppressive soils are heavy, frequently clay or clay-loam in texture (19,27,31,33). The pH is always high, equal to or greater than 8 (18,22). However, no direct relationship exists between physicochemical properties of soils and their level of receptivity to Fusarium wilts. Indeed, all examples of suppressiveness so far studied are fundamentally microbiological in nature; suppressiveness is destroyed by biocidal treatments such as steam, γ-rays, and methyl bromide (2,18,20,22). Generally, soil suppressiveness is expressed against a whole range of forma speciales of *Fusarium oxysporum*, but it does not prevent development of other pathogenic soil fungi (4,25). Finally, one of the most interesting characteristics of suppressiveness is its transmissibility. Even a small proportion (1–10%) of suppressive soil mixed with a conducive, previously heated soil confers suppressiveness to the mixture (11,16,18,22). This series of properties indicates that soil suppressiveness is a function of all or part of the soil microflora.

MECHANISMS OF SUPPRESSIVENESS

Studies of the mechanisms of soil suppressiveness to Fusarium wilts have revealed another characteristic common to all such soils: chlamydospore germination of *Fusarium* is strongly inhibited as well as subsequent development of germ tubes (5,14,24,26). But, despite agreement among authors about these characteristics, interpretations have differed greatly. Although specific antagonistic microbial populations have frequently been demonstrated, we have explained the suppressiveness of Chateaurenard soils in terms of the general phenomenon of nutrient competition between microorganisms (1).

Role of nutrient competition between microorganisms. By comparing microbial interactions in two soils with different receptivity, we have shown that there is a much higher level of fungistasis in suppressive than in conducive soils (5). If glucose is introduced in increasing concentrations to the soils, a larger amount of energy (5–10 times more) is necessary in the suppressive soil to counteract fungistatis, induce chlamydospore germination (Fig. 1), and bring about the development of *Fusarium*. This causes a significant increase in the population of *Fusarium* spp. (Table 1), indicating that the nutrient sink is larger in the suppressive soil. To confirm this, we studied the kinetics of mineralization of small quantities of glucose introduced into suppressive and conducive soils. An initial respiratory rate three times higher in suppressive soil indicates that the initial microbial biomass is about three times higher than in conducive soil. Microbial activity thus develops more rapidly and more intensely in suppressive soils. Consequently, the nutritional stress that accompanies equilibration of the microflora with the new nutritional state of the suppressive soil

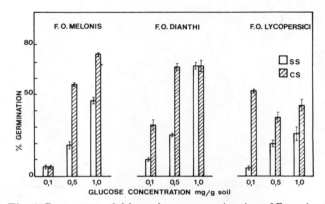

Fig. 1. Percentage of chlamydospore germination of *Fusarium oxysporum* strains after 24 hours of incubation in suppressive (ss) and conducive (cs) soil amended with different concentrations of glucose.

occurs after 12 hr of incubation. After this, microbial activity ceases rapidly. Microbial activity in suppressive soil thus rises and falls suddenly, whereas in conducive soil it is maintained at a high level for 60 hr (Fig. 2).

After amendment with glucose, therefore, *Fusarium* chlamydospores in suppressive soil have only a limited time to germinate and reach the root surface. We believe that an elevated biomass in suppressive soils causes particularly intense competition for nutrients, limiting considerably the possibilities for development of microorganisms—especially of pathogenic *F. oxysporum*. This analysis is supported by Huisman's (13) observation that a chlamydospore of *Fusarium* has only 2–3 hr from the time it is influenced by root exudates to germinate and reach the root apex, where it can penetrate undifferentiated tissue and achieve systemic infection.

Our interpretation explains why modification of the nutritional state of soils induces a modification of their suppressiveness. Amendment of previously infested soils with an energy source lessens the intensity of competition for nutrients and causes an increase in receptivity (Fig. 3). On the other hand, if amendment occurs before infestation of the soil, it induces an increase in the biomass of saprophytes and causes an increase in suppressiveness.

Thus, the suppressiveness of soils to Fusarium wilts is clearly a function of their total microbial activity; but this general mechanism does not explain how a conducive soil can be made suppressive simply by introducing 1–10% suppressive soil. To account for this phenomenon, numerous authors have attempted to demonstrate the presence of specific antagonists in the microbial population.

Role of certain antagonistic microbial populations. Having demonstrated that suppressiveness is microbial in nature, most workers have tried to demonstrate an antagonistic microflora. One of the first hypotheses involved a role for nonpathogenic strains of *F. oxysporum* (26). We demonstrated the same thing in

Table 1. Increase in density of *Fusarium* populations in suppressive and conducive soils amended with different glucose concentrations[a]

Glucose concentration (mg/g)	Suppressive soil				Conducive soil			
	F. oxysporum	*F. roseum*	*F. solani*	Total	*F. oxysporum*	*F. roseum*	*F. solani*	Total
0	100 a[b]	100 a	100 a	100 a	100 a	100 a	100 a	100 a
0.1	97 a	101 a	99 a	99 a	180 b	130 b	106 a	150 b
0.5	103 a	104 a	83 a	101 a	414 c	227 c	254 b	324 c
1	134 b	145 b	112 b	136 b	347 d	255 c	250 b	300 c

[a] Populations estimated 14 days after soil amendment.
[b] Entries with a common letter are not significantly different ($P < 0.05$) based on Duncan's multiple range test.

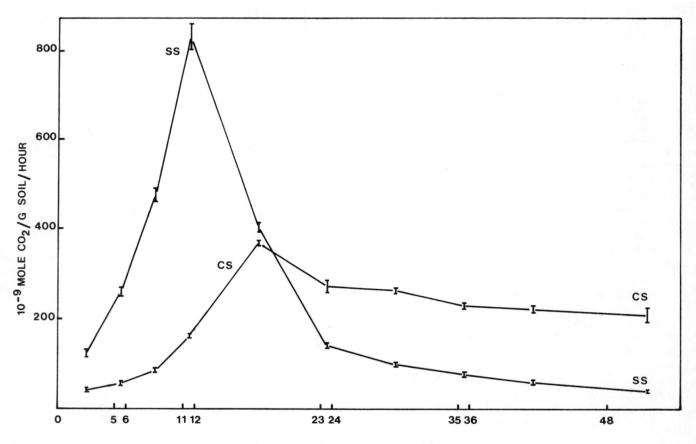

Fig. 2. Kinetics of glucose mineralization in suppressive (ss) and conducive (cs) soil amended with 1 mg of glucose per gram of soil.

the Chateaurenard soils (21), where introduction of nonpathogenic *F. oxysporum* and *F. solani* was sufficient to reestablish suppressiveness in steam-treated soil. This intrageneric competition between ecologically similar forms is much stronger in suppressive soil, where the total biomass is larger and the nutrient sink is greater. Generally, populations of bacteria and actinomycetes are seen as playing a role in the mechanism of soil suppressiveness. For example, Komada and Ezuka (16) in Japan and Arjunarao in India (7,8) claim that actinomycetes antagonistic to *Fusarium* are involved. Smith (24) has shown that in suppressive soils in California certain *Arthrobacter* isolates are associated with the lysis of germ tubes growing from chlamydospores. Finally, Tu et al (29,30) attribute a major role to facultative and anaerobic bacteria, including *Clostridium* spp. and *Bacillus* spp., in suppressiveness of certain soils in Taiwan.

Scher and Baker (22) isolated from a suppressive California soil two strains of fluorescent *Pseudomonas* that established suppressiveness when introduced into a conducive soil. Work by Kloepper et al (15) and subsequently by Scher and Baker (23) suggested that the mode of action with these bacteria may be through competition for iron. Under certain conditions, such as elevated pH, these bacteria produce siderophores that immobilize iron, thus making it unavailable to other microbes. This element would be indispensable to development of *Fusarium* spp. and particularly to elongation of the germ tubes. Certain iron chelates (FeEDDHA) are known to have the same effect as the bacteria, reducing the level of receptivity of the soil.

MANAGEMENT OF SUPPRESSIVENESS AND BIOLOGICAL CONTROL

Ever since it was shown that suppressiveness could be transmitted, workers have attempted to use the technique in biological control practices. Two complementary approaches have been followed: transmission of suppressiveness by mixing the soil into the substrate in which the crop to be protected is growing, and utilization of a specific antagonistic population isolated from suppressive soil.

Direct incorporation of suppressive soil. The suppressiveness of the Chateaurenard and Siagne soils from France and the Salinas and San Joaquin valley soils in California is readily transmitted to other soils that have been steam treated. Introduction of 10% Chateaurenard soil into a "light peat" that was not treated with steam confers a very low level of receptivity on the mixture, equivalent to that of the original suppressive soils (Fig. 4). Using this technique, we have achieved effective protection of crops of carnation, cyclamen, melon, and tomato grown in containers in commercial peats (6). Baker (11) treated carnations in ground beds under commercial greenhouse conditions with Metz sandy loam from the Salinas Valley at the rate of 600 g/m². After 2 years, losses due to Fusarium wilt were reduced by 60% in the treated plots, thus demonstrating the potential benefits of the method.

However, added suppressive soils do carry the risk of transmitting other potentially damaging soilborne pathogens. For this reason, we have tried to make a pathogen-free suppressive soil (PFSS) by incorporating cultures of nonpathogenic fungi into 100 L of steam-treated, suppressive soil from which the fungi were previously isolated. By successive dilution with a disinfected soil, we obtained several cubic meters of PFSS that was used to treat a tomato crop under commercial greenhouse conditions. When incorporated

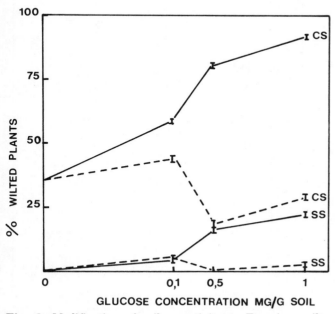

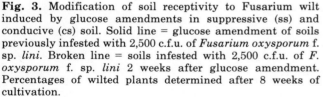

Fig. 3. Modification of soil receptivity to Fusarium wilt induced by glucose amendments in suppressive (ss) and conducive (cs) soil. Solid line = glucose amendment of soils previously infested with 2,500 c.f.u. of *Fusarium oxysporum* f. sp. *lini.* Broken line = soils infested with 2,500 c.f.u. of *F. oxysporum* f. sp. *lini* 2 weeks after glucose amendment. Percentages of wilted plants determined after 8 weeks of cultivation.

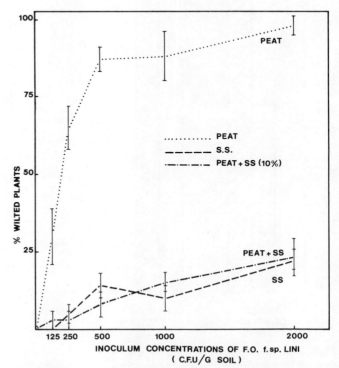

Fig. 4. Transmission of soil suppressiveness to a light peat by addition of 10% suppressive soil from the Chateaurenard region. Percentages of wilted plants determined 6 weeks after infestation with *Fusarium oxysporum* f. sp. *lini* at different concentrations.

into the soil at a rate of 8% by volume to a depth of 20 cm, the PFSS gave complete protection to two successive tomato crops (Fig. 5). These results, along with those of Baker (11), show that transmitted suppressiveness can persist. Thus, the incidence of Fusarium wilts can be considerably reduced by amendment with suppressive soil. But this method is difficult to apply on a large scale, except in the case of particular horticultural crops for which it is sufficient to place a small quantity of suppressive soil on the surface of the substrate to prevent later infections.

Utilization of specific antagonists. The role of a particular population of antagonists in a mechanism of suppressiveness is demonstrated by showing whether the introduction of this population into the soil after steam treatment reestablishes suppressiveness. This simple protocol opens the way to methods of biological control by the introduction of antagonists into soils. Although many studies refer to this type of procedure, it is still difficult to find examples of its use.

Having demonstrated the antagonistic role of certain anaerobic bacteria, Tu and Cheng (29) tried to utilize them against Panama disease of banana. Incorporating 15 L of a concentrated suspension (10^7 bacteria per milliliter) per square meter gave disease control, but only during the first year of the crop. Better results were recorded when the anaerobic bacteria in situ were favored by mixing in a complex amendment and following this with 21 days of flooding.

Scher and Baker (23) showed that the introduction of certain races of fluorescent *Pseudomonas* at the rate of 10^5 c.f.u./g of soil considerably reduced disease severity, even in a conducive soil that was not steam treated. They suggested that this type of intervention could easily be applied on a large scale.

The use of nonpathogenic *Fusarium* to control Fusarium wilts was first suggested in 1944 by Van Koot (32). After Rouxel et al (21) showed the role of *Fusarium* spp. in the suppressiveness of Chateaurenard soil, we attempted to define the conditions under which they could be used for biological control. Introduction of a mixture of nonpathogenic *F. oxysporum* and *F. solani* amended with glucose causes substantial reduction in the incidence of Fusarium wilt of tomato (Fig. 6). This same experiment is currently in progress under commercial greenhouse conditions (12).

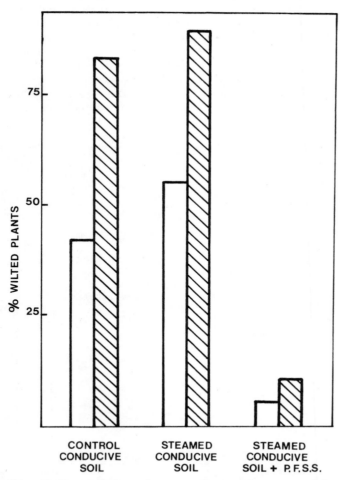

Fig. 5. Transmission of suppressiveness in commercial greenhouse conditions by addition of 8% pathogen-free suppressive soil (PFSS) to a steamed, naturally infested soil. Soil treatments and first plantation of tomatoes in July 1981. Open bars represent disease incidence 20 weeks after the first plantation; shaded bars represent disease incidence 20 weeks after the second plantation of tomatoes in July 1982.

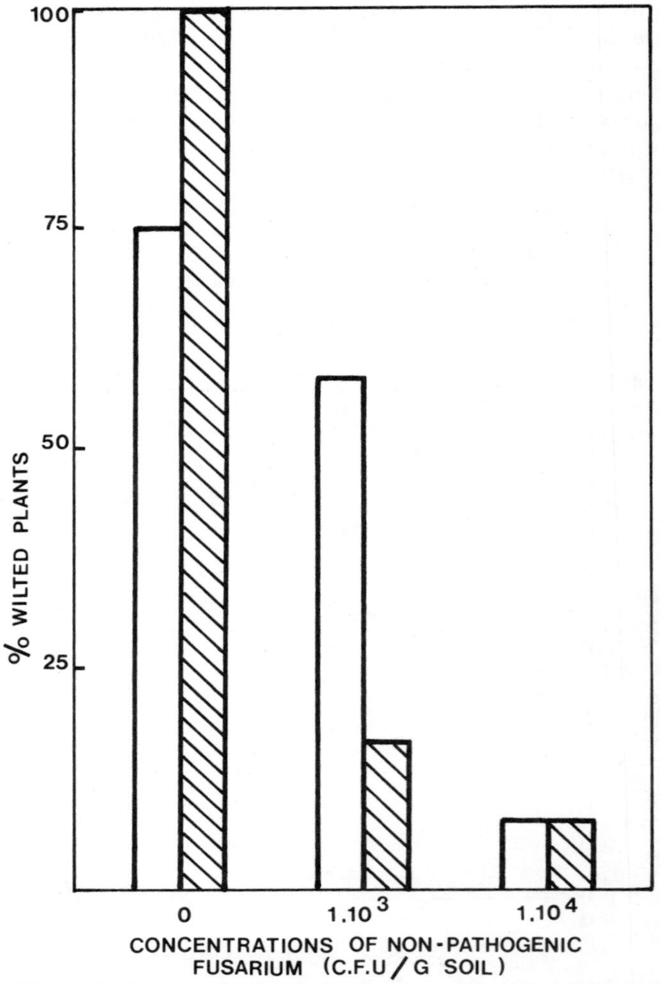

Fig. 6. Induction of suppressiveness in a steamed, conducive soil by addition of nonpathogenic *Fusarium oxysporum* and *F. solani*. Open bars represent addition of nonpathogenic *Fusarium* alone; shaded bars represent soil amended with glucose (5 mg/g of soil) and addition of nonpathogenic *Fusarium* before soil infestation with 500 c.f.u. of *F. oxysporum lycopersici*. Disease incidence determined 11 weeks after the plantation of tomatoes.

CONCLUSIONS

Although current research has not yet resulted in a reliable method of biological control that can be applied on a large scale, progress in understanding the mechanisms of soil suppressiveness to Fusarium wilts is no less important. Two major lines of research have focused attention on the role of the total microbial population acting as a nutrient sink, and thus restricting the opportunity for pathogenic *Fusarium* spp. to develop, and the role of populations of specific antagonists that can inhibit the activity of pathogenic agents.

The two types of mechanisms are not mutually exclusive. On the contrary, they clearly coexist and complement each other in all suppressive soils, and both must be taken into account in developing biocontrol procedures. Although it may seem difficult to base a method of biological control entirely on mechanisms of nutritional competition between microorganism, these same phenomena determine to a large extent the success of procedures based on the massive introduction of a population of antagonists. Indeed, the mass production of microbial antagonists currently poses few serious problems, but the successful establishment of the microbial antagonists in the soil is a more difficult problem.

In the case of nonpathogenic *Fusarium*, Rouxel et al (21) have demonstrated that the antagonistic effects are not manifest in all soil types. Similarly, Scher and Baker (23) have shown that the efficacy of fluorescent *Pseudomonas* is conditioned by soil pH and the level of iron. Thus, it is appropriate to continue studies on the receptivity of soils to soilborne microorganisms in such a way as to define better the environmental conditions essential to the establishment and maximal expression of antagonism by microbial populations selected for biological control.

ACKNOWLEDGMENT

We wish to thank Dr. R. A. A. Morrall for translating and reviewing the manuscript.

LITERATURE CITED

1. Alabouvette, C. 1983. La réceptivité des sols aux Fusarioses vasculaires, rôle de la compétition nutritive entre microorganismes. Doctoral thesis, Université de Nancy. 158 pp.
2. Alabouvette, C., Rouxel, F., and Louvet, J. 1977. Recherches sur la résistance des sols aux maladies. III. Effets du rayonnement γ sur la microflore d'un sol et sa résistance à la Fusariose vasculaire du melon. Ann. Phytopathol. 9:467-471.
3. Alabouvette, C., Rouxel, F., and Louvet, J. 1979. Characteristics of Fusarium wilt-suppressive soils and prospects for their utilization in biological control. Pages 165-182 in: Soil-Borne Plant Pathogens. B. Schippers and W. Gams, eds. Academic Press, New York.
4. Alabouvette, C., Rouxel, F., and Louvet, J. 1980. Recherches sur la résistance des sols aux maladies. VI. Mise en évidence de la spécificité de la résistance d'un sol vis-à-vis des Fusarioses vasculaires. Ann Phytopathol. 12:11-19.
5. Alabouvette, C., Rouxel, F., and Louvet, J. 1980. Recherches sur la résistance des sols aux maladies. VII. Etude comparative de la germination des chlamydospores de *Fusarium oxysporum* et *Fusarium solani* au contact de sols résistant et sensible aux fusarioses vasculaires. Ann. Phytopathol. 12:21-30.
6. Alabouvette, C., Tramier, R., and Grouet, D. 1980. Recherches sur la résistance des sols aux maladies. VIII. Perspectives d'utilisation de la résistance des sols pour lutter contre les fusarioses. Ann. Phytopathol. 12:83-93.
7. Arjunarao, V. 1971. Biological control of cotton wilt. II. In vitro effects of antagonists of the pathogen *Fusarium vasinfectum*. Proc. Ind. Acad. Sci., Sect. B, 74:16-28.
8. Arjunarao, V. 1971. Biological control of cotton wilt. III. In vivo effect of antagonists on the pathogen *Fusarium vasinfectum*. Proc. Ind. Acad. Sci., Sect. B, 74:53-62.
9. Atkinson, G. F. 1892. Some diseases of cotton. Ala. Agric. Exp. Stn. Bull. 41. 65 pp.
10. Baker, K. F., and Cook, R. J. 1974 (original ed.). Biological Control of Plant Pathogens. Reprint ed., 1982. American Phytopathological Society, St. Paul, MN. 433 pp.
11. Baker, R. 1980. Measures to control Fusarium and Phialophora wilt pathogens of carnation. Plant Dis. 64:743-749.
12. Couteaudier, Y., Alabouvette, C., and Louvet, J. 1985. Lutte biologique contre la fusariose vasculaire de la tomate. Résultats en serre de production. Agronomie. In press.
13. Huisman, O. C. 1982. Interrelations of root growth dynamics to epidemiology of root-invading fungi. Annu. Rev. Phytopathol. 20:303-327.
14. Hwang, S. F., Cook, R. J., and Haglund, W. A. 1982. Mechanisms of suppression of chlamydospore germination for *Fusarium oxysporum* f. sp. *pisi* in soils. (Abstr.) Phytopathology 72:948.
15. Kloepper, J. W., Leong, J., Teinzte, M., and Schroth, M. N. 1980. *Pseudomonas* siderophores: A mechanism explaining disease suppressive soils. Curr. Microbiol. 4:317-320.
16. Komada, H., and Ezuka, A. 1970. Ecological study of Fusarium diseases of vegetable crops. I. Survival of pathogenic Fusaria in different soil types. Res. Prog. Rep. Tokai-Kinki Natl. Agric. Exp. Stn. 6.
17. Louvet, J., Alabouvette, C., and Rouxel, F. 1981. Microbiological suppressiveness of some soils to Fusarium wilts. Pages 262-275 in: *Fusarium*: Diseases, Biology and Taxonomy. P. E. Nelson, T. A. Toussoun, and R. J. Cook, eds. Pennsylvania State University Press, University Park. 457 pp.
18. Louvet, J., Rouxel, F., and Alabouvette, C. 1976. Recherches sur la résistance des sols aux maladies. I. Mise en évidence de la nature microbiologique de la résistance d'un sol au développement de la Fusariose vasculaire du melon. Ann. Phytopathol. 8:425-436.
19. Reinking, O. A., and Manns, M. M. 1933. Parasitic and other Fusaria counted in tropical soils. Z. Parasitenkd. 6:23-75.
20. Rouxel, F., Alabouvette, C., and Louvet, J. 1977. Recherches sur la résistance des sols aux maladies. II. Incidence de traitements thermiques sur la résistance microbiologique d'un sol à la Fusariose vasculaire du melon. Ann. Phytopathol. 9:183-192.
21. Rouxel, F., Alabouvette, C., and Louvet, J. 1979. Recherches sur la résistance des sols aux maladies. IV. Mise en évidence du rôle des *Fusarium* autochtones dans la résistance d'un sol à la Fusariose vasculaire du melon.Ann. Phytopathol. 11:199-207.
22. Scher, F. M., and Baker, R. 1980. Mechanism of biological control in a Fusarium-suppressive soil. Phytopathology 70:412-417.
23. Scher, F. M., and Baker, R. 1982. Effect of *Pseudomonas putida* and a synthetic iron chelator on induction of soil suppressiveness to Fusarium wilt pathogens. Phytopathology 12:1567-1573.
24. Smith, S. N. 1977. Comparison of germination of pathogenic *Fusarium oxysporum* chlamydospores in host rhizosphere soils conducive and suppressive to wilts. Phytopathology 67:502-510.
25. Smith, S. N., and Snyder, W. C. 1971. Relationship of inoculum density and soil types to severity of Fusarium wilt of sweet potato. Phytopathology 61:1049-1051.
26. Smith, S. N., and Snyder, W. C. 1972. Germination of *Fusarium oxysporum* chlamydospores in soils favorable

and unfavorable to wilt establishment. Phytopathology 62:273-277.

27. Stotzky, G. 1973. Techniques to study interactions between microorganisms and clay minerals in vivo and in vitro. Bull. Ecol. Res. Comm. Stockholm 17:17-28.

28. Toussoun, T. A. 1975. Fusarium-suppressive soils. Pages 145-151 in: Biology and Control of Soil-Borne Plant Pathogens. G. W. Bruehl, ed. American Phytopathological Society, St. Paul, MN.

29. Tu, C. C., and Cheng, Y. H. 1981. Soil microbial activity in relation to Fusarium wilt-suppressive soil and conducive soil. Pages 1-12 in: Tainan DAIS Scientific Meeting Report.

30. Tu, C. C., Cheng, Y. H., and Chang, Y. C. 1978. Antagonistic effect of some bacteria from Fusarium wilt-suppressive soil and their effect on the control of flax wilt in the field. J. Agric. Res. China 27:245-258.

31. Tu, C. C., Cheng, Y. H., and Chen, M. 1975. Flax Fusarium wilt-suppressive soil in Taiwan. Plant Prot. Bull. Taiwan 17:390-399.

32. Van Koot, Y. 1944. De Fusariumziekte van komkommer en meloen. Meded. Tuinbouwvoorlichtingsdienst 42:1-85.

33. Wensley, R. N., and McKeen, C. D. 1963. Populations of *Fusarium oxysporum* f. sp. *melonis* and their relation to the wilt potential of two soils. Can. J. Microbiol. 9:237-249.

Reduction of Take-all by Mycophagous Amoebas in Pot Bioassays

S. CHAKRABORTY and J. H. WARCUP, Department of Plant Pathology, Waite Agricultural Research Institute, Glen Osmond, South Australia, 5064

The take-all disease of wheat (*Gaeumannomyces graminis* (Sacc.) von Arx & Olivier var. *tritici* Walker) can be suppressed in bioassays using natural soils, including that of the permanent pasture plot (WPP) at the Waite Agricultural Research Institute (20). Although the precise mechanism of suppression is uncertain, the demonstration that suppressive factors are sensitive to heat (10,11,16) and some forms of nitrogen (2), and that suppression can be reestablished by addition of 1% antagonistic soil to fumigated or steamed soil (16,19) or partly reestablished by the recolonization of fumigated soil by airborne contaminants (1), suggests that the suppressive factors are microbial in nature. This has led to the formulation of several hypotheses (12,15), but none unequivocally explains suppression.

Chakraborty and Old (5) and Chakraborty et al (6) have isolated several mycophagous amoebas from the suppressive WPP soil and demonstrated their mycophagy on *G. graminis* var. *tritici*. Chakraborty and Warcup (7) found reduced saprophytic survival of, and a higher association of mycophagous amoebas with, buried mycelium of *G. graminis* var. *tritici* in this soil compared with a nonsuppressive one. Suppressive soils also have higher initial populations of mycophagous and other amoebas (8), and they show consistently higher rhizosphere populations of mycophagous amoebas than nonsuppressive soils (4).

Several organisms including bacteria and fungi have been shown to suppress the disease in field and pot trials (9,14,15). We report the suppression of take-all in pot bioassays using mycophagous amoebas.

MATERIALS AND METHODS

Soil cultures and suspensions of four mycophagous amoebas, *Gephyramoeba* sp., *Saccamoeba* sp., *Thecamoeba granifera* subsp. *minor*, and an unidentified leptomyxid in amoeba saline (13), were used in pot bioassays (16). Soil cultures were prepared by introducing these amoebas singly into autoclaved (121°C for 1 hr on 3 consecutive days) nonsuppressive wheat-field soil. Autoclaved coarse sand was used as the base medium, to which ground oat-grain inoculum of *G. graminis* var. *tritici* was added at the rate of 0.2% and the WPP soil/soil culture at the rate of 1%. In one treatment, a mixture of soil cultures of all four amoebas was added. Stock cultures of all amoebas, except *T. granifera* subsp. *minor*, are maintained in monaxenic cultures with a *Klebsiella* sp.; *T. granifera* subsp. *minor* is maintained with organisms with which it was first isolated. As axenic cultures of these amoebas were not established, the food organisms (*Klebsiella* and the accompanying organisms of *T. granifera* subsp. *minor*) were used as controls.

A total of 350 g of each treatment mixture was added to each 300-cm³ pot, and five wheat seeds of the cultivar Halberg were placed on the surface. The seeds were covered with 50 g of autoclaved coarse sand, and 50 ml of distilled water was added (15% moisture, oven-dry weight). Five replicate pots per treatment were randomized in a growth room (12 hr/day, 15 ± 1°C with a light intensity of 10,760 lux at the height of the plants). Every second day, pots were watered to the original weight. Plants were harvested 4 weeks after seeding, rated for take-all (19), heights were measured, and dry weight of tops determined.

RESULTS AND DISCUSSION

Soil cultures of *Gephyramoeba*, *Saccamoeba*, and *T. granifera* subsp. *minor* significantly ($P<0.01$) reduced disease severity and increased plant height and top dry weight (Table 1). The reduction of disease and the increase in top dry weight of plants where a mixture of amoebas was used was comparable to the suppressive effect of the WPP soil. Soil cultures of the food organisms also significantly reduced disease rating and increased height and top dry weight of plants. The sterilized soil by itself did not have any suppressive effect.

In a further experiment, the effect of a large initial population of the mycophagous amoebas on the severity

Table 1. Disease rating, height, and top dry weight of plants grown in sterile coarse sand with soil cultures of amoebas and inoculum of *Gaeumannomyces graminis* var. *tritici* [a]

Treatment[b]	Disease rating	Plant height (cm)	Top dry weight (mg)
L*Ggt* + WPP soil	1.9	15.9	25.8
L*Ggt* + *Gephyramoeba* sp.	2.2	14.7	24.6
L*Ggt* + *T. granifera* subsp. *minor*	2.0	15.3	23.0
L*Ggt* + *Saccamoeba* sp.	2.8	13.9	22.3
L*Ggt* + Leptomyxid amoeba	3.8	13.3	19.4
L*Ggt* + Mixture of all four amoebas	1.9	15.6	28.7
L*Ggt* + Food organisms	2.2	15.9	23.6
L*Ggt* + Sterile soil	4.2	12.7	16.6
L*Ggt*	4.2	12.7	16.3
D*Ggt* + Sterile soil	0.0	18.9	29.4
D*Ggt*	0.0	19.5	32.2
LSD: $P = 0.01$	0.71	2.37	5.37
$P = 0.05$	0.50	1.65	4.82

[a] Inoculum added at rate of 0.2%; data represent mean of 25 plants per treatment.
[b] L*Ggt* = live and D*Ggt* = dead *G. graminis* var. *tritici*; food organisms = accompanying organisms of *Thecamoeba granifera* subsp. *minor* + *Klebsiella* sp.

Table 2. Disease rating, height, and top dry weight of plants grown in sterile coarse sand with two population densities of mycophagous amoebas and inoculum of *Gaeumannomyces graminis* var. *tritici* [a]

Treatment[b]	Initial population (No./g of coarse sand)	Plant height (cm)	Disease rating	Top dry weight (mg)
L*Ggt* + WPP soil		14.7	2.0	26.8
L*Ggt* + *Gephyramoeba* sp.	2	17.0	2.4	29.2
	20	17.2	1.8	28.8
L*Ggt* + *T. granifera* subsp. *minor*	40	17.1	2.3	24.8
	400	18.1	1.6	26.8
L*Ggt* + *Saccamoeba* sp.	40	16.8	2.6	29.1
	400	17.1	2.4	27.1
L*Ggt* + Mixture of all three amoebas	27	18.2	1.8	27.9
	270	19.8	1.5	33.7
L*Ggt* + Amoeba saline				
5 ml of saline		12.9	4.3	17.4
50 ml of saline		12.6	4.9	16.9
L*Ggt* + Food organisms				
5 ml of organisms		14.1	2.5	22.0
50 ml of organisms (suspension)		14.8	2.1	25.8
D*Ggt* + Amoeba saline				
5 ml of saline		17.4	0.0	29.4
50 ml of saline		18.1	0.0	30.0
LSD: *P* = 0.01		1.69	0.62	4.78
P = 0.05		1.20	0.52	3.39

[a] Inoculum added at rate of 0.2%; data represent mean of 25 plants per treatment.

[b] L*Ggt* = live and D*Ggt* = dead *G. graminis* var. *tritici*; food organisms = accompanying organisms of *Thecamoeba granifera* subsp. *minor* + *Klebsiella* sp.

of take-all was tested. Initial population levels of approximately 2 and 20 per gram of coarse sand for *Gephyramoeba* and 40 and 400 per gram of coarse sand for *Saccamoeba* and *T. granifera* subsp. *minor* were added to 400 g of autoclaved coarse sand containing 0.2% *G. graminis* var. *tritici* inocula. The mixed amoeba treatments consisted of approximately 27 and 270 of these amoebas per gram of coarse sand. The unidentified leptomyxid was not used in this experiment.

Significant (*P*<0.01) increases in height and top dry weight of plants were obtained where a higher initial population of all three amoebas were used together (Table 2). Top dry weight of plants was also significantly increased with an increase in the initial population of the food organisms.

Because both soil culture and saline suspension of the food organisms reduced the disease rating and increased height and top dry weight of plants, attempts were made to determine the composition of the food organisms. In a pot bioassay, using the *Klebsiella* and the accompanying organisms of *T. granifera* subsp. *minor* in separate treatments, it was found that only the accompanying organisms significantly suppressed the disease. Several bacteria, including *Bacillus* and fluorescent *Pseudomonas*, have been shown to suppress take-all in pot and field experiments (3,15,18). The culture of the accompanying organisms of *T. granifera* subsp. *minor* was examined for the presence of fluorescent pseudomonads using a selective medium (17). Fluorescent pseudomonads, as detected by the presence of a yellow-green fluorescence in and around the colonies, were present and may have contributed to the reduced severity of take-all. *Gephyramoeba* and *Saccamoeba* are maintained only with the *Klebsiella* sp., and this bacterium itself did not affect disease levels. The reduction in take-all severity by these amoebas is independent of the effects of the

accompanying organisms of *T. granifera* subsp. *minor*.

Mycophagous amoebas can thus suppress take-all under controlled environmental conditions, although they are unlikely to be the only organisms responsible for suppression. Several other organisms can also reduce take-all severity, at least in pot experiments; therefore, attempts to explain suppression on the basis of specific antagonists may well be futile (12). Several different mechanisms may be functioning simultaneously. This observation is supported by the ability of the accompanying organisms of *T. granifera* subsp. *minor* to suppress take-all and by the apparent lack of a high degree of specificity of suppressive soils (1,19).

LITERATURE CITED

1. Baker, K. F., and Cook, R. J. 1974 (original ed.). Biological Control of Plant Pathogens. Reprint ed., 1982, American Phytopathological Society, St. Paul, MN. 433 pp.
2. Brown, M. E., Hornby, D., and Pearson, V. 1973. Microbial populations and nitrogen in soil growing consecutive cereal crops infected with take-all. J. Soil Sci. 24:296-310.
3. Campbell, R., and Faull, J. L. 1979. Biological control of *Gaeumannomyces graminis*: Field trials and the ultrastructure of the interaction between the fungus and a successful antagonistic bacterium. Pages 603-609 in: Soil-Borne Plant Pathogens. B. Schippers and W. Gams, eds. Academic Press, London. 686 pp.
4. Chakraborty, S. 1984. Population dynamics of amoebae in soils suppressive and non-suppressive to wheat take-all. Soil Biol. Biochem. In press.
5. Chakraborty, S., and Old, K. M. 1982. Mycophagous soil amoeba: Interactions with three plant pathogenic fungi. Soil Biol. Biochem. 14:247-255.
6. Chakraborty, S., Old, K. M., and Warcup, J. H. 1983. Amoebae from a take-all suppressive soil which feed on *Gaeumannomyces graminis tritici* and other soil fungi. Soil Biol. Biochem. 15:17-24.
7. Chakraborty, S., and Warcup, J. H. 1983. Soil amoebae and saprophytic survival of *Gaeumannomyces graminis tritici* in a suppressive pasture soil. Soil Biol. Biochem.

15:181-185.

8. Chakraborty, S., and Warcup, J. H. 1984. Populations of mycophagous and other amoebae in take-all suppressive and non-suppressive soils. Soil Biol. Biochem. 16:197-199.

9. Cook, R. J. 1981. Biological control of plant pathogens: Overview. Pages 23-44 in: Biological Control in Crop Production. G. C. Papavizas, ed. Allanheld, Osmum & Co., London. 461 pp.

10. Cook, R. J., and Rovira, A. D. 1976. The role of bacteria in the biological control of *Gaeumannomyces graminis* by suppressive soils. Soil Biol. Biochem. 8:269-273.

11. Gerlagh, M. 1968. Introduction of *Ophiobolus graminis* into new polders and its decline. Neth. J. Plant Pathol. 74 (Suppl. 2): 1-97.

12. Hornby, D. 1979. Take-all decline: A theorist's paradise. Pages 133-156 in: Soil-Borne Plant Pathogens. B. Schippers and W. Gams, eds. Academic Press, London. 686 pp.

13. Page, F. C. 1967. Taxonomic criteria for limax amoebae, with descriptions of 3 new species of *Hartmannella* and 3 of *Vahlkampfia*. J. Protozool. 14:499-521.

14. Papavizas, G. C., and Lewis, J. A. 1981. Introduction and augmentation of microbial antagonists for the control of soil borne plant pathogens. Pages 305-322 in: Biological Control in Crop Production. G. C. Papavizas, ed. Allanheld, Osmum & Co., London. 461 pp.

15. Rovira, A. D., and Wildermuth, G. B. 1981. The nature and mechanisms of suppression. Pages 385-415 in: Biology and Control of Take-all. M. J. C. Asher and P. J. Shipton, eds. Academic Press, London. 538 pp.

16. Shipton, P. J., Cook, R. J., and Sitton, J. W. 1973. Occurrence and transfer of a biological factor in soil that suppresses take-all of wheat in eastern Washington. Phytopathology 63:511-517.

17. Simon, A., Rovira, A. D., and Sands, D. C. 1973. An improved selective medium for isolating fluorescent pseudomonads. J. Appl. Bacteriol. 36:141-145.

18. Weller, D. M., and Cook, R. J. 1983. Suppression of take-all of wheat by seed treatments with fluorescent pseudomonads. Phytopathology 73:463-469.

19. Wildermuth, G. B. 1977. Studies on suppressive soils in relation to the growth of *Gaeumannomyces graminis* var. *tritici* and other root pathogens of wheat. Ph.D. thesis, University of Adelaide.

20. Wildermuth, G. B. 1980. Suppression of take-all by some Australian soils. Aust. J. Agric. Res. 31:251-258.

Trichoderma as a Biocontrol Agent Against Soilborne Root Pathogens

I. CHET and Y. HENIS, Department of Plant Pathology and Microbiology, Hebrew University of Jerusalem, Faculty of Agriculture, Rehovot 76100, Israel

Biological control of soilborne plant pathogens has recently received considerable attention throughout the world, drawing momentum from the growing public concern regarding the widespread use of hazardous chemicals in pest control (29).

Many Trichoderma isolates are antagonistic to a range of fungi, producing volatile and nonvolatile antibiotics shown to be active against them (7,8). The ability to produce such substances varies between isolates of the same species-aggr. as well as between isolates of different species-aggr. (7). Works dealing with biological control of soilborne plant pathogenic fungi often report that Trichoderma spp. are among the most promising biocontrol agents (11,28,31). Chet and Baker (2) found an isolate of T. hamatum in a Colombian soil naturally suppressive to R. solani. The antagonist was an effective biocontrol agent when applied to infested soil in the greenhouse. Similarly, our group has found a very promising isolate of T. harzianum Rifai capable of controlling Rhizoctonia solani Kuehn and Sclerotium rolfsii Sacc. both in the greenhouse and in the field (15,16,19,20).

To a large extent, the problems involved in the successful application of a biocontrol agent under field conditions stem from lack of knowledge of the conditions leading to the establishment of the agent in the soil and of its ecology, as affected by variability of field soils and climatic conditions. We shall describe here some of our recent findings regarding the application of Trichoderma as a successful biocontrol agent under field conditions, as well as its possible mode of action.

APPLICATION OF
TRICHODERMA PREPARATION

A wheat bran culture of T. harzianum was tested for its capacity to control R. solani in carnation in fields that had been treated with methyl bromide. A linear correlation was obtained between rate of T. harzianum preparation applied to soil and degree of disease control. Disease incidence was reduced by 70% when the T. harzianum preparation was applied broadcast at 150 g (dry weight) per square meter. When the Trichoderma preparation was applied to the soil, direct contact between the roots and the antagonist was random and scarce. We introduced the biological control agent into the root zone by rooting carnation plants as Speedlings (seedlings in trays each in its own compartment) in peat supplemented with T. harzianum preparation (15% by volume). When transferred to soil infested with R.

solani, the treated plants had the lowest infection rate (13%), which was significantly lower than the infection rate in fields receiving T. harzianum as a broadcast. Combining the rooting mixture and the broadcast methods did not result in a further reduction of disease incidence (17). Similarly, the Trichoderma preparation proved efficient when applied to tomato Speedlings against R. solani in naturally infested field plots (18).

Seed or surface coating with protective fungi enables a significant reduction in the dosage of the antagonist's preparation (21). Under greenhouse conditions, the incidence of damping-off disease in cotton caused by R. solani at low temperatures ($20 \pm 2°C$) was 95% as compared with 78.3% at high temperatures ($27 \pm 2°C$). T. hamatum was more effective than T. harzianum as a biocontrol agent (45.4 and 25.0% disease reduction, respectively) at low temperatures; but at the higher temperatures, T. harzianum was superior (18). These findings call for selecting the "right" Trichoderma spp. according to growth conditions.

Planting cotton at depths of 3 and 1 cm resulted in 80 and 16% incidence, respectively, of damping-off caused by R. solani. Though seed coating with Trichoderma spp. reduced disease incidence at both depths, the effect was more pronounced in the shallow planting (18). Application of Trichoderma to seeds was done with Pelgel (methocellulose base) as an adhesive material. Coating with the adhesive material alone did not affect disease incidence and plant stand.

Under field conditions, treatment with T. hamatum reduced cotton damping-off caused by R. solani by 60%, bare patches 23 days after planting by 39%, and significantly increased density of plants (counted 17 days later) by 14%. Spraying with 2.8 kg/ha of pentachloronitrobenzene (PCNB; 75% w.p.) in seed furrows gave similar results, whereas seeds treated with ethylmercury chloride did not differ from the untreated control (18).

INTEGRATED CONTROL

Treatment with either PCNB or soil solarization combined with T. harzianum resulted in a significant improvement in disease control over each treatment alone (4,15,16,19,24).

Trichoderma spp. are among the pioneer colonizers following soil fumigation because they have few competitors and can reproduce rapidly (30). The purpose of the work described in this paper was to achieve an efficient control of soilborne pathogens by integrating fumigation and Trichoderma application. To this end, we studied the ecology of *T. harzianum*, especially its ability to colonize soil that had been fumigated with methyl bromide.

Application of the antagonist after soil fumigation with methyl bromide improved the control of *S. rolfsii* and *R. solani* in a peanut field (16). Although soil fumigation initially reduced the diseases caused by these pathogens, rapid reinfestation by *S. rolfsii* and *R. solani* reduced the effectiveness of the treatment. Application of a preparation of *T. harzianum* prevented reinfestation of the fumigated soil by the pathogens (88% reduction) under both controlled and field conditions. In soil treated with *T. harzianum*, survival of sclerotia was considerably less than in the untreated control. The combined treatment of fumigation and *T. harzianum* applications caused almost total death of sclerotia in soil under both laboratory and field conditions. Applications of *T. harzianum* to the root zone of tomatoes effectively controlled *S. rolfsii* in a field naturally infested with *S. rolfsii* and *R. solani*. Transplanting plants treated with *T. harzianum* into soil fumigated with methyl bromide reduced disease incidence by 93% and increased yield by 160% (16).

The results indicate that the efficiency of fumigation is increased by subsequent application of *Trichoderma* preparations and that such a combination of control methods may enable smaller doses of methyl bromide to be used and may achieve longer-lasting control of soilborne diseases in the field.

MODE OF ACTION OF THE ANTAGONISTIC FUNGUS

The antagonistic interactions that may occur between two fungi are categorized as amensalism, parasitism, and competition. Although most of the work done by our group deals with parasitism and lysis, this does not exclude the possible treatment of the other relationships between *Trichoderma* and its host. The antagonistic ability of *T. harzianum* aggr. against several soilborne plant pathogens was reported by Dennis and Webster (7–9.) Competition for nutrients between germinating sclerotia of *S. rolfsii* and *Trichoderma* provides a possible explanation for the inhibition of sclerotial germination by *Trichoderma* (26,27).

Hyphal interactions between *T. harzianum* and *S. rolfsii* or *R. solani* were observed by light and electron microscopy. Differential interference-contrast light microscopy revealed that hyphae of the mycoparasite reacted to *R. solani* by branching and growing toward it (5). This directed growth was apparently the result of a chemical stimulus released by the *Rhizoctonia*. When contact between the two fungi had been established, the antagonist either grew parallel to its host or coiled around it.

Dennis and Webster (9) studied the response of *Trichoderma* spp. to plastic threads having dimensions similar to those of fungal hyphae. They found that *Trichoderma* never coiled around the threads, which suggested that coiling around other hyphae was not merely due to contact stimulus. Recent evidence (10) points to the presence on *R. solani* hyphae of a lectin that binds to a carbohydrate on *Trichoderma* cell walls. The cell walls of *S. rolfsii* and *R. solani* contain β-1,3-glucan and chitin (1,6), whereas members of the

Oomycetes (e.g., *Pythium* spp.) also contain cellulose (1). *T. harzianum* was able to grow on *R. solani* cell wall as a sole carbon source (20), as well as to attack sclerotia (11). This led us to study the role of the lytic enzymes of the parasite in this host-parasite relationship.

Both scanning and transmission electron microscopy (12,13) revealed that *Trichoderma* spp. attached to the host either by hyphal coils, hooks, or appressoria. Lysed sites and penetration holes were found in the hyphae of the plant pathogenic fungi following removal of parasitic hyphae. High β-1,3-glucanase and chitinase activities were detected in dual agar cultures when *T. harzianum* parasitized *S. rolfsii* compared with the low levels found with either fungus alone. Interaction sites were stained by fluorescein isothiocyanate-conjugated lectins or Calcofluor White M2R New. Appearance of fluorescence indicated the presence of localized cell wall lysis at points of interaction (11) between the antagonist and its host, as well as autolysis of the antagonists's hyphae (22,23).

Elad et al (14) report that *T. harzianum* excreted β-1,3-glucanase and chitinase into the medium when grown on laminarin and chitin, respectively, or on cell walls of the pathogen *S. rolfsii*, as sole carbon source. *T. harzianum* also showed high activity of both enzymes when grown on homogenized *S. rolfsii* sclerotia. Glucanase activity increased by 67% when the fungus was grown on a mixture of laminarin and glucose (3:1 v/v). Similarly, high lytic activity was detected in a wheat bran culture of *S. rolfsii* and in soil when inoculated with *T. harzianum*. Protease and lipase activities were also detected in the medium when the antagonist attacked mycelium of *S. rolfsii*. Protective capacity of *Trichoderma* spp. of bean seedlings against *S. rolfsii* correlated with inhibition of sclerotial germination on soil plates but not with antagonism on agar media. Moreover, sclerotia of different isolates of *S. rolfsii* differed in their susceptibility to the same isolate of *Trichoderma* spp. (25).

Isolates of *T. harzianum* differed in the levels of hydrolytic enzymes produced when mycelia of *S. rolfsii*, *R. solani*, and *Pythium aphanidermatum* were attacked in soil. The phenomenon was correlated with the ability of the *Trichoderma* isolates to control the respective soilborne pathogens.

CONCLUSIONS

The results of experiments conducted over several years have shown that certain isolates of *Trichoderma* spp. may serve as efficient biocontrol agents in naturally infested soils under both greenhouse and field conditions. A significant reduction in crop loss caused by pathogenic fungi could be obtained by maintaining a low inoculum potential for several months. The amount of *Trichoderma* preparation can be reduced by direct application of this material to the root zone of tomatoes and carnations (16,17). This method controlled *R. solani* and *S. rolfsii* as effectively as did broadcasting the biocontrol agent over the rows.

Seed coating may be especially effective for controlling preemergence and postemergence diseases. In such cases the mature plant may be immune, so protection is needed only for a short period at the seedling stage. Therefore, the method of *Trichoderma* application has to be chosen according to the characteristics of disease. Surface coating of seeds was also effective in reducing seedborne *Aspergillus niger* of peanut (*unpublished data*) and of bulbborne *R. solani* of irises (3), thus making biocontrol even more attractive.

Synergistic phenomena are apparently involved in integrated chemical and biological control using *Trichoderma* and fungicides. The pathogen as well as the soil microflora is weakened by the chemical and is therefore better controlled by *Trichoderma* (24,26,27). In addition, the antagonist is favored and even multiplies on dead cells of the pathogen. Combining *Trichoderma* with methyl bromide or soil solarization to prevent reinfestation of the treated soil may prolong the effect of both the chemical and the physical treatment.

The mechanism of antagonism between *Trichoderma* and its hosts follows a sequence of events that includes chemotactic response of *Trichoderma* toward its host, binding to it, and enzymatic degradation. This mechanism is probably only one of many possible interactions between fungi. Antibiotics may also play a role in antagonism (7,8), whereas in natural soil competition is a factor that has to be considered as well.

ACKNOWLEDGMENTS

We would like to thank Dr. Hadar and Dr. Elad for their excellent work and advice.

LITERATURE CITED

1. Bartnicki-Garcia, S. 1973. Fungal cell wall composition. Pages 201-204 in: Handbook of Microbiology, vol. 2. Chemical Rubber Co., Cleveland, Ohio.
2. Chet, I., and Baker, R. 1981. Isolation and biocontrol potential of *Trichoderma hamatum* from soil naturally suppressive of *Rhizoctonia solani*. Phytopathology 71:286-290.
3. Chet, I., Elad, Y., Kalfon, A., Hadar, Y., and Katan, J. 1983. Integrated control of soil-borne and bulb-borne pathogens in iris. Phytoparasitica 10:229-236.
4. Chet, I., Hadar, Y., Elad, Y., Katan, J., and Henis, Y. 1979. Biological control of soilborne plant pathogens by *Trichoderma harzianum*. Pages 585-591 in: Soil-borne Plant Pathogens. B. Schippers and W. Gams, eds. Academic Press, London. 686 pp.
5. Chet, I., Harman, G. E., and Baker, R. 1981. *Trichoderma hamatum*; its hyphal interactions with *Rhizoctonia solani* and *Pythium* spp. Microb. Ecol. 7:29-38.
6. Chet, I., Henis, Y., and Mitchell, R. 1967. Chemical composition of hyphal and sclerotial cells of *Sclerotium rolfsii* Sacc. Can. J. Microbiol. 13:137-141.
7. Dennis, C., and Webster, J. 1971. Antagonistic properties of species-groups of *Trichoderma*. I. Production of nonvolatile antibiotics. Trans. Br. Mycol. Soc. 57:25-39.
8. Dennis, C., and Webster, J. 1971. Antagonistic properties of species-groups of *Trichoderma*. II. Production of volatile antibiotics. Trans. Br. Mycol. Soc. 57:41-48.
9. Dennis, C., and Webster, J. 1971. Antagonistic properties of species-groups of *Trichoderma*. III. Hyphal interaction. Trans. Br. Mycol. Soc. 57:363-369.
10. Elad, Y., Barak, R., and Chet, I. 1983. The possible role of lectins in mycoparasitism. J. Bacteriol. 154:1431-1435.
11. Elad, Y., Barak, R., and Chet, I. 1984. Parasitism of sclerotia of *Sclerotium rolfsii* by *Trichoderma harzianum*. Soil Biol. Biochem. 16:381-386.
12. Elad, Y., Barak, R., Chet, I., and Henis, Y. 1983. Ultrastructural studies of the interaction between *Trichoderma* spp. and plant pathogenic fungi. Phytopathol. Z. 107:168-175.
13. Elad, Y., Chet, I., Boyle, P., and Henis, Y. 1983. Parasitism of *Trichoderma* spp. on *Rhizoctonia solani* and *Sclerotium rolfsii*—scanning electron microscopy and fluorescence microscopy. Phytopathology 73:85-88.
14. Elad, Y., Chet, I., and Henis, Y. 1982. Degradation of plant pathogenic fungi by *Trichoderma harzianum*. Can. J. Microbiol. 28:719-725.
15. Elad, Y., Chet, I., and Katan, J. 1980. *Trichoderma harzianum*: A biocontrol agent effective against *Sclerotium rolfsii* and *Rhizoctonia solani*. Phytopathology 70:119-121.
16. Elad, Y., Hadar, Y., Chet, I., and Henis, Y. 1982. Prevention with *Trichoderma harzianum* Rifai aggr., of reinfestation by *Sclerotium rolfsii* Sacc. and *Rhizoctonia solani* Kühn of soil fumigated with methyl bromide, and improvement of disease control in tomatoes and peanuts. Crop Prot. 1:199-211.
17. Elad, Y., Hadar, Y., Hadar, E., Chet, I., and Henis, Y. 1981. Biological control of *Rhizoctonia solani* by *Trichoderma harzianum* in carnation. Plant Dis. 65:675-677.
18. Elad, Y., Kalfon, A., and Chet, I. 1982. Control of *Rhizoctonia solani* in cotton by seed coating with *Trichoderma* spp. spores. Plant Soil 66:279-281.
19. Elad, Y., Katan, J., and Chet, I. 1980. Physical, biological, and chemical control integrated for soilborne diseases in potatoes. Phytopathology 70:418-422.
20. Hadar, Y., Chet, I., and Henis, Y. 1979. Biological control of *Rhizoctonia solani* damping-off with wheat bran culture of *Trichoderma harzianum*. Phytopathology 69:64-68.
21. Harman, G. E., Chet, I., and Baker, R. 1981. Factors affecting *Trichoderma hamatum* applied to seeds as a biocontrol agent. Phytopathology 71:569-572.
22. Henis, Y., Adams, P. B., Lewis, J. A., and Papavizas, G. C. 1983. Penetration of sclerotia of *Sclerotium rolfsii* by *Trichoderma* spp. Phytopathology 73:1043-1046.
23. Henis, Y., Adams, P. B., Papavizas, G. C., and Lewis, J. A. 1982. Penetration of sclerotia of *Sclerotium rolfsii* by *Trichoderma* spp. (Abstr.) Phytopathology 72:707.
24. Henis, Y., Ghaffar, A., and Baker, R. 1978. Integrated control of *Rhizoctonia solani* damping-off of radish: Effect of successive planting, PCNB, and *Trichoderma harzianum* on pathogen and disease. Phytopathology 68:900-907.
25. Henis, Y., Lewis, J. A., and Papavizas, G. C. 1984. Interactions between *Sclerotium rolfsii* and *Trichoderma* spp.: Relationship between antagonism and disease control. Soil Biol. Biochem. 16:391-395.
26. Henis, Y., and Papavizas, G. C. 1982. Factors affecting susceptibility of *Sclerotium rolfsii* sclerotia to *Trichoderma harzianum* in natural soil. (Abstr.) Phytopathology 72:1010.
27. Henis, Y., and Papavizas, G. C. 1983. Factors affecting germinability and susceptibility to attack of sclerotia of *Sclerotium rolfsii* by *Trichoderma harzianum* in field soil. Phytopathology 73:1469-1474.
28. Liu, S., and Baker, R. 1980. Mechanism of biological control in soil suppressive to *Rhizoctonia solani*. Phytopathology 70:404-412.
29. Mulder, D. 1979. Soil Disinfestation. Elsevier Scientific Publication Co., Amsterdam. 368 pp.
30. Munnecke, D. E., Kolbezen, M. J., Wilbur, W. D., and Ohr, H. F. 1981. Interactions involved in controlling *Armillaria mellea*. Plant Dis. 65:384-389.
31. Wells, H. D., Bell, D. K., and Jaworski, C. A. 1972. Efficacy of *Trichoderma harzianum* as a biocontrol for *Sclerotium rolfsii*. Phytopathology 62:442-447.

Chemical Factors in Soils Suppressive to *Pythium ultimum*

F. N. MARTIN and J. G. HANCOCK, Department of Plant Pathology, University of California, Berkeley, CA 94720, U.S.A.

Inoculum densities of *Pythium ultimum* Trow. fluctuate on a seasonal basis in agricultural soils of the San Joaquin valley of California (3). In fine-textured soils, increases in inoculum density are often the result of saprophytic activity of the fungus and are associated with the incorporation of crop residues (e.g., leaves) at harvest time. As a primary colonizing sugar fungus (2), *P. ultimum* is a poor saprophyte on previously colonized organic matter. Factors that can repress rapid colonization therefore prevent extensive saprophytic activity by *P. ultimum* and reduce subsequent increases in densities of propagules in the soil. This phase of the pathogen life cycle presents a unique opportunity for disease control by managment of inoculum density through reduced saprophytic activity. With less inoculum pressure, disease losses are directly reduced and alternative control measures may have greater success.

Hancock (3,4) has previously reported on the occurrence of several fine-textured soils cropped in a cotton-barley rotation in which *P. ultimum* is present but does not increase in inoculum density upon the addition of cotton leaf debris at harvest time. The soils that are suppressive to inoculum buildup are in marked contrast to adjacent soils that allow *P. ultimum* to increase to high densities. Laboratory investigations with field soil amended with cotton leaf debris gave results similar to those observed in the field.

In the present study, we examine the chemical constituents of these soils and the effects they may have on the ecology of *P. ultimum*, its interrelationships with other soil microbiota, and inoculum production.

MATERIALS AND METHODS

Collection sites previously investigated by Hancock (4) were reevaluated during cotton harvest time for three successive years beginning in 1980. Twenty subsamples from the top 5.0 cm of the furrow shoulder (in 1981, the top 0–20.0 cm was sampled) were bulked, air-dried, mixed, and ground to pass through a 1-mm sieve. The *P. ultimum* ratings of the soils were determined by amending soil (75.0 g) with crushed cotton leaves (0.25 g, 0.83–0.99 mm in diameter) and incubating in polystyrene plastic cups at −0.3 to −1.0 bars for 7 days in a moist chamber at 21°C (3). Subsequent increases in population densities were determined by the soil drop assay method (7), with those soils having net increases greater than 200 propagules per gram of soil termed high *P. ultimum* soils (HPu), those having from 199 to 21 propagules per gram of soil termed moderate *P. ultimum* soils (MPu), and those having fewer than 20 propagules per gram of soil termed low *P. ultimum* soils (LPu) (4).

Saturation extracts and 1 N nitrogen ammonium acetate (pH 7.0) extracts of all soils were performed. Analysis for Ca^{+2}, Mg^{+2}, Na^+, and K^+ was done for both extracts using a Perkin-Elmer 372 atomic absorption apparatus. Analysis for Cl^- and SO_4^{-2} in the saturation extract was done with a Buchler chloridometer and by a barium chloride-gelatin turbimetric procedure (8), respectively. Soil pH was determined by placing electrodes in saturation paste that had equilibrated for 4 hr. Cation exchange capacity was determined by the method of Bowers and Gschwend (1).

Fungal colonization of leaf debris was investigated by amending soil (25.0 g) with crushed cotton leaves (0.08 g, 0.83–0.99 mm in diameter) and adjusting the matric potential with a −0.5 bar ceramic plate extractor in a pressure bomb. After at least a 12-hr equilibration time, the soil was placed in half of a 60-mm-diameter plastic petri dish and covered with polyethylene plastic held in place with a rubber band. These plates were then incubated in a moist chamber at 21°C. At appropriate time intervals, leaf debris was removed from three samples by wet sieving, and fungal colonization was observed by plating on 2.0% water agar amended with 0.1% Tergitol NPX to restrict fungal growth. With this method, there was a strong positive correlation between levels of colonization by *P. ultimum* and net increases in inoculum density ($r = 0.94$). In experiments where HPu soils were amended with Cl^-, the soil was saturated with different concentrations of Cl^- salt solutions prior to adjusting its matric potential.

RESULTS

The *P. ultimum* ratings for soils collected in 1980 and 1981, with few exceptions, agreed with the ratings reported previously (3,4), with the finer-textured Oxalis silty clay and Lethent silty clay loam predominantly LPu soils and the coarser-textured Panoche clay loam predominantly HPu soils. There was no correlation between suppressiveness to *P. ultimum* and concentrations of soluble and exchangeable Ca^{+2}, Mg^{+2}, K^+, and soil pH, which ranged from 7.12 to 7.85. The cation exchange capacity of LPu soils, however, tended to be higher than in HPu soils (ranging from 19.88 to 34.10 meq/100 g). More significant correlations, however, were found between suppressiveness and the concentrations of soluble Na^+, Cl^-, and SO_4^{-2}. With a few exceptions, these ions tended to be higher in LPu soils than in HPu soils, particularly Cl^- (Table 1). In an effort to determine whether Na^+, Cl^-, and SO_4^{-2} were inhibitory to *P. ultimum*, an HPu soil was amended individually with these ions, and the effects on increases in inoculum density were observed. We found that only Cl^- caused significant reductions in net increases of *P. ultimum* inoculum density in field soil (Table 2).

In investigating the colonization of leaf debris in soil, it was found that LPu soils had a significantly higher percentage of leaf colonization by primary colonizing sugar fungi other than *P. ultimum* than did HPu soils and that the frequency of co-colonization of leaf fragments by *P. ultimum* and the other primary colonizing fungi was very low (5). The saprophyte *P. oligandrum* Drechsler was a particularly common member of this mycoflora. Assays of HPu and LPu collection sites for population densities of *P. oligandrum* using a differential medium (Martin and Hancock, *unpublished*) found that LPu soils tended to have higher population densities of *P. oligandrum* than HPu soils (6). When HPu soils were infested with *P. oligandrum* at levels found in LPu soils, there were reductions in degrees of colonization and net increases in inoculum density by *P. ultimum* (Table 3).

By increasing the Cl$^-$ concentrations in an HPu soil, the patterns of leaf colonization by soil saprophytes was altered (Fig. 1). When an HPu soil was amended with Cl$^-$ to concentrations found in LPu soils, there was a 36.3% reduction in colonization; subsequent increases in inoculum density of *P. ultimum* were only 37.2% of the unamended control (Table 4). The reduction in colonization was manifested within the initial 48 hr after leaf amendment and coincided with increased colonization by *P. oligandrum*. In chloride-amended soils, colonization by *P. oligandrum* occurred sooner (40 hr) and to a greater degree (15.2%) than in unamended field soil (53 hr and 7.2%, respectively). The effects of Cl$^-$ on leaf colonization by *P. ultimum* in infested sterile soil were markedly different than in field soil (Fig. 2). Leaf

Table 1. Chloride concentrations in the saturation extract (given in meq of Cl$^-$ per liter) of field soils sampled after cotton harvest[a]

	Collection date	
Soil[b]	14 Oct 1980	25 Oct 1982
HPu		
12	2.5	18.8
13	6.0	16.9
31	···	12.4
33	7.4	9.1
Britz	10.4	···
MPu		
5	···	8.2
15	···	3.0
39	13.9	12.2
100	···	7.9
102	···	9.5
LPu		
2	34.6	21.4
6	···	16.4
7	···	7.5
9	···	9.7
18	···	20.9
19	···	30.9
21	25.7	···
22	55.6	···
24	46.6	28.5
36	15.2	···
37	21.1	···
40	···	8.3
41	18.5	···
Boston	33.3	16.0
X	···	19.4

[a]Soils collected 0–5.0 cm from furrow shoulder.
[b]HPu, MPu, and LPu refer to soils with high, moderate, and low increases in numbers of *Pythium ultimum* propagules.

Table 2. Effect of Na$^+$, Cl$^-$, and SO$_4^{-2}$ on net increases in propagule density of *Pythium ultimum*

Constituent[a]	Propagules/g of soil[b]	Percentage of control
Check	316.7	100.0
NaCl	158.3[c]	50.0
NaH$_2$PO$_4$	355.6	112.3
Check	416.7	100.0
Na$_2$SO$_4$	454.2	109.0

[a]Concentration of Cl$^-$ was 56.6 meq/L for NaCl-amended soil; Na$^+$, 39.0 meq/L for NaH$_2$PO$_4$-amended soil; and SO$_4^{-2}$, 36.5 meq/L for Na$_2$SO$_4$-amended soil.
[b]After 7 days incubation with leaf debris at −0.3 to −0.5 bars.
[c]Statistically significant at $P = 0.05$.

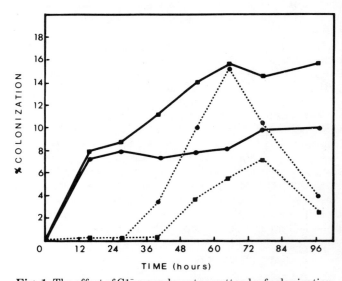

Fig. 1. The effect of Cl$^-$ amendment on cotton leaf colonization by *Pythium ultimum* (solid line) and *P. oligandrum* (dotted line) in a naturally infested, high *P. ultimum* field soil incubated at −0.5 bar and 21°C. The unamended soil (squares) contained 6.2 and the amended soil (circles) contained 44.2 meq of Cl$^-$ per liter, respectively.

Table 3. Effect of *Pythium oligandrum* on leaf colonization and subsequent increases in inoculum density by *P. ultimum*[a]

	Percentage of colonization		*P. ultimum* propagules per gram of soil[b]	Percentage of HPu field soil
Soil	*P. oligandrum*	*P. ultimum*		
12[c]	2.1	20.1	736.8	100.0
12 + *P. oligandrum*[d]	35.5	10.6	498.7	67.7
33[c]	1.4	6.7	163.0	100.0
33 + *P. oligandrum*[d]	9.4	2.7	55.5	34.0

[a]Results in each column for each soil are significant at $P = 0.05$.
[b]After 7 days incubation at −0.1 bars.
[c]Raw HPu field soil.
[d]Amended to 50 propagules per gram of soil with *P. oligandrum* oospores.

colonization by *P. ultimum* in infested sterile soil peaked and then leveled off after 48 hr, whereas colonization was initially low in sterile soil amended with Cl⁻ but gradually increased to nearly the levels found in unamended soils. As for the effects of Cl⁻ on *P. oligandrum* in sterile soil, there was neither a reduction nor a stimulation of colonization with the Cl⁻ concentration used (111.4 meq/L).

At higher Cl⁻ concentrations, *P. oligandrum* was more tolerant to Cl⁻ than *P. ultimum* (Fig. 3). Significant decreases in colonization by *P. oligandrum* did not occur until the Cl⁻ concentrations exceeded 132.7 meq/L, at which point there was more than a 50% reduction in the frequency of colonization by *P. ultimum*. The Cl⁻ concentration (219.8 meq of Cl⁻ per liter) at which colonization by *P. oligandrum* was reduced by 50% (LD_{50}) was nearly the concentration at which *P. ultimum* colonization was inhibited by 75% (214.6 meq of Cl⁻ per liter). The LD_{50} for *P. ultimum* was 127.5 meq of Cl⁻ per liter.

DISCUSSION

P. ultimum is most active as a saprophyte in cotton fields in California after harvest, with the availability of crop residues and the onset of seasonal rains (3). At that time, salinity is also the highest in the soil profile where the fungus is most active (the top 5.0 cm). This results from the cessation of irrigation several months before harvest and the accumulation of soluble salts at the soil surface, deposited there by capillary action and water evaporation. From field observations, it was found that LPu soils tended to have more precipitated salts on the soil surface than did HPu soils. Chemical analyses of these soils confirmed this; and of the various ions present in higher concentrations in LPu than HPu soils, only Cl⁻ reduced the saprophytic activity of *P. ultimum*.

The reduction in saprophytic activity of *P. ultimum* in HPu field soils with increasing Cl⁻ concentrations is manifested in the early phases of leaf colonization. From colonization patterns in field soil, we find that the presence of Cl⁻ reduces the degree of colonization in the first 48 hr and prevents significant increases thereafter. However, in *P. ultimum*-infested sterile soil amended with Cl⁻, colonization patterns are quite different. By excluding competitive microorganisms, colonization by *P. ultimum* continues to increase after 48 hr and approaches the levels found in unamended soils. Therefore, we conclude that the reduction in colonization in field soil is caused by the presence of other competitors and that the delay in colonization by *P. ultimum* caused by Cl⁻ enables these competitors to colonize available substrates. As there is a low frequency of co-colonization of leaf debris by *P. ultimum* and other primary colonizing fungi, the saprophytic activities of *P. ultimum* are effectively suppressed.

One primary colonizing fungus found in LPu field soils that is a strong competitor with *P. ultimum* is the saprophyte *P. oligandrum*. When this fungus is added to HPu soils, there are significant reductions in the frequency of leaf colonization and subsequent increases in inoculum density of *P. ultimum*. In a survey of the collection sites, *P. oligandrum* was found in greater inoculum densities and had a greater saprophytic activity in LPu soils than in HPu soils. One factor that may contribute to its increased saprophytic activity in soils high in Cl⁻ is that it is more tolerant of Cl⁻ than other primary colonizers (i.e., *P. ultimum*), placing it at a

Table 4. Net increases in propagule density of *Pythium ultimum* in an HPu soil amended with NaCl

Cl⁻ (meq/L)	*P. ultimum* propagules per gram of soil[a]	Percentage of control
6.2[b]	508.3	100.0
44.2	188.9	37.2

[a] After 7 days incubation at 21° C and −0.5 bar. Results are significant at $P = 0.05$.
[b] Raw field soil check.

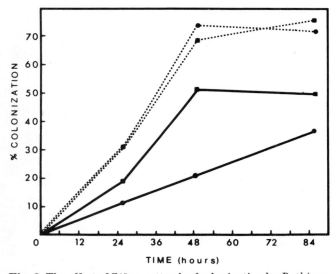

Fig. 2. The effect of Cl⁻ on cotton leaf colonization by *Pythium ultimum* (solid line) or *P. oligandrum* (dotted line) in a sterile, artificially infested, high *P. ultimum* soil incubated at −0.1 bar and 21°C. The unamended soil (squares) contained 3.46–3.71 meq of Cl⁻ per liter, whereas the amended soil (circles) infested with *P. ultimum* or *P. oligandrum* contained 120.6 or 111.4 meq of Cl⁻ per liter, respectively.

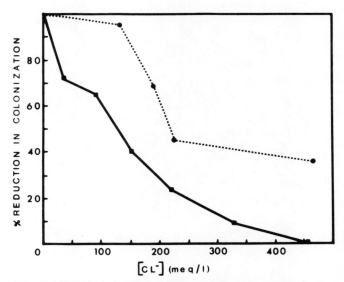

Fig. 3. The effect of different Cl⁻ concentrations on the degree of cotton leaf colonization by *Pythium ultimum* (solid line) and *P. oligandrum* (dotted line) in field soils incubated at −0.1 bar and 21°C.

competitive advantage in soils with high Cl⁻ concentrations.

We conclude that periodic increases in salinity in certain areas of the San Joaquin valley are involved in the suppression of increases in inoculum density of *P. ultimum*. The enhanced soil Cl⁻ concentration directly inhibits the saprophytic activities of *P. ultimum* and alters the ecology of the soil on a long-term basis by allowing the buildup of saprophytic competitors of *P. ultimum*. Because inoculum production in fine-textured agricultural soils by *P. ultimum* depends strongly on its saprophytic activities, any factors that reduce its activities render these soils suppressive to this pathogen.

LITERATURE CITED

1. Bowers, C. A., and Gschwend, F. B. 1952. Exchangeable cation analysis of saline and alkaline soils. Soil Sci. 73:251-261.
2. Garrett, S. D. 1970. Pathogenic Root Infecting Fungi. Cambridge University Press, London.
3. Hancock, J. G. 1977. Factors affecting soil populations of *Pythium ultimum* in the San Joaquin Valley of California. Hilgardia 45:107-122.
4. Hancock, J. G. 1979. Occurrence of soils suppressive to *Pythium ultimum*. Pages 183-189 in: Soil-Borne Plant Pathogens. B. Schippers and W. Gams, eds. Academic Press, London. 686 pp.
5. Martin, F. N., and Hancock, J. G. 1983. Effects of Cl⁻ on the colonization of crop residues by *Pythium ultimum*. (Abstr.) Phytopathology 73:813.
6. Martin, F. N., and Hancock, J. G. 1983. Factors affecting the colonization of crop residues by *Pythium ultimum*. (Abstr.) Phytopathology 73:960.
7. Stanghellini, M. E., and Hancock, J. G. 1970. A quantitative method for the isolation of *Pythium ultimum* from soil. Phytopathology 60:551-552.
8. Tabatabai, M. A. 1974. Determination of sulphate in water samples. Sulphur Inst. J. 10:11-13.

Influence of *Trichoderma* on Survival of *Thanatephorus cucumeris* in Association with Rice in the Tropics

T. W. MEW and A. M. ROSALES, Department of Plant Pathology, The International Rice Research Institute, Los Baños, Laguna, Philippines

Rice sheath blight caused by an aerial form of *Rhizoctonia solani* (*Thanatephorus cucumeris*) is an important rice disease in the tropics (8). Its occurrence is worldwide, ranging from temperate to tropical climates, and from irrigated rice to rain-fed and dryland rice. Sclerotia are the primary source of inoculum (1,5–7). Because of rain splash, at harvest about 20–30% of the sclerotia formed may drop to the ground while the other 70% remained intact with rice straw (1). Viability and number of sclerotia therefore bear a direct relationship to sheath blight incidence. Most of the information, however, is obtained from studies of irrigated rice.

When buried in soil, sclerotia lost their viability (9) because of the surrounding soil fungistatics, soil moisture, and other microbial effects. But their dispersion and thus distribution in a rice field is affected by land preparation, rice cultivation methods, and turbulent wind. Land preparation and rice cultivation methods are associated with rice culture type (2). The influence of factors affecting the viability and infectivity of the sclerotia in relation to the rice culture types is not clear. More sheath blight is often observed in dryland rice than in irrigated rice.

A project has been undertaken to investigate the saprophytic survival of *Thanatephorus cucumeris* and sheath blight development in soil of irrigated and dryland rice fields. In the present study, *Trichoderma harzianum* was isolated from rice straw buried in soil from a dryland rice field. Subsequently, its occurrence in irrigated, rainfed, and dryland rice fields was surveyed,

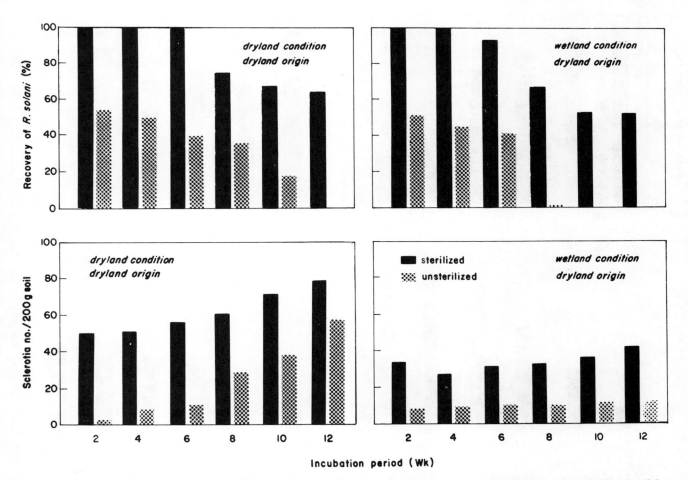

Fig. 1. Survival and sclerotia production of *Rhizoctonia solani* in infected rice straw pieces in sterilized and unsterilized soil from dryland rice field maintained as either dryland or wetland condition.

and its effect on the survival of the pathogen was also studied.

MATERIALS AND METHODS

Recovery of the sheath blight pathogen was determined in sterilized and unsterilized soil samples from irrigated and dryland rice fields. Naturally infected IR-36 rice straw, cut into pieces (2.5 cm long) to include the sheath blight lesion, was washed and buried in both soils. The pathogen was recovered at 2-week intervals for 12 weeks by plating on 2% water agar. Sclerotia production in the infected straw pieces was assessed from 200-g soil samples at 2-week intervals.

Because *T. harzianum* was detected in soil from a dryland rice field, a survey was carried out to determine its occurrence in different rice soils at the International Rice Research Institute experimental farm. Dilution plating from composite soil samples on gallic acid agar medium (3) was done. The effect of *T. harzianum* on decomposition of rice straw infected with sheath blight and on survival of *R. solani* was also determined. Healthy and infected rice straws were infested with a 1-ml spore suspension (1×10^6 spores per milliliter) of *T. harzianum*. Straw weight loss was determined at 2-week intervals for 16 weeks. Subsequently, the effect of the presence of *T. harzianum* on the survival of the *R. solani* in the system was also studied. The cellulolysis adequacy index was used to compare the ability of *Trichoderma* and *R. solani* to decompose rice straw. It was calculated following the method of Garrett (4) by determining the percentage loss in dry weight of inoculated rice straw and filter paper for 7 weeks and the linear growth rate of the fungus (expressed in millimeters per 24 hr) and then by dividing dry-weight loss by mycelial growth rate. Both filter paper and rice straw were used as substrates.

RESULTS AND DISCUSSION

The saprophytic survival of *Thanatephorus cucumeris* in rice straw was affected in unsterilized soil. Recovery of the fungus in straw declined with time. The difference between soils from dryland and irrigated rice fields indicated that survival in both soils was better in dryland than in flooded conditions when soil was unsterilized. In sterilized conditions, although percentage of recovery of the pathogen was lower in irrigated soil than in dryland soil, the pathogen was recovered at the end of the experiment, (i.e., 12 weeks after incubation) (Figs. 1 and 2). The survival of the pathogen in flooded conditions appears related not only to anaerobic conditions but also to other microbial effects.

From 2 to 4 weeks after incubation, sclerotia production in dryland condition was higher in sterilized dryland soil than in sterilized irrigated soil; at 6 weeks

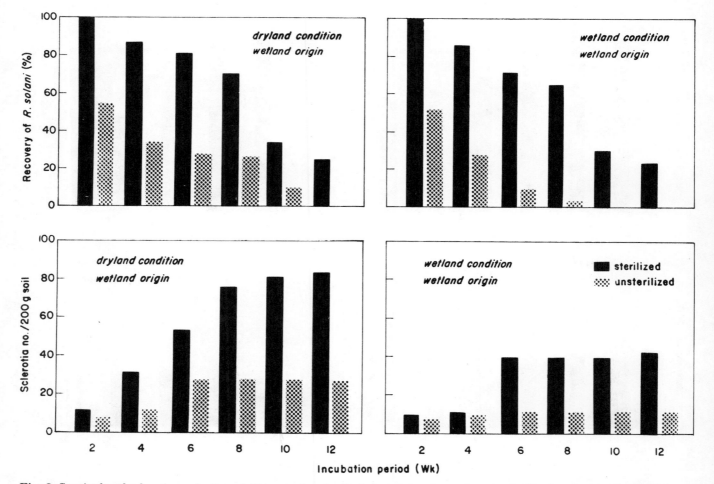

Fig. 2. Survival and sclerotia production of *Rhizoctonia solani* in infected rice straw pieces in sterilized and unsterilized soil from wetland rice field maintained as either dryland or wetland condition.

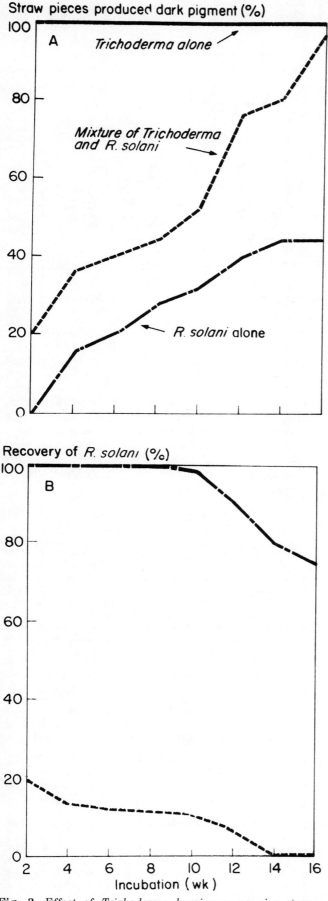

Fig. 3. Effect of *Trichoderma harzianum* on rice straw decomposition **(A)** and on saprophytic survival **(B)** of the sheath blight pathogen.

and thereafter, there was no difference in sclerotia production between the two soils (Figs. 1 and 2). The trend in flooded condition was similar for unsterilized soil, but the number of sclerotia was lower than in sterilized soil. The number of sclerotia was lowest in flooded unsterilized soil. *T. harzianum* colonized the straw pieces buried in dryland rice soil but not those in irrigated soil. The sclerotia produced in such conditions were also colonized by *Trichoderma*.

Rice straw infested with *T. harzianum* alone became dark in color 2 weeks after incubation, indicating decomposition (Fig. 3). Only 40% of the straw pieces inoculated with *Thanatephorus cucumeris* became dark-colored after 16 weeks of incubation. The microbial decomposition of rice straw greatly affected the survival of the sheath blight pathogen (Fig. 3). In the presence of *T. harzianum*, recovery of the sheath blight pathogen was reduced to 20% at 2 weeks after incubation. Rice straw weight losses due to colonization by either or both of the organisms were not significantly different (Fig. 4). *T. harzianum* appeared to be highly competitive; its cellulolysis adequacy index was higher than that of *R. solani* whether the substrate used was filter paper or rice straw (Table 1). The hyperparasitism of *T. harzianum* on *Thanatephorus cucumeris* was confirmed through a light microscope. On agar plate, the mycelial growth of

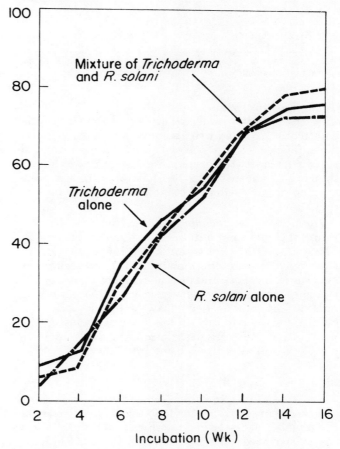

Fig. 4. The weight loss (wet) of rice straw colonized by *Trichoderma harzianum* and *Rhizoctonia solani* at different durations compared with that of rice straw without *T. harzianum* and *R. solani*.

119

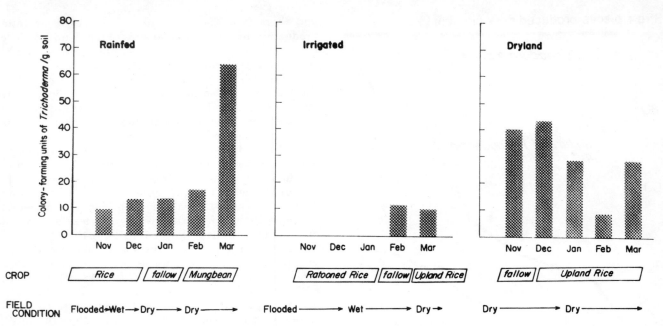

Fig. 5. Occurrence of *Trichoderma harzianum* in soils of rain-fed, irrigated, and dryland rice fields.

Table 1. Cellulolysis adequacy index of *Trichoderma harzianum* and rice sheath blight pathogen

Organism	Cellulolysis adequacy index[a] using	
	Rice straw	Filter paper
T. harzianum	3.40	3.01
Rhizoctonia solani		
From dryland rice	0.64	0.54
From irrigated rice	0.64	0.54

[a]Cellulolysis adequacy index = (percentage loss in dry weight)/(linear growth rate of fungus).

Thanatephorus cucumeris was also inhibited.

In the survey conducted, no *T. harzianum* was detected in irrigated rice soil, nor in rain-fed rice soil when flooded. Occurrence of *T. harzianum* in rain-fed soil when planted to a dryland crop after rice was higher than in soil planted only to rice, but it was highest in the dryland rice field—the field that had never been flooded (Fig. 5).

Thanatephorus cucumeris has a fairly wide host range; it attacks not only rice but other crops as well. The ability of *T. harzianum* to decompose rice straw, thereby affecting the survival of *Thanatephorus cucumeris*, has potential for use in biological control in rice-based cropping systems. Because *T. harzianum* is able to decompose rice straw and has a higher cellulolysis adequacy index than *R. solani*, the possibility of using it to promote dryland crop establishment after rice is being investigated.

CONCLUSION

Rice straw infected with *Thanatephorus cucumeris* was buried in soil from a dryland rice field. Recovery of the fungus was lower in unsterilized than in sterilized soil. In sterilized soil at a water potential of 0.83 bar, the fungus could be recovered from only 50% of the rice straw pieces, and all of the straw was colonized by a *Trichoderma* sp. 4 weeks after incubation. Production of *Thanatephorus cucumeris* sclerotia was affected by the presence of *Trichoderma*, whose ability to decompose rice straw greatly affected the survival of the sheath blight pathogen. When *Trichoderma* and *Thanatephorus cucumeris* were mixed together with rice straw, recovery of the latter was reduced to 20% at 2 weeks after incubation. The cellulolysis adequacy index of the *Trichoderma* on rice straw was higher than that of *Thanatephorus cucumeris*.

LITERATURE CITED

1. Chien, C. C., Jong, S. C., and Chu, C. L. 1963. Studies on the number of sclerotia of rice sheath blight fungus dropped on the paddy field and difference of germinability between natural and cultured ones. (English summary) Inst. Agric. Res. Taiwan 12:7-13.
2. De Datta, S. K. 1981. Principles and Practices of Rice Production. John Wiley, New York. 618 pp.
3. Flowers, R. A., and Littrell, R. H. 1972. Oospore germination of *Pythium aphanidermatum* as affected by casein, gallic acid, and pH levels in a selective agar medium. Phytopathology 62:757.
4. Garrett, S. D. 1966. Cellulose decomposing ability of some cereal foot-rot fungi in relation to their saprophytic survival. Trans. Br. Mycol. Soc. 49:57-68.
5. Hashiba, T., and Mogi, S. 1973. The number and germination ability of *Pellicularia sasakii* (Shirai) S. Ito. in noncultivated paddy soils. (English summary) Plant Prot. Assoc. Hokuriku 21:6-8.
6. Kozaka, T. 1975. Sheath blight in rice plants and its control. Rev. Plant Prot. Res. 8:69-80.
7. Lee, F. N. 1980. Number, viability, and buoyancy of *Rhizoctonia solani* sclerotia in Arkansas rice fields. Plant Dis. 64:298-300.
8. Ou, S. H. 1972. Rice Diseases. Commonw. Mycol. Inst. Kew, Surrey, England. 378 pp.
9. Tsai, W. H., and Yu, C. M. 1977. Epidemiology of rice sheath blight and its effect on rice production. Pages 247-262 in: Diseases and Insect Pests of Rice: Ecology and Epidemiology. R. J. Chiu, ed. Joint Commission on Rural Reconstruction, Taipei. 331 pp.

Biological Control of Fusarium Wilt of Sweet Potato with Cross-Protection by Nonpathogenic *Fusarium oxysporum*

K. OGAWA, Ibaraki Agricultural Experiment Station, Mito, 311-42, and H. KOMADA, Agricultural Research Center, Ministry of Agriculture, Forestry and Fisheries, Yatabe, Tsukuba, 305, Japan

Cross-protection is considered a method of biological control by which a biocontrol agent induces resistance in a host rather than through the mechanism of direct antagonism to the pathogen. Twenty or more examples of cross-protection against soilborne pathogens have been published. However, only a few have been applicable to commercial production, presumably because in most cases protection was induced by an avirulent strain of a pathogen or by an organism pathogenic to other species of crops.

This study was an attempt to use the cross-protection phenomenon to control Fusarium wilt of sweet potato by previous inoculation with *Fusarium oxysporum* isolates obtained from healthy sweet potato plants. The isolates were nonpathogenic not only to sweet potato but also to several other species of major vegetable crops. Preliminary reports of these results have been published (3,4).

MATERIALS AND METHODS

Isolation and inoculation of nonpathogenic *F. oxysporum*. *F. oxysporum* isolates were obtained from the vascular bundles of stems of healthy sweet potato sprouts or from tubers, using Komada's selective medium (2). The pathogenicity of each isolate to sweet potato and several other major vegetable crops was determined by growing the isolates in shake cultures and using bud cells to infest potting soils. Cut sprouts of sweet potato or seeds of other species of crops were planted in the infested soil.

Each nonpathogenic *F. oxysporum* isolate was tested for cross-protection ability against Fusarium wilt by dipping freshly cut ends of sweet potato sprouts into a diluted suspension of bud cells of each isolate produced in 7-day-old shake cultures or by smearing freshly cut ends with a bud-cell suspension paste.

Planting of sweet potatoes. Sweet potato plants were always established from cuttings. Cut ends of the sprouts were inserted diagonally into the test soils. In one case, however, the sprouts were bent and the middle portion of the stems was buried in the soil with the cut ends protruding from the soil.

The volcanic ash soils of the experimental and commercial fields were naturally infested with the pathogen. In the greenhouse experiments, the potting soils were often infested by drenching with bud-cell suspensions of the pathogen or by mixing with a small amount of the pathogen in a culture of soil and wheat bran.

RESULTS

Many *F. oxysporum* isolates were obtained from healthy sweet potato plants. Most were not pathogenic to all the crops tested, such as sweet potato, tomato, cucumber, bottle gourd, melon, radish, and cabbage. Furthermore, some of the nonpathogenic isolates showed cross-protection against Fusarium wilt of sweet potato when sweet potato plants were inoculated with them prior to planting.

Greenhouse experiments. When sweet potato sprouts were preinoculated with a nonpathogenic isolate of *F. oxysporum* and planted in naturally infested soil, or in natural or autoclaved soil artificially infested with the pathogen, a high degree of cross-protection was observed against the wilt (Fig. 1). However, drenching soil with a bud-cell suspension of the fungus was effective only in naturally infested soil in which the inoculum density of the pathogen was assumed to be lowest among the three infestation methods.

Field experiments. In a naturally infested field, cross-protection by preinoculation with nonpathogenic isolates of *F. oxysporum* brought about a marked decrease in wilt incidence and a remarkable increase in yield of sweet potato. The effects were equivalent to those obtained by chemical treatment in which the sprouts were dipped into a benomyl suspension.

Smearing the cut ends of sprouts with a concentrated bud-cell suspension of the protective fungus or dipping the sprouts into a diluted bud-cell suspension also brought about a marked decrease in wilt incidence and a remarkable increase in yield. More yield and less wilt were evident when sprouts were planted as soon as possible after preinoculation. The preinoculation was effective not only against disease caused by the soilborne pathogen but also that caused by tuber-borne inoculum.

The denser the bud-cell suspension and the longer the duration of dipping the sprouts into the bud-cell suspension, the greater was the cross-protective effect (Fig. 2).

Mechanism of cross-protection. Among 12 pathogenic and nonpathogenic fusaria belonging to seven species, only the nonpathogenic isolates of *F. oxysporum* showed cross-protection. No antagonism was observed between the cross-protective isolates and the pathogen in plate cultures. When the cut ends of sprouts were dipped into liquid paraffin, no protection was observed against the disease. This result shows that the cross-protection was not caused by a mechanical plugging of wounds with the bud cells.

Only living bud cells of the isolates caused cross-protection against the disease. Heat-killed bud cells did not cause any cross-protection. Bud-cell germination exudate caused slight cross-protection, but the effect of culture filtrate was indefinite.

Cross-protection was satisfactorily obtained even when the preinoculated sprouts were bent and the bent portion planted. Cross-protection was also observed when the preinoculated sprouts were planted in water and then inoculated by injecting a bud-cell suspension of the pathogen into the stems about 10 cm from the cut ends. If the cut ends were excised 2 days after preinoculation, cross-protection was shown; but no cross-protection was observed if the cut ends were excised immediately after preinoculation. The cross-protective isolates were not pathogenic to sweet potato, as mentioned above; however, colonization of the cut

ends was followed by a remarkable development of local lesions there (Fig. 3). These results suggest that systemic cross-protection was induced mainly by fungal colonization of the wounded cut ends.

Reaction of sweet potato plants against phytotoxic substances produced by the pathogen, resulting in wilting and yellowing of stems and leaves, was nullified by preinoculation with cross-protective isolates.

DISCUSSION

Cross-protection by preinoculation with *F. oxysporum* isolates, which are nonpathogenic to sweet potato as well as to several other species of vegetable crops, shows promise for biological control of Fusarium wilt of sweet potato commercially. It is effective not only against disease caused by soilborne inoculum but also against that transmitted from infested tubers. Further, the

Fig. 1. Cross-protection of sweet potato against Fusarium wilt by preinoculation with a nonpathogenic isolate of *Fusarium oxysporum* in a pot test using artificially infested autoclaved soil (**A**), artificially infested natural soil (**B**), and naturally infested soil (**C**). Left row, preinoculation by dipping cut ends of sprouts into bud-cell suspension of the isolate; center row, preinoculation by drenching soil with a bud-cell suspension of the isolate; right row, uninoculated control. The photograph in (A) is reprinted, by permission, from Ogawa and Komada (3).

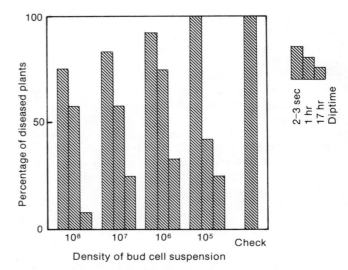

Fig. 2. Influence of dilution of bud-cell suspension of a nonpathogenic isolate of *F. oxysporum* and duration of dipping of sprouts on cross-protection of Fusarium wilt of sweet potato. Reprinted, by permission, from Ogawa and Komada (3).

Fig. 3. Local lesion at the cut end of sweet potato sprout caused by a nonpathogenic isolate of *Fusarium oxysporum* that induces cross-protection against Fusarium wilt of sweet potato.

method does not require the introduction of complicated operations because sweet potatoes are usually planted by cuttings in Japan. Therefore, it will be one of the few examples of biological control using cross-protection in commercial practice.

Research is continuing to clarify the mechanism of the cross-protection phenomenon. It is assumed that cross-protection results from systemic resistance induced by the nonpathogenic *F. oxysporum*, which colonizes and produces local severe infection at the cut ends of sprouts. It is also assumed that the plants react to infection by producing resistance products that are translocated systemically from the basal to the upper stem.

LITERATURE CITED

1. Baker, R., Hanchey, P., and Dottarar, S. D. 1978. Protection of carnation against Fusarium stem rot by fungi. Phytopathology 68:1495-1501.
2. Komada, H. 1975. Development of a selective medium for quantitative isolation of *Fusarium oxysporum* from natural soil. Rev. Plant Prot. Res. 8:114-125.
3. Ogawa, K., and Komada, H. 1984. Biological control of Fusarium wilt of sweet potato by non-pathogenic *Fusarium oxysporum*. Ann. Phytopathol. Soc. Jpn. 50:1-9.
4. Ogawa, K., and Komada, H. 1984. Systemic resistance in cross-protection induced by pre-inoculation with non-pathogenic *Fusarium oxysporum* against Fusarium wilt of sweet potato. Ann. Phytopathol. Soc. Jpn. In press.

Integrated Biological and Chemical Control of Sclerotial Pathogens

J. E. RAHE, Department of Biological Sciences, Simon Fraser University, Burnaby, B.C., V5A 1S6, and R. S. UTKHEDE, Agriculture Canada Research Station, Summerland, B.C., V0H 1Z0, Canada

Chemical control of plant diseases caused by sclerotial pathogens has not been satisfactorily achieved in most cases. The major difficulties arise from the longevity of sclerotial populations in soil (6), the ability of many sclerotial pathogens to attack their hosts over prolonged periods, the belowground infection court of many sclerotial pathogens, and the cost and limited effectiveness of attempts to eradicate sclerotia chemically from soil and crop debris. Optimal control of most plant diseases caused by sclerotial pathogens will probably be achieved only by integrating various approaches aimed at reducing sclerotial inoculum and protecting host plant tissues both chemically and through increased utilization of quantitative resistance. In this paper we are concerned with the potential for integration of chemical and biological approaches for control of diseases caused by sclerotial plant pathogens.

A few notable exceptions notwithstanding, examples of field-effective biological control of plant diseases are limited in number, and examples of applied biological control of diseases caused by sclerotial pathogens are few indeed (14). This situation leaves little to say but much to be learned about integrated biological and chemical control of sclerotial pathogens.

The term *integrated control* describes the use of more than one approach to suppress losses due to pests with greater overall effect than would result from the use of any one of the controls applied singly. The various approaches must interfere minimally with each other, and preferably they should enhance one another. Some of the more promising opportunities for biological control of sclerotial pathogens are considered here with regard to their potential or demonstrated interactions with agrichemicals.

REDUCTION OF INOCULUM VIA DECLINE OF SCLEROTIA

Decline of sclerotial populations occurs via mycoparasitism (3), germination followed by reproductive failure and metabolic exhaustion, which is partly caused by competition with other organisms. Decline is a natural phenomenon that can be enhanced or impeded in various ways (4). Where rates of decline are such that substantial reductions of sclerotial populations occur within a year or two, the phenomenon can be exploited via crop rotations involving nonhost crops. Although crop rotation is considered a cultural practice, the mechanism of inoculum reduction is biological.

Is mycoparasitism of sclerotial plant pathogens in the field a significant factor in their survival? The general lack of demonstrated examples would suggest not; Papavizas (14), however, suggests that this may result from lack of concentrated effort to demonstrate the phenomenon rather than to its absence in nature.

Coniothyrium minitans and *Sporidesmium sclerotivorum* have, to date, shown the most promise for reduction of sclerotial pathogens via mycoparasitism. Both fungi parasitized sclerotia of some but not all species of *Sclerotinia*, *Sclerotium*, and *Botrytis*, and their mycoparasitic ability has been demonstrated in the field for some of these sclerotial species. Both mycoparasites have effected significant and substantial disease reductions in field-grown crops (3). *Sclerotium cepivorum* may be the Methuselah of sclerotial plant pathogens. Little decrease in numbers of viable sclerotia has been observed after burial for 15 years in field plots at Hull, United Kingdom. These plots have recently been determined to be free from *Coniothyrium*, *Sporidesmium*, and *Teratosperma* (a mycoparasite similar to *Sporidesmium*) (5).

Logic suggests that the concentrated nutrients in a sclerotial-sized unit should be an attractive resource in a microbially competitive environment such as soil. Research on *C. minitans* and *Sporidesmium sclerotivorum* notwithstanding, there has been a general lack of attempts to demonstrate sclerotial mycoparasitism in the field. Unavoidably, there is the associated failure to evaluate the effect of chemical treatments and other crop production practices on mycoparasitic activity and survival.

Trichoderma spp. have shown considerable potential for field control of some diseases caused by sclerotial pathogens, but it has yet to be clearly shown that the observed control derives from mycoparasitism (3). *T. harzianum* has reportedly given excellent control of white rot of onions on a commercial scale in Egypt for the past 3 years (1). During this period, 230 ha has been furrow-treated with *T. harzianum* in a barley grain carrier at a rate of 425 kg/ha. Reductions in white rot incidence in treated fields are reported to be 82–87% more than those occurring in controls during the 3-year period. Comparable levels of control were afforded by *T. harzianum* and the fungicide iprodione during the 1981–1982 season, but no attempt to integrate chemical and biological treatments was reported.

Abd-El Moity et al (2) have shown the potential for integration of these approaches by the use of fungicide-tolerant isolates of *T. harzianum* in combination with fungicides. An iprodione-tolerant isolate, applied alone or in combination with iprodione in the furrow, gave better control of white rot on onion transplants in the field than did iprodione alone; and the combination of iprodione with the iprodione-tolerant isolate of *T. harzianum* gave the best control.

Papavizas et al (13) produced a number of apparent mutants of *T. harzianum* resistant to benomyl and other fungicides. Some of these survived in soil with and without amendment with benomyl better than did the wild type parent. Others produced more fungitoxic metabolites effective against radial growth of *S. cepivorum* than did the wild type parent, and one of these reduced the incidence of white rot in transplanted onions grown in sclerotium-infested soil by approximately 78% over both the control and the wild type parent (which was ineffective). Combinations of fungicide and fungicide-resistant antagonists were not tested, but this work illustrates a potential area for integrating chemical and biological control of a sclerotial pathogen.

Stimulation of sclerotial germination in the absence of host plants can reduce sclerotial inoculum, but the practicality of this approach is likely limited to those sclerotial pathogens with limited host range and capacity for repeat germination. Soil applications of onion oil, diallyl disulfide, and related synthetic preparations have resulted in significant reductions of sclerotial populations of *S. cepivorum* and incidence of white rot under field conditions involving a range of soil types (9,12,16). Although this approach appears to have commercial potential, field results have been highly variable.

Only one attempt to combine chemical control with the use of germination stimulants for *S. cepivorum* has been reported (11). All diallyl disulfide plot treatments, with or without subsequent treatment of onions grown on these plots with iprodione, reduced sclerotial populations significantly (approximately 40%) over the control. Incidence of white rot was generally unaffected by these reductions, however: the tests site was heavily infested, and the reduced sclerotial populations surviving the various diallyl disulfide treatments (50–60/kg of soil) were still sufficient to cause severe levels of disease. The site may have been better suited for evaluation of a mycoparasite.

It is important to know the relationship between inoculum density and disease. The net effect of a fungicide is to reduce the amount of new inoculum formed. This reduction tends to keep sclerotial populations low, and biological controls that act by reducing inoculum may be hindered or favored by this effect. Mycoparasites may require a host level above that occurring with the use of effective fungicides in order to become established. In such instances, there would be little potential for effective integration of the two approaches. At low inoculum densities, however, a fungicide may initially provide economic disease reductions but in time allow sclerotial populations to increase to levels where economic control is no longer achieved. Therefore, the integration of a biological control that gives a percentage reduction of inoculum independent of inoculum density (artificial germination stimulants presumably act in this manner) might afford long-term economic control.

The most obvious approach to enhanced decline is the attempt to kill sclerotia directly via fumigation or solarization. Recent successes with solarization suggest that this technique may become a major component of integrated management systems for control of some soilborne fungi; nematodes; weeds; and possibly some bacteria, mollusk, and arthropod pests as well (10). The effective use of solarization is likely to be restricted to agricultural regions of comparatively low latitude, although it has been used successfully as far north as Idaho in the United States (7). Soil fumigation also has a place in pest management for high-value, specialty crop production; but its effectiveness for sclerotial pathogens has been variable and generally not cost-effective.

The common characteristic of solarization and fumigation is that both techniques create at least partial biological vacuums in time and space, and this aspect offers great potential for the integrated use of biological agents. The general failure of attempts at classical biological control for soilborne plant pathogens—the introduction of a foreign organism into an ecosystem with the objective that it will become established there—can be attributed to two factors. First, the introduced organism is often not biologically adapted to the physical environment of the ecosystem in question; second, the ecosystems into which we have attempted to introduce foreign organisms have typically not had any biological vacancies or apartments for rent or sale. The creation of vacant apartments caused by solarization or fumigation affords an enhanced opportunity for an introduced foreign organism to become at least a temporary resident. It still remains for us, as landlords, to select from among the antagonists available ones that will, in a teleological sense, like the physical environment in which we are asking them to settle.

Apparent examples of the practical exploitation of the partial biological vacuums created by soil solarization and fumigation have been reported by Elad et al (8). They observed that *T. harzianum* suppressed *Sclerotium rolfsii* and *Rhizoctonia solani* more effectively when introduced immediately after soil solarization or fumigation with methyl bromide than when introduced alone. The integrated use of *Trichoderma* and solarization gave a longer carryover suppression of *Rhizoctonia* than did solarization alone.

PROTECTION OF HOST TISSUES

Biological agents and chemicals can act as direct protectants of plants. The integration of these approaches requires minimal interference of one with the other. Pesticides with a broad spectrum of activity can potentially interfere with most biological agents used to protect plants. Such interference can be minimized by the selective use of pesticides having a minimum impact on the biological agent in question (exploitation of chemical specificity), selective timing and mode of application of pesticides and biologicals (exploitation of discontinuities in time or space), and the use of strains of biological agents with tolerance to the pesticides selected for integrated use (exploitation of chemical resistance). The use of fungicide-tolerant strains of *T. harzianum* appears to be the only one of these approaches to have received attention thus far with regard to integrated control of sclerotial pathogens.

Experience with the use of *Bacillus subtilis* for biological control of onion white rot illustrates both some potential for integrated chemical and biological approaches and, more important, the many gaps in our knowledge of the interactions involved. Some isolates of *B. subtilis* antagonistic to *S. cepivorum* in dual culture gave significant and substantial, season-long control of onion white rot in field trials conducted in 1978 (15) and 1979 on muck soil and in 1979 on mineral soil when applied as seed treatments (17). These isolates were generally ineffective in muck soil in field trials in 1980 and 1981, but they were effective in greenhouse pot trials using field muck soil in 1980 (17). These results suggest that environmental conditions were probably responsible for their failure to control white rot in the field in 1980 and 1981 (the 1980 and 1981 growing seasons were markedly cooler and wetter than were those of 1979 and 1980).

Combination of bacterial seed treatment with preplant soil treatment with vinclozolin gave either the same or increased levels of control over that afforded by vinclozolin alone in a total of one pot trial and three field trials. The combination of bacterial seed treatment with iprodione gave approximately the same levels of control as did iprodione alone in the three field trials, but in the pot trial the combination treatments were generally less effective than either iprodione or bacterial treatments applied singly (17).

B. subtilis (isolate B2) tolerates the dicarboximide fungicides vinclozolin and iprodione to concentrations of at least 100 ppm in potato-dextrose agar (J. E. Rahe, *unpublished data*). In contrast, thiram is totally inhibitory at concentrations above 1 ppm. This latter finding points to the possibility of interference by thiram seed treatments commonly used on onions with indigenous soil populations of *B. subtilis* and its possible effect on white rot in the field.

B. subtilis appears to meet at least one requirement for integrated use with dicarboximide fungicides for control of white rot. A possibility not yet examined, however, is that *B. subtilis* might degrade one or both of these fungicides and thus reduce their effectiveness. This possibility, if real, could explain the reduced effectiveness of the iprodione + bacteria treatment in the pot trial relative to the effectiveness of iprodione alone.

SUMMARY

It is easy to be discouraged about the potential for integrated chemical and biological control of diseases caused by sclerotial pathogens. Success to date with chemical control has, in general, been less than spectacular, and success with biological control has been even less so. Nevertheless, recent results with *C. minitans*, *Sporidesmium sclerotivorum*, and *Trichoderma* spp. give some cause for optimism. The use of artificial germination stimulants for *S. cepivorum* also appears to be a promising specialized approach. Systematic research to develop practical integrated approaches cannot proceed until field-effective biological controls with known mechanisms of action are developed. We lack knowledge of the effects of agrichemicals on biological balance (4). Antagonism is the biological sine qua non of healthy soil. To what extent do the various agrichemicals used in modern agriculture interfere with this phenomenon? Research on biological agents showing realistic potential for field efficacy should take into account the effects of agrichemicals in general as well as chemicals used specifically for disease control. Biological agents destined to become components of modern agricultural production will unavoidably be exposed to a variety of agrichemicals.

Whether field-effective integration of biological and chemical controls for diseases caused by sclerotial pathogens becomes a reality in the future remains to be seen. It hasn't happened yet.

LITERATURE CITED

1. Abd-El Moity, T. H. 1983. Biological control of white rot of onion in the field. Proc. Int. Workshop Allium White Rot, 2d. Soilborne Diseases Laboratory, ARS/USDA, Beltsville, MD.
2. Abd-El Moity, T. H., Papavizas, G. C., and Shatla, M. N. 1982. Induction of new isolates of *Trichoderma harzianum* tolerant to fungicides and their experimental use for control of white rot of onion. Phytopathology 72:396-400.
3. Ayers, W. A., and Adams, P. B. 1981. Mycoparasitism and its application to biological control of plant diseases. Pages 91-103 in: Biological Control in Crop Production. G. C. Papavizas, ed. Allanheld, Osmun & Co., Totowa, NJ.
4. Baker, K. F., and Cook, R. J. 1974 (original ed.). Biological Control of Plant Pathogens. Reprint ed., 1982. American Phytopathological Society, St. Paul, MN. 433 pp.
5. Coley-Smith, J. R. 1983. Integrated control of *Sclerotium cepivorum*. Proc. Int. Workshop Allium White Rot, 2d. Soilborne Diseases Laboratory, ARS/USDA, Beltsville, MD.
6. Coley-Smith, J. R., and Cooke, R. C. 1971. Survival and germination of fungal sclerotia. Annu. Rev. Phytopathol. 9:65-92.
7. Davis, J. R., and Sorensen, L. H. 1983. Carry-over effects of plastic tarping on Verticillium wilt of potato. Abstr. 627, Int. Congr. Plant Pathol., 4th, Melbourne.
8. Elad, Y., Katan, J., and Chet, I. 1980. Physical, biological, and chemical control integrated for soilborne diseases in potatoes. Phytopathology 70:418-422.
9. Entwistle, A. R., Merriman, P. R., Munasinghe, H. L., and Mitchell, P. 1982. Diallyl disulphide to reduce the numbers of sclerotia of *Sclerotium cepivorum* in soil. Soil Biol. Biochem. 14:229-232.
10. Katan, J. 1981. Solar heating (solarization) of soil for control of soilborne pests. Annu. Rev. Phytopathol. 19:211-236.
11. Merriman, P. R. 1983. Effects of diallyl disulphide and iprodione on sclerotia of *Sclerotium cepivorum* and incidence of white rot in dry bulb onions. Proc. Int. Workshop Allium White Rot, 2d. Soilborne Diseases Laboratory, ARS/USDA, Beltsville, MD.
12. Merriman, P. R., Isaacs, S., MacGregor, R. R., and Towers, G. B. 1980. Control of white rot in dry bulb onions with artificial onion oil. Ann. Appl. Biol. 96:163-168.
13. Papavizas, G. C., Lewis, J. A., and Abd-El Moity, T. H. 1982. Evaluation of new biotypes of *Trichoderma harzianum* for tolerance to benomyl and enhanced biocontrol capabilities. Phytopathology 72:126-132.
14. Papavizas, G. C., and Lumsden, R. D. 1980. Biological control of soilborne fungal propagules. Annu. Rev. Phytopathol. 18:389-413.
15. Utkhede, R. S., and Rahe, J. E. 1980. Biological control of onion white rot. Soil Biol. Biochem. 12:101-104.
16. Utkhede, R. S., and Rahe, J. E. 1982. Treatment of muck soil with onion oil to control onion white rot. Can. J. Plant Pathol. 4:79-80.
17. Utkhede, R. S., and Rahe, J. E. 1983. Chemical and biological control of onion white rot in muck and mineral soils. Plant Dis. 67:153-155.

Yield Depressions in Narrow Rotations Caused by Unknown Microbial Factors and Their Suppression by Selected Pseudomonads

B. SCHIPPERS and F. P. GEELS, Willie Commelin Scholten Phytopathological Laboratory, Baarn; O. HOEKSTRA, J. G. LAMERS, and C. A. A. MAENHOUT, Research Station for Arable Farming and Field Production of Vegetables, Lelystad; and K. SCHOLTE, Department of Field Crops and Grassland Husbandry, Agricultural University, Wageningen, The Netherlands

In different countries and under different economic conditions, crop rotation becomes more and more restricted to a few profitable crops. The reasons for crop intensification, which are primarily economic, have recently been reviewed (13).

Intensification of cropping in agriculture has become possible due to the breeding of new, high-yielding crop varieties with increased resistance to diseases, the development of new pesticides, the increased use of fertilizers, and the invention of various cultural practices that maintain optimal soil fertility.

There is increasing evidence that harmful elements of the root microflora, other than those causing major diseases, set limits to a profitable increase of cropping frequency. This is convincingly demonstrated in detailed, long-term rotational experiments in the Netherlands at the experimental farm "De Schreef" (The Limit) and at the Research Station for Arable Farming and Field Production of Vegetables (PAGV) at Lelystad.

These rotational experiments are located on a heavy sandy clay in one of the youngest polders (East Flevoland) reclaimed from the IJsselmeer in 1958. The rotational experiment at De Schreef was started in 1963 and at the PAGV in 1973. Cooperation in these experiments by crop scientists, soil physicists and chemists, plant pathologists, and economists has not only led to a better understanding of the relationship of actual yields

and costs (economic balance) for a number of crops (6); but it has also thrown light upon the physical, chemical, and biological limits of crop intensification.

NARROW ROTATIONS AND YIELD DEPRESSIONS

Different rotations (14 at De Schreef and 4 at PAGV) are compared with each crop present every year to exclude the annual influence of climate (3,7). Among other variables, four nitrogen dressing rates are applied in subplots that allow comparisons of yield depressions in the different crop frequencies at optimum nitrogen dressing rate. Plot dimensions allow the use of ordinary farm implements.

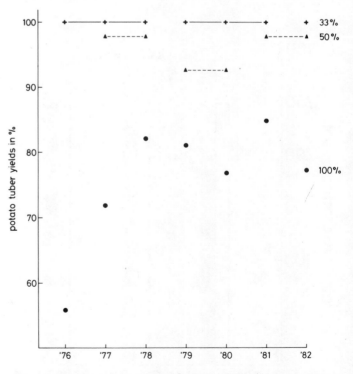

Fig. 2. Development of the relative yield of potato tubers (diameter > 35 mm) at optimal nitrogen dressing rate for 50% potatoes and continuous potato cropping (100%) compared with 33% potato (= 100) in the rotation at PAGV. Lines represent the average yield level over each rotation cycle. Thirty-three percent rotation: sugar beet, winter wheat, and potatoes; 50% rotation: sugar beet and potatoes.

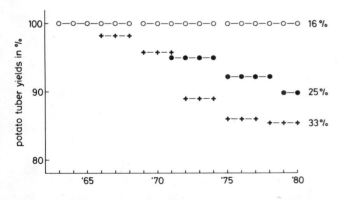

Fig. 1. Development of the relative yield of potato tubers (diameter > 35 mm) at optimal nitrogen dressing rate for 25 and 33% potatoes compared with 16% potatoes (= 100) in the rotation at De Schreef. Lines represent the average yield level over each rotation cycle. Sixteen percent rotation: spring barley, peas, winter wheat, flax, grass seed, and potatoes; 25% rotation: winter wheat, sugar beet, spring barley, and potatoes; 33% rotation: sugar beet, spring barley, and potatoes.

The most serious yield depressions were registered for potato varieties Bintje (Fig. 1) and Saturna (Fig. 2). The losses of yield in the narrow rotations at De Schreef became more severe over the years, but they seem to tend toward their maximum after ± 16 years (sixth cycle of the 33% potato rotation). They are clearly related to the cropping frequency in the rotation (3,7). After 18 years at De Schreef, yield depressions in potato fluctuate around 10% in the 25% potatoes and around 15% in the 33% potatoes in comparison with yields in a 16% frequency of potatoes in the rotation (Fig. 1). After 5 years at the PAGV, yield depressions in potato varied between 20 and 30% in continuous potato cropping compared with yields in a 33% potato cropping frequency. Yield depressions in a 50% potato frequency only account for a few percentage of those in a 33% potato frequency (Fig. 2). Yield depressions in relation to cropping frequency were also registered for winter wheat (Fig. 4) and winter rye.

Physical and chemical soil factors. Detailed physical analysis on pore volume, volume percentage of air (at pF2–0 between 7 and 12 cm), plasticity, workability, and visual value have shown that the soil physical environment changes from year to year but is inconsistent with differences in yield.

The availability of nitrogen, phosphorus, and potassium nutrients in the soil of all rotational plots was demonstrated to be more than sufficient for optimal yield. There was no decline in soil organic matter content (± 2.95) over the years. Physical or chemical soil factors therefore do not account for the observed yield depressions (3,7).

Soil biological factors. *Yield depressions in potatoes.* Yield depressions in high-frequency potato cropping soil were reproducible in pot experiments. Discoloration of the root system was similar to that observed in the field. Treatments of the soil with methyl bromide and pasteurization at 60°C for half an hour eliminated the rotational effects completely (Fig. 3), as well as the differences in uptake of nutrients (Scholte, *personal communication*). Obviously, the rotational effects must be due to the presence of microorganisms that do not form strong resting structures. Nematodes of *Globodera* spp. are not present at detectable population densities at De Schreef. At the PAGV, a potato variety (Saturna) that is resistant to *G. rostochiensis* (pathotype Ro1) was used, and the 50 and 100% potato plots were disinfected every second year with 330 L of metam sodium per hectare. The free-living nematodes *Pratylenchus neglectus* and *Paratylenchus* spp. were not present at detectable levels at PAGV. They were present at De Schreef but did not harm potatoes as far as is known. Nematodes can thus be excluded as the prime causal factor.

At De Schreef, sclerotial densities of *Rhizoctonia solani* on tubers were independent of cropping frequency. Netted (russet) scab (*Streptomyces* spp.) in tubers showed an increase with frequency of cropping, but this was not seen every year. It was never detected at significant levels at PAGV fields, probably because the potato variety Saturna is resistant to this disease. Pentachloronitrobenzene treatments drastically decreased infections by *Rhizoctonia* and *Streptomyces* at De Schreef but did not diminish yield depressions (3). Differences in yield depressions did not change when the potato variety Eigenheimer, insensitive to netted scab, was used. At the PAGV, stem decay caused by *R. solani*, however, was more serious in the continuous cropping but not serious enough to explain the

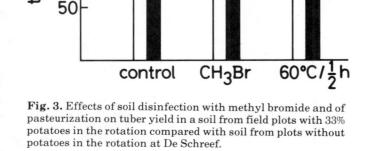

Fig. 3. Effects of soil disinfection with methyl bromide and of pasteurization on tuber yield in a soil from field plots with 33% potatoes in the rotation compared with soil from plots without potatoes in the rotation at De Schreef.

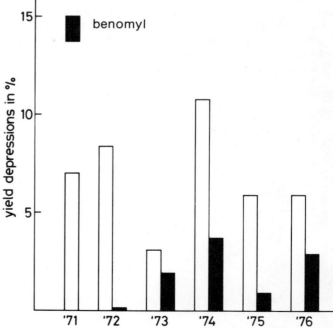

Fig. 4. Relative depression of grain yield of winter wheat at optimal nitrogen dressing rate for 50% cereals with and without benomyl treatment compared with 33% cereals (= 100) in the rotation at De Schreef.

considerable extra yield depressions compared with the 30% potato rotation.

Potato wilt caused by *Verticillium dahliae*, an important disease in warmer and dryer climates, has become of greater importance in temperate countries over the last few years. Wilting symptoms caused by *V. dahliae* have seldom been observed at De Schreef. At the PAGV, microsclerotia were produced in large quantities on diseased stems in both the 50 and 100% cropping frequency. Because microsclerotia of this fungus are known to survive for many years, and because of their production in large quantities in both the 50 and 100% cropping frequencies, *V. dahliae* alone does not seem to account for the enormous difference in yield depressions in monoculture.

Yield depressions in wheat. In winter wheat, yield depressions in rotations with 50% cereals fluctuated around 9% over the last 15 years at De Schreef (Fig. 4) when compared with rotations with 33% cereals. Benomyl applications in the field (GS 31-32, Zadoks scale) reduced eyespot disease (*Pseudocercosporella herpotrichoides*) to a low level, but this only reduced yield depressions by 50% (Fig. 4). At the PAGV, in the rotational experiments with continuous wheat in which eyespot was controlled with benomyl, the yield depressions were estimated to be approximately 10% of the yield in plots with 25% winter wheat.

The yield depressions could be reproduced in pot experiments with soil from the rotation field plots. It was demonstrated that the yield depressions in winter wheat were of microbial origin but not caused by foot rot fungi (Scholte, *unpublished data*).

In continuous wheat plots, the take-all fungus *Gaeumannomyces graminis* var. *tritici* seemed to be controlled by the specific antagonism responsible for take-all decline (4,12). No macroscopic symptoms of this disease have been seen. The pathogen and its lesions, however, can incidentally be found on the roots. This fungus as well as other microorganisms may partly account for the yield depressions.

Yield depressions in winter rye. Although rye has generally been considered to be self tolerant, seed and straw yield depressions up to 30 and 10% respectively in winter rye following winter rye, compared with rye following other crops, have been convincingly demonstrated by Scholte and Kupers (9) in an 18-year rotational experiment on a light sandy soil at optimal nitrogen levels. The yield depressions correlate with a lower uptake of nitrogen and an early degeneration of the root system. Differences in incidence of foot rots could be removed by fungicides, but differences in yield depressions remained. They could be reproduced in pot experiments in soil taken from the rotational field plots and demonstrated to be of microbial origin. A possible role of toxic metabolites from microbial degradation of straw residues could also be excluded (10).

Foot rot diseases in pot experiments were sporadically found despite a careful examination during plant growth. Application of the nematicide oxamyl reduced the numbers of the nematodes *Pratylenchus crenatus* and *P. fallax* but had no effect on the yield depressions nor on the early degeneration of roots. However, application of the fungicides benomyl and captafol reduced yield depressions drastically (Scholte, *personal communication*). For rye after rye (compared with rye after oats), depressions were 13.5% (control), 12.7%

(oxamyl treatment), 4.3% (benomyl + captafol, low dose), and 3.3% (benomyl + captafol, high dose). The fungicides also reduced the fast and early degeneration of the root system. The yield depression of winter rye in high cropping frequency on light sandy soils must therefore be ascribed to pathogenic soil fungi other than the regular pathogens.

YIELD DEPRESSIONS AND MINOR PATHOGENS

The detailed, long-term rotational experiments described here demonstrate, with several important and high-yielding crops, that unknown microbial factors other than regular pathogens can significantly decrease yield and thus set limits to increasing crop intensification.

These microbial factors, which seem to accumulate in narrow rotations, fit the description of minor pathogens given by Salt (8). According to him, parasitism by minor pathogens is confined to root tips, root hairs, and cortical cells, distinguishing them from "major pathogens" that usually penetrate the endodermis and disrupt vascular tissues. The majority of these microorganisms are unspecialized facultative parasites living on root surfaces of many plants or as soil saprophytes, including *Pythium* and *Fusarium* spp. Some affect plant growth by producing toxins but without entering the root, and they usually do not produce distinctive disease symptoms. With most of the yield depressions in potato narrow-rotation soil in field and pot experiments, no distinctive disease symptoms other than a discoloration or restricted development of the root system could be observed in the absence of regular pathogens. It is most likely, therefore, that such minor pathogens are, at least partly, responsible for the observed yield depressions in potato and that they increase with increasing crop frequency.

TREATMENTS WITH SELECTED PSEUDOMONADS

The assumption that certain minor pathogens are at least partly responsible for the observed yield depressions is supported by the elimination or reduction of the yield depressions by seed tuber treatments with siderophore-producing fluorescent pseudomonads. These fluorescent pseudomonads were isolated from the periderm of potatoes harvested from De Schreef and selected for antagonistic activity on an agar medium deficient in Fe^{3+} against a variety of soilborne pathogenic and saprophytic fungi and bacteria of potato (1). The yield depressions in potato could be reproduced in pot experiments using continuous potato cropping soil obtained from the PAGV experimental fields (2): the total plant dry weight was 69% compared with that of potatoes grown in field soil continuously cropped with wheat ($P = 0.05$). Seed tuber treatments with some of the *Pseudomonas* isolates eliminated the yield depressions in potato in pot experiments (Table 1). They also significantly increased yields in some of the narrow potato rotations at the experimental fields in some years (Table 2).

These experiences strongly agree with those of Schroth and Hancock (11), who obtained considerable increases in yield in a variety of tuber crops including

Table 1. Effect of seed tuber treatments with fluorescent pseudomonads on yields of potato in pots in soil continuously cropped with potato or wheat in the field[a]

	Total plant dry weight (%)	
Treatment	Potato soil	Wheat soil
Control	100	100
WCS[b]358	128*[c]	
WCS 365	131***	99
WCS 374	123**	

[a] Soil type = potato.
[b] WCS = Willie Commelin Scholten isolates.
[c] Single asterisk, $P = 0.05$; double, $P = 0.025$; triple, $P = 0.005$.

Table 2. Effect of seed tuber treatments with fluorescent pseudomonads on yields of potato in the field (1981)[a]

Frequency of potato cropping (%)	Yield (tons/ha)	Relative yield (%)			
		Control	WCS 358	WCS 365	WCS 374
67	60	100	104	91	98
33	50	100	111*[b]	104	103
33	51	100	109*	102	100

[a] Soil type = heavy sandy clay from De Schreef.
[b] Asterisk indicates significance at $P = 0.05$.

Table 3. Effect of seed treatments with fluorescent pseudomonads on tuber and total plant weight in soil continuously cropped to radish[a]

Seed treatment	Tuber weight (%)	Total plant fresh weight (%)
Control	100	100
WCS 374	142*[b]	140*
B 10[c]	135*	130*

[a] Pot experiment.
[b] Asterisk indicates significance at $P = 0.05$.
[c] Strain B 10 was provided by Dr. M. N. Schroth (Berkeley).

ACKNOWLEDGMENT

We are grateful to Dr. R. Harling for correcting the English translation of this paper.

potato, beet, and radish by treatments with fluorescent pseudomonads. These increases were demonstrated to be at least partly due to the elimination or suppression of harmful rhizobacteria that inhibit the development of roots (11,14).

The main mechanism of this antagonism by the fluorescent pseudomonads was shown to be iron deprivation by Fe^{3+}-complexing siderophores produced by the fluorescent pseudomonads (5). The increases in yield obtained by seed tuber treatments in narrow rotations could not be obtained in (wheat) soil from fields with no history of potato (Table 1) (2). This indicates that the harmful microorganisms accumulate with crop intensification, which is supported by our experience that considerable increases in radish yields by seed treatments with the selected fluorescent pseudomonads occurred espcially in soil with a history of continuous radish cropping but not in soil without such a history (Table 3) (Geels and Schippers, *unpublished*).

We can conclude that growth stimulation and yield increases following seed treatments are linked to cropping frequency.

CONCLUSIONS

In a variety of crops, considerable yield depressions occur with increasing cropping frequencies. The yield depressions seem to be, at least partly, due to the accumulation of largely unknown, harmful root microorganisms that do not cause distinct symptoms and that have largely been overlooked by plant pathologists. Control of these harmful root microorganisms might impressively increase yields in agriculture. The prospects for controlling these pathogens by seed and seed tuber treatments with certain fluorescent pseudomonads are promising. An interdisciplinary approach to research on this group of microorganisms is essential to a full exploration of their potential.

LITERATURE CITED

1. Geels, F. P., and Schippers, B. 1983. Selection of antagonistic fluorescent *Pseudomonas* spp. and their root colonization and persistence following treatment of seed potatoes. Phytopathol. Z. 108:193-206.
2. Geels, F. P., and Schippers, B. 1983. Reduction of yield depressions in high frequency potato cropping soil after seed tuber treatments with antagonistic fluorescent *Pseudomonas* spp. Phytopathol. Z. 108:207-214.
3. Hoekstra, O. 1981. 15 jaar "De Schreef." Resultaten van 15 jaar vruchtwisselingsonderzoek op het bouwplannenproefveld "De Schreef." Publikatie PAGV 11:1-93.
4. Hornby, D. 1979. Take-all decline: A theorist's paradise. Pages 133-156 in: Soil-borne Plant Pathogens. B. Schippers and W. Gams, eds. Academic Press, New York. 686 pp.
5. Kloepper, J. W., Leong, J., Teintze, M., and Schroth, M. N. 1980. Enhanced plant growth by siderophores produced by plant growth promoting rhizobacteria. Nature 286:885-886.
6. Kupers, L. J. P. 1979. Agronomic aspects of soil-borne pathogens. Pages 357-370 in: Soil-borne Plant Pathogens. B. Schippers and W. Gams, eds. Academic Press, New York. 686 pp.
7. Lamers, J. G. 1981. Continueteelt en nauwe rotaties van aardappelen en suikerbieten. Publikatie PAGV 12:1-65.
8. Salt, G. A. 1979. The increasing interest in "minor pathogens." Pages 289-312 in: Soil-borne Plant Pathogens. B. Schippers and W. Gams, eds. Academic Press, New York. 686 pp.
9. Scholte, K., and Kupers, L. J. P. 1977. The causes of the lack of self-tolerance of winter rye grown on light sandy soils. I. Influences of foot rots and nematodes. Neth. J. Agric. Sci. 25:255-262.
10. Scholte, K., and Kupers, L. J. P. 1978. The causes of the lack of self-tolerance of winter rye grown on light sandy soils. II. Influences of phytotoxins and soil microflora. Neth. J. Agric. Sci. 26:250-266.
11. Schroth, M. N., and Hancock, J. G. 1982. Disease-suppressive soil and root-colonizing bacteria. Science 216:1376-1381.
12. Shipton, P. J. 1975. Yield trends during take-all decline in spring barley and wheat grown continuously. Eur. Mediterr. Plant Prot. Organ. Bull. 5:363-374.
13. Shipton, P. J. 1979. Experimental evidence for the effect on soil-borne diseases of changes in techniques of crop and soil cultivation. Pages 385-397 in: Soil-borne Plant Pathogens. B. Schippers and W. Gams, eds. Academic Press, New York. 686 pp.
14. Suslov, T. V., and Schroth, M. N. 1982. Role of deleterious rhizobacteria as minor pathogens in reducing crop growth. Phytopathology 72:111-115.

Antagonistic Behavior of Root Region Microfungi of Pigeon Pea Against *Fusarium udum*

R. S. UPADHYAY and BHARAT RAI, Centre of Advanced Study in Botany, Banaras Hindu University, Varanasi 221 005, India

Fusarium udum Butler is a soilborne plant pathogen causing wilt disease of pigeon pea (*Cajanus cajan* (L.) Millsp.) in most tropical countries (3), resulting in major economic loss. No successful method of control has yet been evolved, despite increased attention by tropical plant pathologists. The discovery of the perfect state of the pathogen (7) and its possible role in inciting the disease has opened a new avenue for the study of the biology and control of the pathogen. The present study was undertaken to gather information about the antagonistic behavior of other microflora in the root region of pigeon pea with a view to possible biological control.

MATERIALS AND METHODS

The test fungi were *Aspergillus flavus* Link ex Fries, *A. niger* van Tiegh., *A. terreus* Thom, *Cunninghamella echinulata* Thaxter, *Penicillium citrinum* Thom, *Rhizopus nigricans* Ehrenb., *Trichoderma viride* Pers. ex Gray, *Neocosmospora vasinfecta* E. F. Smith, and *Curvularia lunata* (Wakker) Boedijn. These species were frequently isolated from root region of pigeon pea in a separate study (11). The antagonistic potential of these fungi has already been screened on agar (10,12).

To determine the population dynamics of *F. udum* in the root region of pigeon pea in relation to its inoculum amendment with different test microfungi, pure soil-sand inoculum of the test pathogen was prepared by the method described by Garrett (4). The *F. udum* inoculum was amended with soil cultures of the test fungi in the ratio of 20:80, 50:50, and 80:20. Three replicates of earthenware pots (8 cm in diameter and 10 cm high) were each filled with 1 kg of the appropriate soil-sand inoculum mixtures, and the soil moisture of each pot was maintained at 20–25%. Pigeon pea seeds were then sown in each of the pots, which were then placed in culture cabinets at $25 \pm 2°$ C. The control was set by sowing the seeds in the inoculum of *F. udum* diluted in the same ratio as mentioned above with sterilized soil-sand mixture (1:5 ratio) plus 3% maize meal. The population of *F. udum* was estimated in root region (rhizosphere and rhizoplane) of pigeon pea at monthly intervals on the selective medium of Nash and Snyder (6) by the method described by Upadhyay and Rai (11). The incidence of wilt disease was also recorded during the period of each sampling.

In another experiment, pigeon pea seeds were sown in the field. Twenty-day-old seedlings were removed gently from the field and brought into the laboratory. Their root systems were washed in tap water and surface disinfected with 0.5% NaOCl followed by thorough washing by several changes of sterilized distilled water. The root systems of the seedlings were dipped in spore suspension (no. of spores = 1×10^5 per milliliter) of *F. udum* for 5 hr, after which they were transferred into moist chambers separately for 10 hr to activate the spores on the roots. Five such seedlings were transplanted in pots containing pure sand inoculum of different test fungi and were kept under suitable laboratory conditions as described earlier. Control was set with roots dipped in spore suspension of *F. udum* but planted in sterilized soil-sand mixture (1:5 ratio) plus 3% maize meal. The experiment was also conducted with the pure inoculum of *F. udum* in pots but with the roots dipped in spore suspensions of different fungi. However, in this case the control was set by planting seedlings of pigeon pea in pure inoculum of *F. udum* without dipping the roots in the spore suspension of the test fungi. The population of *F. udum* in the rhizosphere and rhizoplane of pigeon pea was estimated on the selective medium by the method mentioned above after 15, 30, 45, and 60 days.

RESULTS

The population of *F. udum* in the soil and root region of pigeon pea was highly suppressed by *A. niger*, *A. flavus*, *A. terreus*, and *T. viride* (cf. control, Table 1). Suppression was highest when *F. udum* inoculum and the test microflora were mixed in the ratio of 20:80, followed by 50:50 and 80:20. These microorganisms exhibited significant inhibitory effect even at the 20:80 ratio of test microfungi to *F. udum*. The inhibitory effect increased with time in the case of *A. niger*. However, with *A. flavus*, *A. terreus*, *P. citrinum*, and *T. viride*, a slight decrease in the rate of reduction of the population was recorded after 2 and 3 months (Table 1). The population of *F. udum* in the root region was not reduced significantly by *Cunninghamella echinulata*, *R. nigricans*, *N. vasinfecta*, and *Curvularia lunata*. On the basis of this study, it may be concluded that the soil amended with *A. niger*, *A. terreus*, *P. citrinum*, and *T. viride* becomes suppressive against *F. udum*.

The incidence of wilt disease of pigeon pea was also recorded (Table 2). The results showed a similar trend as was recorded for population dynamics of the pathogen under the amendments. Soils infested with *F. udum* and then mixed with *A. niger* in different proportions significantly reduced wilt incidence, followed by soils amended with *A. flavus*, *A. terreus*, *P. citrinum*, and *T. viride*. In contrast, soil amended with *Cunninghamella echinulata*, *R. nigricans*, *N. vasinfecta*, and *Curvularia lunata* had little or no effect on disease development.

The population of *F. udum* was affected by the various test fungi when the roots of host seedlings were pretreated with spore suspension of the latter and then planted into soil infested with *F. udum* (Table 3). The colonization of the roots of pigeon pea by the pathogen was suppressed by *A. niger*, *A. flavus*, *A. terreus*, *P. citrinum*, and *T. viride*. However, *Cunninghamella echinulata*, *N. vasinfecta*, *R. nigricans*, and *Curvularia lunata* did not cause any significant effect.

The population of *F. udum* was suppressed on roots pretreated with spore suspension of *F. udum* and grown in mixtures of soil-sand inocula of various fungi

($P = 0.01$, Table 4). In this case, the suppression in the population of *F. udum* was high in comparison with the first experiment described above; but the trend in both experiments was similar.

DISCUSSION

A. niger, *A. flavus*, *A. terreus*, and *P. citrinum* were highly antagonistic against *F. udum* in vitro (10,12). When these fungi were introduced into soil or into root region of pigeon pea they strongly suppressed the population of *F. udum*, thus reducing the incidence of

Table 1. Population of *Fusarium udum* in the rhizosphere of pigeon pea planted in different ratios of *F. udum* inoculum mixed with other test microfungi

Inoculum mixed with	After 1 month			After 2 months			After 3 months		
	80:20[a]	50:50	20:80	80:20	50:50	20:80	80:20	50:50	20:80
Aspergillus flavus	0.21[b]	0.10	0.06	0.30	0.11	0.10	0.24	0.20	0.04
A. niger	0.10	0.08	0.01	0.11	0.10	0.11	0.07	0.04	0.02
A. terreus	0.13	0.11	0.10	0.12	0.11	0.08	0.16	0.16	0.06
Cunninghamella echinulata	2.64	3.00	3.50	3.80	3.50	3.60	3.00	3.20	3.20
Curvularia lunata	2.20	2.00	2.10	3.10	3.70	3.80	4.00	4.10	4.00
Neocosmospora vasinfecta	2.90	2.50	2.20	3.60	3.60	3.10	3.20	3.20	4.10
Penicillium citrinum	0.80	0.20	1.00	1.00	1.50	2.00	2.90	2.70	3.00
Rhizopus nigricans	1.60	2.00	2.80	2.80	2.50	3.20	3.20	3.00	2.60
Trichoderma viride	0.50	0.40	0.40	1.60	1.40	1.50	2.10	2.00	0.50
Control	3.90	3.10	2.80	4.20	4.00	4.00	4.30	4.36	4.25
S.D. (1%)	0.40	0.36	0.29	0.52	0.81	1.22	1.17	1.10	1.12

[a] Ratio of *F. udum* to saprophytes.
[b] Population of *F. udum* per gram of dry rhizosphere soil $\times 10^3$.

Table 2. Monthly wilt incidence of pigeon pea planted in different ratios of soil inocula of *Fusarium udum* and test fungi

Inoculum mixed with	After 1 month			After 2 months			After 3 months		
	80:20[a]	50:50	20:80	80:20	50:50	20:80	80:20	50:50	20:80
Aspergillus flavus	12[b]	10	6	26	18	6	38	20	8
A. niger	2	0	0	6	5	0	7	5	0
A. terreus	17	10	12	40	38	20	50	33	32
Cunninghamella echinulata	28	26	20	45	38	30	63	44	38
Curvularia lunata	16	16	10	30	26	30	65	66	48
Neocosmospora vasinfecta	20	16	8	42	34	30	72	66	58
Penicillium citrinum	10	8	8	36	30	28	37	43	32
Rhizopus nigricans	30	28	26	54	42	36	80	64	40
Trichoderma viride	14	12	10	20	18	14	31	15	9
Control	35	30	22	60	51	48	100	100	100
S.D. (1%)	3.42	5.32	3.71	9.24	11.80	6.36	16.375	10.76	6.62

[a] Ratio of *F. udum* to test fungi.
[b] Percentage incidence of wilting.

Table 3. Population of *Fusarium udum* in the root region (rhizosphere and rhizoplane) of pigeon pea pretreated with spore suspension of various test fungi and planted in soil infested with *F. udum*

Spore suspension	After 15 days		After 30 days		After 45 days		After 60 days	
	Population in rhizo-sphere	Frequency on rhizo-plane (%)	Population in rhizo-sphere	Frequency on rhizo-plane (%)	Population in rhizo-sphere	Frequency on rhizo-plane (%)	Population in rhizo-sphere	Frequency on rhizo-plane (%)
Aspergillus flavus	0.15[a]	45[b]	1.80	50	2.00	66	2.80	65
A. niger	0.12	20	1.60	35	1.80	45	2.00	40
A. terreus	1.20	45	1.80	55	2.10	64	2.30	70
Cunninghamella echinulata	1.50	70	2.80	100	3.50	100	7.50	100
Curvularia lunata	1.50	70	2.00	80	3.20	95	5.80	100
Neocosmospora vasinfecta	1.90	60	2.90	100	3.00	100	6.00	100
Penicillium citrinum	1.60	25	2.60	35	3.10	48	4.60	64
Rhizopus nigricans	1.40	65	3.20	100	4.00	100	8.90	100
Trichoderma viride	0.30	30	1.00	42	1.40	46	2.00	50
Control	2.00	100	5.00	100	6.50	100	7.10	100
S.D. (1%)	0.26	. . .	0.42	. . .	0.51	. . .	0.35	. . .

[a] Population of *F. udum* per gram of dry rhizosphere soil $\times 10^3$.
[b] Percentage frequency of *F. udum* on rhizoplane of pigeon pea.

Table 4. Population of *Fusarium udum* in the root region (rhizosphere and rhizoplane) of pigeon pea pretreated with its spore suspension and then planted in soil-sand inocula of various test fungi

Inoculum of	After 15 days		After 30 days		After 45 days		After 60 days	
	Population in rhizo-sphere	Frequency on rhizo-plane (%)	Population in rhizo-sphere	Frequency on rhizo-plane (%)	Population in rhizo-sphere	Frequency on rhizo-plane (%)	Population in rhizo-sphere	Frequency on rhizo-plane (%)
Aspergillus flavus	0.50[a]	20[b]	1.80	28	1.70	35	1.80	40
A. niger	0.20	15	0.12	24	0.14	30	0.23	32
A. terreus	0.15	20	0.30	28	0.38	35	0.60	40
Cunninghamella echinulata	2.58	80	2.60	85	2.81	100	2.99	100
Curvularia lunata	2.10	75	2.40	82	2.40	85	2.50	85
Neocosmospora vasinfecta	1.60	60	2.10	72	2.60	88	3.00	100
Penicillium citrinum	0.16	30	0.48	50	0.81	66	1.50	80
Rhizopus nigricans	2.80	75	3.20	80	3.50	100	3.00	100
Trichoderma viride	0.80	50	0.90	58	1.36	70	1.58	80
Control	3.00	100	3.60	100	3.60	100	3.40	100
S.D. (1%)	0.26	. . .	0.67	. . .	0.60	. . .	0.89	. . .

[a] Population of *F. udum* per gram of rhizosphere soil $\times 10^3$.
[b] Percentage frequency of *F. udum* on rhizoplane of pigeon pea.

disease development. In cultural experiments, *T. viride* showed a low antagonistic effect against *F. udum* (12); indeed, stimulation of conidial germination and germ tube growth of *F. udum* was recorded in the culture metabolites of *T. viride*. However, soil inoculated with *T. viride* developed a significant level of fungistasis (10,12) and suppressed the population of *F. udum* both in soil and in the root region of pigeon pea. This may be attributed to the fast-growing nature of *Trichoderma*, leading to exhaustion of nutrient. The antagonistic activity of *Trichoderma* against many plant pathogens has been reported by other workers.

Soil inoculated with antagonistic test fungi became suppressive against *F. udum*. The soil infested with these fungi possessed high fungistatic activity against *F. udum* (10,12) and also decreased the ability of *F. udum* to colonize pigeon pea substrate (8). The utility of suppressive soil for biological control of soilborne plant pathogens has been well documented by several workers (1,2,5). This property can be enhanced by the addition of organic substances or by altering other environmental factors. In 1978, we observed that *A. nidulans*, a thermotolerant fungus, when introduced in soil in combination with *F. udum* at $38 \pm 2°C$ suppressed the population of the latter as well as the incidence of wilt caused by it (9). However, at lower temperatures (below 30°C), it has little impact on the pathogen.

Upadhyay and Rai (11) observed that the population of *F. udum* on the roots of healthy pigeon pea was low because of the presence of one or more of the aforesaid antagonistic fungi and vice versa. Therefore, it is likely that *A. niger, A. flavus, A. terreus,* and *T. viride* could play a role in reducing the inoculum of *F. udum* in field soil as well as in the root region of pigeon pea, thus affecting the population dynamics of the pathogen and its disease development. These fungi should be further examined for their capacity to control the wilt disease of pigeon pea caused by *F. udum*.

LITERATURE CITED

1. Alabouvette, C., Rouxel, F., and Louvet, J. 1979. Characterization of Fusarium wilt-suppressive soils and prospects for their utilization in biological control. Pages 165-182 in: Soil-Borne Plant Pathogens. B. Schippers and W. Gams, eds. Academic Press, London.

2. Baker, K. F., and Cook, R. J. 1974 (original ed.). Biological Control of Plant Pathogens. Reprint ed., 1982. American Phytopathological Society, St. Paul, MN. 433 pp.

3. Booth, C. 1971. The Genus Fusarium. Commonw. Mycol. Inst., Kew, Surrey, England.

4. Garrett, S. D. 1963. Soil Fungi and Soil Fertility. Pergamon Press, Oxford.

5. Louvet, J., Alabouvette, C., and Rouxel, F. 1981. Microbiological suppressiveness of some soils to Fusarium wilts. Pages 262-275 in: Fusarium: Diseases, Biology and Taxonomy. P. E. Nelson, T. A. Toussoun, and R. J. Cook, eds. Pennsylvania State University Press, University Park. 457 pp.

6. Nash, S. M., and Snyder, W. C. 1962. Quantitative estimations by plate counts of propagules of the bean root rot Fusarium in field soils. Phytopathology 52:567-572.

7. Rai, B., and Upadhyay, R. S. 1982. Gibberella indica: The perfect state of Fusarium udum. Mycologia 74:343-346.

8. Rai, B., and Upadhyay, R. S. 1983. Competitive saprophytic colonization of pigeon-pea substrate by Fusarium udum in relation to environmental factors, chemical treatments and microbial antagonisms. Soil Biol. Biochem. 15:187-191.

9. Upadhyay, R. S. 1979. Ecological studies on Fusarium udum Butler causing wilt disease of pigeon-pea. Ph.D. thesis, Banaras Hindu University, India.

10. Upadhyay, R. S., and Rai, B. 1978. Tolerance of higher temperature by Aspergillus nidulans in competition with other soil fungi. (Abstr.) Proc. Natl. Acad. Sci. India, 77.

11. Upadhyay, R. S., and Rai, B. 1982. Ecology of Fusarium udum causing wilt disease of pigeon-pea: Population dynamics in the root region. Trans. Br. Mycol. Soc. 74:209-220.

12. Upadhyay, R. S., and Rai, B. 1984. Studies on antagonism between Fusarium udum Butler and root region microflora of pigeon-pea (Cajanus cajan). Acta Mycol. In press.

Control of *Verticillium dahliae* by Coating Potato Seed Pieces with Antagonistic Bacteria

J. A. WADI, Department of Botany, The Islamic University of Gaza, Gaza Strip, Israel, and G. D. EASTON, Department of Plant Pathology, Washington State University, Irrigated Agriculture Research and Extension Center, Prosser, WA 99350, U.S.A.

Control of *Verticillium dahliae* Kleb. on potato is difficult because of the persistence of the fungus in the soil (7). Soil fumigation, long rotations, and planting in virgin land reduce losses from Verticillium wilt. Soil fumigation is expensive and provides only annual control (6). Long rotation with nonsusceptible crops has reduced losses, but these alternate crops give relatively low economic returns. Growing potatoes on virgin land produces high yields, but such land is becoming increasingly less available.

Several workers have used antagonistic microorganisms to control plant pathogens (2,3,8,17). Fluorescent *Pseudomonas* spp. are common inhabitants of the rhizosphere (9,17). Some of these rhizobacteria have been reported to colonize host root systems and increase plant growth when coated on potato seed pieces or radish and sugar beets seeds (13,21).

The investigation reported here sought to determine whether any bacteria isolated from the potato rhizosphere were antagonistic in vitro to *V. dahliae* and to develop methods to test their effectiveness in biological control of *V. dahliae* on potatoes in the greenhouse.

MATERIALS AND METHODS

Soil samples adhering to potato rhizospheres were diluted with sterile distilled water, spread on King's medium B agar in petri plates (11), and incubated 48 hr at 21°C. A spore suspension ($\approx 10^6$ spores per milliliter) of *V. dahliae* that had been cultured on potatoe-dextrose agar for 7 days at 21°C was seeded over the soil dilution on each plate. Zones where the growth of *V. dahliae* was inhibited were noted after 48 hr. Bacterial colonies antagonistic to *V. dahliae* were transferred to fresh King's medium B plates and stored at 4°C.

Rifampicin, streptomycin sulfate, or nalidixic acid mutants (13) of three bacterial isolates numbers 4, 45, and 83 were selected in preliminary studies as the most promising antagonists to *V. dahliae* in vitro. These isolates were compared with a plant-growth-promoting rhizobacterium (PGPR) *Pseudomonas fluorescens* strain B-10 obtained from M. N. Schroth, Department of Plant Pathology, University of California, Berkeley.

Scientific paper 6608, Project 1709, of Washington State University, College of Agriculture, Agricultural Research Center, Pullman, WA 99164. Mention of a product used in these studies does not constitute a recommendation of the product by Washington State University over other products.

Potato seed pieces of the susceptible cultivar Russet Burbank, coated by liquid and dust formulations, were used to assay antagonistic bacteria for control of *V. dahliae* in vivo. Bacteria for liquid formulations were scraped from plates of King's medium B and diluted with 3% methylcellulose solution to give a concentration of $\approx 10^9$ c.f.u./ml. Bacteria for dust formulations were grown as for liquid formulations and added to talc, xanthan gum, and 3% methylcellulose solution to give a concentration of $\approx 10^9$ c.f.u./ml. Treated seed pieces were planted in 15-cm-diameter clay pots containing fertilized field soils naturally infested by *V. dahliae* and additionally infested by adding a *V. dahliae* strain cultured on potato-dextrose agar 15 days at 21°C. The population of *V. dahliae*, determined by plating the soil on ethanol-streptomycin penicillin G agar medium (5), was 10^6 propagules per gram of soil. Untreated seed pieces were also planted in infested and steam-pasteurized field soil as controls.

Each treatment was replicated five times. During all tests, the greenhouse temperatures were 26°C day and 21°C night. The number of wilted plants and weight of tubers in each pot were recorded 9 weeks after planting. Stem segments (1–2 cm in length) from each plant were disinfected with 0.525% sodium hypochlorite and placed on moistened filter paper in petri dishes. After 7 days incubation at 21°C, the stems were examined microscopically for *V. dahliae* conidiophores (14). Populations of bacterial antagonists, the B-10 isolate, and *V. dahliae* in the rhizosphere soil and on the root surfaces were determined by platings on appropriate selective media (5,13).

Isolates 4, 45, and 83 isolated from roots and soil were identified on the basis of phenotypic characteristics and appropriate biochemical tests (4,10,15,16,18–20).

RESULTS

Isolate 45 applied as a dust coating on seed pieces colonized potato roots more abundantly than other bacterial isolates in either dust or liquid coatings (Table 1). Isolate 83 colonized the root surface more when applied in dust, whereas isolate 4 colonized more from a liquid coating. Antagonistic bacteria were not detected on surfaces of roots of plants from noninoculated seed pieces in either infested field soil or pasteurized soil. All bacterial isolates applied in either coating reduced *V. dahliae* propagule populations in rhizospheres (Table 1). All isolates in both coatings except 45 in liquid coating and 83 in dust coating

Table 1. Colonization of potato roots by rhizobacteria and their effects on *Verticillium dahliae* propagules in the rhizosphere, incidence of wilted plants, and plant, root, and tuber weight of potatoes grown in infested soil

| Isolate or soil | Potato seed coating[v] | Bacteria on root surface (c.f.u. × 10³/g root) | *V. dahliae* propagules (× 10³) | | Wilted plants (%) | Plant and tuber weights (g) | | |
| | | | Dried plant tissue | Rhizosphere | | Dry weight | | Tuber weight |
						Tops	Roots	
4[w]	D	0.4 d[x]	90 cd	6 c	66 d	37 c	1.4 cd	33 cd
	L	8.3 c	60 e	8 de	40 e	42 b	1.9 ab	49 ab
45[w]	D	46.6 a	63 de	1 de	13 f	45 a	1.9 ab	56 a
	L	1.1 d	117 abc	4 d	80 b	34 c	1.7 ab	38 c
83[w]	D	18.0 b	111 abc	6 c	100 a	37 c	1.4 cd	41 c
	L	1.0 d	49 e	2 de	100 a	37 c	1.6 c	35 cd
B 10[y]	D	0.6 d	91 cd	11 b	100 a	36 c	1.5 c	33 cd
	L	4.0 cd	95 bcd	9 bc	86 ab	36 c	1.7 ab	35 cd
Control[z]	D	0	148 a	15 a	100 a	29 d	1.2 e	29 e
	L	0	137 ab	15 a	100 a	28 d	1.2 e	30 e
Infested field soil	···	0	144 a	15 a	100 a	28 d	1.2 e	29 e
Pasteurized soil	···	0	0	0	0	46 a	2.0 a	58 a

[v] Bacteria ≈10⁹ c.f.u./ml in a dust coating (D) of talc, xanthan gum, and 3% methylcellulose solution; or in a liquid coating (L) of 3% methylcellulose solution.

[w] Antagonistic to *V. dahliae* in vitro.

[x] Vertical means of all data followed by the same letter are not significantly different according to Duncan's multiple range test ($P = 0.5$).

[y] Plant-growth-promoting rhizobacteria.

[z] Dust and liquid coatings applied without bacteria.

reduced stem infection by *V. dahliae* as measured by incidence of the fungus in dried plant tissue. Only isolates 4 and 45 applied in both coatings reduced wilting of plants of *V. dahliae*. All bacterial isolates, including B-10, from both dust and liquid coatings significantly increased dry top, root, and tuber weights. None of the top or root weights and tuber yields exceeded those obtained in the pasteurized soil control. Treatments with isolate 45 in dust coating and isolate 4 in liquid coating produced tuber weights in *V. dahliae*-infested soil that were equal to those in pasteurized soil. Dust and liquid coatings when applied without bacteria had no effect on plant, root, and tuber weights.

Isolate 4 was identified as *P. fluorescens* biotype C Stanier et al; isolate 45 was identified as *S. flavofungini* Uri and Bekesi; and isolate 83 was identified as *Cellulomonas flavigena* Kellerman and McBeth.

DISCUSSION

Many bacterial strains isolated from naturally infested soil were antagonistic in vitro to *V. dahliae*. Three were identified and compared in the greenhouse studies with a rhizobacterium (*P. fluorescens* strain B-10) previously reported to promote plant growth.

Our results are similar to those of Kloepper and Schroth (12). In their study, application of certain strains of fluorescent *Pseudomonas* spp. to potato seed pieces increased growth and yield of potatoes both in greenhouse experiments and in field plots. Their effective strains were obtained from potato roots and potato periderm and were highly successful as root colonists. Similarly, in our greenhouse studies, several bacterial isolates coated on seed pieces planted in *V. dahliae*-infested field soil resulted in taller plants and greater top, root, and tuber weights. These strains were isolated from potato roots, and the use of a drug resistance technique verified that they were successful as root colonists. Strain B-10, obtained from Kloepper and Schroth, did not reduce wilt caused by *V. dahliae* but did increase root, plant, and tuber weights. Strains 4

and 45 isolated from Washington soils controlled *V. dahliae* and increased tuber weight more than strain B-10. It may be necessary to obtain strains from soils where they will eventually be used to achieve maximum response. The bacterial strains antagonistic to *V. dahliae* and used in this study were isolated from potato rhizospheres in soils cropped previously to potato.

Strains antagonistic to *V. dahliae* in vitro reproduced well in potato rhizospheres and on root surfaces and controlled *V. dahliae* in vivo in pot culture. Previous studies have shown little correlation between antagonism to pathogens in vitro and disease control by these same antagonists in the field (1). *C. flavigena*, *P. fluorescens* biotype C, and *S. flavofungini* coated on potato seed pieces controlled *V. dahliae* and increased plant growth and tuber production in naturally infested field soil. Strain B-10 was less effective than the other three isolates. Results from preliminary field studies we have conducted indicate that these isolates may also reduce *V. dahliae* infection in the field.

LITERATURE CITED

1. Baker, K. F., and Cook, R. J. 1974 (original ed.). Biological Control of Plant Pathogens. Reprint ed., 1982. American Phytopathological Society, St. Paul, MN. 433 pp.
2. Broadbent, P., Baker, K. F., and Waterworth, Y. 1971. Bacteria and actinomycetes antagonistic to fungal root pathogens in Australian soils. Aust. J. Biol. Sci. 24:924-944.
3. Chet, I., and Baker, R. 1981. Isolation and biocontrol potential of *Trichoderma hamatum* from soil naturally suppressive of *Rhizoctonia solani*. Phytopathology 71:286-290.
4. Clark, F. C. 1953. Criteria suitable for species differentiation in *Cellulomonas* and a revision of the genus. Int. Bull. Bacteriol. Nomencl. Taxon. 3:197-199.
5. Easton, G. D., Nagle, M. E., and Bailey, D. L. 1969. A method of estimating *Verticillium albo-atrum* propagules in field soil and irrigation waste water. Phytopathology 59:1171-1172.
6. Easton, G. D., Nagle, M. E., and Bailey, D. L. 1975. Residual effect of soil fumigation and vine burning for control of Verticillium wilt of potato. Phytopathology 65:1419-1422.

7. Green, R. J., Jr. 1969. Survival and inoculum potential of conidia and microsclerotia of *Verticillium albo-atrum* in soil. Phytopathology 59:874-876.

8. Howell, C. R., and Stipanovic, R. D. 1979. Control of *Rhizoctonia solani* on cotton seedlings with *Pseudomonas fluorescens* and with an antibiotic produced by the bacterium. Phytopathology 69:480-482.

9. Katznelson, H., Peterson, E. A., and Rouatt, J. W. 1962. Phosphate dissolving microorganisms on seed and in the root zone of plants. Can. J. Bot. 40:1181-1186.

10. Keddie, R. M. 1974. Genus *Cellulomonas*. Pages 629-631 in: Bergey's Manual of Determinative Bacteriology. 8th ed. R. E. Buchanan and N. E. Gibbon, eds. Williams and Wilkins Co., Baltimore. 1,268 pp.

11. King, E. O., Ward, M. K., and Raney, D. E. 1954. Two simple media for the demonstration of pyocyanin and fluorescein. J. Lab. Clin. Med. 44:301-307.

12. Kloepper, J. W., and Schroth, M. N. 1979. Plant growth promoting rhizobacteria: Evidence that the mode of action involves microflora interactions. (Abstr.) Phytopathology 69:1034.

13. Kloepper, J. W., Schroth, M. N., and Miller, T. D. 1980. Effects of rhizosphere colonization by plant-growth promoting rhizobacteria on potato plant development and yield. Phytopathology 70:1078-1082.

14. McKeen, C. D., and Thorpe, H. J. 1971. An adaptation of a moist-chamber method for isolation and identifying *Verticillium* spp. Can. J. Microbiol. 17:1139-1141.

15. Nonomura, H. 1974. Key for classification and identification of 458 species of the Streptomycetes in ISP. J. Ferment. Technol. 52:78-92.

16. Pridham, T. G., and Tresner, H. D. 1974. Genus *Streptomyces*. Pages 748-829 in: Bergey's Manual of Determinative Bacteriology. 8th ed. R. E. Buchanan and N. E. Gibbon, eds. Williams and Wilkins Co., Baltimore. 1,268 pp.

17. Rouatt, J. W., Peterson, E., Katznelson, A. H., and Henderson, V. E. 1963. Microorganisms in the root zone in relation to temperature. Can. J. Microbiol. 9:227-236.

18. Shirling, E. B. 1972. Cooperative description of type strains of *Streptomyces*. Int. J. System. Bacteriol. 22:265-394.

19. Shirling, E. B., and Gottlieb, D. 1966. Methods for characterization of *Streptomyces* species. Int. J. System. Bacteriol. 16:313-340.

20. Stanier, R. Y., Palleroni, N. J., and Doudoroff, M. 1966. The aerobic pseudomonads: A taxonomic study. J. Gen. Microbiol. 43:159-271.

21. Suslow, T. V., and Schroth, M. N. 1982. Role of deleterious rhizobacteria as minor pathogens in reducing crop growth. Phytopathology 72:111-115.

Application of Fluorescent Pseudomonads to Control Root Diseases

DAVID M. WELLER, Regional Cereal Disease Research Laboratory, U.S. Department of Agriculture, Agricultural Research Service, Pullman, WA 99164, U.S.A.

Fluorescent *Pseudomonas* spp. have received increasing attention in the last few years for their ability to suppress root diseases. Part of the interest reflects the ease with which the pseudomonads can be isolated, cultured, and identified. Several selective media are available (29), and identifying an isolate as *Pseudomonas* is easily done based upon fluorescent pigment production and reactions in several bacteriological tests (29). However, because identification of an isolate to the species level is more difficult, sometimes candidate organisms are reported only as *Pseudomonas* sp. Generally, the fluorescent pseudomonads that have been tested for biocontrol of root diseases fall into the *Pseudomonas fluorescens* and *P. putida* groups (29).

Although the selection of a fluorescent pseudomonad for use against root diseases is often based on considerations such as ease in isolation, identification, and handling, it is their ecological and physiological characteristics that make them strong candidates for biological control. First, the natural habitat for the fluorescent pseuomonads is the soil, but more specifically the particulate organic matter and rhizosphere (20). Many studies have reported isolating pseudomonads from the rhizosphere (20,22), and in some cases fluorescent pseudomonads were an important component of the rhizosphere bacteria (18). Thus, when fluorescent pseudomonads are introduced to the roots they are placed in an environment suitable for their growth and survival. Second, the fluorescent pseudomonads are nutritionally versatile, being able to utilize a large number of organic substrates (29). They are adapted to utilizing root exudates, and the root stimulates their growth as well as that of the other gram-negative bacteria (19,22). Third, the pseudomonads have a fast growth rate relative to other bacteria in the rhizosphere (18). For example, Bowen and Rovira (1) reported that the generation time of pseudomonads on roots of *Pinus radiata* was 5.2 hr, whereas the generation times for the total viable bacteria and *Bacillus* spp. were 7.2 and 39 hr, respectively. Fourth, the fluorescent pseudomonads produce a variety of antibiotics (14) and siderophores (8,16,32) that can inhibit phytopathogenic bacteria and fungi in vitro. If antibiosis is one mechanism of biological control, the ability to produce antibiotics and siderophores could be useful for an antagonist in root colonization or pathogen suppression. Fifth, research on disease-suppressive soils suggests that in some cases fluorescent pseudomonads may provide natural biological control, as in take-all decline (4,28,33), the phenomenon whereby wheat monoculture results in the suppression of take-all

(27). Rovira and Wildermuth (21) showed that gram-negative bacteria proliferate (no doubt fluorescent pseudomonads were present) in lesions on roots of wheat caused by *Gaeumannomyces graminis* var. *tritici* and suggested that this buildup is an integral part of take-all decline. Sher and Baker isolated fluorescent pseudomonads from a Fusarium wilt-suppressive soil and induced suppressiveness (23,24) by adding the bacteria into conducive soil. Finally, and perhaps most important, fluorescent pseudomonads can be successfully introduced into the root system and become established in the rhizosphere and rhizoplane of a variety of crops (2,12,31,33).

EXAMPLES OF CONTROL BY FLUORESCENT PSEUDOMONADS

Fluorescent pseudomonads have been tested against root and seedling diseases of a variety of crops in the greenhouse and field. Biological control has been demonstrated in such diverse crops as potatoes (2,12), sugar beets (31), radish (10,24), wheat (34), cotton (6,7), soybean (13), carnation (15), flax (24), cherry (5), and cucumber (24). Several examples will be discussed. Portions of this topic have previously been reviewed (3,25,26).

Plant growth-promoting rhizobacteria. One of the best examples of biological control of root pathogens by fluorescent pseudomonads is the application of plant growth-promoting rhizobacteria (PGPR) to increase plant growth and yield (25,26).

Burr et al (2) first reported that strains of fluorescent *Pseudomonas* spp. applied to potato seed pieces increased the yield in field tests. These results were confirmed (12) and growth promotion was also demonstrated for sugar beet (31), radish (10), and other crops (26). In field tests, PGPR strains have increased yields of potato 5–33%, of sugar beet 4–8 tonnes per hectare, and root weight of radish 60–144% as compared with untreated plants (26).

The term "rhizobacteria" was coined to describe bacteria that are well adapted to colonizing and growing on roots. The PGPR improve plant growth, but other deleterious or neutral groups of rhizobacteria also occur. More of the beneficial *Pseudomonas* rhizobacteria belong to the *P. fluorescens* and *P. putida* groups, but few fit into recognized biotypes (29). When introduced to seeds or seed pieces, PGPR are thought to promote growth by colonizing the root system and displacing or excluding deleterious fungi and bacteria (30), most of which have not previously been recognized as plant pathogens.

The ability of many PGPR to alter the composition of the root microflora correlates with production of compounds in vitro that inhibit a wide range of organisms. Mutants that lose in vitro antibiosis also lose the ability to enhance plant growth, despite still being able to colonize roots (11). The suppression of the deleterious microorganisms by PGPR appears to be mediated through the production of fluorescent siderophores, Fe^{+3} chelators that have a high affinity for iron and that are involved in its transport (8,26). The siderophores from PGPR strain B10 have been isolated (32). Schroth and Hancock (26) suggested that the siderophores produced by PGPR sequester Fe^{+3} in the rhizoplane, thereby starving other microorganisms for iron, either because they produce siderophores in insufficient quantities or with less affinity for iron than those from PGPR. The siderophore hypothesis is supported by the fact that addition of $FeCl_3$ to King's medium B eliminates siderophore production by PGPR strains and also their ability to inhibit other organisms in vitro. Further, the addition of Fe^{+3} $EDTA^-$ to soil suppressed growth promotion by PGPR (8). Finally, the addition of the siderophore from PGPR strain B10, pseudobactin, to soil increased plant growth.

Suppression of take-all of wheat. The possibility that antagonistic microorganisms are responsible for take-all decline (27) has resulted in studies of the feasibility of applying organisms to simulate the effects of a natural suppressive soil. Cook and Rovira (4) first suggested the possible role of fluorescent pseudomonads in take-all decline; since then, several reports have appeared where bacteria have been applied to plants in the greenhouse or field to control take-all (34). Strains of *Pseudomonas fluorescens* that suppress take-all have been isolated in the state of Washington, USA, from roots of wheat grown in the presence of *G. graminis* var. *tritici* in soil from fields where take-all had declined. Strain 2-79 (NRRL B-15132) when applied as a seed treatment, either alone or combined with 13-79 (NRRL B-15134), has suppressed take-all in experimental field plots (10–27% increased yield) where the take-all fungus was introduced to the seed furrow (34) and in a commercial field (21% increased yield) where the pathogen was naturally present. The treatments have been effective in both fall-sown (winter) and spring-sown (spring) wheat. Tests with treated seed sown in soil fumigated with methyl bromide and with the take-all fungus reintroduced have demonstrated that enhanced plant growth and yield increases resulting from the bacterial treatment are the result of control of take-all rather than direct promotion of plant growth or control of other major or lesser-known pathogens. Studies using an antibiotic-resistant strain of 2-79 indicate that the bacterium is an aggressive root colonist and that root colonization is an important component in disease suppression by the bacteria (33). Unpublished data indicate that for some strains, inhibitory compounds produced by the bacteria are responsible for suppression; for other strains, however, no inhibitory compounds can be detected in vitro even though the bacteria are suppressive.

Suppression of seedling diseases of cotton. Howell and Stipanovic (6,7) reported suppression of seedling damping-off of cotton caused by *Rhizoctonia solani* and *Pythium ultimum* by applying to seed *P. fluorescens,* Pf-5, isolated from the rhizosphere of cotton

seedlings. In *R. solani*-infested soil there was an increase in seedling survival from 30 to 79%. Inhibition of the pathogen was thought to be caused by the production by the bacterium of pyrrolnitrin, an antifungal antibiotic, because the treatment of seeds with the antibiotic alone increased seedling survival from 13 to 70%. In soil infested with *P. ultimum*, strain Pf-5 applied to cotton seeds increased seedling survival from 28 to 71%. Against *Pythium*, the bacterium produced a second antibiotic, pyoluteorin, that duplicated the protection provided by the bacterium.

Suppression of Fusarium wilt diseases. Several strains of fluorescent *Pseudomonas* spp. that suppress Fusarium wilt diseases were isolated from mycelial mats of *Fusarium oxysporum* f. sp. *lini* that had been buried in a Fusarium wilt-suppressive soil (23) or near roots of flax growing in suppressive soil (24). Strain A12, identified as *P. putida*, when added into conducive soil (10^7 c.f.u./g of soil) infested with either *F. oxysporum* f. sp. *lini*, *F. oxysporum* f. sp. *cucumerinum*, or *F. oxysporum* f. sp. *conglutinans* reduced the incidence of wilt in flax, cucumber, and radish to 22, 40, and 61%, respectively, of that in the controls (24). Scher and Baker (24) suggested that disease suppression by *P. putida* resulted from the production of siderophores that bind Fe^{3+} at the rhizoplane, thus limiting the availability of iron to the fungus. This conclusion is supported by the fact that the addition of the chelator EDDHA to conducive soil reduced the incidence of flax wilt. The mechanism is thus similar to that proposed for the suppression of deleterious microorganisms by plant growth-promoting rhizobacteria.

This work demonstrates the feasibility of using fluorescent pseudomonads as soil treatments. Soil application of pseudomonads may be preferable to seed treatments, especially for controlling root diseases of greenhouse-grown crops. Greenhouse soil, when steamed or fumigated to eliminate soilborne pathogens, can become rapidly recolonized by the pathogen if reintroduced. Fluorescent pseudomonads are well adapted to growing in soil following fumigation or sterilization (1,17), and when applied to soil they reduced the chance of reestablishment by the pathogens by filling the biological vacuum or by protecting the roots against infection by a pathogen.

ISOLATION OF EFFECTIVE STRAINS

Currently, no in vitro tests can predict which fluorescent *Pseudomonas* strains from a natural population will be effective in controlling root diseases. Large numbers of strains must be screened in the greenhouse, and these results often do not predict the response of an isolate in the field. The examples of biological control presented in the previous section demonstrate criteria that can aid in the selection of the most effective fluorescent *Pseudomonas* strains.

The fluorescent pseudomonads should be isolated from the environment in which they will be expected to function. Where fluorescent pseudomonads are applied to seeds to protect against root pathogens, the bacteria should be isolated from the rhizosphere or rhizoplane of the target host. Burr et al (2) obtained the original PGPR strains from healthy potato tubers, and other PGPR strains have been isolated from celery, potato, and sugar beet roots. Howell and Sipanovic (6) isolated strain Pf-5 from the cotton rhizosphere, and Weller and Cook (34) isolated take-all suppressive strains from wheat roots.

Little information is available on the specificity of fluorescent pseudomonads that provide biological control for given crops or diseases. The data that are available, however, do not indicate a high degree of specificity. For example, PGPR strain E6 isolated from celery roots enhanced the growth of sugar beet (31), potato (12), and radish (10). Further, PGPR strain B10 from potato suppressed take-all of wheat and Fusarium wilt of flax (9). *P. putida* strain A12 (24) reduced the incidence of Fusarium wilt of cucumber, radish, and flax. Conversely, however, Kloepper et al (12) reported that only one of four PGPR strains that increased radish growth also increased potato growth. Despite the lack of specificity, one might expect a fluorescent pseudomonad to be most effective in protecting the host from which it was isolated because it might be better adapted to utilizing exudates from its original host and thus be more competitive than in the rhizosphere of another plant. Further, bacteria that are present on the root, both in the mucigel and in the intercellular spaces of the cortex, might be even more adapted to a particular plant species because of the close plant-microbe interaction. For example, some of the most suppressive fluorescent pseudomonads of take-all produce a hypersensitive response in tobacco leaves, indicating a possible slight pathogenic relationship with the wheat plant.

The fluorescent pseudomonads should be isolated from roots growing in pathogen-suppressive soil or from inoculum of the pathogen naturally present or introduced into the soil when bacteria are suspected to have a role in disease suppression. The most effective strains selected by Weller and Cook (34) to suppress take-all were from roots of wheat plants grown in soil that had become suppressive to the disease and where the fungus had been introduced. Further, Xing and Weller (*unpublished*) demonstrated that a larger percentage of the fluorescent pseudomonads isolated from roots grown in suppressive soils provided control of take-all in a greenhouse test than those from a conducive soil. Scher and Baker (24) isolated the suppressive strain A12 from mycelium of *F. oxysporum* f. sp. *lini* that was placed in Fusarium wilt-suppressive soil.

In vitro inhibition of a target pathogen or a group of organisms should be given consideration in selecting a potential antagonist. The production of antibiotics and siderophores appears to be an important characteristic of some pseudomonads that suppress disease; one of these compounds was suggested as a mechanism of action in each of the four examples described (6,7,24–26,34). However, because other factors (most of which are unknown) are involved in disease suppression by fluorescent pseudomonads, in vitro antibiosis cannot be used as a sole determinant in the selection process. It is well known that the intensity of in vitro inhibition of a microorganism and its activity in nature do not correlate. Schroth and Hancock (26) pointed out that less than 5% of the bacteria that exhibited antibiosis in vitro enhanced plant growth. Further, strains that protect by other mechanisms can be missed when antibiosis is used as a sole characteristic.

CONCLUSION

The fluorescent *Pseudomonas* group contains a rich source of organisms for use in biological control of root diseases. However, as with many biocontrol agents, application in the field can result in inconsistent results. This should be expected until the ecology of the bacteria and their interaction with the root are better understood. For example, one basic question has not been answered: Where must the bacteria be located to promote disease suppression—rhizosphere, root surface, intercellular spaces of cortex—or is colonization of all critical?

Clearly, more work is needed on the effects of soil physicochemical factors on root colonization, on the host specificity of effective strains, and on the bacteriological characteristics that contribute to the ability of a fluorescent pseudomonad to suppress disease.

LITERATURE CITED

1. Bowen, G. D., and Rovira, A. D. 1976. Microbial colonization of plant roots. Annu. Rev. Phytopathol. 14:121-144.
2. Burr, T. J., Schroth, M. N., and Suslow, T. 1978. Increased potato yields by treatments of seedpieces with specific strains of *Pseudomonas fluorescens* and *P. putida*. Phytopathology 68:1377-1383.
3. Cook, R. J., and Baker, K. F. 1983. The Nature and Practice of Biological Control of Plant Pathogens. American Phytopathological Society, St. Paul, MN. 539 pp.
4. Cook, R. J., and Rovira, A. D. 1976. The role of bacteria in the biological control of *Gaeumannomyces graminis* by suppressive soils. Soil Biol. Biochem. 8:269-273.
5. Cooksey, D. A., and Moore, L. W. 1980. Biological control of crown gall with fungal and bacterial antagonists. Phytopathology 70:506-509.
6. Howell, C. R., and Stipanovic, R. D. 1979. Control of *Rhizoctonia solani* on cotton seedlings with *Pseudomonas fluorescens* and with an antibiotic produced by the bacterium. Phytopathology 69:480-482.
7. Howell, C. R., and Stipanovic, R. D. 1980. Suppression of *Pythium ultimum*-induced damping-off of cotton seedlings by *Pseudomonas fluorescens* and its antibiotic, pyoluteorin. Phytopathology 70:712-715.
8. Kloepper, J. W., Leong, J., Teintze, M., and Schroth, M. N. 1980. Enhanced plant growth by siderophores produced by plant growth-promoting rhizobacteria. Nature 286:885-886.
9. Kloepper, J. W., Leong, J., Teintze, M. and Schroth, M. N. 1980. *Pseudomonas* siderophores: A mechanism explaining disease suppressive soils. Curr. Microbiol. 4:317-320.
10. Kloepper, J. W., and Schroth, M. N. 1978. Plant growth promoting rhizobacteria on radishes. Pages 879-882. in: Proc. Int. Conf. Plant Pathog. Bacter., 4th. Vol. 2. Angers, France.
11. Kloepper, J. W., and Schroth, M. N. 1981. Relationship of in vitro antibiosis of plant growth-promoting rhizobacteria to plant growth and the displacement of root microflora. Phytopathology 71:1020-1024.
12. Kloepper, J. W., Schroth, M. N., and Miller, T. D. 1980. Effects of rhizosphere colonization by plant growth-promoting rhizobacteria on potato plant development and yield. Phytopathology 70:1078-1082.
13. Kommedahl, T., Windels, C. E., Sarbini, G., and Wiley, H. B. 1981. Variability in performance of biological and fungicidal seed treatments in corn, peas, and soybeans. Prot. Ecol. 3:55-61.
14. Leisinger, T., and Margraff, R. 1979. Secondary metabolites of the fluorescent pseudomonads. Microbiol. Rev. 43:422-442.
15. Michael, A. H., and Nelson, P. E. 1972. Antagonistic effect of soil bacteria on *Fusarium roseum* 'Culmorum' from carnation. Phytopathology 62:1052-1056.
16. Misaghi, I. J., Stowell, L. J., Grogan, R. G., and Spearman, L. C. 1982. Fungistatic activity of water-soluble fluorescent pigments of fluorescent pseudomonads. Phytopathology 72:33-36.
17. Ridge, E. H. 1976. Studies on soil fumigation. II. Effects on bacteria. Soil Biol. Biochem. 8:249-253.

139

18. Rouatt, J. W., and Katznelson, H. 1961. A study of the bacteria on the root surface and in the rhizosphere soil of crop plants. J. Appl. Bacteriol. 24:164-171.

19. Rovira, A. D., and Davey, C. B. 1974. Biology of the rhizosphere. Pages 153-204 in: The Plant Root and Its Environment. E. W. Carson, ed. University Press of Virginia, Charlottesville. 691 pp.

20. Rovira, A. D., and Sands, D. C. 1971. Fluorescent pseudomonads—a residual component in the soil microflora? J. Appl. Bacteriol. 34:253-259.

21. Rovira, A. D., and Wildermuth, G. B. 1981. The nature and mechanism of suppression. Pages 385-415 in: Biology and Control of Take-all. M. J. C. Asher and P. Shipton, eds. Academic Press, London. 538 pp.

22. Sands, D. C., and Rovira, A. D. 1971. *Pseudomonas fluorescens* biotype G, the dominant fluorescent pseudomonad in South Australian soils and wheat rhizospheres. J. Appl. Bacteriol. 34:261-275.

23. Scher, F. M., and Baker, R. 1980. Mechanism of biological control in a Fusarium-suppressive soil. Phytopathology 70:412-417.

24. Scher, F. M., and Baker, R. 1982. Effect of *Pseudomonas putida* and a synthetic iron chelator on induction of soil suppressiveness to Fusarium wilt pathogens. Phytopathology 72:1567-1573.

25. Schroth, M. N., and Hancock, J. G. 1981. Selected topics in biological control. Annu. Rev. Microbiol. 35:453-476.

26. Schroth, M. N., and Hancock, J. G. 1982. Disease-suppressive soil and root-colonizing bacteria. Science 216:1376-1381.

27. Shipton, P. J. 1975. Take-all decline during cereal monoculture. Pages 137-144 in: Biology and Control of Soil-Borne Plant Pathogens. G. W. Bruehl, ed. American Phytopathological Society, St. Paul, MN. 216 pp.

28. Smiley, R. W. 1979. Wheat-rhizoplane pseudomonads as antagonists of *Gaeumannomyces graminis*. Soil Biol. Biochem. 11:371-376.

29. Stolp, H., and Gadkari, D. 1981. Nonpathogenic members of the genus *Pseudomonas*. Pages 719-741 in: The Prokaryotes: A Handbook on Habitats, Isolation, and Identification of Bacteria, Vol. I. M. P. Starr, H. Stolp, H. G. Trüper, A. Balows, and H. G. Schlegel, eds. Springer-Verlag, Berlin.

30. Suslow, T. V., and Schroth, M. N. 1982. Role of deleterious rhizobacteria as minor pathogens in reducing crop growth. Phytopathology 72:111-115.

31. Suslow, T. V., and Schroth, M. N. 1982. Rhizobacteria of sugar beets: Effects of seed application and root colonization on yield. Phytopathology 72:199-206.

32. Teintze, M., and Leong, T. 1981. Structure of pseudobactin A, a second siderophore from plant growth promoting *Pseudomonas* B10. Biochemistry 20:6457-6462.

33. Weller, D. M. 1983. Colonization of wheat roots by a fluorescent pseudomonad suppressive to take-all. Phytopathology 73:1548-1553.

34. Weller, D. M., and Cook, R. J. 1983. Suppression of take-all of wheat by seed treatments with fluorescent pseudomonads. Phytopathology 73:463-469.

The Role of Seeds in the Delivery of Antagonists into the Rhizosphere

CAROL E. WINDELS, THOR KOMMEDAHL, G. SARBINI, and H. B. WILEY, Department of Plant Pathology, University of Minnesota, St. Paul 55108, U.S.A.

Seeds have been treated in many ways to protect them from disease, to stimulate their germination, and even to introduce organisms into the rhizosphere (5). In chemical seed treatment, the goal is to protect seeds from seedborne or soilborne pathogens just long enough to enable the seeds to germinate and become established as seedlings, after which their resistance mechanisms are fully active or the cultural methods designed for control are fully implemented. The time needed for protection by the chemical is short and any protection beyond germination is not effective or expected, except for systemics.

The goal of biological seed treatment is the same as for chemical seed treatment; in addition, however, it is expected that colonization of the rhizosphere or root surface occurs to provide protection of roots against pathogens. If an organism applied to seeds can use seed exudates for energy of growth and then enter the rhizosphere (actively or passively), presumably it also can exploit root exudates to maintain continued growth and establishment on the root surface and thereby to provide protection of both seeds and roots (1,5). It may be feasible to protect roots at the proximal but not the distal ends, and this may be relevant to the type of pathogen involved.

Roots generally grow faster than hyphae; therefore roots encounter soilborne fungi before seedborne fungi can colonize root surfaces beyond the immediate vicinity of the seeds. This means that a biocontrol agent must be chosen that can grow rapidly in the different environments of seeds and roots and compete successfully against a variety of soilborne organisms, whether the roots grow fast or slowly.

During the past decade, work has been done at Minnesota on the application of bacteria and fungi to seeds of maize, peas, and soybeans in attempts to control seedling diseases. There is a need for identifying biological, chemical, and physical factors affecting the activity of the antagonists, pathogens, saprophytes, and host plants when testing biological seed treatments for disease protection. This was apparent when we failed to get comparable results in greenhouse and field trials (4). Moreover, in successive field trials, considerable variability in results was encountered (6). This meant that if the seed treatment increased stand (reduced disease) or yield at one time in replicated field trials and did not do so in a second trial, some important factors affecting results were present at one time and not at another (3,8). Some factors such as preparation and application of antagonist can be easily identified and controlled (8).

Organisms applied to seeds seem to be effective mainly in protecting seeds from infection as they germinate (2,7). Analyses made to determine the presence of antagonists in the rhizosphere at various time intervals after their application to seeds indicate that antagonists can be introduced into the rhizosphere (Table 1).

APPLICATIONS TO SEEDS

Maize. When *Chaetomium globosum* was applied to maize seeds, it grew and increased its inoculum concentration manyfold on the pericarp; however, it was not recovered from the rhizospheres of plants growing from treated seeds (Table 2). Instead, the presence of *C. globosum* on seeds resulted in greater populations of several other fungi in the rhizosphere, greater than those found in rhizospheres of nontreated plants. Other species were decreased (Table 2).

When sweet maize seeds were treated with *Aspergillus, Penicillium,* or *Trichoderma* species, these fungi were recovered in populations greater than those in rhizospheres of plants grown from treated than from untreated seeds. For example, spores of *A. flavus* applied at the rate of 15 million per seed were recovered as 500,000 colonies per gram of rhizosphere soil. Also, when spores of *T. harzianum* were applied at 40–50 million per seed, 1.1 million colonies per gram of rhizosphere soil were recovered 10 days after planting (Table 1). Thus it appears that antagonists can be introduced into the rhizosphere of maize by means of seed application but that not all antagonistic fungi behave in the same way. Moreover, it has not yet been shown that the presence of these fungi in any population in the rhizosphere has a protective effect against root pathogens.

Peas. Several fungi and bacteria have been applied to pea seeds to control seedling diseases (1,4,5,7-10). It was clearly established that *Penicillium oxalicum* applied to pea seeds multiplied on those seeds and increased its inoculum potential there (7). It was apparent also that this fungus was present on roots because it could be reisolated in considerable numbers from rhizospheres throughout the growing season of peas, even to harvest (9 and Table 1). For example, when 6–8 million spores of *P. oxalicum* were applied to pea seeds and the rhizospheres were assayed every 2 weeks during the season, from 0.4 to 3.3 million colony-forming units were recovered per gram of rhizosphere soil.

As in work with maize, it was not established that the presence of high populations of *P. oxalicum* had any effect on disease development. Light and electron microscopy revealed that spores had germinated, grown, and produced another crop of spores on the seed surface within 3 days after planting; this was not observed on the root surface (7).

Soybean. Bacteria and fungi were applied to soybean seeds, and these treated seeds were planted in soil known to contain inoculum of *Phytophthora megasperma* var. *sojae*, which causes root rot in soybean. When either *Bacillus* or *Pseudomonas* was applied to seeds of Clay soybeans and then planted in soil, *Pseudomonas* was isolated more frequently than *Bacillus* species from the primary and secondary roots of 4-week-old plants. *Pseudomonas* not only persisted longer on roots but increased nodulation with the seed inoculant *Rhizobium japonicum* mixed in peat, whereas *Bacillus* reduced nodulation.

The application of *Penicillium oxalicum* and *T. harzianum* to soybean seeds resulted in their establishment in the rhizospheres of 10- to 28-day-old plants (Table 1). Nodulation was not adversely affected by these fungi.

In the field, application of inoculum in furrows as soybeans were planted was often more effective than application to seeds in establishing these antagonists in the rhizosphere. This may be explained by the fact that cotyledons of soybean seeds, in contrast to pea seeds, emerge from soil at germination carrying with them the seed-applied inoculum. Furrow applications place inoculum where roots grow.

DISCUSSION

Calculation of propagules in the rhizosphere. When two methods of estimating populations of organisms in the rhizosphere are compared (Table 3), different conclusions are possible. Variation in the soil moisture content, hence the amount of soil clinging to roots, can greatly modify calculations of propagule number. When counts are made per root system, population estimates are more realistic because they are independent of the amount of soil adhering to roots (Table 3).

Potential for introducing seed-applied organisms. Once organisms multiply on seed coats or pericarps and enter the root zones, they encounter a new environment on roots. Whether rhizosphere relationships are established depends upon the type of crop (mono-

cotyledons vs. dicotyledons), crop cultivar, germination habit (epigeal vs. hypogeal), and other host factors.

Crops differ in relative rates of shoot and root growth. Roots grow more slowly than shoots in maize, whereas they develop more rapidly than shoots in peas. This may affect the ability of antagonists to colonize roots of these two crops in the few days after germination and may affect the quantity and quality of root exudates available at any given time.

Organisms applied to seeds of cereal crops probably grow first in rhizospheres of seminal rather than adventitious roots because adventitious roots (the major root system in cereals) grow from the nodes above the seeds. Antagonists applied to legumes may become established along the major root (tap) and its branches. On the other hand, in legumes the pea seed coated with an antagonist remains in the soil (hypogeal), whereas the soybean seed emerges from soil as cotyledons (epigeal), removing seed-applied inoculum from the soil.

There is an inherent advantage in seed applications in that seed exudates can overcome soil fungistasis and thereby enable antagonists to grow and enter the rhizosphere. Seed applications appear to be more strategic for establishing an antagonist in the rhizosphere, but this advantage may not hold for every antagonist, crop, or disease.

In conclusion, when an antagonist moves from the seed, either actively or passively, it encounters a different biological, chemical, and physical environment. No longer is it in the position of being the dominant occupant of a nutrient-rich substrate. Its fate may be death, bare survival, or growth into a suitable niche. The ultimate quest is to identify an antagonist that has the ability to find and then thrive within an appropriate niche and to ward off all competitors before the root encounters or stimulates the growth of a root pathogen.

Table 1. The effectiveness of introducing fungi into the rhizosphere of maize, peas, and soybeans by applying spores to seeds and planting the treated seeds in soil

Antagonist	Crop	Crop age (days)	Colony-forming units per gram of rhizosphere soil ($\times 10^4$)
Aspergillus flavus	Maize	10	50
		20	6
A. niger	Soybean	10	50
		20	0
Penicillium oxalicum	Peas[a]	14	77
		28	80
		42	189
		56	42
	Soybeans	10	60
		20	140
		28	1,200
Trichoderma harzianum	Maize	10	105
		20	135
	Soybeans	10	7
		20	21
		28	980

[a] Two-year average.

Table 2. Effect of coating maize seeds with *Chaetomium globosum* on populations of other fungi in the rhizosphere when coated seeds were planted in soil

Rhizosphere fungus[a]	Colony-forming units per gram of rhizosphere soil	
	Coated seeds	Untreated seeds
Aspergillus	9,000	6,000
Cephalosporium	9,000	13,000
Fusarium	9,000	17,000
Penicillium	57,000	48,000
Trichoderma	17,000	34,000

[a] *Chaetomium globosum* was not recovered from the rhizosphere.

Table 3. Numbers of colony-forming units of *Penicillium oxalicum* present in rhizospheres and roots of peas following planting of seeds coated with spores of *P. oxalicum*

Days after planting	Taproot length (cm)	Colony-forming units[a]	
		Per gram of rhizosphere soil[b]	Per root
3	1.5	777,000	6,300
7	9.7	328,500	36,700
10	18.7	978,400	291,400

[a] Each value is an average of three replicates of five plants per replicate at 3 days and three plants per replicate at 7 and 10 days.
[b] Per gram of oven-dry rhizosphere soil from Little Marvel peas grown in pea field soil in the greenhouse ($21 \pm 4°$ C).

LITERATURE CITED

1. Baker, K. F., and Cook, R. J. 1974 (original ed.). Biological Control of Plant Pathogens. Reprint ed., 1982. American Phytopathological Society, St. Paul, MN. 433 pp.
2. Chang, I., and Kommedahl, T. 1968. Biological control of seedling blight of corn by coating kernels with antagonistic microorganisms. Phytopathology 58:1395-1401.
3. Kommedahl, T., and Mew, I. C. 1975. Biocontrol of corn root infection in the field by seed treatment with antagonists. Phytopathology 65:296-300.
4. Kommedahl, T., and Windels, C. E. 1978. Evaluation of biological seed treatment for controlling root diseases of pea. Phytopathology 68:1087-1095.
5. Kommedahl, T., and Windels, C. E. 1981. Introduction of microbial antagonists to specific courts of infection: Seeds, seedlings, and wounds. Pages 227-248 in: Biological Control in Crop Production. G. C. Papavizas, ed. Allanheld, Osmun & Co., Totowa, NJ. 461 pp.
6. Kommedahl, T., Windels, C. E., Sarbini, G., and Wiley, H. B. 1981. Variability in performance of biological and fungicidal seed treatments in corn, peas, and soybeans. Prot. Ecol. 3:55-61.
7. Windels, C. E. 1981. Growth of *Penicillium oxalicum* as a biological seed treatment on pea seed in soil. Phytopathology 71:929-933.
8. Windels, C. E., and Kommedahl, T. 1978. Factors affecting *Penicillium oxalicum* as a seed protectant against seedling blight of pea. Phytopathology 68:1656-1661.
9. Windels, C. E., and Kommedahl, T. 1982. Rhizosphere effects of pea seed treatment with *Penicillium oxalicum*. Phytopathology 72:190-194.
10. Windels, C. E., and Kommedahl, T. 1982. Pea cultivar effect on seed treatment with *Penicillium oxalicum* in the field. Phytopathology 72:541-543.

Interactions Between Microbial Residents of Cereal Roots

P. T. W. WONG, N.S.W. Department of Agriculture, Agricultural Research Centre, Tamworth, N.S.W., 2340, Australia

Although it is a truism that cereal root-rotting diseases seldom occur alone, research continues to be directed at single pathogens. This paper considers some aspects of microbial interactions between cereal root residents, including saprophytic and parasitic as well as pathogenic microorganisms, that may affect cereal root-rot complexes. In Australia, cereal root-rot complexes involve a number of diseases. These include take-all caused by *Gaeumannomyces graminis* (Sacc.) von Arx & Olivier var. *tritici* Walker; crown rot caused by *Fusarium graminearum* Schwabe and to a lesser extent *F. culmorum* (W. G. Smith) Sacc.; common root rot caused by *Drechslera sorokiniana* (Sacc.) Subram. & Jain and more rarely by *Curvularia ramosa* (Bainier) Boedijn; bare-patch caused by *Rhizoctonia solani* Kuehn; Pythium root rot caused by *Pythium* spp.; as well as others (5). Not all of these soilborne diseases may be present at the one site, but it is not uncommon for two or three of them to occur in the same field. Depending mainly on climatic conditions, one or more of them may become serious in any particular season.

In New South Wales, the most important diseases in the root-rot complex are take-all, crown rot, and common root rot. However, some of the causal fungi show a preference for certain parts of the plant. The take-all fungus attacks the roots whereas *D. sorokiniana* attacks the subcrown internode most readily. *F. graminearum* invades the roots as well as stem tissues but causes most damage in the stem bases (6). Colonization of the seminal roots by these and other fungi is often the springboard for the invasion of the crown roots and other stem tissues. Thus, the prior occupancy of the root cortex by certain fungi has important implications for the emergence of a major pathogen in a root-rot complex. Because cortical cells of wheat roots die within 3 weeks when grown at 15° C (13), this rapid cortical cell death permits the easy ingress of a large variety of rhizosphere and rhizoplane microorganisms. The cortex, then, becomes a major arena for interactions between these root residents.

In recent years, there has been increasing interest in the biological control of soilborne diseases (2); as the goal of biological control appears more attainable, a number of questions arise regarding the disease complex. For example, if take-all is controlled, would there be an upsurge in one or more of the other diseases in the root-rot complex? Or again, if biological control agents are used, could a mixture of microorganisms be developed to protect the roots from several unrelated pathogens?

The first phenomenon is that described by Kreutzer as "disease trading." When one disease is controlled, a "vacuum" is created and a lesser disease assumes greater importance (15). With irrigated wheat in New South Wales, the diseases take-all and crown rot are usually present. In most years, when the full complement of irrigation is applied, take-all is the major disease because it is favored by wet conditions (6). In dry years, when use of irrigation water may be restricted before the end of the season, crown rot becomes the major disease because moisture stress enhances its effects (6). Therefore, in wet or dry years, one disease will take its toll. A similar effect applies to a complex of interacting microorganisms, where the ascendance of one species is contingent on the suppression of the other species.

There is now considerable information on some interactions between *G. graminis* var. *tritici* and various avirulent or hypovirulent fungi that reside in the roots as nonpathogenic parasites (3,8,9,16,28). This cross-protection phenomenon (28) has been shown in pot experiments by having two layers of inoculum, the top layer being the avirulent fungus and the bottom layer the pathogen. When wheat roots grow through the top layer, the roots become colonized by the avirulent fungus. The distal portions of the roots may be infected by the pathogen in the bottom layer, but the disease lesions do not progress toward the crown. The precolonized roots appear to prevent the pathogen from invading the stele in the proximal portions. The fungal antagonists do not appear to suppress the pathogen by hyperparasitism, and (with the exception of some *Phialophora* spp.) they do not produce antibiotics in agar (21). An extensive search for phytoalexins in wheat roots precolonized by *G. graminis* var. *graminis* has yielded no fungitoxic compounds with activity comparable to, say, phaseollin in beans (11). There is evidence that the precolonization of the cortex by *Phialophora* spp. (24) and by *G. graminis* var. *graminis* (Weir and Wong, *unpublished*) induces in the host a more rapid suberization and lignification of the endodermis and xylem tissues. However, these host-induced responses are very localized, and only the portion of the vascular cylinder directly below the cortex colonized by the avirulent fungus is altered in this way. Therefore, for this form of biological control to succeed, a large portion of the root system—and especially the proximal portion of the roots—has to be colonized by the avirulent fungi (28). This appears to have been borne out by pot experiments where different proportions of avirulent fungus and pathogens were used. Thus, a 3:1 ratio of avirulent fungus to pathogen gave more control than a 1:1 ratio, although the amount of pathogen used was the same (9). In field experiments, control was only significant where there were low populations of the pathogen compared with the population of the antagonist (30).

Different genera of saprophytic soil bacteria have been shown to suppress the wheat take-all fungus (7,12,14,26), and bacteria belonging to the *Pseudomonas putida-fluorescens* group suppressed take-all in the glasshouse (7,14,22) and in the field (26). The source of these isolates does not appear to be critical because isolates from a Fusarium-wilt suppressive soil from Salinas Valley, California, were as effective in suppressing take-all as isolates from a take-all suppressive soil (29). Isolates from a take-all suppressive soil also suppressed Fusarium wilt in flax (14).

The mechanisms of suppression by these fluorescent pseudomonads have not been fully elucidated. Although many workers select their bacterial antagonists by evaluating in vitro antibiosis, this does not always correlate with suppression in soil (7,23). The production of antibiotics may have an inhibitory role in certain diseases (17), but other bacterial characteristics that confer "rhizosphere competence" may be more important.

Another mechanism of suppression is the production of siderophores by the fluorescent pseudomonads. Siderophores are low-molecular-weight compounds with a high affinity for Fe^{3+} ions. These Fe-binding siderophores have been found to be produced by a large number of microorganisms, including fungal pathogens (20). The evidence for the role of bacterial siderophores is indirect. The isolation of these compounds from the rhizosphere has not been demonstrated, but fungal siderophores of the hydroxamate class have been detected in many soils.

The evidence for siderophore function consists in showing that soils made suppressive by the addition of the pseudomonads may be rendered conducive by the addition of an excess of Fe^{3+} ions in the form of FeEDTA (Fe ethylenediaminetetraacetate) (2). Further, Kloepper et al (14) have shown that the addition of a semipurified siderophore from the fluorescent pseudomonad called pseudobactin suppressed Fusarium wilt in pot experiments. Recently, Scher and Baker (20) demonstrated that Fusarium wilt could also be suppressed by the addition of FeEDDHA (Fe ethylene di-[O-hydroxyphenylacetate]), which mimics the action of the bacterial siderophore (Table 1). The evidence suggests that the suppression was caused by competition for Fe in the infection court. The availability of Fe in soil is governed by the equilibrium equation:

$$Fe(OH)_3 + 3H^+ \rightleftharpoons Fe^{3+} + 3H_2O$$

Therefore, under alkaline conditions, when iron can become limiting, it is conceivable that microorganisms may be deprived of Fe if they do not have efficient siderophores to scavenge for it. FeEDDHA and the bacterial siderophores have higher Fe stability constants than the siderophores produced by various pathogenic fusaria (20). The stability constants of the bacterial siderophore, EDDHA, the fungal siderophore, and EDTA are $\log_{10} K = 40.0, 33.9, 29.0,$ and 25.0, respectively (20). Hence, the iron-binding abilities of the bacterial siderophore and EDDHA are many orders of magnitude greater than those of the fungal siderophore and EDTA. This explains the inability of FeEDTA to suppress the disease; rather, it supplies the pathogen with iron. Take-all was also suppressed by FeEDDHA whereas FeEDTA had no effect (29). Moreover, although the fluorescent pseudomonads produced the catechol-hydroxamate siderophores in a low-Fe medium as indicated by a peak absorption at about 410 μm, the take-all fungus did not produce any detectable siderophores (Fig. 1) (Wong and Baker, *unpublished*). It appears that siderophores produced by fluorescent pseudomonads are one mechanism of suppression. However, this mechanism only operates in alkaline soils. In soils with pH below 7.0, Fe^{3+} ions are more available to the pathogens, and Fe-limiting conditions do not apply.

Other bacteria such as *Bacillus mycoides* (12) and *B. subtilis* (Huber, *personal communication*) have also reduced the severity of take-all disease in the field. Bednarova-Civinova et al (4) have ascribed their field control of take-all, by pelleting wheat seeds with *P. putida*, to the encouragement of an antagonistic rhizosphere population of *B. cereus* var. *mycoides* in the later stages of wheat growth following the initial colonization of wheat roots by *P. putida*. They suggested that the fluorescent pseudomonads had altered the nutritional status of the rhizosphere to favor the *Bacillus* sp. Stanek (25) has also stressed the role of vitamins and polysaccharides produced by rhizosphere bacteria such as *Agrobacterium* and *Pseudomonas* spp. in stimulating the growth of pathogens like *G. graminis* var. *tritici*.

Misaghi et al (17) have shown that some fluorescent pseudomonads can control *R. solani* in a root-rotting disease of cotton. Can these bacteria suppress *R. solani* in cereals? Can the bacteria that suppress take-all also suppress crown rot or common root rot in the cereal root-rot complex? Indeed, there is evidence that crown

Table 1. Mean incidence of Fusarium wilt when iron chelators (EDTA and EDDHA)[x] were introduced into a conducive soil infested with *Fusarium oxysporum* f. sp. *lini* with or without *Pseudomonas putida*[y]

Soil treatment	Mean incidence of flax wilt (%)
Control	60 b[z]
Control + *P. putida*	10 d
FeEDTA	70 a
FeEDTA + *P. putida*	75 a
FeEDDHA	25 c
FeEDDHA + *P. putida*	5 d

[x] EDTA = ethylenediaminetetraacetate; EDDHA = ethylene di-(O-hydroxyphenylacetate).

[y] After Scher and Baker (20).

[z] Treatments with like letters are not significantly different ($P = 0.05$).

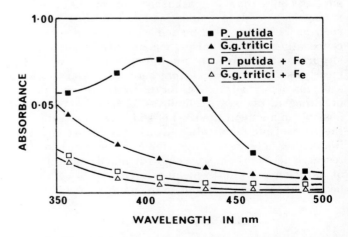

Fig. 1. Absorption spectra for the culture filtrates of *Pseudomonas putida* and *Gaeumannomyces graminis* var. *tritici* in the presence and absence of iron (0.5 g of $FeCl_3$ per liter).

rot may be reduced by the same bacterial isolates that suppressed take-all (Wong, *unpublished*). Atkinson et al (1) have associated the reduction of common root rot with larger populations of antagonistic bacteria in the rhizosphere and laimosphere (soil surrounding the subcrown internode) of some resistant cultivars and breeding lines of wheat. Would these bacteria suppress other cereal root rots?

Do root cortices previously colonized by avirulent fungi reduce infection by cortical invaders such as *Pythium* spp.? There is no information on whether *Phialophora* and *G. graminis* var. *graminis* will do this, but *Microdochium bolleyi* (Sprague) de Hoog and Hermanides-Nijhof reduced *Pythium* infection in wheat and barley when the roots were precolonized by that fungus (19). Reinecke et al (18) have also shown that *M. bolleyi* reduces the numbers of *Fusarium* spp. present in stem bases of wheat and rye. There thus appears to be a significant role for avirulent root-inhabiting fungi, or minor pathogens like *M. bolleyi*, in displacing major pathogens in root systems or in crown tissues.

The differential suppression of pathogens and their antagonists by soil microorganisms should also be explored as a means of control. Deacon and Henry (10) showed that the take-all fungus is susceptible to mycoparasitism by *Pythium oligandrum* and *P. acanthicum*, whereas *G. graminis* var. *graminis* is highly resistant. It may, therefore, be possible to augment the effects of cross-protection using *G. graminis* var. *graminis* by including the *Pythium* mycoparasites. The use of compatible species of bacteria, actinomycetes, and fungi may be one way to counter a suite of root-rotting pathogens that occurs in disease complexes.

CONCLUSIONS

There are few studies like that of Zogg's (31) that examine interactions between a complex of cereal root pathogens. It is becoming apparent, however, that the control of one root pathogen often leads to the ascendance of another and that root rots often occur as complexes that should be managed as a whole. Considerable evidence shows that biological control agents such as the fluorescent pseudomonads can suppress more than one pathogen. If the mechanism of suppression involves competition for a limiting nutrient such as Fe under alkaline soil conditions, then it is conceivable that several pathogens may be similarly suppressed. Therefore, an understanding of the mechanisms of suppression may permit more successful selection of microorganisms for the control of a number of pathogens in disease complexes. Moreover, it may be possible to augment the level of suppression or extend the range of microbial antagonism by the use of compatible species of bacteria, fungi, and actinomycetes. More research needs to be carried out with mixed populations of pathogens if efforts in disease control are not to become as futile as "sniping at an enemy platoon."

LITERATURE CITED

1. Atkinson, T. G., Neal, J. L., Jr., and Larson, R. I 1975. Genetic control of the rhizosphere microflora of wheat. Pages 116-122 in: Biology and Control of Soil-Borne Plant Pathogens. G. W. Bruehl, ed. American Phytopathological Society, St. Paul, MN. 216 pp.

2. Baker, K. F., and Cook, R. J. 1974 (original ed.). Biological Control of Plant Pathogens. Reprint ed., 1982. American Phytopathological Society, St. Paul, MN. 433 pp.

3. Balis, C. 1970. A comparative study of *Phialophora radicicola*, an avirulent fungal root parasite of grasses and cereals. Ann. Appl. Biol. 66:59-73.

4. Bednarova-Civinova, M., Petrikova, V., Stanek, M., and Vancura, V. 1981. Vliv bakterizace semen na *Gaeumannomyces graminis*, rust a vynos psenice. Ochr. Rostlin 17:89-96.

5. Butler, F. C. 1961. Root and foot rot diseases of cereals. N.S.W. Dep. Agric. Sci. Bull. 77. 98 pp.

6. Cook, R. J., and Papendick, R. I. 1972. Influence of water potential of soils and plants on root diseases. Annu. Rev. Phytopathol. 10:349-374.

7. Cook, R. J., and Rovira, A. D. 1976. The role of bacteria in the biological control of *Gaeumannomyces graminis* by suppressive soils. Soil Biol. Biochem. 8:269-273.

8. Deacon, J. W. 1973. Control of the take-all fungus by grass leys in intensive cereal cropping. Plant Pathol. 22:88-94.

9. Deacon, J. W. 1976. Biological control of the take-all fungus, *Gaeumannomyces graminis*, by *Phialophora radicicola* and similar fungi. Soil Biol. Biochem. 8:275-283.

10. Deacon, J. W., and Henry, C. M. 1978. Mycoparasitism by *Pythium oligandrum* and *P. acanthicum*. Soil Biol. Biochem. 10:409-415.

11. Deverall, B. J., Wong, P. T. W., and McLeod, S. 1979. Failure to implicate anti-fungal substances in cross-protection of wheat against take-all. Trans. Br. Mycol. Soc. 72:233-236.

12. Faull, J. L. 1978. Effects of saprophytic bacteria on take-all disease caused by *Gaeumannomyces graminis* var. *tritici*. Ann. Appl. Biol. 89:101-103.

13. Holden, J. 1975. Use of nuclear staining to assess rates of cell death in cortices of cereal roots. Soil Biol. Biochem. 7:333-334.

14. Kloepper, J. W., Leong, J., Teintze, M., and Schroth, M. N. 1980. *Pseudomonas* siderophores: A mechanism explaining disease suppressive soils. Curr. Microbiol. 4:317-320.

15. Kreutzer, W. A. 1965. The reinfestation of treated soil. Pages 495-508 in: Ecology of Soil-borne Plant Pathogens. K. F. Baker and W. C. Snyder, eds. University of California Press, Berkeley. 571 pp.

16. Lemaire, J. M., Doussinault, G., Tivoli, B., Jouan, B., Perraton, B., Dosba, F., Carpentier, F., and Sosseau, C. 1977. La lutte biologique contre le pietin-echaudage des cereales. La monoculture de blé est-elle possible? Institut Technique des Cereales et des Fourrages, 21-32.

17. Misaghi, I. J., Stowell, L. J., Grogan, R. G., and Spearman, L. C. 1982. Fungistatic activity of water-soluble fluorescent pigments of fluorescent pseudomonads. Phytopathology 72:33-36.

18. Reinecke, P., Duben, J., and Fehrmann, H. 1979. Antagonism between fungi of the foot rot complex of cereals. Pages 327-336 in: Soil-borne Plant Pathogens. B. Schippers and W. Gams, eds. Academic Press, London.

19. Salt, G. A. 1979. The increasing interest in "minor pathogens." Pages 289-312 in: Soil-borne Plant Pathogens. B. Schippers and W. Gams, eds. Academic Press, London.

20. Scher, F. M., and Baker, R. 1982. Effect of *Pseudomonas putida* and a synthetic iron chelator on induction of soil suppressiveness to Fusarium wilt pathogens. Phytopathology 72:1567-1573.

21. Sivasithamparam, K. 1975. *Phialophora* and *Phialophora*-like fungi occurring in the root region of wheat. Aust. J. Bot. 23:193-212.

22. Smiley, R. W. 1978. Antagonists of *Gaeumannomyces graminis* from the rhizoplane of wheat in soils fertilised with ammonium or nitrate nitrogen. Soil Biol. Biochem. 10:169-174.

23. Smiley, R. W. 1978. Colonisation of wheat roots by *Gaeumannomyces graminis* inhibited by specific soils, microorganisms and ammonium-nitrogen. Soil Biol. Biochem. 10:175-179.

24. Speakman, J. B., and Lewis, B. G. 1978. Limitation of *Gaeumannomyces graminis* by wheat root responses to

Phialophora radicicola. New Phytol. 80:373-380.

25. Stanek, M. 1979. *Gaeumannomyces graminis* and bacteria in the rhizosphere of wheat. Pages 247-252 in: Soil-borne Plant Pathogens. B. Schippers and W. Gams, eds. Academic Press, London.

26. Weller, D. M., and Cook, R. J. 1981. Control of take-all of wheat with fluorescent pseudomonads. (Abstr.) Phytopathology 71:1007.

27. Wong, P. T. W. 1975. Cross-protection against the wheat and oat take-all fungi by *Gaeumannomyces graminis* var. *graminis*. Soil Biol. Biochem. 7:189-194.

28. Wong, P. T. W. 1981. Biological control by cross-protection. Pages 417-431 in: Biology and Control of Take-all. M. J. C.

Asher and P. J. Shipton, eds. Academic Press, London.

29. Wong, P. T. W., and Baker, R. 1985. Control of wheat take-all and Ophiobolus patch of turfgrass by fluorescent pseudomonads. Pages 151-153 in: Ecology and Management of Soilborne Plant Pathogens. C. A. Parker, A. D. Rovira, K. J. Moore, P. T. W. Wong, and J. F. Kollmorgen, eds. American Phytopathological Society, St. Paul, MN.

30. Wong, P. T. W., and Southwell, R. J. 1980. Field control of take-all of wheat by avirulent fungi. Ann. Appl. Biol. 94:41-49.

31. Zogg, H. 1951. Studien uber die Pathogenitat von Erregergemischen bei Getreidefusskrankheiten. Phytopathol. Z. 18:1-54.

Survival of Fungal Antagonists of *Gaeumannomyces graminis* var. *tritici*

P. T. W. WONG, N.S.W. Department of Agriculture, Agricultural Research Centre, Tamworth, N.S.W., 2340, Australia

The saprophytic survival of the wheat take-all fungus, *Gaeumannomyces graminis* (Sacc.) von Arx & Olivier var. *tritici* Walker, has been studied extensively in the laboratory and in the field (1–3,5,6,9–11). It has been fairly well established that under soil conditions of adequate moisture and moderate temperatures, the wheat take-all fungus does not survive saprophytically for much beyond a year in the field. In recent years, there has been considerable interest in the use of avirulent or hypovirulent root-inhabiting fungi in the biological control of take-all (4,8,13). The ecology of these avirulent fungi, such as *Phialophora graminicola* (Deacon) Walker, *Gaeumannomyces graminis* var. *graminis*, and a lobed hyphopodiate *Phialophora* sp., is not well understood, and there have been few comparable studies on their saprophytic survival in soil (7,11). The present study investigates the survival of pathogenic and avirulent *G. graminis* as well as the avirulent *Phialophora* spp. in two soil types under controlled temperature and moisture conditions in the laboratory.

MATERIALS AND METHODS

Fungi. The fungi used in the experiments are shown in Table 1. They were stored under mineral oil on quarter-strength potato-dextrose agar plus 0.1% yeast extract (Vegemite) at 4°C until ready for use.

Fungal inoculum was in the form of colonized wheat straws as described by Garrett (5). Inocula of two isolates each of *G. graminis* var. *graminis*, *P. graminicola*, and a *Phialophora* sp. (lobed hyphopodia) and an isolate of *G. graminis* var. *tritici* were prepared.

Soils. Two soils were used: a self-mulching black earth and a red brown earth (12). Soils were collected, air-dried, sieved through a 2-mm sieve, and stored in an air-dried condition until use. The black earth was from a wheat field at the Agricultural Research Centre, Tamworth. The red brown earth was from a wheat field at the Agricultural Research Institute, Wagga Wagga, in southwestern New South Wales, where take-all is a serious disease.

Analysis of the soils showed that the black earth (pH 7.4, 0.01M CaCl$_2$) had 2.1% organic matter and 0.9% total nitrogen. The red brown earth (pH 4.7, 0.01M CaCl$_2$) had 1.6% organic matter and 0.18% total nitrogen.

Two hundred grams of soil were placed in a 500-ml, wide-mouthed glass jar, and 50 wheat straws colonized with each fungus were placed in a jar. The jar was shaken so that the straws were buried in the soil. There were two replicates, giving a total of 100 straws per fungal isolate.

Temperature-moisture regimes. Four temperature-moisture regimes were used to simulate different soil conditions: 15°C at −3 bars water potential (moist winter), 30°C at −3 bars (moist summer), 15°C at <−100 bars (dry winter), and 30° at <−100 bars (dry summer). These temperature-moisture regimes, however, were artificial in that the temperatures and moisture levels were constant and did not fluctuate as they would in the field. Survival of the fungi was studied over 6 months because these temperature-moisture regimes would not apply for a longer period.

The jars were placed in incubators at 15°C and 30°C, and the high water potential (−3 bars) was adjusted twice a week. For the treatments with low water potential, the soils were adjusted to −100 bars at the beginning of the experiments; no further adjustment was made in the 6-month period because it was too difficult to equilibrate the soil rapidly following addition of water to the top of the soil.

After incubating for 3 and 6 months, 100 colonized straws of each fungus were recovered from the soils, and the fungi were baited out using wheat seedlings (9). After the wheat seedlings were grown for 4 weeks in a glasshouse at 20–25°C, the root systems were washed free of soil and examined under a dissecting microscope for disease lesions of *G. graminis* var. *tritici*, lobed hyphopodia of *G. graminis* var. *graminis* and the *Phialophora* sp., and the "growth cessation structures" (4) of *P. graminicola*. The root systems and subcrown internodes were scored for presence or absence of the fungus.

In experiment 1, the survival of one isolate each of the four fungi was studied at the four temperature-moisture regimes in the two soil types. In experiment 2, the survival of another isolate each of the three avirulent fungi was studied in the black earth. The other isolate, *G. graminis* var. *tritici*, was unfortunately contaminated during preparation; no data are thus available for comparison with the isolate in experiment 1.

RESULTS

Experiment 1. The survival rates of one isolate each of *G. graminis* var. *tritici*, *G. graminis* var. *graminis*, *P. graminicola*, and a lobed hyphopodiate *Phialophora* sp. were assessed after 3 and 6 months at four temperature-moisture regimes (Tables 2 and 3). In general, the greatest survival occurred in the cool, dry soil (15°C, <−100 bars) followed by the warm, dry soil (30°C, <−100 bars). All the fungi had virtually been eliminated in the warm, moist soils (30°C, −3 bars) after 3 months. Survival was intermediate under cool, moist conditions.

There appears to be better survival of the fungi in the red brown earth in the dry soils compared with the black earth. In cool, moist conditions, survival of the fungi was better in the black earth (Table 2). Survival of *G. graminis* var. *tritici* and *G. graminis* var. *graminis* was comparable under dry conditions, but survival of *G. graminis* var. *graminis* was better than that of *G. graminis* var. *tritici* in the cool, moist soil.

Survival of *P. graminicola* was lower than that of *G. graminis* var. *tritici* under dry conditions but was similar under cool, moist conditions. The lobed hyphopodiate *Phialophora* sp. survived poorly under cool, moist conditions. There was a high survival rate of *G. graminis* var. *graminis* under cool, moist conditions in both soils.

Experiment 2. The survival rates of another isolate each of *G. graminis* var. *graminis*, *P. graminicola*, and the *Phialophora* sp. in the black soil were compared with those in experiment 1 (Table 4). In general, the results are comparable. However, there were considerable differences between the two isolates of a fungus. Thus, whereas one isolate of *G. graminis* var. *graminis* (DAR24167) survived well under cool, moist conditions,

the other isolate (DAR33670) declined rapidly. Similar differences occurred between the *P. graminicola* isolates. Both isolates of *Phialophora* sp. (lobed hyphopodia) survived poorly under cool, moist conditions.

Under dry conditions, the inoculum of *Phialophora* spp. declined more rapidly than that of *G. graminis* var. *graminis*. The two isolates of *G. graminis* var. *graminis* survived at a similar rate under warm, dry and cool, dry conditions.

DISCUSSION

The general trends in saprophytic survival of the three avirulent fungi and the wheat take-all fungus in the two soil types are similar to those found by MacNish (10) for *G. graminis* var. *tritici*. The fungi were virtually eliminated in warm, moist soils (30°C, −3 bars) after 3 months, whereas the best survival occurred under cool, dry conditions (15°C, <−100 bars), followed by warm, dry conditions (30°C, <−100 bars). In general, there were some differences between the survival of the fungi in the two soil types, with greater survival in the black earth than in the red brown earth under cool, moist conditions. The differences in survival of *G. graminis* var. *tritici* in the two soil types may be associated with the nitrogen status of the soil because the black earth had a higher nitrogen content than the red brown earth. It is known that nitrogen prolongs the survival of *G. graminis* var. *tritici* in soil (1,6).

There were also differences in the survival between the two isolates of the same avirulent fungus. This was

Table 1. Fungal isolates used in the survival experiments

Fungus	Isolate	Host	Locality
Gaeumannomyces graminis var. *tritici*	DAR24170	*Triticum aestivum* L.	Carnamah, W.A.
G. graminis var. *graminis*	DAR24167	*Stipa aristiglumis* F. Muell.	Warialda, N.S.W.
G. graminis var. *graminis*	DAR33670	*Ceratochloa unioloides* H.B.K.	Walcha, N.S.W.
Phialophora graminicola	DAR26394	*Agrostis stolonifera* L.	Tamworth, N.S.W.
P. graminicola	DAR27009	Unidentified grass	Cambridge, U.K.
Phialophora sp. (lobed hyphopodia)	DAR32098	*Zea mays* L.	Cambridge, U.K.
Phialophora sp. (lobed hyphopodia)	DAR32099	*Hordeum vulgare* L.	Harpenden, U.K.

Table 2. Survival of *Gaeumannomyces graminis* and *Phialophora* spp. after 3 and 6 months in black earth at four temperature-moisture regimes

Fungus	Sampling time (months)	Percentage survival of fungi at temperature-moisture regime[a]			
		1	2	3	4
G. graminis var. *tritici*	3	40	0	100	63
(DAR24170)	6	7	0	89	43
G. graminis var. *graminis*	3	95	1	99	92
(DAR24167)	6	52	0	97	30
P. graminicola	3	54	0	63	40
(DAR26394)	6	36	0	57	38
Phialophora sp.	3	4	0	87	62
(DAR32098)	6	0	0	34	41

[a] 1 = 15°C, − 3 bars; 2 = 30°C, − 3 bars; 3 = 15°C, <− 100 bars; and 4 = 30°C, <−100 bars.

Table 3. Survival of *Gaeumannomyces graminis* and *Phialophora* spp. after 3 and 6 months in a red brown earth at four temperature-moisture regimes

Fungus	Sampling time	Percentage survival of fungi at temperature-moisture regime[a]			
		1	2	3	4
G. graminis var. *tritici*	3	18	0	100	97
(DAR24170)	6	2	0	87	87
G. graminis var. *graminis*	3	83	1	100	97
(DAR24167)	6	5	0	98	68
P. graminicola	3	18	0	87	39
(DAR26394)	6	7	0	84	15
Phialophora sp.	3	0	0	88	80
(DAR32098)	6	0	0	65	57

[a] 1 = 15°C, − 3 bars; 2 = 30°C, − 3 bars; 3 = 15°C, <− 100 bars; and 4 = 30°C, <− 100 bars.

Table 4. Comparison between the survival of two isolates each of *Gaeumannomyces graminis* var. *graminis* and *Phialophora* spp. after 3 months in black earth at four temperature-moisture regimes

Fungus	Isolate	Percentage survival of fungi at temperature-moisture regime[a]			
		1	2	3	4
G. graminis var. *graminis*	DAR24167	95	1	99	92
	DAR33670	3	0	100	88
P. graminicola	DAR26394	54	0	63	40
	DAR27009	14	0	50	38
Phialophora sp.	DAR32098	4	0	87	62
	DAR32099	2	2	67	27

[a] 1 = 15°C, − 3 bars; 2 = 30°C, − 3 bars; 3 = 15°C, <− 100 bars; and 4 = 30°C, <−100 bars.

most noticeable in *G. graminis* var. *graminis*, where isolate DAR24167 survived well after 3 months under cool, moist conditions whereas isolate DAR33670 was almost eliminated from the soil (Table 4). The two isolates of *P. graminicola* also showed differences in survival under cool, moist conditions. Chambers and Flentje (2,3) and MacNish (10) have reported variation in survival between isolates of the wheat take-all fungus. It was unfortunate in this study that one of the isolates of the pathogen was contaminated during inoculum preparation, preventing comparison between isolates.

Both the isolates of the *Phialophora* sp. survived less well in cool, moist conditions than *P. graminicola* or *G. graminis* var. *tritici*. It appears, therefore, that there is considerable variation in the survival rates of the three avirulent fungi as well as isolates of a fungal species and that this character may be a useful selection criterion for fungal antagonists in the biological control of wheat take-all. The isolate DAR24167 of *G. graminis* var. *graminis* appears to be the best isolate in saprophytic survival among the avirulent fungi studied, being equal to that of the pathogen under warm, dry conditions and surpassing it under cool, moist conditions. The last result has been confirmed in survival experiments in the field (Wong, Southwell, and Murray, *unpublished*).

ACKNOWLEDGMENTS

I thank Mr. C. S. Fang for the *G. graminis* var. *tritici* isolate (DAR 24170) and Mr. R. J. Southwell for technical assistance.

LITERATURE CITED

1. Butler, F. C. 1959. Saprophytic behaviour of some cereal root-rot fungi. IV. Saprophytic survival in soils of high and low fertility. Ann. Appl. Biol. 47:28-36.

2. Chambers, S. C., and Flentje, N. T. 1968. Saprophytic survival of *Ophiobolus graminis* on various hosts. Aust. J. Biol. Sci. 22:1153-1161.

3 Chambers, S. C., and Flentje, N. T. 1969. Relative effects of soil nitrogen and soil organisms on survival of *Ophiobolus graminis*. Aust. J. Biol. Sci. 22:275-278.

4. Deacon, J. W. 1976. Biological control of the take-all fungus, *Gaeumannomyces graminis*, by *Phialophora radicicola* and similar fungi. Soil Biol. Biochem. 8:275-283.

5. Garrett, S. D. 1938. Soil conditions and the take-all disease of wheat. III. Decomposition of the resting mycelium of *Ophiobolus graminis* in infected wheat stubble buried in the soil. Ann. Appl. Biol. 25:742-766.

6. Garrett, S. D. 1940. Soil conditions and the take-all disease of wheat. V. Further experiments on the survival of *Ophiobolus graminis* in infected wheat stubble buried in the soil. Ann. Appl. Biol. 27:199-204.

7. Garrett, S. D. 1974. Cellulose decomposition and saprophytic survival of *Phialophora radicicola*. Trans. Br. Mycol. Soc. 62:622-625.

8. Lemaire, J. M., Doussinault, G., Tivoli, B., Jouan, B., Perraton, B., Dosba, F., Carpentier, F., and Sosseau, C. 1977. La lutte biologique contre le pietin-echaudage des cereales. La monoculture de blé est-elle possible? Institut Technique des Cereales et des Fourrages, 21-32.

9. MacNish, G. C. 1973. Survival of *Gaeumannomyces graminis* var. *tritici* in field soil stored in controlled environments. Aust. J. Biol. Sci. 26:1319-1325.

10. MacNish, G. C. 1976. Survival of *Gaeumannomyces graminis* var. *tritici* in artificially colonized straw buried in naturally infested soil. Aust. J. Biol. Sci. 29:163-174.

11. Slope, D. B., Salt, G. A., Broom, E. W., and Gutteridge, R. J. 1978. Occurrence of *Phialophora radicicola* var. *graminicola* and *Gaeumannomyces graminis* var. *tritici* on roots of wheat in field crops. Ann. Appl. Biol. 88:239-246.

12. Stace, H. C. T., Hubble, G. D., Brewer, R., Northcote, H. K., Sleeman, J. R., Mulcahy, M. J., and Hallsworth, E. G. 1968. A Handbook of Australian Soils. Rellim Tech. Publ., Glenside, South Australia. 435 pp.

13. Wong, P. T. W. 1981. Biological control of take-all by cross-protection. Pages 417-431 in: Biology and Control of Take-all. M. J. C. Asher and P. J. Shipton, eds. Academic Press, London.

Control of Wheat Take-all and Ophiobolus Patch of Turfgrass by Fluorescent Pseudomonads

P. T. W. WONG, N.S.W. Department of Agriculture, Agricultural Research Centre, Tamworth, N.S.W., 2340, Australia, and R. BAKER, Department of Botany and Plant Pathology, Colorado State University, Fort Collins 80523, U.S.A.

It is now well known that certain soils are suppressive to soilborne pathogens, the most notable being take-all decline or suppressive soils (1) and soils suppressive to Fusarium wilt (10). The mechanisms responsible for suppression have not been fully elucidated, but microbial antagonists have been strongly implicated (2,4). Kloepper et al (4) showed that fluorescent pseudomonads from a take-all suppressive soil controlled both take-all in barley and Fusarium wilt in flax. They indicated that fluorescent siderophores (high-affinity iron [Fe^{3+}] ion transport agents) produced by the bacteria were responsible for disease suppression in both these diseases. Scher and Baker (7) confirmed that siderophores produced by fluorescent pseudomonads played a major role in suppressing Fusarium wilt in several hosts, and they showed that a synthetic iron chelator (EDDHA) could also suppress Fusarium wilt. Ophiobolus patch of *Agrostis* turfgrass, a disease similar to wheat take-all and caused by *Gaeumannomyces graminis* var. *avenae* (Turner) Dennis, was controlled in greenhouse experiments by using a layer of soil that was made suppressive by repeated addition of inoculum of the pathogen to the soil (12). The mechanism of suppression, however, was not determined. The present study investigates the suppression of wheat take-all and Ophiobolus patch of *Agrostis* turfgrass by fluorescent pseudomonads from a Fusarium-wilt suppressive soil and possible mechanisms of suppression.

MATERIALS AND METHODS

Fungal and bacterial cultures. The fungi used were *G. graminis* (Sacc.) von Arx & Olivier var. *tritici* Walker (DAR24167) and *G. graminis* var. *avenae* (DAR23143).

The bacteria were isolated from a Fusarium-wilt suppressive soil from Salinas Valley, California, by a baiting technique (6) using mycelial mats of *G. graminis* var. *tritici* as the bait. Fluorescent pseudomonads belonging to the *Pseudomonas putida-fluorescens* group (8) were transferred and stored as lyophilized cultures. Rifampicin-resistant mutants were obtained by streaking 24-hr-old cultures of the bacteria from King's medium B (KB) (3) onto freshly poured plates of KB containing 100 μg of rifampicin (Sigma) per milliliter. Mutants were transferred and lyophilized.

In vitro antibiosis of the bacteria to the fungi was tested on quarter-strength potato-dextrose agar by spotting the bacteria at four positions at the periphery of the plate and placing at the same time a fungal plug at the center. There were three replicates, and the plates were incubated at 25°C.

Control of wheat take-all. Bacterial suspensions in 0.1M $MgSO_4 \cdot 7H_2O$ were prepared by the method of Scher and Baker (7) and thoroughly mixed into the soil infested with ground oat grain inoculum of the pathogen or killed inoculum (0.5% w/w) to give a final bacterial concentration of 10^7 c.f.u./g of soil. Control treatments received an equal amount of 0.1M $MgSO_4 \cdot 7H_2O$. Five hundred grams of a conducive soil (7) were added to 10-cm-diameter plastic pots, and 1 cm of uninfested soil was placed over this. Fifteen wheat seeds (cv. Scout) were sown and covered with a further 1 cm of soil. After emergence, the seedlings were thinned to 10 plants per pot. There were four replicates, and pots were randomized in blocks in a greenhouse (15–23°C). The pots were watered daily until freely draining. After 6 weeks, the root systems were washed out and examined under a dissection microscope for disease symptoms. The tops were cut off above the seed, dried at 80°C for 48 hr, and weighed.

In the second experiment, four isolates of fluorescent pseudomonads (one antibiotic producing, N1R6; and three nonantibiotic producing, S1R3, S2R4, S6R2) were tested for their ability to control wheat take-all. The experiment was carried out as above.

To count bacterial populations in rhizosphere soil, 10 g of soil closely adhering to the wheat roots was shaken in 90 ml of 0.1M $MgSO_4 \cdot 7H_2O$ for 10 min and serially diluted with 0.1M $MgSO_4 \cdot 7H_2O$ and plated out onto KB plates containing 100 μg of rifampicin per milliliter and 100 μg of cycloheximide per milliliter. Bacterial populations were expressed as colony-forming units per gram of oven-dried rhizosphere soil.

Effect of FeEDTA and FeEDDHA on wheat take-all. The experiment was carried out as in the first experiment except that aqueous solutions containing Fe ethylenediaminetetraacetate (FeEDTA) and Fe ethylene di-(O-hydroxyphenylacetate) (FeEDDHA) at 500 μg/ml were added to the soil before sowing the wheat seed. Two weeks later, these chelators were applied again. The wheat plants were harvested after 6 weeks. The root systems were washed out and examined for disease, and tops were oven-dried at 80°C for 48 hr and then weighed.

Control of Ophiobolus patch. Because Ophiobolus patch occurs most commonly in soils that have been fumigated with methyl bromide to control weeds, sterilized soil was used in these experiments. A conducive soil was autoclaved for 1 hr at 121°C on two successive days. *G. graminis* var. *avenae* inoculum (0.5% by weight) or killed inoculum was blended in the soil using a twin-shell blender, and the bacteria (isolates A12R1 and N1R6) were added to half the treatments to give a concentration of 10^7 c.f.u./g of soil. A 1-cm layer of infested soil was placed over this. *Agrostis* seed was sown and covered over with 0.5 cm of uninfested soil. There were four replicates, and the treatments were randomized in blocks in a greenhouse (15–23°C). The pots were watered to field capacity daily. A modified Hoagland's solution was applied weekly after 3 weeks. The tops of the turfgrass were harvested at 6 weeks, dried at 80°C for 48 hr, and weighed.

RESULTS

Control of wheat take-all. The mean dry weight of the tops of the wheat in the treatment where the bacteria and pathogen were present was significantly greater ($P \leqslant 0.01$) than with the pathogen alone, but it was not significantly different from the treatments without the pathogen (Table 1). There was significantly less ($P \leqslant 0.05$) disease in the seminal roots and especially in the crown roots where the bacteria and pathogen were present than in the treatment with the pathogen alone (Table 1). The other treatments were free of disease symptoms, and the dry weights of the tops were not significantly different.

In the second experiment, where an antibiotic-producing and three nonantibiotic-producing isolates of fluorescent pseudomonads were used, all isolates except one nonantibiotic-producing fluorescent pseudomonad significantly ($P \leqslant 0.05$) reduced disease (Table 2). Population densities of the inoculated bacteria at the end of the experiment ranged from 10^5 to 10^6 c.f.u./g of soil (Table 2).

Effect of FeEDDHA and FeEDTA on wheat take-all. The addition of FeEDDHA at 500 μg/g into a conducive soil containing *G. graminis* var. *tritici* significantly reduced ($P \leqslant 0.05$) disease in wheat compared with the treatment with the pathogen but without FeEDDHA (Table 3). The addition of FeEDTA at 500 μg/g of soil did not reduce disease, and the dry weights of tops and root disease ratings were not significantly different ($P \leqslant 0.05$) from the treatment with the pathogen alone. The Fe compounds did not affect the growth of wheat in the absence of the pathogen (Table 3).

Control of Ophiobolus patch. After 4 weeks, the *Agrostis* turf in the treatment with the pathogen alone showed extensive symptoms of chlorosis and necrosis. These symptoms persisted until the end of the experiment at 6 weeks. In the treatment inoculated with the bacteria and pathogen, less than 10% of the pot area was diseased at 4 weeks; the symptoms had disappeared at 6 weeks when the grass was harvested. Dry weights of tops of the grass in treatments inoculated with either of the two isolates of bacteria were significantly greater ($P \leqslant 0.01$) than with the pathogen alone (Table 4). No disease was seen in the control with only killed inoculum of the pathogen.

DISCUSSION

Fluorescent pseudomonads from a Fusarium-wilt suppressive soil reduced take-all in wheat and suppressed Ophiobolus patch in *Agrostis* turfgrass in greenhouse experiments. Both antibiotic-producing and nonantibiotic-producing isolates of fluorescent pseudomonads were equally effective, suggesting that in vitro antibiosis may not be a valuable criterion for screening for effective antagonists of *G. graminis*. It is, therefore, imperative to test the bacterial isolates in soil and preferably in the field to determine their efficacy in disease control.

Fluorescent siderophores have been implicated in the suppression of Fusarium wilt (4,7) and wheat take-all

Table 1. Control of wheat take-all by a fluorescent pseudomonad (N1R6) in a greenhouse experiment

Treatment	Dry weight of tops (grams per pot)	No. of diseased roots per plant	
		Seminal	Crown
Gaeumannomyces graminis			
var. *tritici* alone	0.67	5.0 ± 0.1	2.3 ± 0.2
Plus N1R6	1.22	2.7 ± 0.3	0.4 ± 0.1
N1R6 alone	1.27	0	0
Uninoculated control	1.29	0	0
LSD ($P = 0.01$)	0.12		

Table 2. Effect of four isolates of fluorescent pseudomonads on top weight of wheat with and without take-all

Treatment	Dry weight of tops (grams per pot)	Bacterial population (c.f.u./g of soil)
Gaeumannomyces graminis		
var. *tritici* alone	0.83	0
Plus S1R3	0.89	3.3×10^6
Plus S2R4	1.32	8.1×10^5
Plus S6R2	1.33	6.6×10^6
Plus N1R6	1.36	5.7×10^6
Uninoculated control	1.42	0
LSD ($P = 0.05$)	0.10	

Table 3. Effect of FeEDTA and FeEDDHA on take-all in wheat[a]

Treatment	Dry weight of tops (grams per pot)	No. of diseased roots per plant	
		Seminal	Crown
Uninoculated control	1.31	0	0
Plus FeEDTA	1.41	0	0
Plus FeEDDHA	1.53	0	0
Gaeumannomyces graminis			
var. *tritici* alone	0.47	3.7 ± 0.3	0.9 ± 0.2
Plus FeEDTA	0.37	3.7 ± 0.4	1.1 ± 0.3
Plus FeEDDHA	0.96	2.3 ± 0.3	0.5 ± 0.2
LSD ($P = 0.05$)	0.27		

[a] FeEDTA = Fe ethylenediaminetetraacetate; FeEDDHA = Fe ethylene di-(O-hydroxyphenylacetate).

Table 4. Effect of two isolates of fluorescent pseudomonads on top weight of *Agrostis* with and without *Gaeumannomyces graminis* var. *avenae*

Treatment	Dry weight of tops (grams per pot)	Bacterial population (c.f.u./g of soil)
G. graminis var. *avenae* alone	0.40	0
Plus A12R1	1.26	3.6×10^6
Plus N1R6	1.33	1.9×10^6
Uninoculated control	1.70	0
LSD ($P = 0.01$)	0.13	

(4). The present evidence suggests that siderophores play a part in take-all suppression. The addition of 500 μg of FeEDDHA per gram of soil significantly reduced take-all disease, whereas a similar amount of FeEDTA did not. Scher and Baker (7) suppressed Fusarium wilt with 100 μg of FeEDDHA per gram of soil and suggested that the iron chelator acted in the same way as the bacterial siderophore. Several bacterial isolates and the fungal pathogens were checked for siderophore production in low-Fe media. The fluorescent pseudomonads produced fluorescent siderophores with absorbance peaks at 400–410 nm and at 460 nm (Wong and Baker, *unpublished*). The first peak is similar to that described by Teintze et al (9) for the siderophore pseudobactin. This compound suppressed Fusarium wilt when added to soil (4). However, in our experiments, the pathogens of *G. graminis* var. *tritici* and var. *avenae* did not produce similar siderophores, and the absorbance of their supernatants was similar to that of media amended with 0.5 g of FeCl$_3$ per liter (Wong and Baker, *unpublished*).

FeEDDHA has a high affinity for Fe^{3+} ions and has a stability constant of $10^{33.9}$, which is approaching that of the fluorescent bacterial siderophore (ca. 10^{40}), whereas FeEDTA has a stability constant several orders of magnitude smaller (10^{25}). Therefore, the results suggest that FeEDDHA or the bacterial siderophore was capable of depriving the pathogen of Fe for growth and infection, whereas FeEDTA was not. However, it is unlikely that only one mechanism is involved, and it is perhaps too simplistic to suggest that one or two mechanisms can explain the control of widely differing pathogens like *Fusarium oxysporum* and *G. graminis* by a diverse population of fluorescent pseudomonads.

Weller and Cook (11) have obtained significant control of wheat take-all by pelleting fluorescent pseudomonads from a take-all suppressive soil onto wheat seed. The fluorescent pseudomonads from the Fusarium-wilt suppressive soil have also been shown to suppress take-all in the field (Wong, *unpublished*). However, it is interesting that these bacteria from Salinas Valley have been encouraged in soils cropped almost exclusively to vegetables. It is possible that "specific" suppression may not only apply to one pathogen but may be effective against several plant pathogens.

The use of fluorescent pseudomonads in the field control of Ophiobolus patch may be more easily achieved than for control of take-all because it could be economic to fumigate golf and bowling greens with methyl bromide before introducing the antagonists. It is known that fluorescent pseudomonads rapidly recolonize soils fumigated with methyl bromide (5), and it is conceivable that the introduced fluorescent pseudomonads could dominate the rhizosphere microflora for a sufficiently long period to inhibit the activity of any contaminating inoculum of *G. graminis* var. *avenae*. In time, "general suppression" (2) would develop and succeed the "specific suppression" conferred by the bacteria.

LITERATURE CITED

1. Baker, K. F., and Cook, R. J. 1974 (original ed.). Biological Control of Plant Pathogens. Reprint ed., 1982. American Phytopathological Society, St. Paul, MN. 433 pp.
2. Cook, R. J., and Rovira, A. D. 1976. The role of bacteria in the biological control of *Gaeumannomyces graminis* by suppressive soils. Soil Biol. Biochem. 8:269-273.
3. King, E. O., Ward, M. K., and Raney, D. E. 1954. Two simple media for the demonstration of pyocyanin and fluorescin. J. Lab. Clin. Med. 44:301-307.
4. Kloepper, J. W., Leong, J., Teintze, M., and Schroth, M. N. 1980. *Pseudomonas* siderophores: A mechanism explaining disease suppressive soils. Curr. Microbiol. 4:317-320.
5. Ridge, E. H. 1976. Studies in soil fumigation. II. Effects of bacteria. Soil Biol. Biochem. 8:249-253.
6. Scher, F. M., and Baker, R. 1980. Mechanism of biological control in a Fusarium-suppressive soil. Phytopathology 70:412-417.
7. Scher, F. M., and Baker, R. 1982. Effect of *Pseudomonas putida* and a synthetic iron chelator on induction of soil suppressiveness to Fusarium wilt pathogens. Phytopathology 72:1567-1573.
8. Stanier, R. Y., Palleroni, N. J., and Doudoroff, M. 1966. The aerobic pseudomonads: A taxonomic study. J. Gen. Microbiol. 43:159-271.
9. Teintze, M., Hossain, M. B., Baines, C. L., Leong, J., and van der Helm, D. 1981. Structure of ferric pseudobactin, a siderophore from a plant growth promoting *Pseudomonas*. Biochemistry 20:6446-6457.
10. Toussoun, T. A. 1975. Fusarium suppressive soils. Pages 145-151 in: Biology and Control of Soil-Borne Plant Pathogens. G. W. Bruehl, ed. American Phytopathological Society, St. Paul, MN.
11. Weller, D. M., and Cook, R. J. 1981. Control of take-all of wheat with fluorescent pseudomonads. (Abstr.) Phytopathology 71:1007.
12. Wong, P. T. W., and Siviour, T. R. 1979. Control of Ophiobolus patch in *Agrostis* turf using avirulent fungi and take-all suppressive soils in pot experiments. Ann. Appl. Biol. 92:191-197.

Part IV

Plant Responses and Resistance

Role of Plant Breeding in Controlling Soilborne Diseases of Cereals

P. R. SCOTT and T. W. HOLLINS, Plant Breeding Institute, Cambridge, U.K.

Generalizations about the broad topic covered by this paper are likely to be erroneous, but eyespot (*Pseudocercosporella herpotrichoides* Fron) and take-all (*Gaeumannomyces graminis* (Sacc.) von Arx & Olivier) provide contrasting examples of success and failure in the control of soilborne cereal disease by resistance breeding.

EYESPOT

This splash-dispersed disease of stem bases is characteristic of wet climates where cereals, especially winter wheat, are grown intensively. It also occurs in drier areas, such as the Pacific Northwest of the United States, where very susceptible cultivars are grown. Its lesions reduce yield directly and especially indirectly by causing lodging (7). Plant breeding can contribute greatly to its control in three main ways.

Short stiff straw. Dwarf cultivars generally resist lodging. Scott and Hollins (20) showed that two tall cultivars susceptible to lodging lost much more yield because of eyespot when they were allowed to lodge than when they were kept upright with nets; nets had no effect on two short cultivars that are resistant to lodging.

Shorter straw has been an important objective of wheat breeding programs throughout the world for many decades. An incidental advantage of short straw is that it confers tolerance to eyespot lesions by reducing lodging. This tolerance does not confer resistance to the lesions themselves, which can still reduce yield directly and contribute to the soilborne inoculum affecting a subsequent crop.

Resistance introduced from other cultivars. Limited variation in resistance to eyespot among cultivars had long been known in Europe (6) and North America (28), but it was not until 1950 that more substantial resistance was discovered, in England, in the French winter wheat cultivar Cappelle-Desprez (1,29). Cappelle-Desprez was an outstandingly successful cultivar in several European countries for many years: in Britain, it dominated the wheat crop and was officially recommended throughout the period 1953–1976. At a time when cereal growing was greatly intensifying, the cultivar's eyespot resistance enabled it to be grown even after one or more previous wheat crops.

The resistance is by no means complete: eyespot lesions are still common on Cappelle-Desprez, but they are less frequent and less extensive than on susceptible cultivars, such as Holdfast (Table 1). Although only moderate in degree, the resistance substantially reduces yield loss from eyespot. Application of fungicide may sometimes be advantageous to reduce losses still further.

Much of the resistance of Cappelle-Desprez is determined by one or more genes on a single chromosome, 7A (11). It is readily introduced into new cultivars; in Britain, at least three-quarters of the wheat crop has been protected by this resistance for many years. Furthermore, the resistance has apparently lost none of its effectiveness during more than 30 years of widespread and extensive exploitation in agriculture. It can thus be confidently considered durable (9).

Unfortunately, resistance from Cappelle-Desprez has not been so widely exploited outside Britain, even in its native France. In the Pacific Northwest of the United States, for example, the wheat cultivar Cerco, released for its resistance to eyespot, proved to be nearly as susceptible as Holdfast when we tested it in Britain. Introduction of the moderate but durable resistance of Cappelle-Desprez into locally adapted wheats should control eyespot in a climate that is not highly favorable to the disease (3).

Commercial winter barley cultivars in Britain are usually no more severely attacked than winter wheats that have the resistance of Cappelle-Desprez (25). If necessary, increased resistance could probably be introduced from older cultivars (21).

Resistance from other species. A much more potent source of resistance to eyespot is available in *Aegilops ventricosa*, a wild goat grass of the Mediterranean region. The resistance was transferred into wheat by Maia (13) in France to create the line VPM 1. It is consistently more resistant than Cappelle-Desprez (5); in the moderate epidemics characteristic of the Cambridge area, however, it has not yet been

Table 1. Eyespot score and effect of eyespot inoculation on yield for three winter wheat cultivars in four years[a]

Cultivar	1973	1974	1975	1976	Mean
Eyespot score in inoculated plots (0–100)					
Holdfast	76	66	37	73	63
Cappelle-Desprez	61	31	22	59	43
VPM 1	48	14	11	52	31
LSD (*P* = 0.05)	9	12	5	8	
Yield reduction caused by eyespot inoculation (%)[b]					
Holdfast	17***	19***	7	14**	14
Cappelle-Desprez	−2	8*	6	0	3
VPM 1	4	8	8	3	6

[a] Field trials at Cambridge; six to nine replicates, plots about 3 m².
[b] Significance of yield reductions (not quoted for mean of four trials): * = *P* < 0.05, ** = *P* < 0.01, *** = *P* < 0.001.

possible to demonstrate an advantage over Cappelle-Desprez in terms of yield loss caused by eyespot (Table 1). The resistance is readily transferred to new cultivars in plant breeding programs (2) and, if it remains effective, it could reasonably be expected to render unnecessary the use of fungicides to control eyespot. But the durability of this resistance has not yet been tested by extensive exposure in agriculture.

Potent resistance is also available in rye. This is partially effective in triticale and could perhaps be transferred to wheat (16,22). However, a variant form of *P. herpotrichoides*, characteristic of rye-growing areas and designated R-type, differs from the more usual form, designated W-type, in being equally pathogenic to wheat and rye (10,27). It is therefore unlikely that resistance from rye, if introduced into wheat, would prove durable.

P. herpotrichoides has other variants adapted to *Agropyron repens* and to *Aegilops squarrosa* (4,19). This capacity for adaptation to different cereal species but not to different wheat cultivars (23) must cast some doubt on the durability of resistance introduced into wheat from outside the species. In breeding wheat for resistance to eyespot, we therefore suggest that it is preferable to introduce first the moderate resistance of Cappelle-Desprez, which is of proven durability. Only if this proves inadequate should more potent resistance be introduced, for example from *Aegilops ventricosa*, and then it should be accompanied by resistance from Cappelle-Desprez as a safeguard.

TAKE-ALL

This root disease is present in most except the driest cereal-growing regions. Although species of cereals differ substantially in resistance, resistant cultivars play no part in controlling the disease. However, three approaches to breeding for resistance can be suggested.

Screening cultivars for resistance. It is seldom realized how extensive have been the attempts, since about 1920, to find variation in resistance among cultivars. They have met with negligible success. The mainly negative results have often been published in inconspicuous research reports, but Nilsson (15) surveyed the literature thoroughly and listed more than 100 relevant publications from at least 20 countries, mostly concerned with cultivars of wheat. Scott (18) listed about 30 more, published in the following decade.

A characteristic example of an early study is that of Russell (17), who inoculated 100 wheat cultivars in Canada with take-all in replicated tests, repeated several times. In a given experiment, fourfold differences in response to take-all were sometimes observed between cultivars; but there was much variation in their relative performance from one experiment to another. The results of further experiments on four of the cultivars did not correspond with the first results; furthermore, different seed sources caused more variation than cultivars. Russell concluded that such slight differences in the susceptibility of cultivars as might exist were too small to be used in plant breeding. Unfortunately, many other authors have interpreted more limited data less cautiously, and the literature abounds with claims that cultivars differ in susceptibility, based on isolated observations that

were not, or could not be, confirmed.

From so many negative results, it is clear that cultivars of wheat and barley do not show conspicuous and consistent differences in susceptibility, though large differences often appear in one or two experiments. However, particularly thorough studies by Nilsson (15), Mattsson (14), and Jensen and Jørgensen (8) do show that small but repeatable differences can be found between some cultivars. Mattsson (14) even succeeded in transferring slight resistance from a Russian to a Swedish wheat cultivar. Unfortunately, however, we have to agree with him and with Jensen and Jørgensen (8) that the slight and uncertain advantage to be gained does not justify the very substantial breeding effort required.

After decades of fruitless search, it now seems most unlikely that further screening of cultivars will reveal potent forms of resistance. Indeed, the similarity of cultivars provides evidence that take-all was not a strong selective influence in the evolution of wheat or barley, possibly because the disease was subject to natural biological control by soil microorganisms in the grasslands in which the cereals evolved (18).

Interbreeding populations of cultivars. Even though no source of resistance in wheat or barley is potent enough to be readily selected in conventional breeding programs, a breeding system might nevertheless be devised to recombine, in a single genotype, the putative genes of very minor and perhaps fortuitous effects on take-all from a number of cultivars. If such genes were to act additively, a useful degree of resistance might be expressed: even so, its probably polygenic inheritance would unfortunately make it difficult to transfer to new cultivars, except by similarly laborious means.

We have been trying this speculative approach to the synthesis of resistance that has not occurred naturally by constraining numerous wheat cultivars to interbreed in large populations for a number of years on land heavily infested with take-all (18). Hybrid seed was collected from a part of the population that had inherited male sterility, either through the cytoplasm (21) or the nucleus (24). Only plants with limited take-all lesions on the roots were harvested. The interbreeding population has now been subjected to such selection for 4–5 years. Although neighboring plants are often diseased to substantially different degrees, we have no evidence that this is heritable variation, and it does not appear that the population has shifted markedly toward greater resistance. Whether a slight improvement has been achieved will only be determined when the selected population is compared with its parent population in replicated trials.

Resistance from other species. If a useful degree of resistance cannot be obtained within the species, other species can be considered. The grass species with the highest degree of resistance, for example in the genera *Phleum* and *Poa* (15), do not form fertile hybrids with wheat or barley. Wild species of *Triticum* or *Hordeum* show hardly more variation in resistance than wheat and barley cultivars (8,15). *Aegilops, Agropyron, Secale* (rye), and *Haynaldia* are closely related to *Triticum*, and fertile hybrids can be made with some species. Potent resistance has been reported in all these genera (18); they are very dissimilar from cultivated wheat, however, and the technical difficulty of transferring resistance to wheat while avoiding almost all their other characters should not be underestimated.

Rye, for example, has potent resistance, such that it hardly suffers damage from take-all in agriculture. Unfortunately, recombination of genetic information from wheat and rye is difficult to achieve. They do, however, form a stable hybrid, triticale, which can be grown as a crop with resistance intermediate between that of wheat and rye (26). Transfer of resistance from rye to wheat may also be inhibited by the probably diffuse genetic determination of resistance in rye (12,16).

Like *P. herpotrichoides* and several other necrotrophic parasites of cereals (23), *G. graminis* possesses at least one variant form, var. *avenae*, pathogenically adapted to a particular host species. Adaptation to rye may also occur (26). As with eyespot, there must therefore be some doubt about the durability of resistance derived from other species even if it could be achieved.

CONCLUSIONS

Resistance to foot rot and root rot cereal diseases, most of which are soilborne, is typically more difficult to achieve than resistance to diseases of parts that are above the ground. This may again be a reflection of the tendency for diseases associated with the soil to have been controlled biologically during the evolution of their hosts, under the influence of the abundant soil microflora. The two examples discussed in this paper show the difficulty of generalizing about resistance to soilborne disease. To the cereal breeder, take-all remains one of the most intractable diseases, whereas eyespot provides an outstanding example of successful exploitation of durable resistance.

LITERATURE CITED

1. Batts, C. C. V., and Fiddian, W. E. H. 1955. Effect of previous cropping on eyespot in four varieties of winter wheat. Plant Pathol. 4:25-28.
2. Bingham, J. 1983. Winter wheat. Selection for disease resistance. Rep. Plant Breed. Inst. for 1982, p. 25.
3. Bruehl, G. W., Nelson, W. L., Koehler, F., and Vogel, O. A. 1968. Experiments with *Cercosporella* foot rot (straw breaker) disease of winter wheat. Wash. Agric. Exp. Stn. Bull. 694. 14 pp.
4. Cunningham, P. C. 1965. *Cercosporella herpotrichoides* Fron on gramineous hosts in Ireland. Nature (London) 107:1414-1415.
5. Doussinault, G., and Dosba, F. 1977. An investigation into increasing the variability for resistance to eyespot in wheat. Eyespot variability in the subtribe *Triticinae*. Z. Pflanzenzuecht. 79:122-133.
6. Foex, E., and Rosella, E. 1934. Les piétins du blé. Rev. Pathol. Veg. Entomol. Agric. Fr. 21:9-14.
7. Glynne, M. D. 1944. Eyespot, *Cercosporella herpotrichoides* Fron, and lodging of wheat. Ann. Appl. Biol. 31:377-378.
8. Jensen, H. P., and Jørgensen, J. H. 1976. Screening barley varieties for resistance to the take-all fungus, *Gaeumanno-myces graminis*. Z. Pflanzenzuecht. 76:152-162.
9. Johnson, R. 1978. Practical breeding for durable resistance to rust diseases in self-pollinating cereals. Euphytica 27:529-540.
10. Lange-De La Camp, M. 1966. Die Wirkungsweise von *Cercosporella herpotrichoides* Fron dem Erreger der Halmbruchkrankheit des Getreides. II. Aggressivität des Erregers. Phytopathol. Z. 56:155-190.
11. Law, C. N., Scott, P. R., Worland, A. J., and Hollins, T. W. 1975. The inheritance of resistance to eyespot (*Cercosporella herpotrichoides*) in wheat. Genet. Res. 26:73-79.
12. Linde-Laursen, I., Jensen, H. P., and Jørgensen, J. H. 1973. Resistance of *Triticale*, *Aegilops* and *Haynaldia* species to the take-all fungus, *Gaeumannomyces graminis*. Z. Pflanzenzuecht. 70:200-213.
13. Maia, N. 1967. Obtention de blés tendres résistants au piétin-verse par croisements interspecifiques blés × *Aegilops*. C. R. Seances Acad. Agric. Fr. 53:149-154.
14. Mattsson, B. 1973. Efterforskande av rotdödarresistenta sorter och överföring av resistens till svenskt material. Sver. Utsaedesfoeren. Tidskr. 83:281-297.
15. Nilsson, H. E. 1969. Studies of root and foot rot diseases of cereals and grasses. I. On resistance to *Ophiobolus graminis* Sacc. Lantbrukshoegsk. Ann. 35:275-807.
16. Riley, R., and Macer, R. C. F. 1966. The chromosomal distribution of the genetic resistance of rye to wheat pathogens. Can. J. Genet. Cytol. 8:640-653.
17. Russell, R. C. 1934. Studies in cereal diseases. X. Studies of take-all and its causal organism, *Ophiobolus graminis* Sacc. Can. Dep. Agric. Bull. 170 (N. S.). 64 pp.
18. Scott, P. R. 1981. Variation in host susceptibility. Pages 219-236 in: Biology and Control of Take-all. M. J. C. Asher and P. J. Shipton, eds. Academic Press, London.
19. Scott, P. R., Defosse, L., Vandam, J., and Doussinault, G. 1976. Infection of lines of *Triticum*, *Secale*, *Aegilops* and *Hordeum* by isolates of *Cercosporella herpotrichoides*. Trans. Br. Mycol. Soc. 66:205-210.
20. Scott, P. R., and Hollins, T. W. 1974. Effects of eyespot on the yield of winter wheat. Ann. Appl. Biol. 78:269-279.
21. Scott, P. R., and Hollins, T. W. 1976. Eyespot. Take-all. Rep. Plant Breed. Inst. for 1975, pp. 133-135.
22. Scott, P. R., and Hollins, T. W. 1979. Eyespot. Rep. Plant Breed. Inst. for 1978, pp. 152-153.
23. Scott, P. R., and Hollins, T. W. 1980. Pathogenic variation in *Pseudocercosporella herpotrichoides*. Ann. Appl. Biol. 94:297-300.
24. Scott, P. R., and Hollins, T. W. 1981. Take-all. Rep. Plant Breed. Inst. for 1980, pp. 95-96.
25. Scott, P. R., and Hollins, T. W. 1983. Eyespot. Rep. Plant Breed. Inst. for 1982, pp. 97-98.
26. Scott, P. R., Hollins, T. W., and Gregory, R. S. 1985. Relative susceptibility of wheat, rye, and triticale to isolates of take-all. Pages 180-182 in: Ecology and Management of Soilborne Plant Pathogens. C. A. Parker, A. D. Rovira, K. J. Moore, P. T. W. Wong, and J. F. Kollmorgen, eds. American Phytopathological Society, St. Paul, MN.
27. Scott, P. R., Hollins, T. W., and Muir, P. 1975. Pathogenicity of *Cercosporella herpotrichoides* to wheat, barley, oats and rye. Trans. Br. Mycol. Soc. 65:529-538.
28. Sprague, R., and Fellows, H. 1934. Cercosporella foot rot of winter cereals. U.S. Dep. Agric. Tech. Bull. 428. 24 pp.
29. Vincent, A., Ponchet, J., and Koller, J. 1952. Recherche de variétés de blés tendres peu sensibles au piétin-verse: Résultats préliminaires. Ann. Amelior. Plant. 3:459-472.

Phytophthora drechsleri Causes Crown Rot and the Accumulation of Antifungal Compounds in Cucurbits

A. ALAVI and M. SABER, Plant Pests and Diseases Research Institute, Tehran, Iran; and R. N. STRANGE, Lecturer in the Department of Botany and Microbiology, University College, London, U.K.

Cantaloupes (*Cucumis melo* L. var. *cantalupensis*), melons (*C. melo* var. *inodorus*), cucumbers (*C. sativus* L.), watermelons (*Citrullus lanatus* (Thunb.) Mansf.), pumpkins (*Cucurbita maxima* Duch. ex. Lam.), and marrows (*C. pepo* L.) are grown under irrigation over an area of more than 250,000 ha in Iran (2). They are frequently devastated by a severe wilt known locally as green death (1). *Phytophthora drechsleri* Tucker was isolated from a cantaloupe plant showing symptoms of the disease, and three of Koch's postulates were proved (1). There remained doubt about the first of the postulates because no attempts had been made to show that the fungus was constantly associated with the disease.

Our earlier experiments suggested that species of *Cucumis* (cantaloupe, melon, and cucumber) were susceptible but that species of *Cucurbita* (pumpkins and marrows) were resistant to the isolate previously used (IMI 264494) (2). In attempts to find the cause of this specificity, a search was made for antifungal compounds produced by the plants in response to infection.

MATERIALS AND METHODS

Koch's postulates. Isolates of the fungus were made from cucurbits grown in different parts of Iran (Table 1) by the baiting technique previously described (1). Pathogenicity of the isolates was tested by inoculation of cantaloupe seedlings (1). The fungus was recovered from these plants by surface sterilization and plating on CMA (1).

Antifungal compounds. Cotyledons (10–14 days old) and leaves (from plants 3–4 weeks old) harvested from pot-grown cantaloupes, cucumbers, watermelons, marrows, and pumpkins were arranged in petri dishes and kept moist as previously described for detached leaves of *Vicia faba* (8). Wounds (2–3 mm long and about 1 cm apart) were made in tissues on either abaxial or adaxial surfaces by a needle. Zoospores were produced axenically either in mineral salts solution [Ca(NO₃)₂, 0.1 mM; MgSO₄, 1 mM; KNO₃, 1 mM; and FeSO₄, 0.1 mM] or in water, essentially according to the methods of Halsall and Forrester (4) and Khew and Zentmyer (7). A droplet (20 μl) of inoculum (zoospore suspension in mineral salts solution or in distilled water, 5×10^4 ml⁻¹) was placed on each wound using a microrepette (Jencons Scientific Ltd., Hemel Hempstead, U.K.). Controls received mineral salts solution or distilled water. Plates were wrapped in polyethylene bags and incubated in the dark at 25°C for 48 hr.

After incubation, tissues were homogenized in 70% methyl alcohol (MeOH). The homogenates were filtered through two layers of muslin and then through Whatman No. 1 filter paper. The alcohol was removed on a rotary evaporator at <40°C, and the resulting aqueous extract was partitioned three or four times against equal volumes of ethyl acetate (6). The ethyl acetate fractions were combined, dried by the addition of anhydrous Na₂SO₄, and the volume reduced to give the equivalent of 20 g (fresh weight) of material per milliliter. Both the ethyl acetate and aqueous phases were chromatographed on silica gel thin layer chromatography (TLC) in petroleum ether:ethyl acetate (70:30, v/v). After removal of the solvent the plates were sprayed with a dense conidial suspension of *Cladosporium cucumerinum* (3) and were incubated at 20–25°C in a humid atmosphere for 72 hr.

Areas of TLC plates corresponding to antifungal compounds were scraped and eluted in 70% MeOH. The eludates were dissolved in CH₃CN:H₂O (1:1, v/v) and the compounds separated by reverse phase high-pressure liquid chromatography on a column of Hypersil ODS (250 × 10 mm in diameter). The solvent was CH₃CN:H₂O:CH₃COOH (50:49:1, v/v/v). The effluent from the column was monitored at 354 nm by a Pye Unicam ultraviolet detector (Cambridge, U.K.) and the absorbance recorded on a Teckman chart recorder (TE 200, Teckman Electronics Ltd., Bicester, Oxfordshire, England). Fractions were cut according to the peaks and troughs on the chart and tested for antifungal activity by the *Cladosporium*-TLC assay.

Mass spectra of pure compounds were obtained on a VG 7070 high-resolution, double-focusing mass spectrometer interfaced to a Finnigan INGOS data system.

Table 1. Origins of isolates of *Phytophthora drechsleri*

Province	City	Host	Organ
Bakhtaran	Bakhtaran	Cucumber	Stem
Esfahan	Esfahan	Cucumber	Stem
Esfahan	Esfahan	Melon	Stem
Ilam	Darreh-Shahr	Cantaloupe	Stem
Khorassan	Mashhad	Melon	Stem
Lorestan	Boroojerd	Cucumber	Stem
Lorestan	Khorram-abad	Cucumber	Stem
Markazi	Ghom	Muskmelon	Stem
Markazi	Karadj	Cucumber	Stem
Markazi	Saveh	Cantaloupe	Stem
Markazi	Shemiran	Cucumber	Fruit
Zanjan	Abhar	Cucumber	Stem
Zanjan	Abhar	Melon	Stem

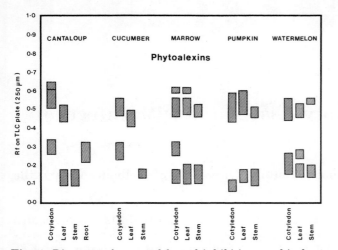

Fig. 1. Diagram of areas of fungal inhibition on thin layer chromatography plates caused by extracts of cucurbits challenged by *Phytophthora drechsleri*. Extracts were chromatographed on silica gel (250 μm thickness) in petroleum ether (40–60 b.p.):ethyl acetate (70:30, v/v) and sprayed with a thick spore suspension of *Cladosporium cucumerinum* in double-strengh Czapek-Dox liquid medium. Shaded areas denote zones of fungal inhibition.

RESULTS AND DISCUSSION

P. drechsleri was isolated from cucurbits with crown rot symptoms from 11 different localities in Iran. In all instances, inoculation of cantaloupes with these isolates gave rise to typical crown rot symptoms, and the same fungus was reisolated from these test plants. These data leave little doubt that *P. drechsleri* is the cause of crown rot of cucurbits in Iran.

In an attempt to discover the reason for differential susceptibility of cucurbits to *P. drechsleri*, the possibility that the fungus might cause the accumulation of antifungal compounds was examined. Cotyledons and leaves of all species tested developed pronounced water-soaking symptoms and lost their turgor when inoculated with droplets containing zoospores of the fungus. In some instances, orange red, necrotic lesions occurred at the site of inoculation. TLC of ethyl acetate extracts of inoculated tissue allowed the separation of up to four compounds that were antifungal (Fig. 1). Aqueous extracts after partitioning with ethyl acetate

were inactive. The compound with an R_f value of 0.45–0.58 was further purified by reverse phase high-pressure liquid chromatography. On mass spectrometry, the compound appeared pure and gave a molecular ion of 318.089155 corresponding to an empirical formula of $C_{20}H_{14}O_4$. There were also prominent fragments at m/z 317, 239, 163, 120, 105, and 77. This compound was found in cotyledons and leaves of cantaloupe, watermelon, marrow, and pumpkin. It clearly differs from coniferyl alcohol found in cucumber plants infected with *Colletotrichum lagenarium* (5). Further work is in progress to obtain sufficient material for proton and ^{13}C NMR spectroscopy so that the structure of the compound may be determined.

ACKNOWLEDGMENTS

We wish to thank M. A. Ahmadi-nejad, Dr. E. Behdad, Mr. B. Esmail-pour, Mr. A. Hashemi, Mr. A. Karimi, and Mr. G. Shirzadi for supplying samples of infected cucurbits and Dr. M. Mruźek for mass spectrometry data.

LITERATURE CITED

1. Alavi, A., and Strange, R. N. 1979. A baiting technique for isolating *Phytophthora drechsleri*, causal agent of crown rot of *Cucumis* species in Iran. Plant Dis. Rep. 63:1084-1086.
2. Alavi, A., Strange, R. N., and Wright, G. 1982. The relative susceptibility of some cucurbits to an Iranian isolate of *Phytophthora drechsleri*. Plant Pathol. 31:221-227.
3. Bailey, J. A., and Burden, R. S. 1973. Biochemical changes and phytoalexin accumulation in *Phaseolus vulgaris* following cellular browning caused by tobacco necrosis virus. Physiol. Plant Pathol. 3:171-177.
4. Halsall, D. M., and Forrester, R. I. 1977. Effects of certain cations on the formation and infectivity of *Phytophthora* zoospores. 1. Effect of calcium, magnesium, potassium and iron ions. Can. J. Microbiol. 23:994-1001.
5. Hammerschmidt, R., and Kuć, J. 1982. Lignification as a mechanism for induced systemic resistance in cucumber. Physiol. Plant Pathol. 20:61-71.
6. Keen, N. T., Sims, J. J., Erwin, D. C., Rice, E., and Partridge, J. E. 1971. 6a-hydroxyphaseollin: An antifungal chemical induced in soybean hypocotyls by *Phytophthora megasperma* var. *sojae*. Phytopathology 61:1084-1089.
7. Khew, K. L., and Zentmyer, G. A. 1973. Chemotactic response of zoospores of five species of *Phytophthora*. Phytopathology 63:1511-1517.
8. Strange, R. N., Deramo, A., and Smith, H. 1978. Virulence enhancement of *Fusarium graminearum* by choline and betaine and of *Botrytis cinerea* by other constituents of wheat germ. Trans. Br. Mycol. Soc. 70:201-207.

Changes in Root Tissue Permeability Associated with Infection by *Phytophthora cinnamomi*

DAVID M. CAHILL and GRETNA M. WESTE, School of Botany, University of Melbourne, Parkville, Victoria, 3052, Australia

Phytophthora cinnamomi Rands causes root rot resulting in severe disease and death of many species in Victorian dry sclerophyll forests. In a study to ascertain how the pathogen actually kills plants, the cellular damage caused by this pathogen to aseptically grown seedling roots was examined by measuring electrolyte loss from inoculated roots. The leakage is primarily caused by increased permeability resulting from membrane damage and can be measured as changed conductivity in a fluid bathing the roots. Changes in tissue permeability are an important early indication of pathogenesis (2,8,11). Alterations in temperature of the root and its environment have been shown to affect susceptibility to *P. cinnamomi* (2,3). Changes in leakage were therefore measured over a range of temperatures for both susceptible and field-resistant species.

The presence of microorganisms around a growing root stimulates greater root exudation (7). To test whether observed leakage associated with *P. cinnamomi* was caused by the presence of the fungus and not by its action directly on root cells, a nonpathogen was inoculated onto host roots. Culture filtrates of *P. cinnamomi* were also tested for their capacity to increase leakage from roots (2).

MATERIALS AND METHODS

Seedlings of 13 species (Table 1) were grown in a controlled environment and roots were inoculated with 10–20 zoospores of *P. cinnamomi* from an axenically prepared suspension (4). Zoospores encysted on the root surface and germinated within 0.5 hr; the germ tubes penetrated root tissue within 2 hr. Damage to cell membranes was measured by the change in conductance of the fluid bathing the roots, using a Phillips conductivity meter (PW 9501/01) as previously described (9). Roots were severed 1 cm behind the tip, washed to remove electrolytes released during excision, and placed in 5 ml of distilled water in a water bath at 21 ± 0.5°C and continuously shaken. Changes in conductance were measured 2, 8, 16, and 24 hr after inoculation. Uninoculated roots served as controls. Measurements were made immediately after washing the roots and at half-hourly intervals thereafter for 4 hr. Results were converted to μmho per gram (fresh weight) of tissue.

To examine the effect of temperature, the root systems of 8-month-old seedlings of *Eucalyptus marginata* (susceptible) and *E. calophylla* (field resistant) growing in a sand and vermiculite mix were subjected to one of three different temperatures—14, 24, and 28°C—while the tops were maintained at 24°C. Roots were inoculated with approximately 20 zoospores, then excised 24 hr later and placed in glass distilled water (pH 6.8). Changes in conductivity of the water were measured over the succeeding 4 hr. Lesion lengths were recorded, and all roots were plated onto agar at the conclusion of the experiment. Uninoculated controls were treated similarly.

For tests with the nonpathogen, spores from a culture of *Mortierella ramanniana* were inoculated onto host roots and conductivity measured on excised roots 24, 48, and 72 hr after inoculation.

Culture filtrates were derived from a continuously shaken culture of *P. cinnamomi* grown in the medium of Chee and Newhook (1). Roots were immersed directly into the filtrate.

RESULTS

A progressive increase in conductance of the fluid bathing the roots was observed in most inoculated species compared with uninoculated controls. This is illustrated for jarrah, *E. marginata* Sm. (susceptible), and marri, *E. calophylla* (field resistant) (Fig. 1A and B). Other species, notably the resistant species of Table 1, showed little if any electrolyte loss over the period of analysis (Fig. 1C and D). Both the amount of leakage and its time of occurrence varied to some extent with host susceptibility (Table 1). When the nonpathogenic *M. ramanniana* was inoculated onto host roots, there was no leakage when compared with controls up to 48 hr after inoculation. Similarly, the culture filtrate prepared

Table 1. Time in hours after inoculation with *Phytophthora cinnamomi* at which significant changes in conductance occurred

Species	Hours
Susceptible	
Eucalyptus marginata	8
E. sieberi	16
Themeda australis	8
Xanthorrhoea australis	16
X. resinosa	16
Field resistant	
Acacia melanoxylon	2
A. pulchella	2
E. calophylla	2
E. maculata	8
Gahnia radula	2
Resistant	
Juncus bufonius	>24
Triticum sp.	>24
Zea mays	>24

from *P. cinnamomi* induced little change in conductance when compared with controls.

Leakage of electrolytes from roots of the two *Eucalyptus* spp. held at both 14 and 24°C was of a similar magnitude and ranged from 20 to 30% of the total available, calculated on a freeze-thaw basis inoculation on roots grown at 14°C (Table 2). *E. marginata* lost significantly greater quantities of electrolytes at both temperatures. The fungus was readily reisolated from the roots at 14°C and was, therefore, responsible for the leakage although the lesion observed at higher temperatures was absent.

Leakage from roots of plants grown and inoculated at 28°C was much greater. Lesion length was more than twice that of those grown at 24°C. The significantly greater leakage compared with roots at 24°C probably reflects the amount of membrane damage. The optimum temperature for growth of the particular isolate of *P. cinnamomi* is around 28°C, and hence bigger lesions and greater electrolyte loss may be expected. Again, *E. marginata* lost a significantly greater quantity of electrolytes than *E. calophylla*.

To test whether the results obtained could be reproduced mechanically, healthy roots of both species were artificially wounded by gentle forceps compression and changes in conductivity measured. Leakage was

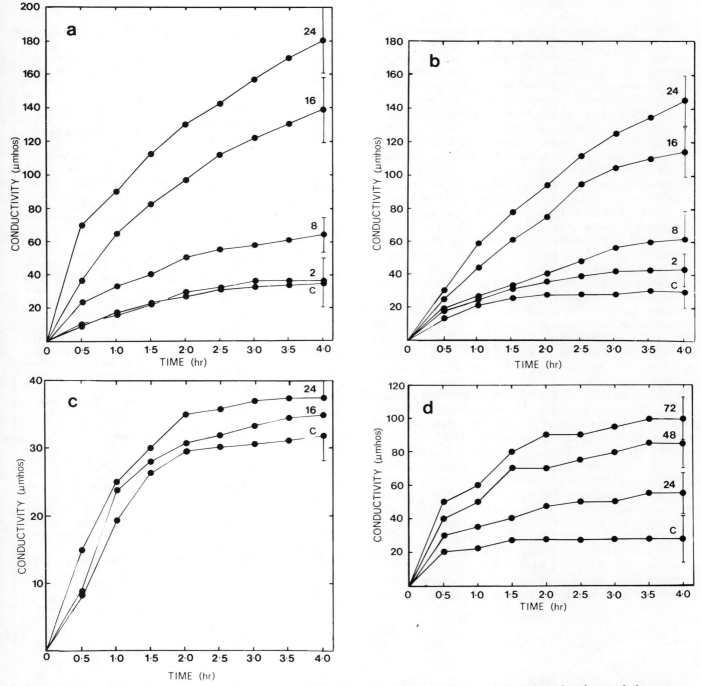

Fig. 1. Conductance of the fluid bathing roots either inoculated with *Phytophthora cinnamomi* or uninoculated controls for a range of postinoculation periods (16, 24, 48, or 72 hr). **(A)** *Eucalyptus marginata* (susceptible), **(B)** *E. calophylla* (field resistant), **(C)** wheat (resistant), and **(D)** *Juncus bufonius* (resistant). Vertical bars represent 95% confidence limits.

Table 2. Lesion length 24 hr after inoculation[a]

Length (mm)	Eucalyptus calophylla			E. marginata		
	14°C	24°C	28°C	14°C	24°C	28°C
Mean lesion length	0	9	22	0	18	33
Range	0	8–11	18–27	0	14–26	26–39

[a]Fungus was reisolated in all cases.

Table 3. Percentage of total electrolytes lost from inoculated and wounded roots after 4 hr in distilled water

Temperature (°C)	Eucalyptus calophylla[a]		E. marginata[b]	
	Inoculated	Wounded	Inoculated	Wounded
14	20	⋯	28	⋯
24	19	28	33	27
28	40	30	50	32

[a]Field resistant.
[b]Susceptible.

similar in both species and was of the magnitude obtained for inoculated roots maintained at 24°C. There was no difference between the susceptible and field resistant species (Table 3).

DISCUSSION

Observations of the 13 species tested have shown that the roots of all species attract and are penetrated by zoospores of *P. cinnamomi* whether they are susceptible, field resistant, or resistant. Any resistance is expressed after penetration (6). These experiments demonstrated that in all species examined, the tissue was penetrated within 2 hr; leakage from damaged cell membranes also occurred within 2 hr. Rates of tissue penetration and of significant leakage were interrelated, varying to a certain extent with host susceptibility. In resistant species, penetration was slow and did not extend beyond a defined lesion. The time after inoculation before a significant amount of leakage occurred was more than 24 hr. Field-resistant species reacted rapidly to pathogen invasion, but there was less total leakage compared with susceptible roots. In the latter, penetration, damage, and hence leakage were slow, extensive, and continuous.

Leakage may be a consequence of membrane wounding by the pathogen; obviously, however, wounding cannot explain either the increased leakage at higher temperatures or the difference in response between susceptible and field-resistant species. The latter processes may be a function of plant response to the pathogen, and hence they vary with temperature and species.

Changes in permeability have been recorded as very early symptoms of fungal invasion, being observed before increases in respiration (11) and before visible symptoms. Water-soaking is normally the first visible sign of leakage from infected roots (5); it was observed 24–48 hr after inoculation with *P. cinnamomi*, whereas conductivity changes were often marked at 2+ hr after zoospore inoculation. Changes in membrane permeability associated with disease have usually been directly linked to the action of a fungal toxin or enzyme (10). There is to date no direct evidence that *P. cinnamomi* produces a toxin. Woodward et al (12,13) have isolated a 1,3-β-glucan from *P. cinnamomi* culture filtrates that caused wilting in high concentrations when applied to cut stems. In experiments described in

this paper, culture filtrates did not induce leakage. This result may be compared with leakage induced by culture filtrates of *Fulvia fulva* (Cooke) Ciferri reported by Dow and Callow (2).

LITERATURE CITED

1. Chee, K. H., and Newhook, F. J. 1965. Nutritional studies with *Phytophthora cinnamomi* Rands. N. Z. J. Agric. Res. 8:523-529.
2. Dow, J. M., and Callow, J. A. 1979. Leakage of electrolytes from isolated leaf mesophyll cells of tomato induced by glyco-peptides from culture filtrates of *Fulvia fulva* (Cooke) Ciferri (syn. *Cladosporium fulvum*). Physiol. Plant Pathol. 15:27-34.
3. Grant, B. R., and Byrt, P. N. 1984. Root temperature effects on the growth of *Phytophthora cinnamomi* in the roots of *Eucalyptus marginata* and *E. calophylla*. Phytopathology 74:179-184.
4. Hwang, S. C., Ko, W. H., and Aragaki, M. 1975. A simplified method for sporangial production by *Phytophthora cinnamomi*. Mycologia 67:1233-1234.
5. Newton, H. C., Maxwell, D. P., and Sequeira, L. 1973. A conductivity assay for measuring virulence of *Sclerotinia sclerotiorum*. Phytopathology 63:424-428.
6. Phillips, D. P., and Weste, G. 1984. Field resistance in three native monocotyledon species which colonize indigenous sclerophyll forest after invasion by *Phytophthora cinnamomi*. Aust. J. Bot. 32:339-352.
7. Rovira, A. D. 1969. Plant root exudates. Bot. Rev. 35:35-57.
8. Thatcher, F. S. 1939. Osmotic and permeability relations in the nutrition of fungal parasites. Am. J. Bot. 26:449-458.
9. Weste, G., and Cahill, D. 1982. Changes in root tissue associated with infection by *Phytophthora cinnamomi*. Phytopathol. Z. 103:97-108.
10. Wheeler, H. E. 1976. Permeability alterations in diseased plants. Pages 413-429 in: Physiological Plant Pathology. R. Heitefuss and P. H. Williams, eds. Encyclopedia of Plant Physiology, vol. 4. Springer-Verlag, Berlin.
11. Wheeler, H. E., and Black, H. S. 1963. Effects of *Helminthosporium* and victorin upon permeability. Am. J. Bot. 50:683-695.
12. Woodward, J. R., Keane, P. J., and Stone, B. A. 1979. β-glucans and β-glucan hydrolases in plant pathogenesis with special reference to wilt-inducing toxins from *Phytophthora* species. Pages 113-141 in: Chemistry and Biology of Fungal Polysaccharides. Advances in Chemistry. A.C.S. Symp. Ser.
13. Woodward, J. R., Keane, P. J., and Stone, B. A. 1980. Structures and properties of wilt-inducing polysaccharides from *Phytophthora* species. Physiol. Plant Pathol. 16:439-454.

Stability of Verticillium Resistance of Potato Clones and Changes in Soilborne Populations with Potato Monoculture

J. R. DAVIS, J. J. PAVEK, D. L. CORSINI, and L. H. SORENSEN, University of Idaho Research and Extension Center, Aberdeen 83210, U.S.A.

Verticillium wilt (*Verticillium dahliae* Kleb.) of potato (*Solanum tuberosum* L.) is recognized as a major limiting factor of potato production in several areas of the world, including the United States. Because of this, a considerable portion of our efforts has been directed toward the development of potato cultivars with high degrees of resistance to Verticillium wilt. From these efforts we have obtained a collection of germ plasm with genes for *V. dahliae* resistance. Data describing the stability of Verticillium resistance with the continuous cropping of potato are lacking in the literature. We have investigated the stability of resistance with monoculture of the potato genotypes included in this study and have related results to yield and quality. When potato genotypes were compared during continuous cropping, the effects on soilborne populations of *Verticillium* spp. were also evaluated. The results of these studies are described in this paper. A portion of this work has been published (3).

MATERIALS AND METHODS

Resistant potato genotypes A66107-51 (-51), A68113-4 (-4), and Targhee (T) were compared with commonly grown susceptible cultivars Russet Burbank (RB) and Butte (B) by continuously cropping each genotype for five consecutive years (1977–1981) in the same respective field plots. In 1982, the cultivar RB was uniformly cropped over all plot locations and comparisons of wilt severity and yield were made between cropping sites.

The field area of this investigation consisted of a Declo loam soil at the University of Idaho Research and Extension Center, Aberdeen, Idaho. For more than 30 years this field had been previously cropped with barley, wheat, and potato and had a known history of Verticillium wilt. Field plots were designed as a randomized block with four replicates per treatment. Plot dimensions were 7.2 × 22.5 m with a 0.9-m spacing between rows. During the 6-year period of this study, cultural management practices of the University of Idaho involving fertility, irrigation, and insect and weed control were followed. Assays for *V. dahliae* colonization of potato stem tissue were made each year, and the survival of *Verticillium* spp. in soil was evaluated by previously described methods (1,2,4). Each year of this investigation, the dates of planting (20–23 May) and harvest (25 September–7 October) were similar. During the spring of each year (May–June), soil samples for assay of *Verticillium* spp. were randomly collected from the uppermost 23 cm of soil profile. Stem samples were collected during the last week of August each year for *V. dahliae* assays, and during a similar time period (28 August–2 September) plants were routinely evaluated for Verticillium wilt as previously described (4). Potatoes were harvested each year between 25 September and 7 October from 12 m of row in each plot. Following determinations of yield and grade, the colony-forming units of *Verticillium* spp. and disease severities were compared and these factors were correlated to evaluate relationships. Duncan's multiple range test was routinely used for comparisons between treatments. All analyses were made in accordance with appropriate experimental designs.

RESULTS AND DISCUSSION

The results of 5 years of continuous cropping demonstrate the stability of field resistance to Verticillium wilt for three potato genotypes (Table 1). During continuous cropping for 5 years, the resistant potato clones (T, -51, -4) have remained resistant. Because symptoms of Verticillium wilt may not always be reliable for an accurate diagnosis (5), the degree of *V. dahliae* colonization in potato tissue was taken into

Table 1. Stability of Verticillium wilt resistance of five potato genotypes after 5 years of continuous cropping in the same respective plot sites[y]

Genotypes	Stems with severe wilt, 31 August (%)	Log$_{10}$ of *V. dahliae* (c.f.u. +1) per gram of apical stem tissue (24 August)
Russet Burbank	59 a[z]	3.4321 a
Butte	58 a	3.8056 a
Targhee	5 b	2.5186 b
A68113-4	1 b	0.1653 c
A66107-51	0 b	0.1653 c

[y]Data collected at Aberdeen, Idaho, in 1981.

[z]Different letters denote significant differences (*P* = 0.01) by Duncan's multiple range test.

Table 2. Influence of 5 years of continuous cropping with several potato genotypes upon soilborne populations of *Verticillium* spp.

Genotypes	Colony-forming units per gram of air-dried soil	
	V. dahliae	*V. tricorpus*
Russet Burbank	315 a[z]	1 a
Butte	228 b	16 ab
Targhee	199 b	15 ab
A68113-4	81 c	61 c
A66107-51	85 c	53 bc

[z]Different letters denote significant differences (*P* = 0.05) by Duncan's multiple range test.

account. Our results with *V. dahliae* assays from stem tissue (Table 1) confirm that these clones remain resistant to *V. dahliae* colonization with continuous cropping.

During each year of this study (3), wilt severities for both -51 and -4 have remained negligible (1.0% or less). Wilt severities of T have ranged from 5 to 13% with continuous cropping, and the incidence of wilt for either RB or B has consistently exceeded 39% by late August.

Table 2 shows relative soil population levels of *V. dahliae* and *V. tricorpus* Isaac after 5 years of monoculture with each genotype. *V. dahliae* populations were significantly lower in locations where B, T, -51, and -4 had been grown, compared with sites where RB had been cropped. In locations where the most resistant clones (-51 and -4) had been grown, *V. dahliae* populations were the lowest. In contrast to this, an inverse relationship was evident with *V. tricorpus*. In plot sites of the most resistant potatoes, *V. tricorpus* populations were higher than in RB sites.

When RB was cropped over all plot locations in 1982, the $\log_{10}$ values of *V. dahliae* populations in soil were found to be highly correlated with the incidence of wilt severity ($r = 0.69$, $P = 0.001$), whereas *V. tricorpus* populations were negatively correlated ($r = -0.51$, $P = 0.05$). When RB was grown over all plot sites in 1982, the incidence of wilt was significantly less (on 20 August) by 46–75% in plot sites where -51 and -4 had been grown for 5 years than in RB sites. These results were further verified by assays of *V. dahliae* in potato stem tissue. Potato stems collected on 19 August had significantly fewer colony-forming units of *V. dahliae* in RB stems grown over plot locations with a history of -51 and -4 than from plot sites with a history of RB cropping.

Throughout these investigations, wilt severity has been closely correlated with yield. When RB was cropped over all plot sites in 1982, the incidence of wilt was negatively correlated with tuber yield ($r = -0.47$, $P = 0.05$) and positively correlated with undersized tubers ($r = 0.67$, $P = 0.01$). Potato genotypes with the highest degree of *V. dahliae* resistance (-51 and -4) have consistently maintained the highest yields along with acceptable quality. Yields of -51 have ranged from 26 to 110% higher than RB, whereas yield increases of -4 have ranged from 38 to 142% higher.

LITERATURE CITED

1. Butterfield, E. J., and Devay, J. E. 1977. Reassessment of soil assays for *Verticillium dahliae*. Phytopathology 67:1073-1078.
2. Davis, J. R., and McDole, R. E. 1979. Influence of cropping sequences on soilborne populations of *Verticillium dahliae* and *Rhizoctonia solani*. Pages 399-405 in: Soil-borne Plant Pathogens. B. Schippers and W. Gams, eds. Academic Press, London.
3. Davis, J. R., Pavek, J. J., and Corsini, D. L. 1983. Development of new potato cultivars with high Verticillium resistance and increased yield potential by the year 2000. Page 71 in: Research for the Potato in the Year 2000. W. J. Hooker, ed. Proc. C.I.P. Decennial Congr., Lima, Peru.
4. Davis, J. R., Pavek, J. J., and Corsini, D. L. 1983. A sensitive method for quantifying *Verticillium dahliae* colonization in plant tissue and evaluating resistance among potato genotypes. Phytopathology 73:1009-1014.
5. Isaac, I., and Harrison, J. A. C. 1968. The symptoms and causal agents of early dying disease (Verticillium wilt) of potato. Ann. Appl. Biol. 61:231-244.

Field Resistance of Wheat Cultivars to Crown Rot (*Fusarium graminearum* Group 1)

R. L. DODMAN and G. B. WILDERMUTH, Queensland Department of Primary Industries, Toowoomba, and T. A. KLEIN and F. W. ELLISON, I. A. Watson Wheat Research Centre, University of Sydney, Narrabri, N.S.W., 2390, Australia

Crown rot is an important disease of wheat in northern New South Wales and southern Queensland. The disease is caused by *Fusarium graminearum* Schwabe, group 1, a different pathotype from that which causes head scab in wheat and stalk and cob rot in maize. Aspects of infection, symptoms, epidemiology, and control have been reviewed by Burgess et al (1).

Differences in the susceptibility of wheat cultivars to crown rot were first observed by McKnight and Hart (2), who noted a lower incidence of "whiteheads" in some cultivars. In detailed investigations on cultivar resistance in both the glasshouse and the field, Purss (3) found no correlation between seedling blight and the behavior of cultivars in the field. He concluded that inoculation of seed with a spore suspension of *F. graminearum* was a reliable method of producing disease and that whitehead production and crown symptoms just prior to maturity gave a suitable assessment of cultivar resistance. Two cultivars with some resistance to crown rot were found and several more were reported by Wildermuth and Purss (4) after screening some 400 additional cultivars. The results of further investigations are reported in this paper.

MATERIALS AND METHODS

Isolates of *F. graminearum* from diseased wheat tillers were stored in liquid nitrogen and cultured on Czapek-Dox agar. Inoculum was prepared by colonizing a mixture of sterilized wheat and barley grain for 14 days with single-spore cultures of *F. graminearum*. Spores were obtained from this grain by shaking with water and decanting; seed was then soaked in the spore suspension for 1 hr and dried before planting. Colonized grain was also used as inoculum by blending it with water and pouring the suspension alongside plants at the late-tillering stage of growth (8 weeks after seeding). Alternatively, the colonized grain was dried, ground to pass a 2-mm sieve, and then placed in the furrow with the seed at planting.

Cultivars were sown directly into fields containing natural inoculum at Bellata and Edgeroi in New South Wales. Both sites had grown wheat for at least 2 years, with the previous crops being cultivars Songlen at Bellata and Timgalen at Edgeroi. Each site had 25% of plants severely affected by crown rot in the previous year. The soil at Bellata was a grey-black, self-mulching clay and at Edgeroi was a black, self-mulching clay.

At Millmerran in Queensland, plots with natural inoculum were obtained by planting a susceptible cultivar in an area previously inoculated with colonized grain, growing the plants to maturity, and cultivating the stubble into the soil. Trials were planted in these areas in the following season. The soil was a heavy, black, cracking clay.

Disease severity was assessed by pulling plants at maturity, separating the tillers, and removing the lower leaf sheaths. Tillers were divided into diseased and healthy classes based on the presence or absence of a honey-brown discoloration of the tiller base. Plants with at least one diseased tiller were rated as diseased. From a sample of at least 50 plants (usually in excess of 200 tillers), the percentage of diseased tillers was determined.

The bread wheat cultivars Banks, Cook, Gala, Gamut, Oxley, Puseas, and Songlen were assessed in conjunction with the durum wheat cultivar Durati.

RESULTS

When seed was inoculated with a spore suspension of *F. graminearum*, the level of crown rot was increased and differences between cultivars were expressed (Table 1). Both methods of inoculation with colonized grain increased crown rot severity, with the greatest difference between susceptible and resistant cultivars occurring when inoculum was spread alongside the growing plants (Table 2).

At the two sites in New South Wales, cultivars grown in fields with natural inoculum showed differences in resistance (Table 3). Crown rot was more severe with early sowing, whereas Cook was consistently the most resistant cultivar. At Millmerran in Queensland, high levels of disease in susceptible cultivars (Oxley had 85.0% and Puseas had 81.3% diseased tillers) were also obtained with natural inoculum. The cultivar Gala showed excellent resistance (19.1% diseased tillers).

DISCUSSION

Seed inoculation of wheat can increase the incidence of crown rot and can be used to screen cultivars for resistance to this disease. These findings complement the results of Purss (3), who demonstrated that seed inoculation can provide disease ratings of cultivars that parallel those from natural infection. The major deficiency of seed inoculation in disease screening studies is that a considerable number of plants may die from "seedling blight."

Other methods of inoculation have been examined to determine their effectiveness in increasing disease

Table 1. Disease severity in two cultivars either uninoculated or seed inoculated with a spore suspension of *Fusarium graminearum*

	Percentage of diseased plants	
Cultivar	Uninoculated	Inoculated
Gamut (susceptible)	23.3 a[z]	51.6 c
Gala (resistant)	18.5 a	29.2 b

[z] Means followed by the same letter do not differ significantly ($P = 0.05$).

Table 2. Disease severity in two cultivars either uninoculated or inoculated in two ways with grain colonized with *Fusarium graminearum*

	Percentage of diseased tillers[y]		
Cultivar	Uninoculated	Inoculum alongside plants	Inoculum in furrow
Oxley (susceptible)	46.6 c[z]	92.9 e	78.3 d
Gala (resistant)	7.2 a	25.6 b	32.2 bc

[y] Arcsin transformation applied to data for analysis; values given are equivalent means.

[z] Means followed by the same letter do not differ significantly ($P = 0.05$).

Table 3. Disease severity in five cultivars grown at Bellata and Edgeroi in 1979 and sown either in May or July

	Percentage of diseased tillers[y]			
	Bellata		Edgeroi	
Cultivar	May	July	May	July
Durati	94 a[z]	86 a	83 a	60 b
Songlen	94 a	81 a	82 a	51 b
Banks	85 a	83 a	81 a	51 b
Oxley	85 a	86 a	69 a	81 a
Cook	79 a	53 b	59 b	38 b

[y] Square root transformation applied to data for analysis; values given are equivalent means.

[z] Means within the same column followed by the same letter do not differ significantly ($P = 0.05$).

levels and reducing disease escapes. The use of colonized grain avoids the problem of seedling blight and provides consistent and reliable results. The ability to induce high levels of crown rot is of particular significance in disease-screening studies because it provides flexibility in the choice of screening sites. This can be important where it is desirable to avoid confounding influences caused by other root and crown diseases such as common root rot (*Bipolaris sorokiniana*).

Resistant cultivars consistently have a lower proportion of tillers with basal browning. This criterion is the most reliable means of identifying resistance. Tillers without symptoms are often infected by *F. graminearum*, but there appears to be little effect on final grain production. Yield losses with more resistant cultivars are thus lower than with highly susceptible cultivars. Crown rot has occurred in all cultivars assessed, indicating that resistance is not complete. Disease severity can vary considerably, with different levels of disease occurring in the same cultivar based on changes in seasonal conditions, inoculum level, time of planting, soil type, and soil nutrient status.

The screening procedures described have been successfully applied to the selection of resistant lines from breeding material. Segregating material has been inoculated, grown to maturity, and individual healthy plants selected. Subsequent generations have been screened as lines with further selection for maximum resistance. Lines with resistance equal to that of Gala have been obtained from the cross Gala (resistant) $\times$ Puseas (highly susceptible).

Potential cultivars are screened for resistance to crown rot before being released to commercial growers. This information can be used by growers in areas prone to crown rot. The use of cultivar resistance, in combination with rotations, long fallowing (18 months), and stubble burning, provides an effective integrated approach to crown rot control.

LITERATURE CITED

1. Burgess, L. W., Dodman, R. L., Pont, W., and Mayers, P. 1981. Fusarium diseases of wheat, maize and grain sorghum in eastern Australia. Pages 64-76 in: *Fusarium*: Diseases, Biology and Taxonomy. P. E. Nelson, T. A. Toussoun, and R. J. Cook, eds. Pennsylvania State University Press, University Park.
2. McKnight, T., and Hart, J. 1966. Some field observations on crown rot disease of wheat caused by *Fusarium graminearum*. Queensl. J. Agric. Anim. Sci. 23:373-378.
3. Purss, G. S. 1966. Studies on varietal resistance to crown rot of wheat caused by *Fusarium graminearum* Schw. Queensl. J. Agric. Anim. Sci. 23:475-498.
4. Wildermuth, G. B., and Purss, G. S. 1971. Further sources of field resistance to crown rot (*Gibberella zeae*) of cereals in Queensland. Aust. J. Exp. Agric. Anim. Husb. 11:455-459.

Variability in *Phytophthora cactorum* in India

V. K. GUPTA, K. S. RANA, and N. M. MIR, Department of Mycology and Plant Pathology, Himachal Pradesh Agriculture University, Nauni, Solan-173230, India

Rootstocks are known to differ in their reactions to *Phytophthora cactorum* (3). When the problem was discussed in the workshop on *P. cactorum* held in British Columbia, Canada, in 1974, it was concluded that information about variability in the pathogen was lacking (8), especially with respect to Indian isolates. The physiology of the pathogenicity of different pathotypes needs further study, and the present study is an endeavor in this direction.

MATERIALS AND METHODS

Isolation of *P. cactorum*. Soil samples were collected from apple orchards located in the states of Himachal Pradesh and Jammu and Kasmir, India. Each sample contained soil from the collar region of the infected plant to a depth of 8 in. *P. cactorum* was isolated by using hard, raw apple fruits as bait. Soil samples were transferred to plastic containers 6 in. deep, and water was poured into these containers to bring its level half an inch above the soil. Apple fruits were placed in the containers, which were kept at room temperature (19–25°C). Fruits developing hard brown lesions were utilized for isolating *P. cactorum*. Potato-dextrose agar having benomyl (20 ppm), streptomycin sulphate (60 ppm), penicillin G (60 ppm), and brassicol (100 ppm) was used for isolation of the fungus. The petri plates were incubated at 23 ± 1°C for 3–5 days.

Reaction of *P. cactorum* isolates on apple. Dormant basal portions of 10 rootstocks were cut into pieces 5–6 cm long. Four-millimeter cylinders of bark were removed with a cork borer and replaced with freshly growing *P. cactorum* cultures. These were placed in moist chambers in petri plates that were incubated at 23 ± 1°C for 10 days to record the lesion size. There were four replicates.

Preparation of enzymes. The previous year's growth of five rootstocks was inoculated with five isolates of *P. cactorum*, and infected bark was taken after 10 days for enzyme assay. The collected plant tissues were crushed in a pestle and mortar using 1.5 ml of 0.2 M phosphate buffer (pH 6.6) per gram of tissue. The extract was filtered through several layers of muslin cloth, centrifuged at 2,000 rpm for 20 min (13), and the supernatant used as a crude enzyme source.

Assaying the enzymes. *Polyphenol oxidase.* The method described by Matta and Dimond (7) was used, where 0.01 M catechol served as substrate. Change in absorbance was measured at 495 nm in a Bausch and Lomb Spectronic 20 colorimeter. The results were expressed as change in optical density per gram of fresh material per minute. An increase in absorbance by 0.01 in 1 min was taken as one unit (2).

Peroxidase. Modified Hampton's (5) method was employed, using 0.05 M pyrogallol as the substrate. Change in absorbance was measured at 420 nm in the same colorimeter as above, and the activity was expressed as change in optical density per gram of fresh tissue. Increase in absorbance by 0.01 in 1 min was taken as one unit (2).

Production of macerating enzymes. The cultures were grown in glucose asparagine solution by adding enzyme inductor substances (1.5% carboxy methyl cellulose [CMC] for cellulase and 1.5% pectin for pectolytic enzymes). Fifty milliliters of medium in 250-ml flasks was inoculated with uniform-sized culture pieces and incubated at 23 ± 1°C for 15 days. After the filtrate was centrifuged at 2,000 rpm for 20 min, a few drops of toluene were added in the clear suspension and the filtrate was stored below 4°C in a refrigerator.

Assaying. The activity of polygalacturonase (PG), polymethyl galacturonase (PMG), and cellulase (CX) was determined by measuring loss of viscosity (6) using the Fenske-Ostwald viscometer in a water bath at 30 ± 1°C. The substrates were 1% solutions each of CMC, sodium polypectate, and pectin maintained at pH 5.2 in acetate buffer. The reaction mixture consisted of 5 ml of the respective substrate, 2.5 ml of buffer, and 2.5 ml of enzyme extract. The initial reading (Vo) for each substrate was taken with the reaction mixture in which active enzyme extract was replaced by an equal amount (2.5 ml) of boiled enzyme extract. The viscosity of different reaction mixtures was determined after 120 min of reaction time (Vt). The viscosity of distilled water was recorded as Vw. The percentage of activity was calculated as $(Vo - Vt)/(Vo - Vw) \times 100$.

The activity of polygalacturonate transeliminase (PGTE) and pectin methyl transeliminase (PMTE) was also assayed viscometrically as above using 1% pectin and 1.2% sodium polypectate in borate buffer at pH 8.7.

RESULTS

The reactions of seven isolates of *P. cactorum* on the basal parts of 10 apple rootstocks have been categorized as trace, moderately susceptible, susceptible, and highly susceptible depending upon the type and lesion size (Table 1). The isolates clearly had different reactions on different rootstocks. MM104 and M2 behaved as susceptible rootstocks to all seven isolates. MM105 was moderately susceptible or susceptible to isolates P1–P6 and was resistant to P7. Rootstocks M4, M26, and MM115 had trace infections that did not increase further with most of the isolates, but isolates P4 on M4 and P5 on MM115 were virulent. MM106, MM110, and MM111 had mixed reactions with different isolates. It is clear from this data that the reaction with different isolates varied in the same rootstock, thereby showing that variability exists in *P. cactorum* isolates.

Reaction of young growth of apple rootstocks. Data on lesion size on the previous year's growth of five rootstocks reveal different reactions to different isolates (Table 2). M4 was resistant to all isolates except P4. MM104 was highly susceptible, but the lesion size varied with different isolates. P4 was the most and P1 the least virulent in these rootstocks. On MM114, P5 was more virulent in comparison with other isolates. MM101 and MM110 had mixed reactions.

Polyphenol oxidase activity. Polyphenol oxidase activity in both healthy and infected twigs was assayed after 10 days of inoculation (Table 3). Of the healthy rootstocks, MM104 and MM101 had less activity of polyphenol oxidase in comparison with M4 and MM114; MM110 had the maximum activity. There was an increase in the activity of this enzyme in all rootstocks after infection. The degree of increase in activity varied in different rootstocks with fungal isolates. The rootstocks that were susceptible to a particular isolate had less activity of this enzyme after infection as compared with the resistant one. P1 isolate (least virulent) increased the activity of this enzyme to a maximum level in all the rootstocks. Maximum enzyme activity with this isolate was observed in MM110, followed by M4 and MM114 (resistant rootstocks) and last by MM101 (susceptible).

Similarly, the P2 isolate had maximum activity with MM114 (resistant) and minimum with MM104 (susceptible). Isolate P5 was less virulent on M4, on which the

highest activity of polyphenol oxidase was observed; rootstocks MM101 and MM104 were susceptible to this isolate and had lesser activity. P3 and P4 isolates also brought about similar changes in the enzyme activity in different rootstocks depending upon the reaction.

Peroxidase activity. The activity of peroxidase enzyme was assayed after 10 days of incubation of the rootstocks at $23 \pm 1°C$ (Table 4).

Of the healthy rootstocks, MM104 and MM101 had less activity of peroxidase compared with M4, MM114, and MM110; there was an increase in its activity in all rootstocks after infection. The degree of increase or decrease in activity varied in different rootstocks in relation to infection with different isolates. Susceptible rootstocks had less peroxidase activity after inoculation than resistant ones.

Maximum peroxidase activity was observed in MM114 with P4 isolate. Least activity was observed in MM101 (susceptible), followed by MM104 (susceptible) with P2. With P1 isolate, maximum activity was found in M4 and MM110 and minimum in MM101. With P2 isolate, maximum activity was observed in MM114 followed by M4 and minimum in MM104. With P3 isolate, maximum activity was found in M4 and minimum in MM101.

Production of macerating enzymes by different isolates in vitro. The activity of all the pectolytic enzymes (Table 5) was maximum in P5 isolate. Activity of PG was significantly more in P5 as compared with other isolates, where it was almost the same; activity was lowest in P3.

The activity of PGTE was also maximum in P5, and it

Table 1. Reaction of different isolates of *Phytophthora cactorum* on apple rootstocks[a]

Rootstock	Fungus isolate						
	P1	P2	P3	P4	P5	P6	P7
M2	S	HS	MS	S	HS	S	HS
M4	T	T	T	MS	T	R	T
M7	T	T	T	MS	MS	MS	T
M26	T	T	T	T	T	T	T
MM104	MS	S	S	HS	HS	S	HS
MM105	MS	HS	MS	S	S	S	R
MM106	T	S	S	S	T	S	S
MM110	MS	S	S	MS	S	MS	MS
MM111	T	T	T	MS	MS	T	T
MM115	T	T	T	T	MS	T	T

[a]R = no infection; T = infection established, but lesion size did not increase; MS = lesion size from 6 to 15 mm, but restricted; S = lesion size from 15 to 30 mm; and HS = lesion larger than 30 mm.

Table 2. Reaction of different isolates of *Phytophthora cactorum* on twigs of different rootstocks of apple

Fungus isolate	Mean lesion length (mm) on rootstocks					
	M4	MM101	MM104	MM110	MM114	Mean
P1	7.0	7.01	11.25	7.66	7.00	7.98
P2	7.0	36.66	41.66	11.38	7.02	20.74
P3	7.01	14.13	22.33	10.75	7.00	12.24
P4	11.37	10.25	40.87	7.50	7.50	15.49
P5	7.00	27.75	30.75	12.25	19.25	19.40
Mean	7.87	19.16	29.35	9.89	9.55	15.16

SD (5% level) based on:
Rootstocks:	0.275
Isolates:	0.275
Rootstocks × isolates:	0.308

Table 3. Activity of polyphenol oxidase in healthy and inoculated rootstocks

Fungus isolate	Activity (units per gram of tissue) in rootstocks					
	M4	MM101	MM104	MM110	MM114	Mean
P1	88.25	70.75	82.16	110.83	98.50	90.09
P2	72.33	33.83	32.08	48.50	103.33	58.01
P3	54.50	28.83	36.75	28.00	91.66	47.94
P4	47.91	58.08	9.66	76.33	89.50	56.29
P5	103.50	17.25	30.41	53.91	29.75	46.96
Mean	73.30	41.75	38.20	63.51	82.55	59.86
Control	9.90	7.91	3.75	16.71	11.00	9.85

SD (5% level) based on:
Rootstocks:	0.209
Isolates:	0.209
Rootstocks × isolates:	0.514

Table 4. Activity of peroxidase in different healthy and inoculated rootstocks

Fungus isolate	Activity (units per gram of tissue) in rootstocks					
	M4	MM101	MM104	MM110	MM114	Mean
P1	537.60	316.87	417.60	459.20	467.80	439.81
P2	577.60	273.60	186.93	180.80	652.80	374.34
P3	504.00	265.60	339.20	299.20	436.60	368.92
P4	344.00	516.80	201.60	497.60	552.80	422.56
P5	377.60	177.60	196.80	268.80	315.20	267.20
Mean	468.16	310.09	268.43	341.12	485.09	374.56
Control	382.20	330.40	300.00	486.40	515.00	402.80

SD (5% level) based on:
Rootstocks:	11.73
Isolates:	12.84
Rootstocks × isolates:	28.72

Table 5. Production of macerating enzymes by different isolates of *Phytophthora cactorum*

Fungus isolate	Percentage reduction in viscosity after 2 hr of incubation				
	PG	PGTE	PMG	PMTE	CX
P1	37.50 (37.67)[a]	24.40 (29.60)	36.95 (37.41)	39.56 (38.82)	27.46 (31.58)
P2	37.70 (37.88)	27.20 (31.44)	37.25 (37.58)	43.40 (41.21)	20.64 (27.01)
P3	35.40 (36.51)	26.60 (31.04)	34.36 (35.85)	37.64 (37.80)	32.58 (34.90)
P4	37.70 (37.88)	17.03 (24.35)	32.00 (34.42)	30.98 (33.79)	27.32 (31.49)
P5	46.76 (43.15)	34.90 (36.23)	43.01 (40.96)	43.72 (41.37)	27.72 (31.82)
SE (± 1)	0.207	0.851	0.540	0.540	0.550
SD (5%)	0.434	1.787	1.134	1.134	1.155

[a]Figures in parentheses are transformed values.

was almost double that of P4. P1, P2, and P3 had less activity than P5 but significantly more than P4. The activity of PMG was also maximum in P5, followed by P2, P1, and P3. It was almost equal in P1 and P2 and less in P3. Least activity of this enzyme was observed in P4. The activity of PMTE was maximum in P5 and was almost equal to that in P2, followed by P1 and P3; activity was minimum in P4.

Cellulase activity also varied in different isolates. It was maximum in P3, almost the same in P1, P4, and P5, and minimum in P2.

DISCUSSION

Marked differences have been observed in the susceptibility of the same rootstock to *P. cactorum* in different parts of the world (3,4,8,11,12). This has been attributed to the existence of different pathotypes in the fungus, but conclusive evidence is still lacking. Seven isolates of *P. cactorum* (P1–P7) from different parts of India varied in their reactions with a set of 10 apple rootstocks (M2–26; MM104–115). On a single rootstock, variability was indicated by the different reactions given by different isolates. This is also true on the young twigs of the same rootstocks. Sewell and Wilson (12) and Aldwinckle et al (1) have also shown the existence of variability in *P. cactorum* isolates on the basis of differential reactions on apple rootstocks and cultivars. This point was discussed earlier in the *P. cactorum* workshop in 1974 (8). Mircetich (10) has shown M26 rootstock to be susceptible to *P. cactorum* in California, and this rootstock is resistant to all the isolates in our studies. This gives a clear indication that Indian isolates are different from California isolates and that variability exists in the pathogen.

Infection with *P. cactorum* isolates increased the activity of polyphenol oxidase enzyme in all infected rootstocks, but the increase was much more in resistant reactions than in susceptible ones regardless of rootstock and isolate involved. With infection, the activity of peroxidase enzyme also differed in apple rootstocks. Maximum increased activity was observed in resistant reactions, and there was a slight decrease in susceptible reactions. Mir (9) has also found that the activity of polyphenol oxidase and peroxidase in apple rootstocks differs in reaction with *P. cactorum*. Increased activity of polyphenol oxidase is found in resistant rootstock (M4) and slightly less than this in

susceptible (MM104). Similarly, peroxidase activity increased in resistant lines. These isolates also differed in their capacity to produce pectolytic and cellulolytic enzymes. It can thus be concluded from this study that the seven isolates of *P. cactorum* differed among themselves.

LITERATURE CITED

1. Aldwinckle, H. S., Polach, F. J., Molin, W. T., and Pearson, R. C. 1975. Pathogenicity of *Phytophthora cactorum* isolates from New York apple trees and other sources. Phytopathology 65:989-994.
2. Farkas, G. L., and Kiraly, Z. 1962. Role of phenolic compounds in the physiology of plant diseases and disease resistance. Phytopathol. Z. 55:105-150.
3. Fitzpatrick, R. E., Mellor, F. C., and Welsh, M. F. 1944. Crown rot of apple trees in British Columbia root stock and scion resistant trials. Sci. Agric. 22:533-544.
4. Gupta, V. K., and Rana, K. S. 1983. Differential reaction of apple root stocks and cultivars to Pythiaceous fungi associated with collar rot. Abstr. Natl. Symp. Adv. Tree Sci. 60 pp.
5. Hampton, R. E. 1963. Activity of some soluble oxidases in carrot slices infected with *Thielaviopsis basicola*. Phytopathology 53:497-499.
6. Hancock, J. G., and Miller, R. L. 1965. Relative importance of polygalacturonate trans-eliminase and other pectolytic enzymes in southern anthracnose, spring black stem, and Stemphylium leaf spot of alfalfa. Phytopathology 55:346-355.
7. Matta, A., and Dimond, A. E. 1963. Symptoms of Fusarium wilt in relation to quantity of fungus and enzyme activity in tomato stem. Phytopathology 53:574-578.
8. McIntosh, D. L. 1975. Proceedings of the 1974 APDW workshop on crown rot of apple trees. Can. Plant Dis. Surv. 55:109-116.
9. Mir, N. M. 1983. Studies on the variability and control of *Phytophthora cactorum* (Leb. and Cohn.) Schroet. causing collar rot of apple. Ph.D. thesis, HPKVV, COA, Solan. 157 pp.
10. Mircetich, S. M. J. 1982. Phytophthora root and crown rot of deciduous fruit trees in California. Proc. Deciduous Tree Fruit Disease Workers. Bad Apple, Salt Lake City, Utah. 2(2):1-9.
11. Schmidle, A. 1957. On inoculation experiments of apple tree with *Phytophthora cactorum* (Leb. and Cohn.) Schroet., the agent of collar rot. Phytopathol. Z. 28:329-342.
12. Sewell, G. W. F., and Wilson, J. F. 1959. Resistance trials of some apple root stock varieties to *Phytophthora cactorum* (L. and C.) Schroet. J. Hortic. Sci. 34:51-58.
13. Sridhar, R., Chandra Mohan, D., and Mahadevan, A. 1969. The role of the parasite and its metabolites triggering host physiology. Phytopathol. Z. 64:21-27.

Glasshouse Test for Tolerance of Wheat to Crown Rot Caused by *Fusarium graminearum* Group 1

T. A. KLEIN, University of Sydney, Plant Breeding Institute, I. A. Watson Wheat Research Centre, Narrabri, N.S.W., 2390; C. M. LIDDELL and L. W. BURGESS, Department of Plant Pathology and Agricultural Entomology, University of Sydney, N.S.W., 2006; and F. W. ELLISON, University of Sydney, Plant Breeding Institute, I. A. Watson Wheat Research Centre, Narrabri, N.S.W., 2390, Australia

Crown rot, caused by *Fusarium graminearum* Schwabe group 1, is a disease occurring in wheat crops throughout the eastern wheat belt of Australia (1,2). Losses have been more pronounced on the heavy black and grey clay soils in southern Queensland and northern New South Wales. Differences in susceptibility to crown rot between wheat cultivars have been observed (4–6). The use of cultivars with tolerance to crown rot is one of the means available to growers to reduce losses from this disease. The procedures developed to assess adult plants in the field for tolerance to crown rot are both time consuming and labor intensive. These two factors preclude the possibility of active screening and selection of early generation material in a program breeding for tolerance to crown rot. A test for assessing seedling tolerance of wheat cultivars has been developed. The following paper presents the procedures involved and preliminary findings of two experiments, one in Narrabri and one in Sydney.

MATERIALS AND METHODS

Experiment 1 (Narrabri). Fifteen seeds of each of four wheat (*Triticum aestivum*) cultivars and one oat (*Avena sativa*) cultivar were sown in separate aluminum canisters (diameter of 50 mm, height of 200 mm). Each canister contained 350 g of black self-mulching subsoil with a water potential of -0.03 MPa. The moisture characteristic of this soil was determined by the filter paper method (3). Soil water potential was subsequently determined gravimetrically. Cultivars Coolibah (oats), Cook, and Timgalen have adult plant tolerance to crown rot, whereas cultivars Banks and Songlen are susceptible to crown rot as adult plants. Air-dry soil (45 g) was added to a depth of 20 mm above the seed.

Inoculum was prepared by growing *F. graminearum* group 1 on sterilized grain for 3 weeks. Colonized grain was dried and milled through a 2-mm screen. A layer of inoculum (0.6 g) was spread uniformly over the soil surface of each canister. Air-dry soil (45 g) was then added to a depth of 20 mm above the layer of inoculum, and water was added to the surface to bring the surface water potential to either -0.03 MPa (treatment 1) or -0.1 MPa (treatment 2). Soil water potential was maintained in the range of -0.03 to -0.1 MPa in treatment 1 and in the range of -0.1 to -0.3 MPa in treatment 2 by adding water to the soil surface.

Four replicates of each cultivar in water treatment 1 and water treatment 2 were placed in a growth cabinet at 22°C. A further four replicates of each cultivar in each water treatment were placed in a glasshouse with a diurnal temperature range of 10 to 20°C. The growth cabinet and the glasshouse had a 12-hr photoperiod. The number of healthy seedlings was counted 25 and 31 days after sowing.

Experiment 2 (Sydney). Ten plants of each of six wheat cultivars were grown in pasteurized black earth (70 kg wet) in a 60-L container (diameter of 400 mm, height of 500 mm). Seeds were sown in a circle (diameter of 300 mm). Cultivars Cook, Kite, and SUN 69A have adult plant tolerance to crown rot; cultivars Songlen, Banks, and SUN 41A-24 are susceptible to crown rot as adult plants.

Seeds were sown on the subsoil, and a 20-mm layer of soil was added. Inoculum was applied in a layer of finely ground chaff (700 μm), colonized saprophytically by *F. graminearum* group 1, and a further 20 mm of soil was added. Soil was maintained at a constant water potential (-0.1 MPa) by watering the soil surface as required. Plants were grown in a glasshouse with temperatures of 11°C at night and 20°C during the day. The number of healthy plants was recorded 39, 64, and 102 days after sowing.

RESULTS

Experiment 1 (Narrabri). There was a higher level of seedling death in seedlings grown at 22°C than those grown at 10–20°C. The difference in death rates between cultivars was also greater at 22°C. The oat cultivar, Coolibah, had a higher proportion of healthy plants than did any of the wheat cultivars (Table 1). Cultivars Cook and Banks were more tolerant to crown rot as seedlings than cultivars Songlen and Timgalen. The mean proportion of seedlings remaining healthy, in all cultivars, was the same in both surface soil water treatments.

Experiment 2 (Sydney). There was a difference between the relative tolerance of cultivars to crown rot at the three times of assessment (Table 2). Cultivars Kite, SUN 69A, and Cook had a similar proportion of dead plants at each assessment. The cultivars Banks and SUN 41A-24 displayed a degree of tolerance at both 39 and 64 days after sowing, but they had a high death rate 102 days after sowing. Cultivar Songlen was the most susceptible cultivar at all times of assessment. The

Table 1. Mean number of healthy plants growing in soil infested with *Fusarium graminearum* group 1 and assessed 25 and 31 days after seeding (experiment 1)[a]

Cultivar	At 25 days[b]		At 31 days[b]		At 31 days[c]	
	10–20°C	22°C	10–20°C	22°C	10–20°C	22°C
Coolibah (oats)	1.5	0.7	1.8	2.1	0.5	0.7
Cook	1.7	4.8	2.9	6.7	4.3	6.2
Banks	3.3	4.7	4.0	7.5	3.0	5.7
Timgalen	4.9	7.2	7.2	8.6	7.2	9.8
Songlen	4.9	7.2	6.6	9.4	7.1	9.3
LSD (5%)	1.7	1.1	1.5	1.2	2.2	0.6

[a] Mean of four replicates each of 15 plants. Each value transformed [$\sqrt{(100 - x + 0.5)}$ where x is the percentage of healthy plants] for statistical analysis.
[b] Water potential = −0.03 MPa.
[c] Water potential = −0.10 MPa.

Table 2. Mean number of healthy plants 39, 64, and 102 days after seeding into soil containing a layer of chaff colonized by *Fusarium graminearum* group 1 (experiment 2)[a]

Cultivar	At 39 days	At 64 days	At 102 days
SUN 69A	2.5	2.5	7.0
Kite	2.5	2.5	7.4
Cook	2.7	2.7	7.1
SUN 41A-24	2.3	5.6	8.3
Banks	3.8	4.7	9.1
Songlen	5.6	7.9	9.4
LSD (10%)	1.7	2.1	1.5

[a] Mean of four replicates each of 10 plants. Each value transformed [$\sqrt{(100 - x + 0.5)}$ where x is the percentage of healthy plants] for statistical analysis.

difference between the cultivars with the highest proportion of healthy seedlings and those with the lowest proportion was greatest 64 days after sowing and lowest 39 days after sowing.

DISCUSSION

Cultivars differ in their tolerance as seedlings to crown rot. The oat cultivar Coolibah and the wheat cultivars Cook, Banks, Kite, and SUN 69A have a greater degree of seedling tolerance than the cultivars Songlen, Timgalen, and SUN 41A-24.

Seedling tolerance was correlated with adult plant tolerance in six of the eight cultivars assessed. The oat cultivar Coolibah may possess true seedling resistance to crown rot. Adult plants of oats are not often severely affected in the field. Cultivars Cook, Kite, and SUN 69A have both seedling and adult plant tolerance to crown rot. Cultivar SUN 41A-24 has intermediate seedling and adult plant tolerance to crown rot, whereas cultivar Songlen is susceptible at both growth stages. Cultivar Banks appears to be losing its seedling tolerance after 64 days and is exhibiting its adult plant susceptibility 102 days after sowing. In contrast, cultivar Timgalen has little seedling tolerance but has a degree of adult plant tolerance.

Screening wheat cultivars as seedlings for tolerance to crown rot is affected by temperature. The differentiation between cultivars is more rapid if the temperature is increased to 22°C. This temperature is similar to the optimum temperature for growth of *F. graminearum* group 1 in vitro.

The two surface soil water potential treatments used at Narrabri did not affect the relative proportions of healthy seedlings in the glasshouse or in the growth cabinet. The development of symptoms in adult plants in the field is generally more severe if a moisture stress occurs at anthesis or during grain fill. Therefore, further examination of seedling tolerance is warranted to determine whether manipulation of soil water potential may assist in screening for seedling tolerance to crown rot.

The results of screening cultivars were consistent between the two experiments. The fact that consistent results were obtained at two separate locations (Narrabri and Sydney), and by two different operators, increases confidence in the feasibility of screening wheat cultivars as seedlings for tolerance to crown rot. Purss (5) could find no correlation between seedling and adult plant reaction to crown rot, but he did suggest that a seedling test "could be of value if true resistance (to crown rot) was found." The procedures described in this paper will allow a greater number of wheat cultivars to be screened for seedling tolerance to crown rot, perhaps allowing "true resistance" to be identified in the future.

LITERATURE CITED

1. Burgess, L. W., Dodman, R. L., Pont, W., and Mayers, P. 1981. Fusarium diseases of wheat, maize and grain sorghum in eastern Australia. Pages 64-76 in: *Fusarium*: Diseases, Biology and Taxonomy. P. E. Nelson, T. A. Toussoun, and R. J. Cook, eds. Pennsylvania State University Press, University Park. 457 pp.
2. Burgess, L. W., Wearing, A. H., and Toussoun, T. A. 1975. Survey of fusaria associated with crown rot of wheat in eastern Australia. Aust. J. Agric. Res. 26:791-799.
3. Fawcett, R. G., and Collis-George, N. 1967. A filter paper method for determining the moisture characteristics of soil. Aust. J. Exp. Agric. Anim. Husb. 7:162-167.
4. McKnight, T., and Hart, J. 1966. Some field observations on crown rot disease of wheat caused by the fungus *Fusarium graminearum*. Queensl. J. Agric. Anim. Sci. 23:373-378.
5. Purss, G. S. 1966. Studies of varietal resistance to crown rot of wheat caused by *Fusarium graminearum* Schw. Queensl. J. Agric. Anim. Sci. 23:475-498.
6. Wildermuth, G. B., and Purss, G. S. 1971. Further sources of field resistance to crown rot (*Gibberella zeae*) of cereals in Queensland. Aust. J. Exp. Agric. Anim. Husb. 11:455-459.

Development of Inoculation Technique for *Rhizoctonia solani* and Its Application to Screening Cereal Cultivars for Resistance

H. J. McDONALD and A. D. ROVIRA, CSIRO Division of Soils, Glen Osmond, S.A., 5064, Australia

Rhizoctonia solani Kuehn is the causal agent of a seedling disease of cereals known as "bare patch" in alkaline soils with low rainfall and is considered a major pathogen of cereals in South Australia (1). Natural populations of *R. solani* are highly variable (9), and infectivity is affected by soil disturbance (2).

These factors make it difficult to conduct field and laboratory studies on the effects on plants of known populations of *R. solani*. To overcome some of these problems, an inoculation technique was developed to provide a range of disease symptoms similar to those in the field, while also ensuring reproducible results in glasshouse, controlled environment, and field experiments. In this paper we describe this technique and its use to screen cereal cultivars for resistance to *R. solani* under controlled environmental conditions.

MATERIALS AND METHODS

An isolate of *R. solani* was obtained from roots of wheat seedlings with severely truncated, spear-tipped roots from a field site in South Australia. The perfect stage of this isolate, identified as *Thanatephorus cucumeris* (Frank) Donk, was induced using the soil-over-culture method (8).

White millet seed was chosen to provide the base for the inoculum because the uniformly sized seeds (about 1 mm in diameter) provided well-defined propagules when colonized by the fungus. The inoculum was prepared by soaking γ-irradiated millet seed for 16 hr, draining off water, and autoclaving at 121°C for 1 hr on each of three consecutive days. After each autoclaving the flasks were shaken to prevent packing of the seed. *R. solani* cultures grown on potato-dextrose agar were cut into 5-mm cubes, added to the flasks of sterilized millet, and incubated at 25°C to allow colonization of the seed; after 3 weeks, the inoculum was dried in a laminar flow unit.

In the initial experiment, the effect of incubation time and inoculum quantity on disease severity was assessed using unsterilized, air-dry, calcareous sandy loam (pH 8) described as Dy 5.43 (solodized solonetz or solodic planisol, calcic phase) (5). Propagules were mixed through 1-kg lots of soil at rates of 0, 8, 16, and 32 per kilogram, and 10% water (w/w) was then added. Duplicate 1-kg samples of soil were placed in individual plastic containers, covered, and incubated at 10°C for 0, 2, and 4 weeks. Six wheat seeds (cv. Condor) were planted per container. The soil moisture content was adjusted to 10% (w/w) and the containers placed into a controlled environment cabinet at 10°C with 8 hr of light per day.

After 21 days, shoot height and seminal root length were measured and the roots rated for disease on a 0–5 scale, with 0 = no disease; 1 = <25% of primary roots with severe truncation and necrosis; 2 = 25–50% of primary roots with severe truncation and necrosis; 3 = 50–75% of primary roots with severe truncation and necrosis; 4 = >75% of primary roots with severe truncation and necrosis; and 5 = 100% of primary roots severely truncated, top stunted and moribund.

This technique, with 2 weeks of incubation, was then used to ascertain whether barley and oat varieties differed in their susceptibility to *R. solani*. Seven barley and 10 oat cultivars were tested against three rates of *R. solani* (0, 8, and 16 propagules per kilogram of soil) with four replicates per treatment. *R. solani* root damage was assessed on the five-point scale.

In a separate experiment, the effect of soil disturbance on *R. solani* damage to roots was studied by setting up eight replicate soil samples each with 8 and 16 propagules per kilogram of soil in individual containers; after 2 weeks, four were left undisturbed at each disease level; four were transferred to plastic bags and returned to their original containers. Condor wheat was planted into these soils, and after 3 weeks roots were assessed for root damage by *R. solani*.

RESULTS

Both the number of propagules per kilogram of soil and incubation had significant effects on seminal root length and disease rating (Fig. 1). Shoot height was significantly affected by the number of propagules. The F ratio shows that disease rating was the most sensitive

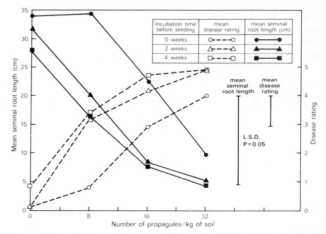

Fig. 1. The effect of population density of *Rhizoctonia solani* and incubation time on root disease rating and root length of wheat grown under controlled environmental conditions.

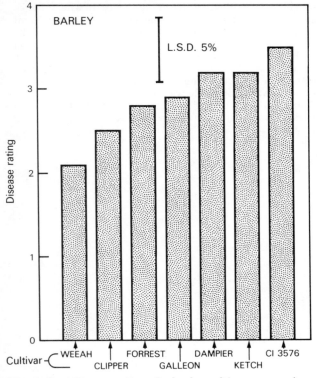

Fig. 2. Root disease ratings for barley cultivars grown in soil containing *Rhizoctonia solani* at 16 propagules per kilogram of soil.

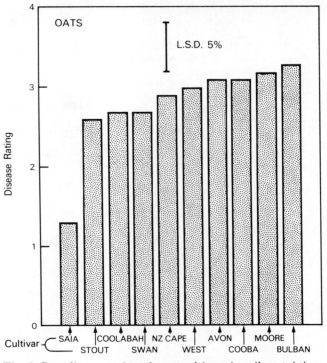

Fig. 3. Root disease ratings for oat cultivars in soil containing *Rhizoctonia solani* propagules at 16 propagules per kilogram of soil.

parameter in assessing disease severity.

This experiment indicates that millet seeds colonized by *R. solani* produced a range of disease ratings and symptoms similar to those found in field plants and also that the severity of *R. solani* attack could be manipulated. Because incubation times of 2 or 4 weeks produced no significant difference in disease ratings, a 2-week period was used in later studies.

When the technique was applied to barley and oat cultivars, there were significant differences between cultivars in root damage caused by 16 propagules per kilogram (Figs. 2 and 3). Differences between cultivars at 8 propagules per kilogram were significant at $P < 0.1$, with Saia oats and Weeah barley the most resistant (or tolerant).

Disturbing the soil reduced the root disease rating on wheat from 2.3 to 1 for 8 propagules per kilogram and 2.5 to 1.25 for 16 propagules per kilogram. The effect was significant at $P < 0.01$.

DISCUSSION

Although different methods of inoculum preparation have been used (6), the problems of propagule size (3) and age of inoculum (4) affecting disease potential have made it difficult to relate results obtained in pot experiments to the disease in the field. The development of uniformly sized propagules (i.e., millet seed colonized by *R. solani*), together with a known inoculum age and incubation time, solves some problems and enhances experimental reproducibility.

The technique described in this paper overcomes some of the difficulties associated with *Rhizoctonia* research—the difficulty of simulating the range of field infection with artificial inoculum; the variations in

disease levels in field soils; and the effect of disturbance when field soils are collected, transported, and placed in pots.

The preliminary trials with barley and oats have shown one cultivar of each cereal to be more resistant to, or tolerant of, the pathogen than most other cultivars. Each of these resistant (or tolerant) cultivars has characteristics that distinguish it from the other cultivars: Weeah barley has a more vigorous early rooting habit and is less sensitive to manganese deficiency; Saia oats is diploid, whereas the remaining oat varieties are hexaploid.

The technique has considerable potential for screening cereal varieties for resistance, for testing chemical and biocontrol agents, and for studying the effect of soil type and conditions upon *R. solani*.

The reduction in root damage caused by *R. solani* after soil disturbance may, in part, explain the lower incidence of the pathogen in wheat grown in cultivated soil than when grown in a no-till system (7). The mechanisms involved in reduction of damage by *R. solani* following soil disturbance have yet to be studied, but one could be that hyphae that have ramified through the soil no longer get the energy they need from the original propagules. Fragmented hyphae may be more prone to attack by soil microorganisms and thus less infective.

LITERATURE CITED

1. Banyer, R. J. 1966. Cereal root diseases and their control. Part III. J. Agric. 415-417.
2. Dubé, A. J. 1971. Studies on the growth and survival of *Rhizoctonia solani*. Ph.D. thesis, University of Adelaide.
3. Henis, Y., and Ben-Yephet, Y. 1970. Effect of propagule size of *Rhizoctonia solani* on saprophytic growth, infectivity, and virulence on bean seedlings. Phytopathology 60:1351-1356.

4. Kamal, M., and Weinhold, A. R. 1967. Virulence of *Rhizoctonia solani* influenced by age of inoculum in the soil. Can. J. Bot. 45:1761-1765.

5. Northcote, K. H., Hubble, G. D., Isbell, R. F., Thompson, C. H., and Bettenay, E. 1975. A Description of Australian Soils. CSIRO Australia.

6. Papavizas, G. C., and Lewis, J. A. 1979. Integrated control of *Rhizoctonia solani*. Pages 415-424 in: Soil-borne Plant Pathogens. B. Schippers and W. Gams, eds. Academic Press, London. 686 pp.

7. Rovira, A. D., and Venn, N. R. 1985. Effect of rotation and tillage on take-all and Rhizoctonia root rot in wheat. Pages 255-258 in: Ecology and Management of Soilborne Plant Pathogens. C. A. Parker, A. D. Rovira, K. J. Moore, P. T. W. Wong, and J. F. Kollmorgen, eds. American Phytopathological Society, St. Paul, MN.

8. Tu, C. C., and Kimbrough, J. W. 1975. A modified soil-over-culture method for inducing basidia in *Thanatephorus cucumeris*. Phytopathology 65:730-731.

9. Weinhold, A. R. 1977. Population of *Rhizoctonia solani* in agricultural soils determined by a screening procedure. Phytopathology 67:566-569.

Phytophthora cinnamomi: A Study of Resistance in Three Native Monocotyledons That Invade Diseased Victorian Forests

DARREN PHILLIPS and GRETNA WESTE, Botany School, University of Melbourne, Parkville, Victoria, 3052, Australia

Phytophthora cinnamomi Rands causes severe disease in Victorian dry sclerophyll forests. The death of up to 75% of the understory results in increased bare ground (8), which is colonized by apparently resistant sedges such as Gahnia radula (R. Br.) Benth. and Lepidosperma laterale R. Br. and by grasses such as Poa sieberana Spring. These species show no visible secondary symptoms of disease. They were studied to determine whether they were completely resistant to penetration and infection or whether they were only field resistant, perhaps providing a source of infection and disease extension within native forest.

MATERIALS AND METHODS

Young plants of L. laterale, G. radula (both Cyperaceae), and Poa sieberana (Poaceae) were collected from disease-free forest. Plants, roots, and soil were tested by lupin baiting for the presence of P. cinnamomi (3), then were washed free of soil and placed in root boxes prepared from black polyvinyl chloride. The boxes were divided into two separate compartments with a sealed divider, and plants were placed so that one main root was directed into each compartment. The root tips were inoculated with an axenically prepared suspension of zoospores of P. cinnamomi. Plants were inoculated either with zoospores or with sterile distilled water (controls). Some plants in the divided root boxes were zoospore-inoculated on one of the two main roots, with the other serving as a control. Subsequent root growth, lesion length, and extension were recorded directly on polyacetate sheets with 1-cm^2 grid placed on the acrylic plastic cover of each root box. Roots were measured and mapped daily for 4 days after inoculation and then every 48 hr for another 6 days. Twenty-four plants were studied, including 12 of L. laterale, 10 of Poa sieberana, and 2 of G. radula. Ten days after inoculation, roots were cut into segments 2–3 mm long and either surface sterilized and plated into P_{10} VP agar (5) or sectioned and stained for microscopic examination.

RESULTS

Necrotic lesions formed in roots of all species examined 24–48 hr after inoculation (Fig. 1). The pathogen was therefore shown to infect and cause primary symptoms in these species. When lesions formed, root growth ceased and the root tips subsequently disintegrated. Lateral roots usually formed immediately behind the necrotic tissue 4–6 days after inoculation. Lesion extension was rapid in the three species for the first 3 days, but it ceased within 6 days. The longest lesions, which were formed in Poa sieberana roots, were nearly twice as long as those of L. laterale. No secondary symptoms were exhibited by any of the inoculated plants. There was no chlorosis, wilting, or dieback of shoots.

The growth rate of uninfected control roots remained constant after false inoculation with sterilized distilled water: Poa sieberana, 1.11 cm per day; L. laterale, 0.8 cm per day. However, in all infected roots growth ceased 24–48 hr after inoculation.

In one plant each of L. laterale and G. radula, a new root tip actually grew directly from the necrotic, previously terminated root at 9 and 5 days after inoculation respectively. Infected plants usually produced fewer new roots than the uninfected controls. In Poa sieberana, the number of new roots produced in the infected root systems was reduced by 50% as compared with the control 10 days after inoculation. P. cinnamomi was reisolated from all inoculated roots of G. radula, from 75% of L. laterale, and from 50% of Poa sieberana. Less than 20% of the total length of the main root was invaded by the pathogen in each of the three species; hyphal invasion was generally confined to the lesion.

Microscopic examination showed that infected roots contained chlamydospores, hyphal vesicles, and hyphae that grew along between the cells and also penetrated them. Some cells, throughout the cortex of the lesion area, were filled with fungal cytoplasm. A characteristic constriction of hyphae was observed at points of intracellular wall penetration.

No sporangia were observed on the root surface in L. laterale and Poa sieberana. In one single-rooted L. laterale, three encysted zoospores were seen scattered among fine root hairs of laterals adjacent to the infected and necrotic main root tip 11 days after inoculation.

The amount of fungal material declined markedly with distance from the lesion and was absent after 10 mm from the inoculation point in L. laterale and 18 mm in Poa sieberana in the roots examined. No fungal material was observed in control or uninfected roots.

Associated cellular disruption was also progressively reduced with distance from the point of inoculation. Deposits probably containing callose were observed in root tissue when stained with decolorized aniline blue. The small papillae fluoresced bright yellow and were observed in cell walls adjacent to invading intercellular hyphae in heavily infected tissue of L. laterale in the cortex and stele, but not in Poa sieberana. Cahill and Weste (2) also observed papillae in G. radula and Juncus bufonius (rush), but not in Themeda australis R. Br. (grass).

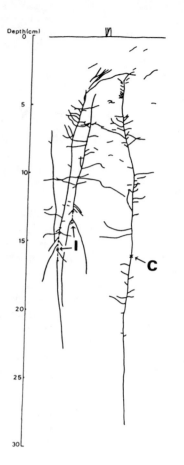

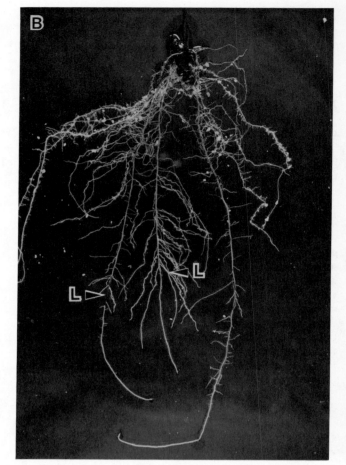

Fig. 1. *Gahnia radula* grown in divided root box. **(A)** Root system, as mapped on transparent overlay, showing sites of inoculation (I) with zoospores on the two main roots. Control root (C) was sham inoculated. **(B)** Whole root system, washed free of soil at end of experiment, showing lesions (L) that developed 24–48 hr after inoculation.

DISCUSSION

The three understory species studied are common colonizers of native dry sclerophyll forests, utilizing ground bare of vegetation because of disease caused by *P. cinnamomi*. In the absence of visible or secondary symptoms, they were considered to show resistance (8).

These investigations have demonstrated that the pathogen successfully infected the roots of each species studied. The formation of lesions extending from the site of inoculation demonstrated that the pathogen penetrated the roots and produced a typical primary symptom: necrosis of the root tissue. Therefore these species were not resistant to penetration, and resistance was expressed after penetration.

Lesion extension, which has been used to indicate susceptibility (9), was rapid for 3 days after inoculation but had ceased by 6 days. In fully susceptible species, necrosis may extend throughout the whole root system (8). In resistant species, lesions are restricted, as in *J. bufonius* (7). In each species investigated, the lesions were contained in size and limited in time. No lesions extended more than 18 mm or continued to extend more than 6 days after inoculation. The limited extent of hyphal colonization and the lack of extension of the fungus into lateral roots are evidence of resistance.

The pathogen persisted within the contained lesion for at least 10 days when it was reisolated. This persistence demonstrates incomplete resistance and may provide a source of inoculum for secondary infection and disease extension.

The infected root ceased growth 24–48 hours after inoculation. Lateral root formation occurred immediately behind the lesion 4–6 days after infection.

The capacity to replace dead roots via lateral root formation immediately behind the necrotic zone is an important response to infection and is considered to be a major factor in the resistance of these three species (1,4).

In *G. radula* and *L. laterale*, but not in *Poa sieberana*, new root tips grew out of the infected and necrotic root tissue. This remarkable capacity to renew infected root tips is probably a highly significant factor in the survival capacity of plants growing in infested soil.

The resistance shown by the sedges and grass may be compared with that previously recorded for *J. bufonius* (7), for *T. australis* (2), and for wheat (6). Evidently there is a wide range of response to *P. cinnamomi*. Wheat, inoculated at germination, contained viable pathogen at harvest. No secondary symptoms were observed and no plants were killed, but yield was reduced a significant 47% (6).

LITERATURE CITED

1. Batini, F. E., and Cameron, J. N. 1975. The effects of temperature on the infection of New Zealand blue lupins by *Phytophthora cinnamomi*. West. Aust. For. Dep. Res. Pap. 25.

2. Cahill, D., and Weste, G. 1983. Formation of callose deposits as a response to infection with *Phytophthora cinnamomi*. Tr. Br. Mycol. Soc. 80:23-29.
3. Chee, K., and Newhook, F. J. 1965. Improved methods for use in studies on *Phytophthora cinnamomi* and other *Phytophthora* species. N. Z. J. Agric. Res. 3:88-95.
4. Halsall, D. M. 1978. A comparison of *Phytophthora cinnamomi* infection in *Eucalyptus sieberi*, a susceptible species, and *Eucalyptus maculata*, a field resistant species. Aust. J. Bot. 26:643-656.
5. Tsao, P. H., and Ocana, G. 1969. Selective isolation of species of *Phytophthora* from natural soils on an improved antibiotic medium. Nature 223:636-638.
6. Weste, G. 1978. Comparative pathogenicity of six root parasites towards cereals. Phytopathol. Z. 93:41-55.
7. Weste, G., and Cahill, D. 1982. Changes in root tissue associated with infection by *Phytophthora cinnamomi*. Phytopathol. Z. 103:97-108.
8. Weste, G., and Taylor, P. 1971. The invasion of native forest by *Phytophthora cinnamomi*. I. Brisbane Ranges, Victoria. Aust. J. Bot. 19:281-294.
9. Zentmyer, G. A. 1980. *Phytophthora cinnamomi* and the diseases it causes. American Phytopathological Society Monograph 10.

Relative Susceptibility of Wheat, Rye, and Triticale to Isolates of Take-all

P. R. SCOTT, T. W. HOLLINS, and R. S. GREGORY, Plant Breeding Institute, Cambridge, U.K.

Wheat crops are highly susceptible to take-all. Rye is much more resistant (see references in Scott [5]): rye crops are probably seldom damaged by the disease even in areas where rye is cropped repeatedly on the same land, as on the sandy soils of Norfolk and Suffolk, England. Triticale, the hybrid between wheat and rye, has often been reported to be closer to wheat than to rye in susceptibility (2,6,9); however, this is based on observations of young seedlings inoculated in the glasshouse.

In eight field trials, by contrast, we repeatedly found

that the degree of root infection on adult plants of hexaploid triticale was approximately midway between the substantial infection on hexaploid wheat and the very limited infection on rye (Fig. 1). With one exception (1982 in Fig. 1), this result held true over a wide range of disease severities in different trials covering a period of 6 years. One trial (1981a in Fig. 1) included uninoculated plots: the effect of inoculation on the yield of triticale (reduction of 24%, $P <0.001$) was also intermediate between the effects on wheat (53%, $P <0.001$) and rye (3%, $P <0.05$). Our field results confirmed those of Jensen and Jørgensen (1).

Triticale thus appears to inherit a valuable degree of resistance from rye, expressed in adult plants in field trials but not in seedlings in the glasshouse. In a glasshouse test with inoculated adult plants (five plants per plot, 10 replicates), scores of take-all on roots (0–100 scale) were 82 on wheat, 76 on triticale (difference not significant, $P >0.05$), and 37 on rye (difference from triticale, $P <0.001$). This suggests that the field environment, and not merely the adult growth stage, is necessary for the expression of resistance in triticale.

VARIATION IN SUSCEPTIBILITY OF TRITICALE

Octoploid triticale (with 21 pairs of chromosomes from hexaploid wheat and seven pairs from diploid rye) inherits a higher proportion of its genome from wheat than hexaploid triticale (14 pairs from tetraploid wheat and seven pairs from rye). As might be expected, we found (Table 1) that octoploid lines of triticale were more susceptible than hexaploid lines but less susceptible than wheat in two inoculated field trials. However, at a naturally infested site with more moderate disease, octoploids and hexaploids were similar (Table 1). Differences between octoploid and hexaploid triticales have not been detected in glasshouse tests at the seedling stage (2,3).

Wheat and rye cultivars show negligible variation in

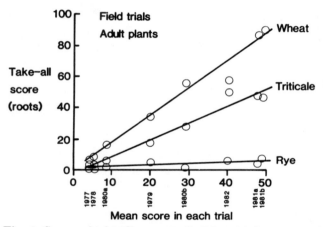

Fig. 1. Scores of take-all on roots (0–100 scale) for groups of wheat, hexaploid triticale, and rye lines in eight field trials. For each trial, scores for the three cereal species are plotted against the mean score for all entries in the trial. For each cereal species, a regression line is fitted to the values from the eight trials. In 1977, 1978, 1979, 1980b, and 1981b: inoculated trials with six replicates, plots 0.1 m²; mean of 5 wheats, 7–34 triticales, 3–4 ryes. In 1981a and 1982: inoculated trials with six replicates, plots 3 m²; mean of 2 wheats, 4–6 triticales, 2 ryes. In 1980a: naturally infected trial with eight replicates, plots 6 m²; mean of 2 wheats, 4 triticales, 2 ryes.

Table 1. Scores of take-all on roots (0-100 scale) for groups of wheat, triticale, and rye lines in three field trials

Cereal	1980b[w]	1981b[w]	1980a[x]
Wheat	57 (5)[y] ***[z]	91 (5) ***	18 (2) ***
Octoploid triticale	42 (7) ***	65 (7) ***	7 (2)
Hexaploid triticale	29 (7) ***	47 (31) ***	7 (4) ***
Rye	2 (4)	8 (4)	2 (3)

[w]Inoculated trials with six replicates; plots = 0.1 m².
[x]Naturally infected trial with eight replicates; plots = 6 m².
[y]Number of entries per group shown in parentheses.
[z]Significant difference ($P < 0.001$) between vertically adjacent values.

Table 2. Scores of take-all on roots (0–100 scale) for six lines of hexaploid triticale in two field trials[z]

Line	1977	1979
1975/61	2.0	11.2
1975/39	1.8	13.2
1975/71	2.5	16.0
1975/23	7.1	19.2
1975/26	9.5	28.5
1975/13	11.6	31.1
LSD ($P = 0.05$)	4.8	15.7

[z]Inoculated trials with six replicates; plots = 0.1 m². Correlation coefficient between 1977 and 1979 = 0.97 ($P <0.001$).

susceptibility to take-all (5,8). Surprisingly, we found that some hexaploid triticale breeding lines differed consistently in susceptibility in field trials. This is illustrated in Table 2 for six lines chosen for retesting in 1979 because of their apparent variation in 1977. Some triticale lines do not possess the full complement of seven pairs of rye chromosomes, so deficiency in one or more pairs could have accounted for the greater susceptibility of some lines. However, A. G. Seal and M. D. Bennett (Plant Breeding Institute) kindly checked four of the lines (1975/23, 26, 39, and 61) and found them all to have the full chromosomal complement. The genetic origin of the variation in susceptibility therefore remains in doubt.

PATHOGENICITY OF ISOLATES

We collected 136 isolates of *Gaeumannomyces graminis* from cereal crops in Britain in 1982 and found that, in the glasshouse, some caused nearly as much root infection on rye seedlings as on wheat seedlings; others caused much more infection on wheat than on rye. We used the differences (Δ) between the disease scores (0–5 scale) on wheat and rye as a measure of adaptation to rye. The frequency distribution of Δ for the 136 isolates (Fig. 2) was skewed and possibly bimodal but showed no absolute distinction between rye-adapted isolates and others. However, isolates selected in one test for their

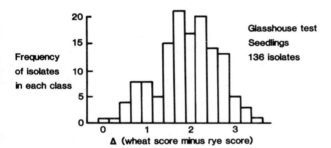

Fig. 2. Frequency distribution of Δ (difference between disease scores on wheat and rye) for 136 isolates of *Gaeumannomyces graminis* from cereal crops in Britain. Low values of Δ indicate strong adaptation to rye.

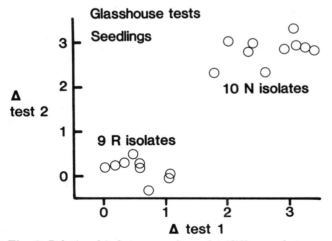

Fig. 3. Relationship between values of Δ (difference between disease scores on wheat and rye) in two tests for 19 isolates of *Gaeumannomyces graminis*. Nine isolates (R) selected for low Δ (strong adaptation to rye) in test 1; 10 isolates (N) selected for high Δ in test 1.

low values of Δ (here designated R isolates) or for their high values of Δ (designated N isolates) performed similarly in a second test (Fig. 3). The 136 isolates were obtained from wheat, barley, and rye crops in diverse regions of Britain, including numerous areas where rye is not grown.

DISCUSSION

The moderate resistance to take-all of adult plants of triticale grown in the field is of great potential benefit to cereal growers. Although less resistant than rye, triticale may provide an improved alternative crop to wheat where yield is limited by take-all. It could also substitute for barley, which is nearly as susceptible to take-all as wheat (5).

Unexpectedly, variation in resistance among lines of triticale may provide some scope for further improvement. However, although this variation appears greater than that found within any other cultivated cereal (8), it is probably of marginal value beside the substantially improved resistance, relative to wheat, of triticale as a species.

The greater susceptibility of octoploid than of hexaploid triticale might suggest that resistance is determined by the ratio between wheat and rye chromosomes in the triticale genome, rather than by individual genes from rye. Genetically diffuse determination of resistance in rye is also consistent with the negligible effect of individual pairs of rye chromosomes in wheat-rye addition lines (2,4,7). Although this suggest that the transfer of rye resistance to wheat cultivars would be difficult, we consider the attempt worth making in view of the paucity of resistance in wheat (8). Our results show that selection for resistance would need to be applied to adult plants in the field.

The discovery of rye-adapted isolates of *G. graminis* was unexpected. We do not know whether they may jeopardize the resistance of triticale, or even whether they are damaging to adult rye plants in the field. Our limited evidence suggests that R isolates are not strongly selected by the cultivation of rye: they do not appear to cause severe infection on rye where the crop is grown intensively. Nevertheless, their effect on triticale requires careful study.

ACKNOWLEDGMENT

We thank J. R. Paine for valuable practical assistance.

LITERATURE CITED

1. Jensen, H. P., and Jørgensen, J. H. 1973. Reactions of five cereal species to the take-all fungus (*Gaeumannomyces graminis*) in the field. Phytopathol. Z. 78:193-203.
2. Linde-Laursen, I., Jensen, H. P., and Jørgensen, J. H. 1973. Resistance of *Triticale, Aegilops* and *Haynaldia* species to the take-all fungus, *Gaeumannomyces graminis*. Z. Pflanzenzuecht. 70:200-213.
3. Mielke, H. 1974. Untersuchungen über die Anfälligkeit verschiedener Getreidearten gegen den Erreger der Schwarzbeinigkeit, *Ophiobolus gramini* Sacc. Mitt. Biol. Bundensanst. Land Forstwirtsch. Berlin-Dahlem 160. 61 pp.
4. Riley, R., and Macer, R. C. F. 1966. The chromosomal distribution of the genetic resistance of rye to wheat pathogens. Can. J. Genet. Cytol. 8:640-653.
5. Scott, P. R. 1981. Variation in host susceptibility. Pages

219-236 in: Biology and Control of Take-all. M. J. C. Asher and P. J. Shipton, eds. Academic Press, London.

6. Scott, P. R., and Hollins, T. W. 1979. Take-all. Rep. Plant Breed. Inst. for 1978, pp. 153-154.

7. Scott, P. R., and Hollins, T. W. 1983. Take-all. Rep. Plant Breed. Inst. for 1982, pp. 98-99.

8. Scott, P. R., and Hollins, T. W. 1985. Role of plant breeding in controlling soilborne diseases of cereals. Pages 157-159 in: Ecology and Management of Soilborne Plant Pathogens. C. A. Parker, A. D. Rovira, K. J. Moore, P. T. W. Wong, and J. F. Kollmorgen, eds. American Phytopathological Society, St. Paul, MN.

9. Skou, J. P. 1975. Studies on the take-all fungus *Gaeumannomyces graminis*. V. Development and regeneration of roots in cereal species during the attack. Arsskr. K. Vet. Landbohoejsk. Copenhagen, pp. 142-160.

New Inoculation Technique for *Gaeumannomyces graminis* var. *tritici* to Measure Dose Response and Resistance in Wheat in Field Experiments

A. SIMON and A. D. ROVIRA, CSIRO Division of Soils, Glen Osmond, S.A., 5064, Australia

The characteristic patchiness of take-all in wheat (2), caused by the fungus *Gaeumannomyces graminis* var. *tritici*, makes field experiments difficult. It is thus often necessary to manipulate the level of disease with artificial inoculum.

Oat kernels colonized by *G. graminis* var. *tritici* had been used before 1982, but a recent study by Simon (*unpublished*) has shown that different-sized fractions of the ground inoculum produced different levels of disease; also, when whole oat grains are used, the fungus develops from a propagule with a larger energy-rich base than is usual for natural field inoculum.

In this study, we evaluated annual ryegrass seed and white millet seed colonized by *G. graminis* var. *tritici* as alternative forms of inoculum.

MATERIALS AND METHODS

Annual ryegrass (*Lolium rigidum*) seed and white millet (*Panicum miliaceum*) seed was γ-irradiated (2.5 megarads) and soaked in water for 16 hr. The excess water was drained off and the seed autoclaved in 1-L Erlenmeyer flasks for 1 hr on each of three successive days. This sterilized seed was inoculated with 5-mm cubes of *G. graminis* var. *tritici* culture on potato-dextrose agar and incubated at 25°C for 4 weeks before air-drying.

The inocula were used in two field experiments in 1982, one designed to obtain dose-response curves and the other to screen wheat cultivars for resistance to *G. graminis* var. *tritici*. The experiments were conducted on a calcareous sandy loam, pH 8.0, and classified as Gcl.12; solonized brown soil; calcic xerosol (3).

Dose-response experiment. Ryegrass and millet seed propagules were applied at 0, 30, 60, 120, and 240 propagules per kilogram of soil. The soil (about 3 kg) in each microplot was removed to a depth of 10 cm using a post-hole digger 20 cm in diameter. The inoculum was mixed through two-thirds of the soil, which was returned to the hole. Ten wheat seeds (cv. Condor) were sown centrally within a 20-cm-diameter circle and covered with the remaining soil. Five replicate plots of each treatment were arranged in a randomized complete block design. Twelve weeks after seeding, two plants were removed from each plot for assessment of disease on the roots and measurement of shoot weight. Grain yield was determined on the remaining eight plants at maturity.

Screening for resistance. The above procedure was used in this trial with ryegrass propagules at rates of 0, 16, and 64 propagules per kilogram of soil. Grain yield

was determined on 33 cultivars of wheat—from the Australian Wheat Collection, from Centro Internacional de Mejoramiento de Maiz y Trigo at Temuco in southern Chile, and from commercial cultivars—in a split-plot design with three replicates. The cultivars from the Australian collection were selected as resistant from 4,000 cultivars screened in the field since 1977 using a variety of inoculation techniques. Inoculum of *G. graminis* var. *tritici* at 16 propagules per kilogram did not significantly reduce the grain yield of any cultivar. This treatment was combined with the uninoculated treatment to calculate the resistance index (RI) (Table 1):

$$RI = \frac{\text{Grain yield with 64 propagules per kg of soil}}{\text{Mean grain yield with 0 and 16 propagules per kg of soil}} \times 100.$$

Table 1. Range of resistance in wheat to *Gaeumannomyces graminis* var. *tritici*

Cultivar	Resistance index[a]
Aus 7875[b]	110
Temu 272-72[c]	92
Aus 3294	90
Aus 600	79
Aus 5938	78
Aus 310	76
Temu 89-72	74
Halberd	72
Aus 5559	69
Aus 5945	68
Aus 1080	67
Aus 774	62
Aus 5915	61
Aus 2564	55
Aus 4245	55
Aus 7303	54
Warigal	51
Temu 248-72	43
Aus 7795	38
Aus 2036	37
Aus 3766	36
Aus 481	33
Aus 12435	32
Aus 19117	31
Kite	31
Aus 7775	22
Temu 243-73	18
Condor	15
LSD ($P < 0.05$):	48

[a] $RI = \dfrac{\text{Grain yield with 64 propagules per kg of soil}}{\text{Mean grain yield with 0 and 16 propagules per kg of soil}} \times 100.$

[b] Aus = Australian Wheat Collection.

[c] Temu = Centro Internacional de Mejoramiento de Maiz y Trigo collection in Chile.

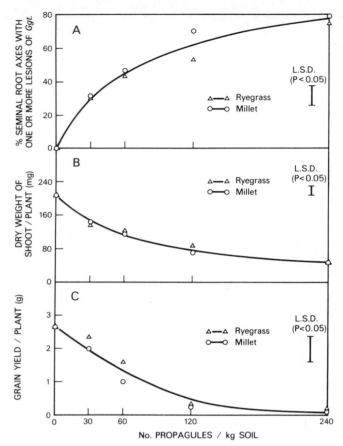

Fig. 1. Effect of increasing population densities of two forms of *Gaeumannomyces graminis* var. *tritici* inoculum in field microplots. (**A**) Incidence of pathogen on seminal root axes, (**B**) dry weight of plant shoots, and (**C**) grain yield of wheat (A and B measured 12 weeks after seeding).

Because 1982 was a drought year (136 mm of rain from April to October), an additional 58 mm of water was added during August–September.

RESULTS

In the dose-response experiment, increasing numbers of propagules of both forms of inoculum gave an increase in disease incidence on the roots, a decrease in shoot weight at 12 weeks after seeding, and a decrease in grain yield. There was no significant difference between the two forms of inoculum, although the millet seed inoculum consistently gave more disease (Fig. 1).

The results of the cultivar screening experiment in Table 1 show a range of resistance; Condor is the most susceptible and Aus 7875 the most resistant.

DISCUSSION

Use of ryegrass and millet seed colonized by *G. graminis* var. *tritici* permitted the numbers of propagules of uniform size to be manipulated in unsterilized soil naturally low in *G. graminis* var. *tritici*. Uniform propagule size is important in dose-response studies, as Hornby (1) has shown that infectivity of propagules is proportional to their size. The larger size of the millet seed (5.3 mg) compared with that of the ryegrass seed (1.8 mg) could therefore account for the consistently higher disease severity with millet seed inoculum.

Propagules of ryegrass and millet seed are small enough to be used without grinding, and their food reserves are sufficient for the propagules to remain infective in moist soil for at least 4 weeks (Heather J. McDonald, *unpublished*).

In the screening experiment, the wheat seed was placed within an area of smaller diameter than the inoculum. The aim was to reduce the chance of roots escaping infection and thus to reduce the variability in yield response discussed by Scott (4) and experienced by us in previous inoculation experiments. However, grain yield between replicates in this experiment varied considerably; this raises the question whether yield is the best criterion by which to assess resistance. We believe that both ryegrass and millet inocula could be used to obtain dose-response curves for cultivars being screened for resistance and to indicate the population densities at which resistance breaks down. It is feasible that, in screening for resistance, the most susceptible cultivars could be identified at low rates and the most resistant cultivars at high rates of inoculum.

In field experiments involving large plots, both forms of inoculum could be readily applied either with the seed into the drill-row or incorporated through the soil to study such topics as the effect of soil type, fertilizer, fungicides, and biological control agents. Simon (*unpublished*) is using ryegrass inoculum to screen soils for biological suppression of *G. graminis* var. *tritici*.

ACKNOWLEDGMENTS

We are indebted to Mr. R. Manley, Avon, S.A., for the use of his land on which these experiments were conducted and to the Wheat Industry Research Council for financial support.

LITERATURE CITED

1. Hornby, D. 1975. Inoculum of the take-all fungus: Nature, measurement, distribution and survival. EPPO Bull. 5:319-333.
2. MacNish, G. C., and Dodman, R. L. 1973. Incidence of *Gaeumannomyces graminis* var. *tritici* in consecutive wheat crops. Aust. J. Biol. Sci. 26:1301-1307.
3. Northcote, K. H., Hubble, G. D., Isbell, R. F., Thompson, C. H., and Bettenay, E. 1975. A Description of Australian Soils. CSIRO, Australia. 170 pp.
4. Scott, P. R. 1981. Variation in host susceptibility. Pages 219-236 in: Biology and Control of Take-all. M. J. C. Asher and P. J. Shipton, eds. Academic Press, London. 538 pp.

Part V
Influence of Soil, Environment, and Nutrition

Soil as an Environment for the Growth of Root Pathogens

D. M. GRIFFIN, Department of Forestry, Australian National University, Canberra, A.C.T., 2601, Australia

An understanding of the soil as an environment for root pathogens properly forms a part of ecology, yet research on these pathogens appears to have become increasingly separated from general ecology in recent decades. In this contribution, some concepts concerning the ecological niches of soil microbes are discussed. In particular, attention is focused on the dimensions of the niche hypervolume in relation to nutritional factors because these are both particularly poorly known and of high potential importance.

During its saprophytic phase, the soilborne root pathogen shares the same terrestrial environment as other soil microorganisms. Lynch (16) has provided a general survey of the characteristics of this environment; in recent years there have been many reviews of special aspects, such as my own on the role of water in microbial ecology (9–11). Rather than repeat what has already been done, I have decided to be quite frankly speculative. Lately, there has been a remarkable renaissance in microbial ecology, as witnessed by the books edited by Lynch and Poole (17) and Slater et al (33) and the article by Schlegel and Jannasch (30); but this is scarcely reflected as yet in most of the work on soilborne pathogens. I have selected one of the major themes within the renaissance and shall speculate on its significance for a better understanding of the ecology of soilborne pathogens.

I shall approach my main topic by first considering the concept of the niche. There is now a large evolving literature relating to the term "ecological niche," and much of this is confusing because of the variety of definitions adopted by different workers (e.g., Odum [21], Tribe and Williams [38], Pianka [23], Schlegel and Jannasch [30], and Wimpenny et al [42]). For my purpose, however, the ideas of Hutchinson (13), already discussed in a mycological context by Naughton (19), are particularly useful. In Hutchinson's treatment, the ability of an organism to grow, reproduce, and contribute to the maintenance of the species, i.e., its fitness, is affected by a wide range of factors. Thus each organism has cardinal points in relation to temperature, and so temperature may be considered as a dimension affecting fitness. If there are n-dimensions affecting an organism's fitness, then its niche is that hypervolume in n-dimensional space the boundaries and shape of which are determined by the response curves of fitness to change in each dimension and by interactions between the various dimensions. The concept may be further refined. The fundamental niche can then be seen as determined by the organism's own genetic properties, as "a hypothetical, idealized niche in which the organism encounters no 'enemies,' such as competitors or predators..." (24). The realized niche, however, is the one that actually exists, being determined by genetic properties as expressed in the presence of other organisms.

For most higher organisms, it is the realized niche that is best known, by means of information gained from field observations. The fundamental niche is known in respect of few of its dimensions. The reverse is the situation with microbes. Extensive laboratory experimentation has defined the characteristics of the fundamental niche in respect of many dimensions for many microbes. The difficulties of observing the microbe and simultaneously evaluating the environment (all without disturbance) in the microbe's natural habitat have greatly inhibited the gaining of an understanding of realized microbial niches. This is particularly so in the case of soil microbiology, because of the opaque, heterogeneous nature of soil. Given this situation, there is an obvious and inevitable temptation to predict the form of the realized niche from a partial knowledge of the fundamental niche (37). Too often, however, the results of laboratory experiments have been poor predictors of behavior in soil in the field. Usually, this has been because the total environment for growth in the laboratory has borne little resemblance to that in soil, even in regard to easily controlled factors such as temperature. Sometimes, however, the failure has been caused by the inevitable simplifications inherent in most experimental work, including the reduction of interactions between environmental variables. Nonetheless, pure cultures studies, or studies under artificially well-defined conditions, are important in clarifying concepts and frameworks, even if one must agree with those who have argued that eventually it is absolutely necessary to perform studies on soil microbial ecology in soils. Only in this way can realized niches be determined.

Let me now focus on one dimension (or group of dimensions) of the Hutchinsonian hypervolume—that of response to nutrient availability. There now seems to be overwhelming evidence that, in soil, most microbes are hungry most of the time (15). This must apply as much to pathogenic fungi during their saprophytic phase as to obligate saprophytes. It is almost axiomatic that for such chemoheterotrophs as fungi, the nutrients in most critical shortage are energy-yielding, organic compounds like carbohydrates and amino acids. In some instances, when abundant, mature plant residues are being decomposed, utilizable nitrogen compounds may be the limiting factor.

It is therefore surprising that so little is known about the relationship between nutrient (substrate) concentration and fungal activity. Many have pointed out the inappropriateness of standard laboratory media because of their high substrate concentrations, yet this has not led to significant work on fungal pathogens growing in nutritionally realistic systems. In these respects, the dimensions of the fundamental niche are poorly known for the vast majority of fungi, and those of the realized niche are almost entirely unknown—and these are the most critical niche dimensions for soilborne fungi! Rapid advances have been made recently in regard to nutritional aspects of bacterial ecology, mainly through use of chemostats, and Slater (32) has provided an excellent outline of the principles involved. Although comparable work with fungi poses additional problems, largely derived from their filamentous form, it is to be expected that a comparable revolution in fungal ecology will occur in the next decade.

Let me now refer to some of the basic concepts concerning nutrients and growth. The fundamental relationship between rate of reaction and substrate concentration in first-order, enzyme-mediated reactions is given by the familiar Michaelis-Menten equation:

$$q = q_{max}s/(K_m + s)$$

where q = rate of reaction; q_{max} = maximum rate of reaction, i.e., unlimited by substrate availability; s = substrate concentration; and K = Michaelis constant, i.e., value of s when $q/q_{max} = 0.5$.

More generally, it is often found empirically that the growth rate of an organism relates to the concentration of a limiting substrate by the Monod equation of similar form:

$$\mu = \mu_{max}s/(K_s + s)$$

where μ = specific growth rate (gram per gram [dry weight] per hour); μ_{max} = maximum specific growth rate, i.e., unlimited by substrate availability; s = substrate concentration; and K_s = saturation constant, i.e., value of s when $\mu/\mu_{max} = 0.05$.

Consider first an organism in an environment of abundant nutrient availability, or where s is large. Then μ largely depends upon μ_{max}. The success of an organism in such an environment will therefore depend in large part on its maximal specific growth rate. Consider now an organism in an environment where nutrient availability is almost vanishingly small, or where s is very small. (The available solutes in many natural systems are in fact present in nanomolecular concentrations [14]). Two strategies can lead to relatively high specific growth rates. The first is for the organism to increase the capacity of the transport system or the concentration of the enzyme as substrate concentration declines. This effectively increases μ_{max} in the Monod equation and thus the value of μ in the given conditions. However, the generation of increased transport system capacity or enzyme concentration will usually be energy demanding and therefore a poor strategy in a situation of developing stress. A preferable alternative is for K_s to be very small, i.e., the affinity of the organism for the substrate is very high. This is true whether the scavenging function for the limiting substrate is an active membrane transport system or the first enzyme in a metabolic sequence.

Organisms that characteristically live in environments of high nutrient availability and that have high values of μ_{max} have been called copiotrophs, whereas those that characteristically live in impoverished environments and that have low values for K_s are oligotrophs (25,26). The copiotrophs and oligotrophs of Poindexter bear a certain similarity to r- and K-strategists of MacArthur and Wilson (18) and other population biologists (22). Here r and K arise from the venerable Verhulst-Pearl logistic equation, with r-selection indicating that higher population growth rates and higher productivity are favored and K-selection indicating that a more efficient utilization of resources, say by a closer cropping of the food supply, is favored. These terms have been little used in a microbiological context (3,12,31).

Hirsh (12) has tabulated many characteristics of nutrient uptake and utilization and their consequences for model oligotrophs. Only two characteristics beyond those discussed above will be noted here. The first is the maintenance coefficient, which is derived from the equation

$$ds/dt = (1/y_g)\,(dx/dt) + mx$$

where x = mass of cells per volume; t = time; y_g = yield coefficient (growth), i.e., (mass organism produced)/(mass substrate utilized in growth); and m = maintenance coefficient (the weight of substrate used per dry weight of cells to maintain cell viability without growth). The second is μ_{min}, the slowest growth rate sustainable by the organism at limiting nutrient concentrations. It would seem likely that a low maintenance coefficient and a low μ_{min} would assist growth and survival in impoverished environments.

Remarkably few data exist for the values of K_s, μ_{max}, μ_{min}, and m for fungi, and scarcely any exist for plant pathogens. Such fungal data as I have found are presented in Table 1, along with a few representative data for bacteria. Note that the values for K_s, μ_{max}, and m for the copiotrophic bacteria *Escherichia coli* and *Pseudomonas* sp. are higher than those for the oligotrophic *Arthrobacter* and *Spirillium* sp. The value of m is, however, undoubtedly very sensitive to environmental factors; for instance, it increases tenfold in *Saccharomyces cerevisiae* as the concentration of sodium chloride in the broth increases by 1 M (41). For bacteria in soil it has been estimated to be as low as 0.0025 (15).

What of the fungi? It is difficult to say much except that their values for K_s, μ_{max}, μ_{min}, and m differ greatly. Within one species—*Neurospora crassa*—there are two uptake systems with very different affinities, the high-affinity system being induced, or more probably derepressed, only at low concentrations of substrate. On the limited data available, *Fusarium aquaeductuum* would quality as an oligotroph and *Saccharomyces cerevisiae* as a copiotroph in accord with their known habitats. *Mucor hiemalis* is confusing because of its low μ_{max} yet high m. Here perhaps an explanation can be found in the differing growth strategies of fungal colonies, hyphal density and rates of radial extension often being inversely related. Thus the ratio of colony radial growth rate (micrometers per hour) to specific growth rate (grams per gram per hour) for *Aspergillus nidulans*, *Penicillium chrysogenum*, and *M. hiemalis* are 986, 432, and 4,296 respectively (39).

As can be seen, there are virtually no data relating to

Table 1. Some values for the saturation constant (K_s glucose), maximum (μ_{max}), and minimum (μ_{min}) specific growth rates (dry weight basis) and maintenance constant (m; glucose unless otherwise stated) of fungi and bacteria[a]

Organism	K_s (μM)	μ_{max} (per hr)	μ_{min} (per hr)	m (g/g per hr)
Fungi				
Achlya bisexualis		0.81	<0.04	
Agaricus sp.		0.26		
Aspergillus nidulans	444–666	0.148 (25° C)	0.015 (25° C)	0.018
		0.360 (37° C)		
Fusarium aquaeductuum	1.6	0.13		
F. graminearum		0.28		
Geotrichum candidum	5.5	0.61		
Graphium sp.		0.187		
Mucor hiemalis		0.099		0.04
Neurospora crassa	(i) 8,000	0.35		
	(ii) 10			
Penicillium chrysogenum		0.28	0.014	0.022
Saccharomyces cerevisiae	600	0.45		0.036
Bacteria				
Arthrobacter globiformis	16	0.37	0.01	<0.01
Escherichia coli	21	0.96	0.008	0.076
Pseudomonas sp. (lactate)	20	0.64	0.05	0.066
Spirillum sp. (lactate)	5.8	0.35	0.01	0.016

[a] Data derived from the literature (1,2,4–6,25,27–29,34,36,40).

the nutritional dimensions of either the fundamental or realized niches of plant pathogenic fungi. The scant data available for other microbes make it likely, however, that the physiological parameters discussed do have profound ecological significance. The armchair biologist can think of many possibilities. Do fungal pathogens with a prolonged saprophytic survival phase (7,8) possess such a double uptake system as that already discovered in a number of organisms, using the low-affinity system when growing actively as a parasite in almost pure culture in the relatively high nutrient concentrations found within the living host, and then switching to the high-affinity system during the phase of saprophytic survival, with negligible growth within the dead host? During the survival phase, are μ_{min} and m exceptionally low? Are pioneer colonizers characterized by high linear growth rates (allowing them to spread rapidly over a substrate) but high values of K_s and low values of μ_{max} (making them poor competitors in crowded sites)? How do these important parameters differ between root-invading fungi and their saprophytic competitors in the rhizosphere? Small differences in the value of parameters are likely to be unimportant in natural situations, however, because of the slow rate at which weaker competitors are probably eliminated (35). During the requisite passage of time, the environment may well have changed to favor the previously weaker competitor.

If a considerable diversity of values could be demonstrated for soilborne plant pathogens, the data would significantly aid in understanding their ecology, not least in regard to models such as that of Newman and Watson (20). This model is very successful in predicting microbial abundance and activity in the rhizosphere from the premises that growth at each point is controlled by the concentration of soluble organic substrate and that changes in substrate concentration result from production by roots and subsequent diffusion, production by microbial breakdown of insoluble substrates, and use by microorganisms. Values of K_s, μ_{max}, and m are required, and generalized values from the literature were used in the model. If values for a given pathogen were known, it is likely that

the model could be developed to predict the chances of success for the pathogen within the rhizosphere under different conditions.

LITERATURE CITED

1. Bull, A. T., and Bushell, M. E. 1976. Environmental control of fungal growth. Pages 1-31 in: The Filamentous Fungi. Vol. 2. J. E. Smith and D. R. Berry, eds. Edward Arnold, London.
2. Bull, A. T., and Trinci, A. P. J. 1977. The physiological and metabolic control of fungal growth. Adv. Microb. Physiol. 15:1-84.
3. Carlile, M. J. 1980. From prokaryote to eukaryote: Gains and losses. Symp. Soc. Gen. Microbiol. 30:1-40.
4. Dawes, E. A. 1976. Endogenous metabolism and the survival of starved prokaryotes. Symp. Soc. Gen. Microbiol. 21:19-54.
5. Dow, C. S., Whittenburg, R., and Carr, N. R. 1983. The 'shut-down' or 'growth precursor' cell—an adaptation for survival in a potentially hostile environment. Symp. Soc. Gen. Microbiol. 34:187-247.
6. Forage, A. J., and Righelato, R. C. 1979. Biomass from carbohydrates. Pages 289-313 in: Economic Microbiology. Vol. 4. A. H. Rose, ed. Academic Press, London.
7. Garrett, S. D. 1970. Pathogenic Root-Infecting Fungi. Cambridge University Press, Cambridge. 294 pp.
8. Garrett, S. D. 1981. Soil Fungi and Soil Fertility. 2nd ed. Pergamon Press, Oxford. 150 pp.
9. Griffin, D. M. 1978. Effect of soil moisture on survival and spread of pathogens. Pages 175-197 in: Water Deficits and Plant Growth. Vol. 5. T. T. Kozlowski, ed. Academic Press, New York.
10. Griffin, D. M. 1981. Water potential as a selective factor in the microbial ecology of soils. Pages 141-151 in: Water Potential Relations in Soil Microbiology. J. F. Parr, W. R. Gardner, and L. F. Elliott, eds. Soil Sci. Soc. Am. Spec. Publ. 9.
11. Griffin, D. M. 1981. Water and microbial stress. Adv. Microb. Ecol. 5:91-136.
12. Hirsch, P. 1979. Life under conditions of low nutrient concentrations. Pages 357-372 in: Strategies of Microbial Life in Extreme Environments. M. Shilo, ed. Verlag Chemie, Weinheim.
13. Hutchinson, G. E. 1958. Concluding remarks. Cold Spring Harbor Symp. Quant. Biol. 22:415-427.
14. Konings, W. N., and Veldkamp, H. 1983. Energy transduction and solute transport mechanisms in relation to

environments occupied by microorganisms. Symp. Soc. Gen. Microbiol. 34:153-186.

15. Lockwood, J. L., and Filonow, A. B. 1981. Response of fungi to nutrient-limiting conditions and to inhibitory substances in natural habitats. Adv. Microb. Ecol. 5:1-61.

16. Lynch, J. M. 1979. The terrestrial environment. Pages 69-71 in: Microbial Ecology: A Conceptual Approach. J. M. Lynch and N. J. Poole, eds. Blackwell, Oxford.

17. Lynch, J. M., and Poole, N. J., eds. 1979. Microbial Ecology: A Conceptual Approach. Blackwell, Oxford. 266 pp.

18. MacArthur, R. H., and Wilson, E. O. 1967. The Theory of Island Biogeography. Princeton University Press, Princeton, N. J. 203 pp.

19. Naughton, S. J. 1981. Niche: Definition and generalizations. Pages 79-88 in: The Fungal Community. D. T. Wicklow and G. C. Carroll, eds. Marcel Dekker, New York.

20. Newman, E. I., and Watson, A. 1977. Microbial abundance in the rhizosphere: A computer model. Plant Soil 48:17-56.

21. Odum, E. P. 1971. Fundamentals of Ecology. 3rd ed. Sanders, Philadelphia. 574 pp.

22. Pianka, E. R. 1970. On r- and K-selection. Am. Nat. 104:592-597.

23. Pianka, E. R. 1976. Competition and niche theory. Pages 114-141 in: Theoretical Ecology. R. M. May, ed. Blackwell, Oxford.

24. Pianka, E. R. 1978. Evolutionary Ecology. 2d ed. Harper and Row, New York. 307 pp.

25. Poindexter, J. S. 1981. Oligotrophy: Fast and famine existence. Adv. Microb. Ecol. 5:63-89.

26. Poindexter, J. S. 1981. The caulobacters: Ubiquitous unusual bacteria. Microbiol. Rev. 45:123-179.

27. Righelato, R. C. 1975. Growth kinetics of mycelial fungi. Pages 79-103 in: The Filamentous Fungi. Vol. 1. J. E. Smith and D. R. Berry, eds. Edward Arnold, London.

28. Righelato, R. C. 1979. The kinetics of mycelial growth. Pages 385-401 in: Fungal Walls and Hyphal Growth. J. H. Burnett and A. P. J. Trinci, eds. Cambridge University Press, Cambridge.

29. Righelato, R. C., Trinci, A. P. J., Pirt, S. J., and Peat, A. 1968. The influence of maintenance energy and growth rate on the metabolic activity, morphology and conidiation of Penicillium chrysogenum. J. Gen. Microbiol. 50:399-412.

30. Schlegel, H. G., and Jannasch, H. W. 1981. Prokaryotes and their habitats. Pages 43-82 in: The Prokaryotes. Vol. 1. M. P. Starr, H. Stolp, H. G. Truper, A. Balows, and H. G. Schlegel, eds. Springer Verlag, Berlin.

31. Seifert, R. P. 1981. Applications of a mycological data base to principles and concepts of population and community ecology. Pages 11-23 in: The Fungal Community. D. T. Wicklow and G. C. Carroll, eds. Marcel Dekker, New York.

32. Slater, J. H. 1979. Microbial populations and community dynamics. Pages 45-63 in: Microbial Ecology: A Conceptual Approach. J. M. Lynch and N. J. Poole, eds. Blackwell, Oxford.

33. Slater, J. H., Whittenburg, R., and Wimpenny, J. W. T., eds. 1983. Microbes in Their Environment. Symp. Soc. Gen. Microbiol. 34. Cambridge University Press, Cambridge.

34. Solomons, G. L. 1975. Submerged culture production of mycelial biomass. Pages 349-364 in: The Filamentous Fungi. Vol. 1. J. E. Smith and D. R. Berry, eds. Edward Arnold, London.

35. Taylor, J., and Williams, P. 1975. Theoretical studies on the coexistence of competing species under continuous flow conditions. Can. J. Microbiol. 21:90-98.

36. Tempest, D. W., and Neijssel, O. M. 1978. Eco-physiological aspects of microbial growth in aerobic nutrient-limited environments. Adv. Microb. Ecol. 2:105-155.

37. Tempest, D. W., Neijssel, O. M., and Zevenboom, W. 1983. Properties and performance of microorganisms in laboratory culture; their relevance to growth in natural ecosystems. Symp. Soc. Gen. Microbiol. 34:119-152.

38. Tribe, H. T., and Williams, P. A. 1967. Investigations into the basis of microbial ecology in soil, illustrated with reference to growth of soil diphtheroids and Azotobacter in a model system. Can. J. Microbiol. 13:467-480.

39. Trinci, A. P. J. 1969. A kinetic study of the growth of Aspergillus nidulans and other fungi. J. Gen. Microbiol. 57:11-24.

40. Trinci, A. P. J., and Thurston, C. F. 1976. Transition to the non-growing state in eukaryotic micro-organisms. Symp. Soc. Gen. Microbiol. 26:55-80.

41. Watson, T. G. 1970. Effects of sodium chloride on steady-state growth and metabolism of Saccharomyces cerevisiae. J. Gen. Microbiol. 64:91-99.

42. Wimpenny, J. W. T., Lovitt, R. W., and Coombs, J. P. 1983. Laboratory model systems for the investigation of spatially and temporally organised microbial systems. Symp. Soc. Gen. Microbiol. 34:67-117.

Lethal Temperatures of Soil Fungi

G. J. BOLLEN, Laboratory of Phytopathology, Agricultural University, Wageningen, The Netherlands

Data on heat resistance of soil fungi are indispensable for reliably predicting results of such measures against pathogens as selective heat treatments, solar heating of soils, and composting of plant residues. Heat treatment of soil, seed, and plant propagation material was introduced almost a century ago. It has always been practiced since that time, although its role has been taken over in part by chemicals. In disinfestation of seeds and propagation material, heat treatment was largely replaced by fungicide use. Its use for disinfestation of glasshouse soils was surpassed by fumigants.

For various reasons there is a renewed interest in heat treatments for disease control. Solar heating of moistened soil under plastic tarps is recommended for controlling pathogens in tropical and subtropical areas (12,14). In The Netherlands, pollution problems associated with fumigation of glasshouse soil with methyl bromide have recently led to severe restrictions in its use, and it will probably be banned in the near future. Soil steaming is the most effective alternative.

Another application of heat treatment is associated with composting of plant refuse. Fungal pathogens do not survive composting of residues of infested crops (11,16,17). Their elimination is attributed both to heat evolved during the process and to the formation of toxic products.

A knowledge of heat resistance of pathogens and their antagonists is indispensable for understanding the effects of heat treatments. Many workers have reported on thermal sensitivity of fungi (2–5,12,14). This paper adds data on pathogens that were used in experiments on composting and on selective heat treatments of glasshouse soil.

MATERIALS AND METHODS

Sources of inoculum. Heavily infested soils with diseased roots or tubers were used for assessment of lethal temperatures of obligate pathogens and fusaria. The inoculum of *Olpidium brassicae* was a suspension of air-dried root material of *Solanum villosum* in water. Soil cultures were pure cultures grown in autoclaved potting soil (organic matter, 68%; pH-KCl 5.5) enriched with oatmeal (2%), and these were used for most of the fungi.

Soil heat treatments. Three methods were used. Quantities up to 30 L were heated in a modified type of the soil pasteurization apparatus designed by Aldrich and Nelson (1). Samples up to 1 L were treated with a laboratory soil pasteurizer (4). Soil moisture was 50% of moisture-holding capacity or more. Soil cultures were heated in a thermostatically controlled water bath. Six cubic centimeters of soil was placed in test tubes and adjusted with sterile water to 10 cm³. Thermocouples were placed in the soil suspension. The tubes were plugged and partially immersed in the water bath. Temperatures varied less than 0.2°C of the adjusted values. Treatments lasted 30 min.

Viability tests. Survival of obligate pathogens and some nonobligates was assessed by using test plants. In most experiments, the soil was replanted up to three times to obtain optimal detection. Survival of fungi grown in soil cultures was tested by plating out soil particles on potato-dextrose agar (PDA; pH 5.6), by soil dilution plates using the same medium, or by adding 10 ml of the treated soil to test plants. For the analysis of the fungal flora of heat-treated field soil, PDA with oxgall (5 mg/ml) and oxytetracyline (50 μg/ml) was used.

RESULTS AND DISCUSSION

Obligate pathogens. Among the fungi tested, only *O. brassicae* survived treatment at 60°C and then only in two of the four experiments (Table 1). Its thermal death point (TDP) decreased by about 5°C when the root material was kept moist for 4 days before treatment. Its

Table 1. Lethal temperatures of obligate pathogens subjected to 30-min heat treatments

Pathogen	Test plant[a]	Number of experiments	Number of plants per treatment	Lethal temperature[b] (°C)
Olpidium brassicae[c]				
Not activated	*Solanum villosum*	1	30	55.0–60.0
Not activated	*S. villosum*	2	40	60.0–62.5
Activated	*S. villosum*	1	40	55.0–57.5
Plasmodiophora brassicae	Chinese cabbage	2	100	50.0–55.0
	Chinese cabbage	1	100	55.0–60.0
Synchytrium endobioticum	Potato	1	12	50.0–60.0
	Potato	1	16	55.0–60.0

[a] The test plants were of the same species as the host plant from which the isolates were obtained, except for the isolate of *O. brassicae* that originated from tulip.
[b] Range given is temperature of treatment that permitted survival followed by that which was lethal.
[c] Air-dried roots with resting spores either activated or not activated by incubation under moist conditions for 4 days before treatment.

resistance resulted from resting spores, as zoospores are reported to be highly sensitive to heat (8). Campbell and Grogan (7) found survival of resting spores after treatments up to 65°C for 10 min.

Lethal temperatures of *Plasmodiophora brassicae* were close to 55°C. *Synchytrium endobioticum* survived a 55°C treatment, but with a severe loss of virulence. The fungus could still infect sprouts of the highly susceptible potato cultivar Deodara, but the minute lesions never developed into actual warts. The reaction of the susceptible cultivar to the heat-treated inoculum was similar to the reaction of resistant cultivars to untreated inoculum.

Oomycetes. These fungi are rather sensitive to heat (Table 2). The most resistant species was *Pythium aphanidermatum*. The isolate tested formed oospores abundantly in soil cultures. *P. sylvaticum* and *Phytophthora capsici* were chosen to study the relationship of heat resistance to the ability to form oospores. Both species are heterothallic; therefore, oospores were only formed in soil cultures that had been inoculated with compatible mating types. The cultures to be treated were 5–8 weeks old and contained numerous oospores. Formation of oospores increased heat resistance of *P. capsici*. This observation suggests that it is important to know whether one or both mating types are present when heat treatments are used in marginal situations for controlling heterothallic pathogens. Such marginal situations occur in applying thermotherapy to sensitive planting material, soil solarization at marginal temperatures, or heat treatment of heavy soils with steam-air mixtures.

***Fusarium* species.** The formae speciales of *F. oxysporum* were among the most resistant pathogens (Table 3). Two of the four isolates of *F. oxysporum* f. sp. *dianthi* that were tested even survived at 60°C. The pathogen is known for its survival in soil locations that received poor steam penetration during treatment. Heat resistance was determined by treatments of infested glasshouse soils as well as pure cultures in soil of isolates obtained from the infested soils. Killing the fungi in soil cultures required an increment of at least 2.5°C (one stage in the series of treatments) more than that required to eliminate the pathogen from infested soils. This was probably a result of the extremely high densities of inoculum attained in soil cultures (viable counts up to $5.10^7/g$). The higher the number of spores the greater is the chance of resistant spores among them. The resistance of some strains of *Fusarium* species was also noticed by other workers (9,10). This limits the possibilities of utilizing thermotherapy with planting stock.

In spite of the close relationship between *Cylindrocarpon* and *Fusarium*, the former was appreciably more heat sensitive. Although the soil cultures contained numerous chlamydospores, *C. destructans* did not survive treatment at 47.5°C. The results apply to only two isolates, and more data are needed before a general conclusion is made. An isolate from cyclamen also did not survive 50°C (4).

Other pathogens. The low TDP of *Colletotrichum coccodes* was unexpected. An analysis of the mycoflora of a potato field soil revealed the fungus in samples treated up to 60°C. The TDP of *Verticillium dahliae* was

Table 2. Lethal temperatures of oomycetes subjected to 30-min heat treatments

Pathogen	Source	Lethal temperature	
		With test plants[a]	By plating[b]
Phytophthora capsici A₁	Pepper	42.5–45.0	40.0–42.5
P. capsici A₂	Pepper	42.5–45.0	42.5–45.0
P. capsici A₁ + A₂	Pepper	>50.0	47.5–50.0
P. cryptogea	Gerbera	40.0–42.5	40.0–45.0
Pythium aphanidermatum	Cucumber	>52.5	50.0–52.5
P. sylvaticum male	Lettuce	47.5–50.0	42.5–45.0
P. sylvaticum female	Lettuce	45.0–47.5	42.5–45.0
P. sylvaticum male + female	Lettuce	47.5–50.0	42.5–45.0

[a] Test plants were cucumber seedlings for all pathogens except *P. cryptogea*, where Chinese aster was used. Each treatment included 100 seedlings in separate 100-ml pots. The treated inoculum was mixed (1:10) with steamed potting soil.
[b] For each treatment, 50 particles of the treated soil culture were plated.

Table 3. Lethal temperatures of nonobligate pathogens estimated by 30-min heat treatments of soil cultures

Pathogen	Source	Lethal temperature[a]
Colletotrichum coccodes (2)[b]	Potato, tomato	45.0–50.0
Cylindrocarpon destructans (2)	Lilium speciosum	45.0–47.5
Fusarium avenaceum (1)	Carnation	57.5–60.0
F. oxysporum f. sp. *dianthi* (2)	Carnation	55.0–57.5
F. oxysporum f. sp. *dianthi* (2)	Carnation	60.0–65.0
F. oxysporum f. sp. *gladioli* (1)	Freesia	57.5–60.0
F. oxysporum f. sp. *lycopersici* (4)	Tomato	57.5–60.0
F. oxysporum f. sp. *melongenae* (2)	Eggplant	57.5–60.0
F. redolens f. sp. *dianthi* (1)	Carnation	55.0–57.5
F. solani f. sp. *phaseoli* (1)	Bean	45.0–50.0
Phomopsis sclerotioides (2)	Gherkin	45.0–50.0
Verticillium dahliae (2)	Potato, tomato	45.0–47.5
V. dahliae (1)	Sugar beet	40.0–45.0

[a] Estimated by using soil dilution plates; density of viable counts in the suspension of untreated soil cultures varied from 2.10^5 to 1.10^8 per milliliter (= 0.3 g of soil). Lower limit of detection was 2 viable counts per milliliter of suspension.
[b] Number of isolates tested.

lower than that of isolates tested earlier (4) but was in accordance with the results of the comprehensive study by Pullman et al (14). Potato plants grown on 50°C-treated soil from infested fields did not show symptoms, whereas those on untreated soil were seriously affected. Sensitivity of *V. dahliae* also appears from the effectiveness of solar heating in controlling the pathogen (12).

Saprophytes. Lethal temperatures of most species were in the same range as those of parasites. A treatment at 70°C eliminated most species, including *Trichoderma* spp., which are known for their antagonism toward pathogens.

Among resistant fungi, cleistothecia-forming penicillia and aspergilli were the most common species. Few of them are antagonists of pathogens. An example is *Talaromyces flavus* (13), a common fungus in glasshouse soils (6).

Spore germination of distinctly resistant fungi is typically activated by a heat shock. Breaking of dormancy was reported for spores of heat-resistant ascomycetes (15). It was also very characteristic of the spores of the two most resistant fungi recorded in a survey of more than 250 soils from The Netherlands—the conidia of *Gilmaniella humicola*, a hyphomycete, and the chlamydospores of *Tephrocybe carbonaria*, a basidiomycete. Both fungi survived a treatment of 90°C in moist soils. Germination of chlamydospores of *T. carbonaria* was still activated when 8-year-old soil cultures were heated: treatment at 85°C for 30 min yielded 336 times as many colonies as untreated soil. In cultures of *G. humicola* stored for the same period, germination was not enhanced by heat. Apparently, constitutive dormancy of the spores had been terminated. Activation of germination by treatments at sublethal temperatures was not observed for pathogens used in the present study.

LITERATURE CITED

1. Aldrich, R. A., and Nelson, P. E. 1969. Equipment for aerated steam treatment of small quantities of soil and soil mixes. Plant Dis. Rep. 53:784-788.

2. Baker, K. F., and Roistacher, C. N. 1957. Heat treatment of soil. Pages 123-137 in: The U.C. System for Producing Healthy Container-grown Plants Through the Use of Clean Soil, Clean Stock and Sanitation. K. F. Baker, ed. Calif. Agric. Exp. Stn. Man. 23. 332 pp.

3. Barran, L. R. 1980. Effect of heat, freeze-thawing and desiccation on the survival of *Fusarium sulphureum* spores. Trans. Br. Mycol. Soc. 75:425-427.

4. Bollen, G. J. 1969. The selective effect of heat treatment on the microflora of a greenhouse soil. Neth. J. Plant Pathol. 75:157-163.

5. Bollen, G. J. 1974. Fungal recolonization of heat-treated glasshouse soils. Agro-Ecosystems 1:139-155.

6. Bollen, G. J., and Van der Pol-Luiten, B. 1975. Mesophilic heat-resistant soil fungi. Acta Bot. Neerl. 24:254-255.

7. Campbell, R. N., and Grogan, R. G. 1964. Acquisition and transmission of lettuce big-vein by *Olpidium brassicae*. Phytopathology 54:681-690.

8. Campbell, R. N., and Lin, M. T. 1976. Morphology and thermal death point of *Olpidium brassicae*. Am. J. Bot. 63:826-832.

9. Grouet, D. 1967. Recherches sur les fusarioses. III. Mise au point du traitement à l'eau chaude de la fusariose du glaïeul. Ann. Epiphyt. 18:285-304.

10. Hargreaves, A. J., and Fox, R. A. 1978. Some factors affecting survival of *Fusarium avenaceum* in soil. Trans. Br. Mycol. Soc. 70:209-212.

11. Hoitink, H. A. J., Herr, L. J., and Schmitthenner, A. F. 1976. Survival of some plant pathogens during composting of hardwood tree bark. Phytopathology 66:1369-1372.

12. Katan, J. 1981. Solar heating (solarization) of soil for control of soilborne pests. Annu. Rev. Phytopathol. 19:211-236.

13. Marois, J. J., Johnston, S. A., Drum, M. T., and Papavizas, G. C. 1982. Biological control of Verticillium wilt of eggplant in the field. Plant Dis. 66:1166-1168.

14. Pullman, G. S., DeVay, J. E., and Garber, R. H. 1981. Soil solarization and thermal death: A logarithmic relationship between time and temperature for four soilborne plant pathogens. Phytopathology 71:959-964.

15. Warcup, J. H., and Baker, K. F. 1963. Occurrence of dormant ascospores in soil. Nature 197:1317-1318.

16. Wijnen, A. P., Volker, D., and Bollen, G. J. 1983. De lotgevallen van pathogene schimmels in een composthoop (The fate of pathogenic fungi in a compost heap). Gewasbescherming 14:5.

17. Yuen, G. Y., and Raabe, R. D. 1979. Eradication of fungal plant pathogens by aerobic composting. Phytopathology 69:922.

Relation Between Root Infection with *Phytophthora cinnamomi* and Water Relations in Susceptible and Field-resistant *Eucalyptus* Species

PETER DAWSON and GRETNA WESTE, Botany School, University of Melbourne, Parkville, Victoria, 3052, Australia

The actual cause of death in susceptible species of *Eucalyptus* infected with *Phytophthora cinnamomi* Rands has puzzled plant pathologists since first reported (8). The pathogen commonly invades unsuberized roots and rarely extends throughout the major, suberized roots. The appearance of infected, susceptible *Eucalyptus* trees resembles that caused by drought because diseased forest trees show wilting of shoots; dry, graying foliage; and branch dieback. All susceptible plant species appear severely water stressed as a result of infection (12). The experiments described in this paper were designed to compare the relationship between infection with *P. cinnamomi* and subsequent water relations of susceptible with field-resistant *Eucalyptus* species.

MATERIALS AND METHODS

Four-month-old seedlings of *E. seiberi* L.A.S. Johnson (susceptible) and *E. maculata* Hook. (field resistant) were grown in pots cut vertically in half to allow the attachment of an acrylic plastic front for observation of root growth and lesion development. Plants were inoculated on the root tips with an axenic suspension of zoospores from *P. cinnamomi*. At the completion of the experiments, roots were surface sterilized in 70% alcohol and segments were plated onto a selective agar medium (6) to enable calculation of the percentage of the root system infected. During the 15 days following inoculation, lesion extension and the component factors in plant water relations were measured. Transpiration was recorded from daily loss in weight and calculated as milligrams of water per square millimeter of leaf area per day; the water lost was subsequently replaced by syringe. Leaf water vapor was measured with a vapor diffusion porometer, both surfaces of three leaves being measured daily at noon. Leaf water potential was determined with a pressure bomb (9) and leaf relative water content was also measured (1). Root and stem hydraulic conductivities were measured using a pressure chamber (4). The root system or stem was excised at lower stem level, washed gently, and placed in a large test tube filled with nutrient solution inside the pressure bomb. Bomb pressure was maintained at 5 bars, and the flow rate of exudate was measured with a pipette attached to the stem.

Root and stem hydraulic conductivities were determined from the equation (5): $J_v = L\Delta P x^{-1}$, where J_v is the hydraulic flux, based on the cross-sectional area of stem or root ($mm^3/mm^2/sec$) of root exudate; ΔP is the pressure potential applied across stem or root; x is the length of stem or root (mm); and L is the hydraulic conductivity ($mm^2/kPa/sec$) (5).

Osmotic potentials of exuded sap and nutrient solution were ignored because they represented less than 0.01% of the total water potential applied across the root system.

RESULTS

The first symptom observed in both species was a dark brown lesion 7–10 mm long that formed overnight behind the tips of inoculated roots. Growth of inoculated roots ceased. In *E. maculata*, neither fungus nor the initial lesion extended farther than 10 mm, and no other symptoms were observed. In *E. sieberi*, the lesions extended 12 mm a day on each of the three roots for 2–3 days, during which they were visible in the root boxes. Secondary symptoms of shoot wilting were observed in *E. sieberi* after a period that varied from 3 to 15 days after inoculation.

Symptom development in *E. sieberi* occurred in three clearly defined, successive stages. Stage 1 was lesion formation, approximately 10 hr after inoculation; stage 2, lesion extension associated with reduced root hydraulic conductivity; stage 3, water stress in shoots that followed the continued and increasing reduction in root conductivity.

Inoculated *E. maculata* plants displayed only stage 1 in symptom development. Root conductivity of inoculated *E. sieberi* plants (Fig. 1) showed a marked reduction during stages 2 and 3 of disease development when compared with controls. These differences of hydraulic conductivity were not the result of changes in root length. The decreased root conductivity is particularly remarkable in that the fungus was isolated from only 8–15% of the total root system of *E. sieberi* at stage 3 of disease. At stage 2 of symptom development, the root hydraulic conductivity of *E. sieberi* was only 50% of the control (Fig. 1). By stage 3, it had fallen to 9% of the control value. There was no significant reduction in hydraulic conductivity of *E. maculata* roots, and only 2–4% of the root systems were infected.

Water stress did not appear in shoots of *E. sieberi* until root conductivity was significantly less than that of uninfected controls. At stage 3, wilting was observed in shoots; measurements showed that leaf conductance and leaf xylem water potential, transpiration, and leaf relative water content (Table 1) had declined significantly but still followed the same pattern with noon maxima. Leaf conductance was reduced 75%. Leaf xylem water potential changed from −0.4 to −1.2 MPa. The relative water content of the leaves fell from 88 to 54%, and shoot transpiration rate was only 26% of that of the uninfected controls. These changes were absent in infected *E. maculata* plants.

Stem conductivity of *E. sieberi* did not change with infection. In both control and infected plants, the conductivity was three orders of magnitude greater than control root conductivity. Stems therefore did not contribute significantly to the reduced hydraulic conductivity measured in infected root systems.

DISCUSSION

Despite an adequate water supply in the soil, all infected *E. sieberi* plants became severely water stressed. These experiments demonstrated that in every case, water stress was associated with gross impairment of root hydraulic conductivity. Infected *E. maculata*

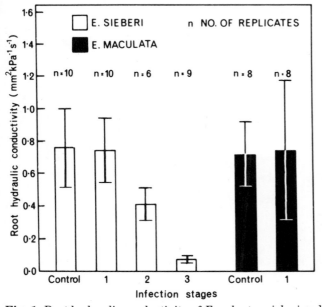

Fig. 1. Root hydraulic conductivity of *Eucalyptus sieberi* and *E. maculata* for uninfected control plants and for plants infected by *Phytophthora cinnamomi* at different stages of disease development. (Confidence levels of 95% are shown. Mean is represented by bar on histograms.)

plants were not water stressed, and root conductivity was scarcely affected. In *E. sieberi*, shoot water relations (Table 1) were not affected at stage 2 in symptom development when hydraulic conductivity was reduced by 50% (Fig. 1); but shoot water relations were severely disrupted by stage 3 when root hydraulic conductivity was only 9% of the control. The onset of changes in root conductivity was associated with lesion extension and was observed to 2–14 days after inoculation in all plants. The range in time taken for failure in water transport probably reflects variability in seedling vigor and in susceptibility to the pathogen.

Water transport through the roots was reduced by 91%, or practically eliminated, when less than one-sixth of the root system was infected. Once water flow within the roots failed, water stress symptoms quickly developed in the shoots. The disproportion between the amount of root infected and the reduction in hydraulic conductivity indicates that the failure of water transport cannot be caused directly by decay of the relatively small diseased portion of the root. Previous research demonstrated large and highly significant reduction in leaf relative water content, leaf xylem water potential, and transpiration in 1- to 2-year-old container-grown plants of *Isopogon ceratophyllus* (2). Measurements recorded from mature forest trees of *E. obliqua* L. Herit. showed that plants infected by *P. cinnamomi* had lower leaf xylem water potential than uninfected trees (11). *P. cinnamomi* is reported to have similar effects on the water status of *E. marginata* Sm. growing on laterites (10). The sudden death syndrome recorded on trees of *E. baxteri* (Benth.) Maiden & Blakeley and *E. macrorhyncha* F. Muell. ex Benth. infected with *P. cinnamomi* in the Brisbane Ranges of Victoria (12) may have been caused by the failure of root water conductivity, thus preventing an adequate water supply to the crown during periods of rapid transpiration.

Decreased root hydraulic conductivity has been demonstrated previously for cotton plants infected by *Phymatotrichum omnivorum* (7) and for safflower infected by *Phytophthora cryptogea* (3). None of these infected plants wilted until root conductivity was reduced to a small fraction of that in uninfected controls. The hydraulic conductivity measured for infected *E. sieberi* roots is comparable with that calculated on a leaf area basis for infected safflower roots (3).

Although the exact mechanism remains unknown, the experiments described in this paper have demonstrated failure in root water transport in a

Table 1. Transpiration, leaf relative water content (RWC), and leaf xylem water potential of *Eucalyptus sieberi* and *E. maculata* plants inoculated with *Phytophthora cinnamomi*

Stage of infection	Replicates (seedlings)	Leaf xylem water potential (−MPa)	Transpiration[a] (mg/mm^2/hr × 10^2)	RWC[a] (%)
E. sieberi				
Control	10	0.3 ± 0.1	34 ± 13	87 ± 3
Infection stage				
Root lesions	10	0.5 ± 0.1	28 ± 12	85 ± 2
Root conductance affected	6	0.4 ± 0.1	37 ± 9	88 ± 4
Shoot symptoms	9	1.2 ± 0.1	9 ± 4	54 ± 7
E. maculata				
Control	8	0.4 ± 0.1	62 ± 18	90 ± 3
Infection stage[b]				
Root lesions	8	0.5 ± 0.1	57 ± 14	91 ± 4

[a] Mean and 95% confidence limits.
[b] Later infection stages absent in *E. maculata*.

susceptible species, *E. sieberi*, when infected with *P. cinnamomi*. Further investigations are in progress to determine whether this xylem dysfunction is caused by embolism in vessels associated with plugging and tyloses, by hormone imbalance, or by high-molecular-weight compounds produced either from the fungus (as in enzymic degradation) or from host interaction.

LITERATURE CITED

1. Barrs, H. D., and Weatherley, P. E. 1962. A re-examination of the relative turgidity technique for estimating water deficits in leaves. Aust. J. Biol. Sci. 15:413-428.
2. Dawson, P., and Weste, G. 1982. Changes in water relations associated with infection by *Phytophthora cinnamomi*. Aust. J. Bot. 30:393-400.
3. Duniway, J. M. 1977. Changes in resistance to water transport in safflower during the development of *Phytophthora* root rot. Phytopathology 67:331-337.
4. Mees, G. S., and Weatherley, P. E. 1957. The mechanism of water absorption by roots. I. Preliminary studies on the effects of hydrostatic pressure gradients. Proc. R. S. London B 147:367-380.
5. Milburn, J. A. 1979. Water Flow in Plants. Longmans, London. 225 pp.
6. Ocana, G., and Tsao, P. H. 1966. A selective agar medium for the direct isolation and enumeration of *Phytophthora* in soil. (Abstr.) Phytopathology 56:893.
7. Olsen, M. W., Misaghi, I. J., Goldstein, D., and Hine, R. B. 1983. Water relations in cotton plants infected with *Phymatotrichum*. Phytopathology 73:213-216.
8. Podger, F. D. 1972. *Phytophthora cinnamomi*, a cause of lethal disease in indigenous plant communities in Western Australia. Phytopathology 62:972-981.
9. Scholander, P. F., Hammel, H. T., Bradstreet, E. D., and Hemmingsen, E. A. 1965. Sap pressure in vascular plants. Science 148:339-346.
10. Shea, S. R., Shearer, B., and Tippett, J. 1982. Recovery of *Phytophthora cinnamomi* Rands from vertical roots of jarrah (*Eucalyptus marginata* Sm.). Australas. Plant Pathol. 11:25-28.
11. Weste, G. 1980. Vegetation changes as a result of invasion of forest on krasnozem by *Phytophthora cinnamomi*. Aust. J. Bot. 28:139-150.
12. Weste, G., and Taylor, P. 1971. The invasion of native forest by *Phytophthora cinnamomi*. I. Brisbane Ranges, Victoria. Aust. J. Bot. 19:291-294.

Effects of Soil Temperature, Moisture, and Timing of Irrigation on Powdery Scab of Potatoes

R. F. DE BOER, Plant Research Institute, Department of Agriculture, Burnley, Victoria, 3121; P. A. TAYLOR and S. P. FLETT, Irrigation Research Institute, Department of Agriculture, Tatura, Victoria, 3616; and P. R. MERRIMAN, Plant Research Institute, Department of Agriculture, Burnley, Victoria, 3121, Australia

Powdery scab of potatoes (*Solanum tuberosum* L.) caused by *Spongospora subterranea* (Wallr.) Largerh. f. sp. *subterranea* Tomlinson has been reported from most potato-growing districts in Australia.

In Victoria, the severity and frequency of outbreaks of powdery scab have increased in recent years. Consignments of affected seed and ware potatoes are often downgraded, and scab on potatoes in cool stores can increase the frequency of storage rots caused by other pathogens (5). The situation is thought to have been caused, in part, by the recent introduction of irrigation practices whereby growers are encouraged to irrigate crops, especially during tuber set, to ensure that yield-reducing water deficits do not occur. Tuber set is thought to be significant because studies by Hughes (4) suggest that tubers are susceptible to infection by *S. subterranea* only at an early stage of growth. Cool, wet soils are reported to favor the development of powdery scab (1,3–6,11–13), and observations in Victoria indicate that such conditions occur after irrigation. Therefore, the yield benefits derived from irrigation may be offset by losses due to powdery scab.

The objective of this study was to examine in more detail the effect of soil temperature and moisture on disease development and to determine whether withholding irrigation during the early stage of crop growth could influence the incidence of powdery scab.

EFFECT OF SOIL TEMPERATURE AND MOISTURE

Soil temperature. A layer of autoclaved sand in free-draining containers (170 mm × 110 mm × 90 mm) was inoculated with a suspension (1 g in 20 ml of water) of spore balls of *S. subterranea*. Two presprouted potato tubers, cultivar Kennebec, were placed in the sand in each container and covered with more autoclaved sand.

Each container was covered with a fine plastic mesh to reduce shoot growth. Containers were incubated at constant temperatures of either 10.0, 12.5, 15.0, 17.5, or 20.0° C and watered with distilled water. After 3 months, the numbers of infected tubers and powdery scab pustules per tuber were recorded.

The percentage of tubers with scab and number of pustules per tuber were greatest at 12.5° C (Table 1). No disease occurred at 10.0° C.

Soil matric water potential. Two presprouted potato tubers, cultivar Sebago, were planted into a clay loam of pH 5.5 in plastic cups (110 mm in diameter and 80 mm deep). A small hole was made in the base of each cup to allow aeration, and cups were then inverted onto ceramic plates connected to a water tension apparatus. The cups were kept in darkness at 17.5–20° C, and soil was maintained at matric water potentials of either −0.001, −0.01, −0.02, −0.03, or −0.1 bars. When the first roots appeared 1 week after planting, a suspension (1 g in 20 ml of water) of spore balls of *S. subterranea* was injected into each cup. Four weeks after inoculation, segments of root with a uniform distribution of root hairs were selected. Segments were stained and

Table 1. Effect of soil temperature on the incidence and severity of powdery scab on potato tubers

Temperature (° C)	Tubers with scab (%)	Number of scab pustules per tuber
10.0	0.0 ± 0.0[y]	0.0 ± 0.0[y]
12.5	65.3 ± 30.6	51.1 ± 39.0
15.0	41.0 ± 4.1	16.7 ± 16.7
17.5	7.6 ± 8.1[z]	2.0 ± 3.1[z]
20.0	10.3 ± 5.7	4.5 ± 0.7

[y] Standard errors of the means.
[z] No disease recorded in two replicates.

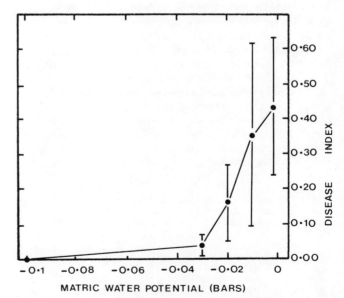

Fig. 1. Effect of matric water potential on the infection of root hairs of potatoes by *Spongospora subterranea*. Disease index = sum of individual root hair infection ratings divided by total number of microscopic fields examined. Root hair infection ratings per microscopic field: 1 = 1–5, 2 = 6–15, 3 = 16–30, and 4 = more than 30 sporangia.

examined microscopically for sporangia of *S. subterranea*. The assessment of sporangial infection in root hairs was expressed as a disease index (2) (Fig. 1).

The disease index was greatest at −0.001 bars and decreased with decreasing matric water potentials (Fig. 1). Sporangia of *S. subterranea* were not detected on roots from treatments of −0.1 bars.

EFFECT OF TIMING OF IRRIGATION

The effect of timing of furrow irrigation on disease on the cultivar Kennebec was investigated under controlled and field experiments. In each experiment, plots were irrigated at planting. Subsequent treatments were normal irrigation (6- to 9-day cycle in controlled experiment, 12- to 14-day cycle in field experiment); irrigation withheld until tuber set, followed by normal irrigation; and irrigation withheld until 3 weeks after tuber set, followed by normal irrigation. Tuber set was designated as the time when the tips of 50% or more of stolons had swollen to at least 5 mm in diameter. The number of diseased tubers was assessed at harvest.

An excavated site of 8.5 m × 1.3 m × 0.3 m in a polyethylene igloo was filled with a sandy clay loam of pH 5.8. A cubic meter of soil and potato debris infested with *S. subterranea* was then added and incorporated with a rotary hoe. Tubers were planted at 9-cm intervals in rows 1 m long and spaced 9 cm apart, with a single guard row between each two-row plot. The igloo was air-conditioned, and temperatures ranged between 9 and 25°C. The temperature in the soil was maintained between 10 and 18°C (averaging 15°C) by circulating cold water (10°C) through pipes buried below the surface. Tuber set was recorded 7 weeks after planting, and plots were harvested 9 weeks later.

The field experiment was established at a site where five successive crops of potatoes had been affected with powdery scab. The soil type was a clay loam of pH 5.5. Tubers were planted at 25-cm intervals in two rows per 3-m plot with an inter-row spacing of 0.8 m and two guard rows between each plot. Aluminum foil was laid on the soil before plants emerged to minimize effects of rainfall and maintain favorable temperatures for infection. Tuber set occurred 7 weeks after planting, and plots were harvested 22 weeks later.

The level of disease in the controlled experiment was high, and the percentage of tubers with scab was progressively reduced by delaying irrigation until tuber set and 3 weeks after tuber set (Table 2). Similarly, the percentage of tubers with scab in the field experiment was significantly reduced by delaying irrigation.

Table 2. Effect of timing of irrigation on the incidence of powdery scab in controlled and field experiments

	Tubers with scab (%)	
Irrigation treatment	Controlled experiment	Field experiment
Normal (6- to 9-day cycle, controlled experiment; 12- to 14-day cycle, field experiment)	80.9 a[z]	21.5 a
Delayed until tuber set	44.7 b	6.6 b
Delayed until 3 weeks after tuber set	4.2 c	4.4 b

[z] In each column, mean values followed by the same letter are not significantly different from each other at $P = 0.05$ according to Duncan's multiple range test.

DISCUSSION

The results from the controlled experiment on irrigation, which showed that tubers were susceptible to *S. subterranea* only during their early stage of growth, have important practical applications. In soils infested with *S. subterranea*, it may be possible to minimize the development of scab on tubers by withholding irrigation during tuber set. This concept of disease control by modified irrigation practices has been successfully developed for control of common scab (*Streptomyces scabies* (Thaxt.) Waks. and Henrici) (7–10). These experiments defined the time of susceptibility of tubers to infection by *S. scabies* and the influence of water potential on this disease. The information is used as a basis for control whereby high soil moisture levels are maintained by regular irrigation during the first weeks of tuber set.

To assess the potential for controlling powdery scab by withholding irrigation during tuber set, we need to define the conditions required for the development of disease on tubers. Further studies are therefore required to determine the time of susceptibility of tubers and the time taken for infection of tubers by *S. subterranea* under a given set of conditions, especially those of moisture and temperature.

LITERATURE CITED

1. Bhattacharyya, S. K. 1974. Studies on powdery scab. Pages 67-69 in: 25th Annual Scientific Report of the Central Potato Research Institute, Simla, India. 162 pp.
2. Flett, S. P. 1983. A technique for detection of *Spongospora subterranea* in soil. Trans. Br. Mycol. Soc. 81:424-425.
3. Hims, M. 1976. The weather relationships of powdery scab disease of potatoes. Proc. Br. Plant Pathol. Ann. Appl. Biol. 84:274-275.
4. Hughes, I. K. 1980. Powdery scab (*Spongospora subterranea*) of potatoes in Queensland: Occurrence, cultivar susceptibility, time of infection, effect of soil pH, chemical control and temperature relations. Aust. J. Exp. Agric. Anim. Husb. 20:625-632.
5. Karling, J. S. 1968. The Plasmodiophorales. 2d ed. Hafner Publishing Company, New York. 256 pp.
6. Kole, A. P. 1954. A contribution to the knowledge of *Spongospora subterranea* (Wallr.) Lagerh., the cause of powdery scab of potatoes. Tijdschr. Plantenziekten. 60:1-65.
7. Lapwood, D. H., and Hering, T. F. 1970. Soil moisture and the infection of young potato tubers by *Streptomyces scabies* (common scab). Potato Res. 13:296-304.
8. Lapwood, D. H., Wellings, L. W., and Hawkins, J. H. 1971. Irrigation as a practical means to control potato common scab (*Streptomyces scabies*). Plant Pathol. 20:157-163.
9. Lapwood, D. H., Wellings, L. W., and Hawkins, J. H. 1973. Irrigation as a practical means to control potato common scab. (*Streptomyces scabies*): Final experiment and conclusions. Plant Pathol. 22:35-41.
10. Lewis, B. G. 1970. Effects of water potential on the infection of potato tubers by *Streptomyces scabies* in soil. Ann. Appl. Biol. 66:83-88.
11. Melhus, I. E., Rosenbaum, J., and Schultz, E. S. 1916. *Spongospora subterranea* and *Phoma tuberosa* on the Irish potato. J. Agric. Res. 7:213-253.
12. Ramsey, G. B. 1918. Influence of moisture and temperature upon infection by *Spongospora subterranea*. Phytopathology 8:29-31.
13. Würzer, B. 1965. Ergänzende Untersuchungen über den Pulverschorf der Kartoffel und dessen Erreger *Spongospora subterranea* (Wallr.) Lagerh. Arb. Landwirtschaftlichen Hochschule Hohenheim 34:1-104.

Influence of Depleted Oxygen Supply on Phytophthora Root Rot of Safflower in Nutrient Solution

A. D. HERITAGE, CSIRO Centre for Irrigation Research, Private Mail Bag, Griffith, N.S.W., 2680, Australia, and J. M. DUNIWAY, Department of Plant Pathology, University of California, Davis, 95616, U.S.A.

Phytophthora cryptogea Pethybr. & Laff. is an economically important root rot pathogen of safflower (*Carthamus tinctorius* L.). The severity of infection in field crops has been found to increase with both frequency and intensity of irrigation (8,40). In fact, the occurrence of Phytophthora root rots in a wide range of other crops is associated with wet soils (3,15,32,36,39). Given favorable soil conditions of moisture, temperature, and organic matter, appreciable decreases in soil oxygen levels may result after only 24 hr of flood irrigation on fine-textured soils (29). Under these conditions, oxygen levels in soil air have been shown to fall to between 0.5 and 1% (10,25).

Consequences of reduced soil oxygen levels for plant growth have been widely reported (1,4,6,11,12,18,29,31). Fungal growth appears to be less sensitive to reductions in oxygen levels than is plant root growth. A selection of soil fungi have been tested for their response to reduced O_2 levels (19). The majority was insensitive except at most extreme treatments. Species of *Phytophthora* maintain growth in both liquid and solid media down to O_2 levels of 1% (24). However, species of *Phytophthora* vary in their response to low O_2 levels (17,24): whereas most grow more slowly in low O_2 than in air, a few species grow faster in 1% O_2 than in air (21%).

Interactions between depleted oxygen in the root zone and plant diseases have been examined in a few plant pathogen associations and have been discussed in review articles (6,11,28). Low soil-oxygen levels have been associated with increased Phytophthora root rot in citrus (32) and avocado (33). However, an earlier report had demonstrated lower incidence of root rot in avocado under these conditions (5), whereas infection of *Pinus* spp. by *P. cinnamomi* was unaffected by depleted O_2 levels (37). More recently, bean root rot caused by *Fusarium solani* was reported to be aggravated by soil atmospheres that were depleted in oxygen (21–23).

The aim of the present study is to examine the effect of depleted oxygen levels in nutrient solution on the incidence of safflower root rot caused by *P. cryptogea*.

MATERIALS AND METHODS

Growth of safflower plants. Two cultivars of safflower were used in these experiments: Biggs, a highly resistant cultivar (35); and Gila, a cultivar having some elements of resistance (27) and favorable agronomic characteristics.

Seedlings were raised in vermiculite and transferred to 1-L preserving jars containing Hoagland's solution (half strength) after 1 week. Each jar held one seedling, which was supported by a plastic foam collar fitted into a hole in the jar lid. Except during the application of a specified gas treatment, safflower roots were aerated at 15 L/hr via a glass tube. Seedlings were raised for a further 14 days in a glasshouse (15–30°C, night and day temperatures). Before removal of seedlings to a controlled atmosphere chamber for inoculation under varying oxygen treatments, the nutrient solution was renewed. Constant temperature conditions (25 ± 0.1°C) were maintained in the chamber while 85 W/m² (wavelength of 300–700 nm) of fluorescent and incandescent light was applied for 14 hr/day. After this treatment, plants were returned to the glasshouse where air bubbling was resumed during symptom development. Disease severity was recorded 7 days after inoculation. Symptoms in the shoots and roots of plants were rated on a scale of 1 to 7. Final fresh weights of shoots and roots were determined after the roots were blotted dry in a standardized manner.

Fungal inoculum. The culture of *P. cryptogea* used in all experiments was the A2 mating type, which was originally isolated from safflower in California (isolate number P201). Zoospores used for inoculation were obtained by incubating disks of fungal mycelium in autoclaved sand maintained at a constant water potential and subsequently removing the disks to water (20).

Safflower plants were inoculated when 3 weeks old by pipetting 1 ml of zoospore suspension containing 5×10^4 spores into each jar. Gentle mixing of the solution in the jars was maintained by constant bubbling with air (15 L/hr) for 2 hr, after which the desired gas treatment was applied.

Preparation of gas mixtures. Gases having oxygen concentrations lower than air (21% O_2) were obtained by mixing compressed air and nitrogen (high-purity grade, 99.9997% pure) through capillary tubes under constant pressure. The final composition of oxygen in the mixture was checked against standard gas mixtures using gas chromatography. Gas mixtures were regulated to individual jars by capillaries (15 L/hr).

In one experiment, a gas mixture containing 0.5% O_2 was applied to both inoculated and uninoculated safflower roots for varying periods. In another experiment, gas mixtures containing varying concentrations of oxygen (in the range of 0 to 21%) were bubbled for a period of 24 hr after inoculation.

RESULTS

Effect of prolonged O_2 depletion. The effect of prolonged periods of depleted O_2 on the development of safflower root rot was tested using cultivars Biggs and Gila. Uninoculated plants of both cultivars developed wilting symptoms in their shoots and evidence of root discoloration after prolonged exposure to 0.5% O_2 (Table 1). Gila cultivar appeared to be less severely affected than Biggs. For example, Gila plants that had been subjected to 48 hr of low O_2 grew to 83% of control plants (grown in air), whereas Biggs plants treated for the same time grew to only 40% of control plants. When the low O_2 treatment was extended to 72 hr, shoot weights of both cultivars were similarly affected. Plant shoot tips but not wilted leaves in all uninoculated plants were observed to recover during the 7-day posttreatment period.

Phytophthora root rot symptoms were progressively worse with increasing duration of depleted O_2 (Table 1). Significant reductions in shoot weights were observed in plants having only 6–12 hr of treatment. Cultivar Biggs was more severely affected than Gila by the combination of prolonged O_2 depletion and infection with *P. cryptogea*. Shoot fresh weights of Gila plants having 24 hr of treatment, for example, developed to 27.3% of untreated plants, whereas Biggs plants having a similar treatment weighed only 8.4% of untreated control plants. Prolonged O_2 depletion similarly increased the severity of visible symptoms and decreased the fresh weight of roots of infected plants (Table 1). Neither wilted leaves nor shoot tips of inoculated plants recovered during the 7-day posttreatment period.

Effect of varying oxygen concentrations. The effect of varying the oxygen content of the gas mixtures supplied to safflower roots on root rot development was tested with cultivar Gila. Roots were exposed to oxygen percentage levels of 21 (air), 5, 2, 0.5, and 0 (N_2) for 24 hr following the 2-hr inoculation period.

Uninoculated plants were unaffected by the three gas mixtures having the highest O_2 concentrations (Table 2). A slight but significant ($P < 0.05$) reduction in shoot fresh weight was observed at 0.5% O_2, whereas plants whose roots were bubbled with 0% O_2 (N_2) grew to less than half the fresh weight of the controls (air).

Phytophthora root rot was more severe in plants whose roots were subjected to those gas mixtures containing O_2 concentrations less than air (Table 2). Final shoot fresh weights of plantings growing in jars that were bubbled with 5 and 2% O_2 were reduced to about 50% of uninoculated controls. Progressively more severe reductions in fresh weight of plant parts resulted with decreasing O_2 levels, with fresh shoots weighing 35 and 12% of control plants at 0.5 and 0% O_2, respectively.

DISCUSSION

Depletion of oxygen in the root zone influences root growth and the severity of root rot development caused by *P. cryptogea* in safflower.

In the absence of the disease organism, cultivar Gila was superior to Biggs in its tolerance of prolonged periods of oxygen depletion. This result confirms observations of these cultivars in fields that are subjected to excessive waterlogging (C. A. Thomas, *personal communication*). Cultivar selection for tolerance to waterlogging may be necessary where irrigation practices are not sophisticated or where prolonged rainfall may be experienced in conjunction with crops growing on poorly drained soils.

Susceptibility of safflower to infection by *P. cryptogea* increased both with prolonged low O_2 levels (0.5%) and with decreased levels of O_2 (0–5%) applied for 24 hr after inoculation. This observation supports field observations with safflower (40) and highlights the possibility of low soil oxygen conditions as a critical soil factor influencing the severity of root rot in this crop. These results, however, were carried out in solution culture and need to be repeated under soil conditions before they can be used to interpret field observations.

Table 1. Effect of prolonged O_2 depletion (0.5%) on Phytophthora root rot of safflower

Inoculated[x]	Time (hr)	Fresh weight (g) Shoots	Roots	Symptoms[y] Shoots	Roots
Biggs					
No	0	23.0 ab[z]	7.6 defg	1	2
	6	23.3 ab	7.6 defg	1	2
	12	24.9 a	7.2 defg	1	2
	24	21.6 ab	7.4 defg	1	2.7
	48	9.2 cdef	4.0 fg	1.3	4
	72	13.3 cd	6.2 defg	1.3	3.2
Yes	0	16.2 bc	5.8 defg	2	3.5
	6	12.0 cde	4.7 efg	1.7	4.7
	12	2.2 fg	1.3 fg	3.4	6.2
	24	1.9 fg	1.4 fg	5.5	6.5
	48	1.5 fg	1.0 g	6.2	7
	72	1.1 g	0.6 g	6.0	6.7
Gila					
No	0	27.9 a	8.9 d	1	1
	6	30.0 a	8.2 d	1	1
	12	23.9 b	6.0 def	1	1.7
	24	23.9 b	6.9 de	1	1.2
	48	22.4 b	7.8 de	1	2.2
	72	13.9 c	6.4 de	1.7	2
Yes	0	22.4 b	5.3 defg	1	4
	6	15.2 c	4.3 efgh	1	5
	12	15.9 c	5.5 defg	1.7	4.5
	24	7.7 de	2.7 fgh	4.8	5.5
	48	2.5 fgh	1.2 h	5.2	6.7
	72	1.8 gh	0.7 h	4.7	6.7

[x] Inoculation with zoospores of *Phytophthora cryptogea* before treatment.
[y] Visible symptom rating of 1–7 on an increasing scale of disease severity.
[z] Weights within the same experiment that are followed by different letters are significantly different by Duncan's multiple range test, $P = 0.05$.

Table 2. Effect of varying O_2 concentrations on Phytophthora root rot of safflower (cv. Gila)

Inoculated[x]	O_2 (%)	Fresh weight (g) Shoots	Roots	Symptoms[y] Shoots	Roots
No	21	30.2 a[z]	7.6 defg	1	1.2
	5	25.0 ab	7.5 defg	1	1
	2	26.9 ab	7.5 defg	1	1.5
	0.5	21.2 ab	6.4 efg	1	1.2
	0	13.0 cd	4.7 efg	1	2.5
Yes	21	27.4 a	4.7 efg	1	3
	5	13.2 cd	7.0 defg	2.4	5.2
	2	15.3 c	4.1 fg	2.4	4.5
	0.5	10.3 cde	4.2 fg	2	5.7
	0	3.4 fg	1.2 g	5.2	5.5

[x] Inoculation with zoospores of *Phytophthora cryptogea* before treatment.
[y] Visible symptom rating of 1–7 on an increasing scale of disease severity.
[z] Weights within the same experiment that are followed by different letters are significantly different by Duncan's multiple range test, $P = 0.05$.

The normally highly resistant cultivar Biggs was severely infected when subjected to low oxygen conditions. Previous reports have demonstrated a reduction in the resistance of Biggs to Phytophthora root rot when subjected to high levels of inoculum, low light intensities, high temperatures (16), or prior water stress (7).

The mechanism by which low oxygen levels were able to exacerbate root damage caused by *P. cryptogea* was not examined in this study. Release of root exudates has been implicated in the chemotaxis of *Phytophthora* zoospores toward roots (2,14,38). Increases in root exudation under depleted soil oxygen have been reported in other plant species (9,13,26). However, because this experiment was carried out in stirred solution culture, zones of accumulated root exudates are not likely to have occurred on the safflower roots. In fact, in these experiments, the low O_2 stress was imposed after the infection period, suggesting that a change in the root's resistance to the fungus had occurred. Under low O_2 levels, reduced production of oxidized polyphenols has been reported to account for increased development of Phytophthora rot of cocoa pods (30). More recently, a resistance mechanism involving two antifungal polyacetylene compounds (safynol and dehydrosafynol) has been reported in Biggs, the safflower cultivar that is highly resistant to root rot (34). The function of this and other possible resistance mechanisms under reduced oxygen concentrations is worthy of future research.

ACKNOWLEDGMENTS

This study was carried out at Davis and was supported by the Australian Oilseeds Research Committee. We thank Carol Lindquist and Shirley Heritage for technical assistance and Xavier Flores, Department of Vegetable Crops, Davis, for help with gas mixing.

LITERATURE CITED

1. Bradford, K. J., and Yang, S. F. 1981. Physiological responses of plants to waterlogging. HortScience 16:3-8.
2. Chang-Ho, Y., and Hickman, C. J. 1970. Some factors involved in the accumulation of phycomycete zoospores on plant roots. Pages 103-108 in: Root Diseases and Soil-borne Pathogens. T. A. Toussoun, R. A. Bega, and P. E. Nelson, eds. University of California Press, Berkeley.
3. Cook, R. J., and Papendick, R. I. 1972. Influence of water potential of soils and plants on root diseases. Annu. Rev. Phytopathol. 10:349-374.
4. Curtis, D. S., and Zentmyer, G. A. 1949. Effect of oxygen supply in nutrient solution of avocado and citrus seedlings. Soil Sci. 67:253-260.
5. Curtis, D. S., and Zentmyer, G. A. 1949. Effect of oxygen supply on Phytophthora root rot of avocado in nutrient solution. Am. J. Bot. 36:471-474.
6. Drew, M. W., and Lynch, J. M. 1980. Soil anaerobiosis, microorganisms and root function. Annu. Rev. Phytopathol. 18:37-66.
7. Duniway, J. M. 1977. Predisposing effect of water stress on the severity of Phytophthora root rot in safflower. Phytopathology 67:884-889.
8. Erwin, D. C. 1952. Phytophthora root rot of safflower. Phytopathology 42:32-35.
9. Fulton, J. M., and Erickson, A. E. 1964. Relation between soil aeration and ethyl alcohol accumulation in xylem exudate of tomatoes. Soil Sci. Soc. Am. Proc. 28:610-614.
10. Furr, J. R., and Aldrich, W. W. 1943. Oxygen and carbon-dioxide changes in the soil atmosphere of an irrigated date garden on calcareous very fine sandy loam soil. Proc. Am.

Soc. Hortic. Sci. 42:46-52.
11. Grable, A. R. 1966. Soil aeration and plant growth. Adv. Agron. 18:57-106.
12. Greenwood, D. J. 1970. Soil aeration and plant growth. Rep. Prog. Appl. Chem. 55:423-431.
13. Grineva, G. M. 1961. Excretion by plant roots during brief periods of anaerobiosis. Sov. Plant Physiol. (Engl. Transl.) 8:549-552.
14. Halsall, D. M. 1976. Zoospore chemotaxis in Australian isolates of *Phytophthora* species. Can. J. Microbiol. 22:409-422.
15. Hickman, C. J., and English, M. P. 1951. Factors influencing the development of red core in strawberries. Trans. Br. Mycol. Soc. 34:223-236.
16. Johnson, L. B., and Klisiewicz, J. M. 1969. Environmental effects on safflower reaction to *Phytophthora drechsleri*. Phytopathology 59:469-472.
17. Klotz, L. J., Stolzy, L. H., and DeWolfe, T. A. 1963. Oxygen requirements of three root-rotting fungi in a liquid medium. Phytopathology 53:302-305.
18. Letey, J., Stolzy, L. H., and Blank, G. B. 1962. Effect of duration and timing of low soil oxygen content on shoot and root growth. Agron. J. 54:34-37.
19. Macauley, B. J., and Griffin, D. M. 1969. Effects of carbon dioxide and oxygen on the activity of some soil fungi. Trans. Br. Mycol. Soc. 53:53-62.
20. MacDonald, J. D., and Duniway, J. M. 1978. Influence of the matric and osmotic components of water potential on zoospore discharge in *Phytophthora*. Phytopathology 68:751-757.
21. Miller, D. E., and Burke, D. W. 1975. Effect of soil aeration on Fusarium root rot of beans. Phytopathology 65:519-523.
22. Miller, D. E., and Burke, D. W. 1977. Effect of temporary excessive wetting on soil aeration and Fusarium root-rot of beans. Plant Dis. Rep. 61:175-179.
23. Miller, D. E., Burke, D. W., and Kraft, J. M. 1980. Predisposition of bean roots to attack by the pea pathogen, *Fusarium solani* f. sp. *pisi*, due to temporary oxygen stress. Phytopathology 70:1221-1224.
24. Mitchell, D. J., and Zentmyer, G. A. 1971. Effects of oxygen and carbon dioxide tensions on the growth of several species of *Phytophthora*. Phytopathology 61:787-791.
25. Reuther, W., and Crawford, C. L. 1947. Effects of certain soil and irrigation treatments on citrus chlorosis in a calcareous soil. II. Soil atmosphere studies. Soil Sci. 63:227-240.
26. Rittenhouse, R. L., and Hale, M. G. 1971. Loss of organic compounds from roots. II. Effect of O_2 and CO_2 tension on release of sugars from peanut roots under axenic conditions. Plant Soil 35:311-321.
27. Rubis, D. D., and Black, D. S. 1958. Gila—a new safflower variety. Ariz. Exp. Stn. Bull. 301. 5 pp.
28. Schoeneweiss, D. F. 1975. Predisposition, stress, and plant disease. Annu. Rev. Phytopathol. 13:193-211.
29. Scott Russell, R. 1977. Plant root systems. McGraw-Hill, London. 298 pp.
30. Spence, J. A. 1961. Black-pod disease of cocoa. II. A study of host-parasite relations. Ann. Appl. Biol. 49:723-734.
31. Stolzy, L. H., and Letey, J. 1964. Measurement of oxygen diffusion rates with the platinum microelectrode. III. Correlation of plant response to soil oxygen diffusion rates. Hilgardia 35:567-576.
32. Stolzy, L. H., Letey, J., Klotz, L. J., and Labanauskas, C. K. 1965. Water and aeration as factors in root decay of *Citrus sinensis*. Phytopathology 55:270-275.
33. Stolzy, L. H., Zentmyer, G. A., Klotz, L. J., and Labanauskas, C. K. 1967. Oxygen diffusion, water and *Phytophthora cinnamomi* in root decay and nutrition of avocados. Proc. Am. Soc. Hortic. Sci. 90:67-76.
34. Thomas, C. A., and Allen, E. H. 1971. Light and antifungal polyacetylene compounds in relation to resistance of safflower to *Phytophthora drechsleri*. Phytopathology 61:1459-1461.
35. Thomas, C. A., and Zimmer, D. E. 1970. Resistance of Biggs safflower to Phytophthora root rot and its inheritance. Phytopathology 60:63-64.

36. Wager, V. A. 1942. *Phytophthora cinnamomi* and wet soil in relation to the dying-back of avocado trees. Hilgardia 14:519-532.
37. Zak, B. 1961. Aeration and other soil factors affecting southern pines as related to littleleaf disease. U.S. Dep. Agric. For. Serv. Tech. Bull. 1248. 30 pp.
38. Zentmyer, G. A. 1970. Tactic responses of zoospores of *Phytophthora*. Pages 109-111 in: Root Diseases and Soil-borne Pathogens. T. A. Toussoun, R. V. Bega, and P. E. Nelson, eds. University of California Press, Berkeley.
39. Zentmyer, G. A., and Richards, S. J. 1952. Pathogenicity of *Phytophthora cinnamomi* to avocado trees, and the effect of irrigation on disease development. Phytopathology 42:35-37.
40. Zimmer, D. E., and Urie, A. L. 1967. Influence of irrigation and soil infestation with strains of *Phytophthora drechsleri* on root rot resistance of safflower. Phytopathology 57:1056-1059.

Pea Root Pathogen Populations in Relation to Soil Structure, Compaction, and Water Content

J. M. KRAFT, U.S. Department of Agriculture, Agricultural Research Service, Irrigated Agriculture Research and Extension Center, Prosser, WA 99350, and R. R. ALLMARAS, U.S. Department of Agriculture, Agricultural Research Service, Columbia Plateau Conservation Research Center, Pendleton, OR 97801, U.S.A.

In eastern Washington and northeastern Oregon, peas (*Pisum sativum* L.) are grown primarily in rotation with fall-planted cereals where rainfall adequately supports annual cropping. In this area, wheat yields have increased 300% in the last 50 years; in contrast, average processing pea yields are static. Field-to-field and year-to-year variations are large. One of the most important and recurring yield constraints of peas is root disease caused primarily by *Fusarium solani* (Mart.) Sacc. f. sp. *pisi* and *Pythium ultimum* Trow (3–5). Yield reductions of peas from root diseases range from 10 to 50% (3,6).

Early, vigorous pea growth is essential for an economical yield, primarily because soilborne diseases are directly influenced by the vigor of the developing crop (5). Such stress factors as inadequate fertility and moisture, soil compaction, and occasional high air temperatures exacerbate soilborne diseases. In some field sites, tractor wheels and tillage implements create compacted soil layers that are not readily penetrated by infected pea roots. These infected roots are usually severely rotted, reduced in volume, and unable to grow into sources of water and nutrients necessary for optimum yields (5).

Our study was conducted in the green pea and wheat producing area of Walla Walla County, Washington, and Umatilla County, Oregon. The dominant Walla Walla silt loam in this area is a typic Haploxeroll, with a minimum of horizonation and an annual dry season. In cultivated soils there is a significant layering of abiotic properties primarily because of tillage practices. Currently, wheat yields range from 4,000 to 8,000 kg/ha with straw residues from 8,500 to 16,000 kg/ha. Furthermore, long-term use of nitrogen fertilizers has reduced soil pH from 6.9 to values of 5.3 or less in the upper 20 cm. Conservation tillage is being developed for wheat culture in this area and will most likely increase straw residues on or near the soil surface. In addition, compaction patterns induced by implement use and traffic will change. The effects of this altered tillage system on pea production and root diseases are unknown. To understand the changes that will occur, our objectives were to determine the extent and depth of soil compaction in a crop rotation of peas and wheat; the depth and population distribution of *F. solani* f. sp. *pisi* and *P. ultimum* in several different growing sites; and the effect tillage pans and wheat straw residues have on

Second author now with the U.S. Department of Agriculture, Agricultural Research Service, University of Minnesota, St. Paul, MN 55108, U.S.A.

pea growth, soil water infiltration, and root pathogen distributions.

MATERIALS AND METHODS

For about 4 years, we have been measuring the abiotic environment of the pea root associated with pathogen populations and distribution and with root disease. Twenty-two field sites were assayed in 7.5- or 15-cm increments to a depth of 60 cm. Soil measurements included dry bulk density, organic carbon, pH, weight of coarse organic fragments (mainly undecomposed wheat residues), and representative water content distributions with soil depth. In addition, soil samples were assayed to determine distribution and populations of *P. ultimum* and *F. solani* f. sp. *pisi*.

Dry bulk density was determined by weight and volume measurements of soil cores obtained with a modified Oakfield soil tube (2-cm diameter). Depth increment was 7.5 cm in earlier work and was finally reduced to 2 cm; the depth of sampling was 60 cm. At least 15 random cores provided a bulk density mean with a standard error of 50 kg/m^3. Recently incorporated crop residue (coarse organic fragments) was first determined by sieving the air-dry soil plus residue on a 0.5-mm-mesh screen and then correcting for soil contaminants by carbon dilution. Organic carbon was determined by dry combustion. Soil pH was determined in a mixture of one part soil and two parts 0.01M CaCl$_2$. Soil water content vs. depth and soil water movement were both determined by field measurements with tensiometers and neutron moisture probes above and below the tillage pan.

Propagule numbers and depth of distribution of *P. ultimum* and *F. solani* f. sp. *pisi* were determined from soil samples taken at various times during the crop year and rotation. Soils were serially diluted in 0.3% water agar and assayed on selective media (7,8). Later in the study, dry bulk density, coarse organic matter, and propagule numbers were determined on the same sample to improve precision.

RESULTS AND DISCUSSION

A definite tillage pan in all pea, wheat, or wheat-fallow field sites sampled was evident regardless of soil type and water regime (dry land or supplementally irrigated). This tillage pan was 20 ± 5 cm deep and about 7 cm thick. In addition, the straw turned down after a wheat crop (moldboard plow) was still present and largely undecomposed during the succeeding pea crop. Preliminary measurement indicated that the straw mat

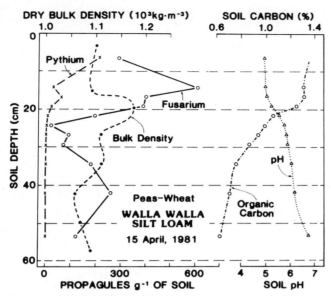

Fig. 1. Typical relationship between pea root pathogen abundance and related soil properties indicative of long-term and recent tillage history in a Walla Walla silt loam (pea-wheat rotation).

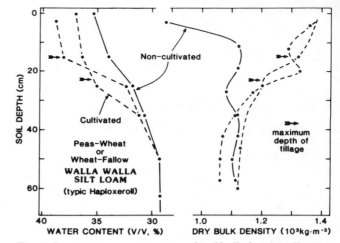

Fig. 2. Long-term management and soil bulk-density effects on soil water content vs. depth in a soil profile undergoing steady-state drainage. Arrows indicate maximum depth of tillage in both a pea-wheat or a wheat-fallow cropping system. The wheat-pea cultivation system had the deepest tillage. Note the relative absence of any restriction of water flow in the noncultivated soil.

was located within 10 cm above the maximum depth of tillage. Estimated amounts of incorporated straw ranged from 2,000 to 7,000 kg/ha. The tillage pan is significant because compaction directly influences the severity and extent of pea root disease (6). The impact of the combined tillage pan and superimposed straw layer has not yet been determined. In a typical soil profile, *P. ultimum* was characteristically found in the upper 20 cm and was absent below the plow layer. *F. solani* f. sp. *pisi* propagules were found throughout the upper 60 cm of soil, but their frequency was always low in the tillage pan under the plow layer (Fig. 1).

The sparsity of *F. solani* f. sp. *pisi* propagules in the tillage pan and their presence below it are related to the impaired drainage from compaction in the tillage pan (Fig. 2) and the saprophytic survival of *F. solani* f. sp. *pisi* under dry soil conditions. In fields not cropped to peas for 5 or more years, *F. solani* f. sp. *pisi* was not detected in the plow layer but was always recovered below it. Long-term cultivation in this area has produced an environment beneath the plow layer that is favorable for long-term survival of *F. solani* f. sp. *pisi* (Fig. 2). Slower drying and associated delay of optimum moisture for field operations are other effects of long-term cultivation (1).

Long-term use of ammoniacal fertilizers in wheat has also lowered the pH from about 6.9 to as low as 4.9 in the plow layer. This lowered pH (Fig. 1) is detrimental to pea growth and is favorable for *F. solani* f. sp. *pisi* (Allmaras and Kraft, *unpublished*).

Propagule abundance is consistent with the soil moisture optima for the two pathogens (2). The permanent tillage pan restricts internal drainage during winter and spring as typified in the steady-state water-movement relation in Fig. 2. The excessively wet plow layer in winter and spring favors *P. ultimum* in both the saprophytic and pathogenic modes, but the relatively dry subsoil even during winter recharge favors *F. solani* f. sp. *pisi*. Pea roots excavated to depths greater than 60 cm showed diseased root patterns at various depths

readily explained by the observed populations and their depth distribution (Allmaras and Kraft, *unpublished*).

CONCLUSIONS

A tillage pan was present at about 20 cm in all 22 pea and wheat fields surveyed. The pan was detected by an increase in bulk density just below the deepest presence of undecomposed wheat residue. *F. solani* f. sp. *pisi* propagules were present everywhere in the top 60 cm except in the tillage pan, and *P. ultimum* propagules were located only in the tillage layer. These propagule distributions were related to the moisture environment and comparative optima for the organism.

F. solani f. sp. *pisi* propagules are present to infect a pea root throughout its life cycle and are found as deep as 60 cm; *P. ultimum* propagules are present in the plow layer to infect the developing pea root, especially when it encounters concentrations of undecomposed straw. Thus an integration of improved tillage, varieties, fertilizers, and pest management practices is needed to reduce disease and other stresses on peas.

Because compaction is unavoidable, research is also needed to develop tillage and planting equipment that breaks up the pressure pan just below the seed row. In addition, cultural practices are needed to assure that the developing pea root system does not grow into the incorporated straw residues. These considerations are especially critical because of the changes in harvesting techniques and development of new conservation tillage systems.

LITERATURE CITED

1. Allmaras, R. R., Ward, K., Douglas, C. L. Jr., and Ekin, L. G. 1982. Long-term cultivation effects on hydraulic properties of a Walla Walla silt loam. Soil Tillage Res. 2:265-279.
2. Baker, K. F., and Cook, R. J. 1974 (original ed.). Biological Control of Plant Pathogens. Reprint ed., 1982. American Phytopathological Society, St. Paul, MN. 433 pp.
3. Kraft, J. M., and Berry, J. W., Jr. 1972. Artificial infestation of large field plots with *Fusarium solani* f. sp. *pisi*. Plant Dis. Rep. 56:398-400.

4. Kraft, J. M., and Burke, D. W. 1971. *Pythium ultimum* as a root pathogen of beans and peas in Washington. Plant Dis. Rep. 55:1056-1060.

5. Kraft, J. M., Burke, D. W., and Haglund, W. A. 1981. Fusarium diseases of beans, peas, and lentils. Pages 142-156 in: *Fusarium*: Diseases, Biology, and Taxonomy. P. E. Nelson, T. A. Toussoun, and R. J. Cook, eds. Pennsylvania State University Press, University Park. 457 pp.

6. Kraft, J. M., and Giles, R. A. 1979. Increasing green pea yields with root-rot resistance and subsoiling. Pages 407-413 in: Soil-borne Plant Pathogens. B. Schippers and W. Gams, eds. Academic Press, London.

7. Mircetich, S. M., and Kraft, J. M. 1973. Efficiency of various selective media in determining *Pythium* populations in soil. Mycopathol. Mycol. Appl. 50:151-161.

8. Nash, S. M., and Snyder, W. C. 1962. Quantitative estimations by plate counts of propagules of the bean root rot *Fusarium* in field soils. Phytopathology 52:567-572.

Wax Layers for Partitioning Soil Moisture Zones to Study the Infection of Wheat Seedlings by *Fusarium graminearum*

C. M. LIDDELL and L. W. BURGESS, Fusarium Research Laboratory, Department of Plant Pathology and Agricultural Entomology, University of Sydney, Sydney, New South Wales, 2006, Australia

Crown rot caused by *Fusarium graminearum* Schwabe group 1 (4) is a serious disease of dryland wheat on the northern plains of the eastern wheat belt of Australia (1). It is common on alkaline (pH 6.5–8.5), deep, cracking gray clay and black earth soils, causing significant losses in crops that suffer late moisture stress. These losses result from premature tiller death caused by rotting of the crown and tiller bases, which may be infected directly or subsequently colonized after infection of the coleoptile, subcrown internode, or scutellum. Infection of either the seminal or crown roots is rare (even if artificially inoculated) and of little consequence (1). The overseasoning inoculum of the fungus is mainly hyphae within crop residues, which may remain infective for at least a year in the surface layer of the soil (6). This layer corresponds to the tillage zone of the seedbed, which has been fully described for a black earth by Collis-George and Lloyd (3). It consists of a dense consolidated layer overlain by 10–20 cm of loose, friable soil. Under normal conditions the consolidated layer remains moist for some time after rain, whereas the surface soil dries rapidly. The soil has a high clay content (40–80%) and excellent structure and fertility. Seed is normally sown in or onto the consolidated layer.

Field observations indicate that dry surface soil may reduce the severity of the disease by limiting seedling infection despite high levels of the fungus. This aspect of the epidemiology of the disease is being studied by experiments designed to evaluate the role of soil moisture in the process of infection at the seedling stage. A layered pot is being developed for this purpose, and this communication reports on a preliminary experiment.

MATERIALS AND METHODS

The principal design feature of the experiment was the layered pot (Fig. 1), which has the subsoil (moist consolidated layer) of the field separated from the topsoil by a thin wax layer. This design permits the water content of the topsoil to be varied while maintaining an optimal moisture content in the root zone (subsoil). A second wax layer over the soil surface is designed to stop water loss by evaporation. Tapered 125-mm plastic pots were used, and the soil was obtained from the A horizon of a black earth (pH 7.8) on the I. A. Watson Wheat Research Centre, Narrabri, New South Wales.

The wax partitions were modeled on those of Jensen and Kirkham (5). Petroleum jelly and hard paraffin wax (mp 52–54° C) were melted and blended together (ratio of 9:1 by volume), and the mixture was cooled to 42° C before being poured onto the soil surface. It was allowed to set to the required thickness, and the excess was poured off. This method was used for both upper and lower wax layers, giving a thin, even, smooth layer that sealed off the soil surface completely.

The subsoil was wet up to $\Theta g = 0.31$ g/g (approximately −0.1 MPa water potential) by aspirating weighed amounts of water onto the soil in sealed plastic bags up to $\Theta g = 0.24$ g/g; it was then mixed and stored for 14 days with daily kneading to facilitate

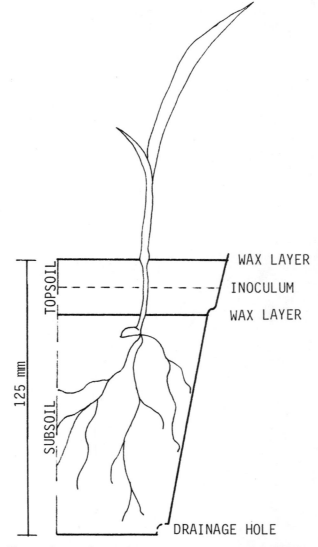

Fig. 1. Layered pot showing the location of seedlings in relation to the wax layers and fungal inoculum layer.

equilibration. At the end of this time, 870 g of this soil was placed as subsoil into each of 30 pots and packed to a bulk density of 1.1 g/cm³. Fifteen seeds of *Triticum aestivum* L. cv. Timgalen were placed 1 cm below the subsoil surface of each pot. Sufficient water was added to bring the subsoil water content up to $\Theta g = 0.31$ g/g, and the pots were covered with plastic film for 24 hr to allow redistribution of the water. The surface of the subsoil was then sealed with a thin wax layer (1–2 mm).

The topsoil for five pots was prepared at each of the following gravimetric water contents (approximate water potential in parentheses): 0.26 g/g (−0.3 MPa), 0.24 g/g (−0.5 MPa), 0.23 g/g (−0.7 MPa), 0.21 g/g (−1.0 MPa), and 0.20 g/g (−1.5 MPa) by aspirating the full amount of water into sealed plastic bags containing the soil as described above. One treatment of $\Theta g = 0.31$ g/g (−0.01 MPa) was prepared in the same manner as the subsoil.

Four of the five pots were inoculated with 0.1 g of powdered (1-mm mesh) colonized wheat chaff (2) in a band midway between the upper and lower wax layers. The pathogen was isolated from a plant collected in September 1981 from a crown rot sampling site in the Darling Downs, Queensland. The fifth pot acted as a control, and no chaff was added. The topsoil surface of all pots was sealed with a thick (2–3 mm) wax layer and placed in a glasshouse with a 12°C night/18°C day air temperature regime. Thermistors were placed in one pot at each water content and connected to a continuous chart recorder. The plants were grown for 21 days from emergence, and the subsoil was watered through the base by capillary action when required.

At the termination of the experiment, all the plants were carefully removed and washed under running tap water. They were then surface sterilized in 1.8% sodium hypochlorite for 1 min and allowed to dry in a sterile airflow. Sections of the scutellum, coleoptile, subcrown internode, leaf sheath bases, and vascular tissue of the crown were plated on potato-dextrose agar containing streptomycin sulphate at 1.0 g/L. The plates were incubated under fluorescent lights (2) for 4 days.

RESULTS

The moisture content of the topsoil had no effect on the emergence of the seedlings (Table 1), which was high and in the usual range for the seed lot used. The seedlings were normal, although some had a thin coating of wax on the coleoptile because the wax

becomes sticky at temperatures above 20°C and adheres to the growing coleoptile. All plants grew normally up to the end of the experiment, and the inoculated plants began to display symptoms after 10 days. The only aboveground symptom was an early, rapid necrosis of the coleoptile; however, the subcrown internode and scutellum were also discolored in many plants. An analysis of variance showed that there was a significant influence of topsoil water content on infection (Table 1). No control plants were infected by *F. graminearum* group 1.

Many fungi and bacteria were isolated from all plants, but a lower diversity and abundance came from plants grown with dry (0.23 g/g, 0.21 g/g, 0.20 g/g) topsoil. *Bipolaris sorokiniana* (Sacc. ex Sorok.) Shoemaker and *F. equiseti* (Corda) Sacc. were common in all pots; but *Rhizopus* sp. and numerous bacteria dominated the isolation plates from the pots at 0.31 g/g, 0.26 g/g, and 0.24 g/g. Most of these were isolated from the scutellum and coleoptile, but *F. graminearum* was isolated from all tissue and was the only organism isolated from the subcrown internode and crown. The temperature of the topsoil during the day was usually in the range of 22 to 27°C, but it reached a maximum of 34°C on three occasions; at night it was a constant 13°C. The daily air temperature was never above 20°C and decreased to 12°C at night. The gravimetric water content of the topsoil of all pots decreased by 0.01–0.02 g/g over the whole experiment (21 days).

DISCUSSION

The results of this experiment show that statistically valid data can be obtained using standard analysis of variance techniques; however, experimental variation was greater than desirable, resulting in a high LSD value in Table 1. Increased replication should reduce the effect of the variation on statistical analyses, but we are trying to isolate the variation experimentally to keep the future experiments to a manageable size. Although the variability may result from a number of factors (including the soil or fungal inoculum), experiments prepared with strict attention to homogeneity have displayed a high level of unexplained variation. The host and pathogen behaviors are also possibilities. The host seed was from a breeding selection trial and should have been quite uniform. This is not a strictly controlled source, however, and may have been variable. The pathogen, being grown from a single-spore source culture, should have been uniform; but this fungus is well known for morphological variation in culture and may reflect this in pathogenesis. The experimental system itself may be responsible, but this is unlikely because some variation has been a common problem with most experiments conducted on this disease by many workers and is to be expected in an experiment like this. Furthermore, plants have been grown for up to 4 weeks in layered pots without inoculum, and there is no obvious effect on growth. The duration of experiments is kept as short as possible to help minimize these effects.

Control of the water content of the topsoil is a critical factor. Of particular concern, therefore, was the slight yet significant drying of the topsoil during the experiment. It was quite serious because the topsoil was

Table 1. Percentage emergence of wheat seedlings and percentage of 21-day-old seedlings yielding *Fusarium graminearum* group 1 after growing in infested topsoil at six moisture contents

Seedlings	Initial moisture content (g/g)					
	0.31	0.26	0.24	0.23	0.21	0.20
	Approximate water potential (MPa)					
	−0.1	−0.3	−0.5	−0.7	−1.0	−1.5
Emergence						
Control[a]	100	100	93.4	93.3	93.3	93.3
Inoculated[b]	93.3	91.7	93.3	91.7	98.3	93.3
Infected[c]	17.3	24.2	21.2	37.0	10.3	3.5

[a] One sample of 15 seeds per treatment.
[b] Mean of four replicates of 15 seeds, LSD (0.05) = 9.2.
[c] Yielding pathogen from crown vascular tissue. Mean of four replicates, LSD (0.05) = 23.2.

prepared on the wetting curve of the hysteresis envelope, and any drying would immediately put the soil on the drying curve. For this soil, it could mean a rapid lowering of the water potential by -0.3 to -0.5 MPa. This problem may have contributed to the variation and should be borne in mind when viewing the data.

The results show that the infection of wheat seedlings by *F. graminearum* is strongly inhibited by dry soil (Θg = 0.21, 0.20 g/g; $\Psi \approx -1.0, -1.5$ MPa). Wearing and Burgess (7) demonstrated that *F. graminearum* could grow saprophytically at almost its maximum rate at these water potentials on sterile soil. Obviously, therefore, there are factors operating at certain water potentials to reduce the infection of plant tissue by this pathogen in nonsterile soil. Both the host and the soil microflora may be involved and are possibly the most important of the factors. The host may have a physiological defense mechanism that is influenced by soil water potential, or the host coleoptile may alter the soil structure in its immediate vicinity as it grows and thus make conditions unsuitable for fungal growth and infection at certain water potentials. Microbiological antagonists would be expected to suppress the pathogen at the wettest conditions; indeed, the results did show an insignificant reduction in the infection rate at Θg = 0.31, 0.26, and 0.24 g/g ($\Psi \approx -0.1, -0.3,$ and -0.5 MPa, respectively).

Future work will be directed at more precise control of the topsoil water potential and limiting the variation. The most recent experiments are being conducted on single plants in small growth tubes with a larger number of replicates to aid statistical analysis. Control of the topsoil water potential is being achieved by the use of a blend of purified yellow beeswax and paraffin oil as the wax layers. The water potential itself is now being established on the drying curve by pressure membrane apparatus, isopiestic equilibration, or mixing with dry soil. The effect of other factors (such as pH and soil type) on the infection of wheat by *F. graminearum* group 1 will be examined using this system.

LITERATURE CITED

1. Burgess, L. W., Dodman, R. L., Pont, W., and Mayers, P. 1981. Fusarium diseases of wheat, maize and grain sorghum in eastern Australia. Pages 64-76 in: *Fusarium*: Diseases, Biology and Taxonomy. P. E. Nelson, T. A. Toussoun, and R. J. Cook, eds. Pennsylvania State University Press, University Park.
2. Burgess, L. W., and Liddell, C. M. 1983. Laboratory Manual for *Fusarium* Research. Department of Plant Pathology and Agricultural Entomology, University of Sydney. 162 pp.
3. Collis-George, N., and Lloyd, J. E. 1979. The basis for a procedure to specify soil physical properties of a seed bed for wheat. Aust. J. Agric. Res. 30:831-846.
4. Francis, R. G., and Burgess, L. W. 1977. Characteristics of two populations of *Fusarium roseum* 'Graminearum' in eastern Australia. Trans. Br. Mycol. Soc. 68:421-427.
5. Jensen, C. R., and Kirkham, D. 1963. Oxygen-18 as a tracer in soil aeration studies. Soil Sci. Soc. Am. Proc. 27:499-501.
6. Wearing, A. H., and Burgess, L. W. 1977. Distribution of *Fusarium roseum* 'Graminearum' group 1 in eastern Australian wheat belt soils and its mode of survival. Trans. Br. Mycol. Soc. 69:429-442.
7. Wearing, A. H., and Burgess, L. W. 1979. Water potential and the saprophytic growth of *Fusarium roseum* 'Graminearum.' Soil Biol. Biochem. 11:661-667.

Effect of Frost on Fusarium Root Rot of Alfalfa and Possibility of Double Trait Selection

C. RICHARD, R. MICHAUD, C. WILLEMOT, M. BERNIER-CARDOU, and C. GAGNON, Station de Recherche, Agriculture Canada, Sainte-Foy, Québec, G1V 2J3, Canada

It has been demonstrated that Fusarium root rot and Fusarium wilt diseases reduce survival of alfalfa after a freeze test and that freezing the plants before inoculation enhances the development of the disease, reducing survival and yield (5). On the other hand, some progress has been made in selecting alfalfa for resistance to Fusarium root and crown rot by inoculating the fungus directly on the cut end of the taproot (4). We have therefore investigated the possibility of selecting alfalfa simultaneously for frost hardiness and Fusarium resistance by freezing alfalfa plantlets in an inoculated soil.

MATERIALS AND METHODS

Plants, fungal pathogens, and inoculation techniques. Alfalfa seedlings (cv. Saranac AR) were grown for 7 weeks in a growth room at 24°C with a 16-hr photoperiod and a light intensity of 350 μE/sec/m^2 in a pasteurized garden soil-vermiculite mixture (9:1, v/v) preinoculated with three fusaria isolated from rotted roots of alfalfa (3), *Fusarium equiseti* (Corda) Sacc., *F. oxysporum* Schlecht., and *F. solani* (Mart.) Sacc. After hardening for 3 weeks at 1°C, the plants were frozen at −4°C to enhance root penetration by fungi and then returned to initial growth conditions for 3 weeks.

LT$_{50}$. After a second 3-week hardening period at 1°C, plants were frozen at five different temperatures (−10, −12, −14, −16, and −18°C) and then returned to initial growth conditions. Survival counts were made after 3 weeks of regrowth. Probit analysis was used to estimate the temperature at which 50% of the population was killed (LT$_{50}$).

Screening. Healthy plants that survived at temperatures lower than or equal to the LT$_{50}$ were intercrossed by honeybees in the greenhouse. An equal amount of seed from each of the 87 selected plants was bulked to form a population, thereafter named SM-601, that was tested for frost and root rot resistance.

Frost resistance, survival, and root rot resistance. To evaluate the progress of the selection for the two traits (frost and root rot resistance), we compared the selected population (SM-601) in a series of three experiments with the unselected control (Saranac AR) and with a population derived from two cycles of selection for Fusarium resistance in Saranac AR (SM-17) (5). In each experiment, a completely randomized design was used; each population was represented by 16

pots of five plants each for each test temperature.

Experiments. Frost resistance of the three populations was evaluated in experiment A by subjecting 10-week-old plants to a freeze test. The LT$_{50}$ was then estimated as described above. In experiment B, the ability of the selected population to support *Fusarium* colonization and recover after a freeze stress was tested by freezing unhardened 7-week-old plants at −4°C in a preinoculated soil with the three fusaria used for selection. After 3 weeks of regrowth, survival counts were made and the top growth was determined after drying the surviving seedlings at 80°C for 24 hr. Their roots were evaluated for root rot using the 0–11 grade Horsfall-Barratt scale (2). In experiment C, the resistance to root rot was assessed for the three populations by inoculating 10-week-old seedlings with *F. acuminatum* on the cut end of the taproot (4). After 3 weeks of incubation at 24°C, root rot was evaluated by measuring the percentage and length of vertical discoloration in the split taproot.

RESULTS

Subjecting seedlings to subfreezing temperatures enhanced penetration of roots by fusaria from the pre-inoculated growth medium (Fig. 1). *F. equiseti, F. oxysporum,* and *F. solani* were reisolated from diseased roots. At 1°C, there was almost no root rot; the lower the freezing temperature, the more developed was the root rot.

In experiment A, frost resistance (LT$_{50}$) of the double-trait selected population (SM-601) was −12.6°C as compared with −11.4°C for the unselected population (Saranac AR) (Table 1). The LT$_{50}$ of SM-17 (Fusarium resistant) was not different from that for unselected Saranac AR (Table 1).

After the freeze stress at −4°C (unhardened plants) in experiment B, survival after 3 weeks of regrowth was

Table 1. Frost resistance (LT$_{50}$) of double-trait selected population (frost and root rot resistance) as compared with unselected cultivar (Saranac AR) and population selected for Fusarium resistance only (SM-17)

| Population | LT$_{50}$ (°C)[y] | Confidence limits | |
		Lower	Upper
SM-601	−12.6 b[z]	−13.0	−12.3
Saranac AR	−11.4 a	−11.7	−11.2
SM-17	−11.4 a	−11.6	−11.1

[y] Temperature at which 50% of the population is killed.
[z] Classification based on a relative potency test.

Present address of third author: Food Research Institute, Central Experimental Farm, Ottawa, Ontario, Canada K1A 0C6

much higher (60.2%) for the selected population than for the unselected Saranac AR (21.4%) and the Fusarium-resistant population (SM-17) (14.8%). Dry matter yield (per plant) and root rot of the survivors, however, were not statistically different for the three populations (Table 2).

Although less root rot was observed in the double-trait selected populations (SM-601) in experiment C, mean rot index and vertical rot were not statistically different from the other two populations (Table 3).

DISCUSSION

Winter stresses predispose alfalfa to root colonization by soil microorganisms (3), and plants injured by freez-ing are more susceptible to pathogens (5). This phenomenon is clearly illustrated in Figure 1, where the number of sites and the degree of infection increase with decreasing temperature. It is the first time that this relationship between freezing injury, root penetration by fungi, and subsequent rot development has been documented.

Frost resistance was improved by selecting surviving plants after a freeze test. The survival of the SM-601 population was better than the unselected cultivar after a freeze stress without hardening. In addition, after a cold acclimatation period, the LT_{50} of the selected population was 1.2° C less than that of the unselected cultivar. Gasser and Willemot (1) reduced the LT_{50} of cv. Angus by 1.4° C in one cycle of selection under artificial freeze tests.

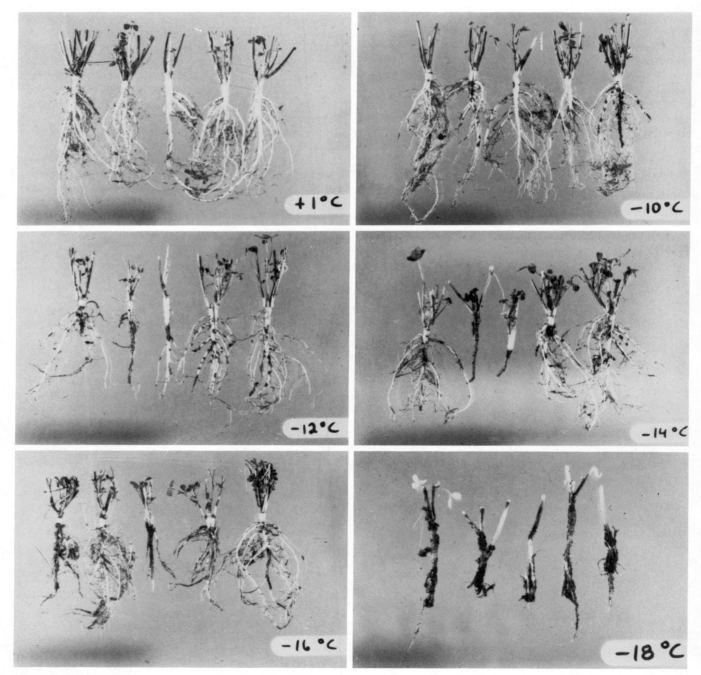

Fig. 1. Effect of freezing temperatures on root rot of alfalfa grown in a soil contaminated with *Fusarium equiseti, F. oxysporum,* and *F. solani* as compared with unfrozen root (+1° C) in the same substrate.

Table 2. Survival, dry matter yield, and rot index of double-trait selected population as compared with single-trait selected population and with unselected cultivar 3 weeks after freeze stress ($-4°$C) in a soil preinoculated with *Fusarium equiseti*, *F. oxysporum*, and *F. solani*

Population	Survival (%)[w]	Dry matter yield (g/plant)[x]	Rot index (%)[y]
SM-601	60.2 (48.9, 70.8) b[z]	0.15 ($\pm$ 0.04) a	64.8 ($\pm$ 8.9) a
Saranac AR	21.4 (13.5, 30.9) a	0.06 ($\pm$ 0.05) a	56.2 ($\pm$ 11.1) a
SM-17	14.8 (8.2, 23.0) a	0.06 ($\pm$ 0.05) a	65.0 ($\pm$ 11.8) a

[w] Figures in parentheses represent confidence limits (68%).
[x] Figures in parentheses represent standard deviation.
[y] Percentage of root tissue per plant.
[z] Classification obtained by *t* test ($\alpha = 0.05$).

Table 3. Root rot of double-trait selected population as compared with single-trait selected population and with unselected cultivar 3 weeks after inoculation with *Fusarium acuminatum* on cut end of taproot

Population	Vertical rot (mm)	Rot index (%)[y]
SM-601	14.0	20.1
Saranac AR	16.3	24.7
SM-17	15.6	29.4
Standard deviation	$\pm$ 0.84	$\pm$ 3.5
P[z]	0.14	0.19

[y] Calculated on a per-plant basis.
[z] Probability of a greater F value under equality of the populations.

Although the difference was not significant ($P = 0.14$ and 0.19), the selected population showed less vertical rot (-14%) and had a smaller rot index (-32%). As a result of selecting resistant alfalfa plants inoculated with *F. acuminatum* on the cut end of the taproot (4), less progress was made for rot index (-18%); however, a more significant and substantial progress was obtained for vertical rot (-20%).

In experiment B, the double-trait selected population SM-601 had three times the survival of the control (Saranac AR) after a freezing stress in a contaminated soil. This selected population can then be characterized by its ability to recover from freeze damage and root invasion by fungi, an ability that was considered to be a component of winter hardiness in barley (6). This may also be the case for alfalfa.

CONCLUSIONS

Subjecting alfalfa roots to freezing temperatures favors penetration of *Fusarium* into roots and subsequent rot development. Descendants from healthy plants surviving freeze tests in a *Fusarium*-infected medium are more resistant to frost and have a better recovery in the presence of root colonizers.

LITERATURE CITED

1. Gasser, H., and Willemot, C. 1974. Amélioration génétique de la luzerne à la résistance au froid aprés une génération de sélection. Can. J. Plant Sci. 54:833-834.
2. Horsfall, J. G., and Barratt, R. W. 1945. An improved grading system for measuring plant diseases. (Abstr.) Phytopathology 35:655.
3. Richard, C. 1981. Examen de la microflore endoracinaire de la luzerne en fonction de l'âge, de l'état sanitaire et de l'emplacement dans la racine. Phytoprotection 62:67-78.
4. Richard, C., Michaud, R., Freve, A., and Gagnon, C. 1980. Selection for root and crown rot resistance in alfalfa. Crop Sci. 20:691-695.
5. Richard, C., Willemot, C., Michaud, R., Bernier-Cardou, M., and Gagnon, C. 1982. Low-temperature interactions in Fusarium wilt and root rot of alfalfa. Phytopathology 72:293-297.
6. Smith, M. N., and Olien, C. R. 1978. Pathological factors affecting survival of winter barley following controlled freeze tests. Phytopathology 68:773-777.

Reduction in Infection of Wheat Roots by *Gaeumannomyces graminis* var. *tritici* with Application of Manganese to Soil

A. D. ROVIRA, CSIRO Division of Soils, Glen Osmond, S.A., 5064; and R. D. GRAHAM and J. S. ASCHER, Agronomy Department, Waite Agricultural Research Institute, Glen Osmond, S.A., 5064, Australia

Although micronutrient nutrition has been shown to affect the resistance of plants to disease (2), there is little information on the importance of micronutrients to soilborne root diseases. Manganese (Mn) addition has reduced potato scab (5), Fusarium wilt of cotton (11), and take-all in wheat (8). An observation by the first author that wheat growing in a calcareous sand had fewer whiteheads caused by *Gaeumannomyces graminis* (Sacc.) von Arx & Olivier var. *tritici* Walker where the crop had been sprayed with Mn than in unsprayed parts of the field led us to investigate under controlled environment conditions whether Mn deficiency predisposes wheat to infection by *G. graminis* var. *tritici*.

MATERIALS AND METHODS

Two factorial experiments were conducted with sieved Mn-deficient calcareous sand classified as Uc1.11 or calcic regosol (7) from Eyre Peninsula, South Australia.

Experiment 1 was a $3 \times 3 \times 5$ factorial experiment with three levels of *G. graminis* var. *tritici* (0, 0.05, and 0.1%, w/w) as ground oat kernel inoculum, three rates of Mn (0, 1, and 10 mg/pot with 300 g of soil), and five replicates. The inoculum of *G. graminis* var. *tritici* was prepared by growing the pathogen for 4 weeks at 25°C on autoclaved whole oats that had imbibed water for 24 hr before autoclaving. The colonized oat kernels were ground and the 250–500 μm fraction used in this experiment.

Experiment 2 was a $3 \times 2 \times 3$ factorial experiment with three levels of *G. graminis* var. *tritici* (0, 3, and 6

colonized dead ryegrass seeds [12] per 300 g of soil), two levels of Mn (0 and 10 mg/pot with 300 g of soil), and three replicates. In both experiments, the inoculum was mixed through dry soil and the Mn was then added to each pot as $MnSO_4$ solution. This was mixed through the soil, and water was added to 25% w/w (−0.3 bar). At the start of each experiment, the following nutrients were added and mixed through the soil for each pot: 105 mg of $Ca(NO_3)_2 \cdot 4H_2O$, 13 mg of K_2SO_4, 7.4 mg of $MgSO_4 \cdot 7H_2O$, 16.5 mg of KH_2PO_4, 1.5 mg of $FeSO_4 \cdot 7H_2O$, 0.5 mg of H_3BO_3, 0.8 mg of $CuSO_4 \cdot 5H_2O$, 1.3 mg of $ZnSO_4 \cdot 7H_2O$, 0.1 mg of $CoSO_4 \cdot 7H_2O$, 1.5 mg of NaCl, and 0.1 mg of $H_2MoO_4 \cdot H_2O$.

Five germinated seeds of wheat, *Triticum aestivum* cv. Condor, were sown 1 cm deep in each pot. The plants were grown at 15°C with a 12-hr day under lamps giving a photon flux density of 600 μM/m^2 per second. Pots were watered to full weight every 2 days and rerandomized within each block every 4 days until harvest after 26 days.

The roots were washed out of the soil with deionized water; after removing free water, fresh weights were recorded for both roots and shoots. Dry weights of roots were obtained after assessment of disease. Roots and shoots were analyzed for Mn, Zn, and Cu by atomic absorption spectrophotometry following nitric-perchloric acid digestion.

The severity of *G. graminis* var. *tritici* infection was assessed by the number of seminal roots per plant with black stelar lesions, the number of stelar lesions per plant, and the total length of lesions (mm) per plant. All data were assessed by analysis of variance appropriate to the factorial design.

Table 1. Vegetative yields of 26-day-old wheat plants in manganese-deficient soil in response to Mn and *Gaeumannomyces graminis* var. *tritici*

Manganese supplied (mg/300 g of soil)	Fresh weight (mg/plant) at inoculum level of				Dry weight (mg/plant) at inoculum level of			
	0	0.05%	0.1%	Mean	0	0.05%	0.1%	Mean
			Shoots[a]					
0	496	486	470	484	70	66	60	65
1	506	500	500	502	84	80	80	81
10	536	532	526	531	102	100	100	101
Mean	512	506	499		85	82	80	
			Roots[b]					
0	846	778	680	768	52	44	40	45
1	890	878	820	816	62	57	50	56
10	982	990	868	947	67	62	58	62
Mean	906	882	789		60	54	49	

[a] LSD ($P < 0.05$) values for fresh weight: Mn, 18; pathogen, interaction not significant. For dry weight: Mn, 4.1; pathogen, interaction not significant.
[b] LSD ($P < 0.05$) values for fresh weight: Mn, pathogen, 57; interaction not significant. For dry weight: Mn, pathogen, 4.3; interaction not significant.

RESULTS

Experiment 1. The tops showed no visible evidence of Mn deficiency nor of take-all, with only a 10% increase in fresh weight of shoots due to manganese. However, there was a 50% increase in top dry weight with Mn (Table 1). The direct effect of Mn on physiological processes in the host plant, such as photosynthesis and lignin synthesis, could account for the different ratios of fresh weight to dry weight in Mn-sufficient and Mn-deficient plants. Infection by *G. graminis* var. *tritici* was similar to that encountered in moderately infested crops—about a 26–44% incidence; these levels of infection reduced weights of tops and roots but produced no visible top symptoms.

Plants at 0 and 1 mg of Mn per pot were below the critical minimum tissue concentration of about 20 μg/g required for optimal growth of young wheat (Table 2) (3). Infection with *G. graminis* var. *tritici* had no effect on the Mn levels in roots or tops.

At each level of *G. graminis* var. *tritici*, Mn affected each of the parameters for assessing infection by the take-all fungus (number of seminal roots with lesions,

number of lesions per plant, and total length of lesions per plant; see Table 3). Even the lower rate of Mn reduced take-all consistently, although this reduction was statistically significant only for total length of lesions per plant with 0.05% concentration of *G. graminis* var. *tritici* inoculum.

Experiment 2. The plants in soil not amended with Mn showed acute deficiency symptoms, but the results are consistent with those in experiment 1 in that the addition of Mn increased both top weight and Mn concentration (Table 4). The use of propagules of standard size produced slightly higher levels of disease than the ground oat inoculum of experiment 1, but the reduction of *G. graminis* var. *tritici* infection with Mn (Table 5) was consistent with that found in experiment 1.

In the Mn-deficient plants, both levels of *G. graminis* var. *tritici* produced the same high incidence of infected roots; other studies (Rovira, *unpublished*) have shown that 70% incidence appears to be a "saturation" level of disease, with little increase in the number of infected roots with increasing inoculum. The results in experiment 2 indicate that this saturation level of root infection was achieved with three *G. graminis* var. *tritici* propagules per 300 g of soil when Mn supply was

Table 2. Concentration of manganese (μg/g) in shoots of wheat in Mn-deficient soil in response to added Mn and *Gaeumannomyces graminis* var. *tritici*

Manganese supplied (mg/300 g of soil)	Concentration of Mn in shoots[a] at inoculum level of			
	0	0.05%	0.1%	Mean
0	9.6	9.9	9.6	9.7
1	12.7	12.3	12.4	12.5
10	34.8	35.8	33.6	34.6

[a]LSD (*P* <0.05) values: Mn, 1.3; pathogen and pathogen-Mn interaction, not significant.

Table 3. Effect of manganese supply and *Gaeumannomyces graminis* var. *tritici* on severity of infection of seminal roots of wheat in Mn-deficient soil

Manganese supplied (mg/300 g of soil)	Inoculum level[a]			
	0	0.05 %	0.1%	Mean
Number of seminal roots per plant with lesions[b]				
0	0	2.2	3.1	2.65
1	0	2.1	2.8	2.45
10	0	1.8	2.0	1.90
Mean	0	2.0	2.6	
Number of lesions per plant[c]				
0	0	3.3	4.9	4.10
1	0	2.9	4.0	3.45
10	0	2.0	2.7	2.35
Mean	0	2.7	3.9	
Total length (mm) of lesions per plant[d]				
0	0	6.7	8.2	7.45
1	0	4.5	7.0	5.75
10	0	3.0	4.0	3.50
Mean	0	4.7	6.4	

[a]Ground oat kernel inoculum as percentage w/w.
[b]Plants averaged seven seminal roots each. LSD (*P*<0.05) values: Mn, pathogen, 0.49; pathogen-Mn interaction not significant. Values adjusted conservatively for number of zeros.
[c]LSD (*P*<0.05) values: Mn, pathogen, 0.92; pathogen-Mn interaction not significant.
[d]LSD (*P*<0.05) values: Mn, pathogen, 1.9; pathogen-Mn interaction not significant.

Table 4. Effect of manganese and *Gaeumannomyces graminis* var. *tritici* on dry weight of tops and roots and Mn content of wheat seedlings

Mn supplied (mg/300 g of soil)	Propagules per 300 g of soil			
	0	3	6	Mean
Dry weight tops (mg/plant)[a]				
0	49	42	39	44
10	90	85	76	83
Mean	69	64	58	
Dry weight roots (mg/plant)[b]				
0	83	48	42	57
10	162	119	136	138
Concentration of Mn in tops (μg/g)[c]				
0	10.6	11.9	13.3	11.9
10	48.0	53.1	41.0	47.4
Concentration of Mn in roots (μg/g)[d]				
0	20.1	24.4	24.6	23.1
10	47.3	50.1	55.9	51.1

[a]LSD (*P*<0.05) values: Mn, 7.7; pathogen, 6.3; pathogen-Mn, interaction not significant.
[b]LSD (*P*<0.05) values: Mn, 30; pathogen, pathogen-Mn interaction not significant.
[c]LSD (*P*<0.05) values: Mn, 7.3; pathogen, pathogen-Mn interaction not significant.
[d]LSD (*P*<0.05) values: Mn, 11.3; pathogen, pathogen-Mn interaction not significant.

Table 5. Effect of manganese nutrition on percentage of roots with lesions caused by *Gaeumannomyces graminis* var. *tritici*[a]

Mn supplied (mg/300 g of soil)	Propagules per 300 g of soil[b]			
	0	3	6	Mean
Percentage of roots with lesions				
0	0.1	74	76	50
10	0	30	50	27
Mean	0	52	63	

[a]Based on an average of 36 roots from five plants in each 300-g pot of soil, all treatments in triplicate.
[b]LSD (*P*<0.05) values: Mn, pathogen, 15; interaction not significant.

low; with adequate Mn, however, even six propagules did not reach the saturation level.

DISCUSSION

We consider it significant that effects of this magnitude were measured in seedlings that were only beginning to show measurable effects of Mn deficiency. The implications of these results are considerable. In South Australia, there is widespread incipient and transitory (season-dependent) Mn deficiency on solonized brown soils and on siliceous and calcareous sands that have problems with take-all. Thus, correction of Mn deficiency should reduce take-all on a field scale enabling the healthier roots to make better use of soil water and nutrients that are often limiting in these soils.

Many factors that affect the level of take-all disease in the field also affect the release of soluble Mn^{2+} ions in the soil. For example, ammonium fertilizers and decreased soil pH, which alleviate take-all, can also release Mn^{2+}; whereas liming (4,10) and increasing pH above 6.5 with or without changing the calcium level (9) increases take-all and can reduce the level of soluble Mn^{2+} in the soil. Further studies are required on this more general link between Mn^{2+} availability and take-all.

Manganese may exert its influence on the severity of root infection by *G. graminis* var. *tritici* by several mechanisms. Mn^{2+} may be toxic to the fungus in acid soils, as shown for *Streptomyces scabies* of potato (6). This is consistent with the recent report (8) that take-all was decreased when Mn supply was increased from the equivalent of 1 to 4 × Hoagland's solution in sand culture: the higher concentration of Mn may have been toxic to the saprophytic survival of the fungus. However, our results have been obtained in the deficiency range with less Mn^{2+} present in the soil, suggesting that Mn may be acting through physiological processes in the plant. This hypothesis is supported by our observations of less take-all on plants sprayed with Mn in the field. Improved Mn nutrition increases the rate of photosynthesis, which could increase exudation of soluble organic compounds by roots. These exudates influence the rhizosphere microflora and possibly the ectotrophic growth of the take-all fungus (1). A further mechanism may be that ligneous materials are involved in defence against *G. graminis* var. *tritici* in the form of lignitubers (13), and these structures may be more poorly developed in manganese-deficient plants because lignin production is controlled by Mn-activated enzyme systems.

ACKNOWLEDGMENTS

We are grateful for the support of the Wheat Industry Research Committee of South Australia and the Australian Wheat Industry Research Council.

LITERATURE CITED

1. Cook, R. J., and Rovira, A. D. 1976. The role of bacteria in the biological control of *Gaeumannomyces graminis* by suppressive soils. Soil Biol. Biochem. 8:269-273.
2. Graham, R. D. 1983. Effects of nutrient stress on susceptibility of plants to disease, with special reference to the trace elements. Adv. Bot. Res. 10:221-276.
3. Graham, R. C., and Loneragan, J. F. 1981. The critical level of manganese for wheat. Pages 95-96 in: Proceedings National Workshop in Plant Analysis, Goolwa, South Australia, February 1981. Dep. Agric. S.A.
4. Huber, D. M. 1981. The role of nutrients and chemicals. Pages 317-341 in: Biology and Control of Take-all. M. J. C. Asher and P. J. Shipton, eds. Academic Press, London. 538 pp.
5. McGregor, A. J., and Wilson, G. C. S. 1966. The influence of manganese on the development of potato scab. Plant Soil 25:3-16.
6. Mortvedt, J. J., Berger, K. C., and Darling, H. M. 1963. Effect of manganese and copper on the growth of *Streptomyces scabies* and the incidence of potato scab. Am. Potato J. 40:96-102.
7. Northcote, K. H., Hubble, G. D., Isbell, R. F., Thompson, C. H., and Bettenay, E. 1975. A Description of Australian Soils. CSIRO Publ., Australia. 170 pp.
8. Reis, E. M., Cook, R. J., and McNeal, B. L. 1982. Effects of mineral nutrition on take-all of wheat. Phytopathology 72:224-229.
9. Reis, E. M., Cook, R. J., and McNeal, B. L. 1983. Elevated pH and associated reduced trace-nutrient availability as factors contributing to increased take-all of wheat upon soil liming. Phytopathology 73:411-413.
10. Rovira, A. D. 1981. The microbiology and biochemistry of the rhizosphere in relation to root disease of wheat and direct drilling in Brazil. Pages 259-277 in: The Soil/Root System in Relation to Brazilian Agriculture. R. S. Russell, K. Igue, and Y. R. Mehta, eds. IAPAR Publ., Londrina, Brazil. 372 pp.
11. Sadasivan, T. S. 1965. Effect of mineral nutrients on soil microorganisms and plant disease. Pages 460-470 in: Ecology of Soil-borne Plant Pathogens. K. F. Baker and W. C. Snyder, eds. University of California Press, Berkeley. 571 pp.
12. Simon, A., and Rovira, A. D. 1985. New inoculation technique for *Gaeumannomyces graminis* var. *tritici* to measure dose response and resistance in wheat in field experiments. Pages 183-184 in: Ecology and Management of Soilborne Plant Pathogens. C. A. Parker, A. D. Rovira, K. J. Moore, P. T. W. Wong, and J. F. Kollmorgen, eds. American Phytopathological Society, St. Paul, MN.
13. Skou, J. P. 1981. Morphology and cytology of the infection process. Pages 175-179 in: Biology and Control of Take-all. M. J. C. Asher and P. J. Shipton, eds. Academic Press, London. 538 pp.

Effect of Parent Materials Derived from Different Geological Strata on Suppressiveness of Soils to Black Root Rot of Tobacco

E. W. STUTZ and G. DÉFAGO, Institut für Phytomedizin; R. HANTKE, Institut für Geologie; and H. KERN, Institut für Phytomedizin, Eidgenössische Technische Hochschule, ETH-Zentrum, 8092 Zürich, Switzerland

Suppressive soils are known to exist in Australia (2), France (6), and the United States (10). Such a soil has recently been discovered in a tobacco field in Switzerland where, after 24 years of monoculture and despite the presence of the causal agent *Thielaviopsis basicola* (Berk. & Br.) Ferr., no black root rot was found (3). The aim of the present study was to determine whether this suppressiveness results from the decline effect, caused by tobacco monoculture, or from a general phenomenon of this region (near Payerne, western Switzerland) related to geological features through the parent materials from which the soils were derived.

METHOD

Soil samples were taken from 96 fields located in a 22-km^2 area between the Broye plain and Lake Neuchâtel. Each sample consisted of four subsamples (about 1 kg each) taken from the top 20 cm of the perimeter of a 4-m square located inside the field. Each sample was air-dried for 2 days, thoroughly mixed, sieved (mesh size of 0.8 cm), filled in 20 pots of 50 cc, and

immediately inoculated with *T. basicola* (10^4 endoconidia per cubic centimeter of soil, ETH-strain D 127, highly virulent to tobacco). After 2 weeks of incubation at 20°C, one tobacco seedling (*Nicotiana glutinosa* L.) at the four-leaf stage was planted in each pot. Three weeks later, the degree of root disease was estimated on an eight-class scale and calculated in percentage of darkened area overgrown with chlamydospores. The sample collection and the experiment were repeated five times. Twenty soil samples were analyzed for N, P, K, and Mg; pH and particle size were recorded for each soil sample (1).

RESULTS

Certain fields were found to be highly suppressive, some less so, and some not at all. Suppressiveness was not associated with N, P, K, Mg, pH, crop, or crop rotation; nor could any clear relation to particle size be established (Table 1). Suppressiveness of the soils was lost after heat treatment (121°C, 30 min) and reduced by air drying (20°C, 4 weeks; Fig. 1); it was transmitted to a conducive sandy soil by addition of 5% suppressive soil

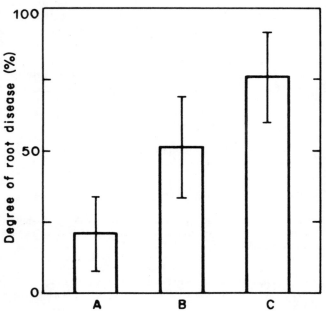

Fig. 1. Effect of heat treatment and air drying on suppressiveness. A = suppressive soil without treatment, B = dried by air (20°C, 4 weeks), and C = treated by heat (121°C, 30 min). Average data of five repetitions and standard deviation of the mean. Treatment A is significantly different (*P* = 0.05, Duncan's multiple range test) from B and C.

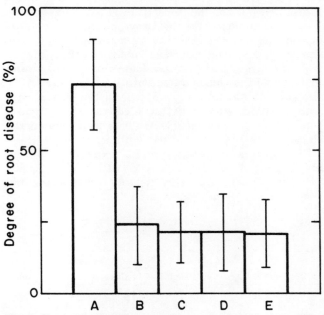

Fig. 2. Transmission of suppressiveness to a conducive soil by addition of different amounts of suppressive soil. Volume ratio of conducive to suppressive soil: A, 100:0; B, 95:5; C, 90:10; D, 75:25; E, 0:100. Average data of five repetitions and standard deviation of the mean. Ratio A is significantly different (*P* = 0.05, Duncan's multiple range test) from the others.

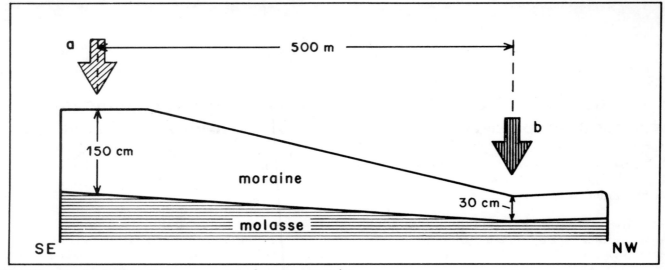

Fig. 3. Longitudinal soil profile: a = suppressive, b = not suppressive.

and an incubation time of 6 weeks (Fig. 2). The suppressive soils were located exclusively on the hill between the Broye plain and Lake Neuchâtel; no suppressive fields were found in the plain or on the shore of the lake. On the hill, the fields showed differences in their suppressive capacity. Both vertical and longitudinal soil profiles showed that the depth of moraine can be correlated to the degree of suppressiveness. Fields with a moraine thickness of more than 1 m were found to be highly suppressive, those with about 60 cm were less so, and those with thicknesses less than 20 cm not at all (Fig. 3).

The geological survey showed that the hill consists of molasse, deposited in the Aquitanian. The molasse is covered by a soil layer composed of ground moraine material from the Rhone glacier, which retreated about 16,000 years ago (4,7,9). The thickness of the soil layer varies from a few centimeters to 2.5 meters (on average, 30–100 cm), and the topsoil is often mixed with weathered molasse. In the Broye plain, the moraine deposits lie under younger lake or steam sediments. The shore of Lake Neuchâtel was found to be lake chalk (9).

The molasse consists of sandstone layers containing feldspar and is consolidated with lime and marl. The clay content is generally low. The sandstones are mostly fine grained, but sometimes coarser (11). These sandstones are impermeable. Water is drained along the single layers and does not accumulate in pools (8).

The moraine contains material transported by the glacier from the Swiss Alps over a distance of 150–200 km. The soil is medium textured and consists of loamy sands, sandy loams, and loams. The clay content is

between 10 and 25%, the silt content about 25–40%. Gravel and stone content is low.

DISCUSSION

This work confirmed the presence of soils suppressive to *T. basicola* in Switzerland (3) and showed that suppressiveness is found in a 22-km² area. Suppressiveness seems to result from biological factors, but not from a decline effect. It is transmissible to a conducive sandy soil. The suppressiveness is eliminated by heat treatment and reduced by air drying. These results are comparable with those of earlier authors (2,6,10,12) who proved the biological nature of suppressiveness. Hsi (5) showed that acidified soil, crop sequence, and high air and soil temperatures suppressed *T. basicola*. In our experiments, no correlation with pH or crop rotation could be found, nor a relationship with the quantity of organic matter as reported by Broadbent and Baker (2).

To our knowledge, no earlier report has considered the geological history of suppressive soils. The suppressiveness reported here was found in weathered morainic layers. It seems that only this type of soil is able to support the biological principle of suppressiveness.

ACKNOWLEDGMENT

This research was supported by the Swiss Federal School of Technology, Grant 0.330.083.50/2.

Table 1. Characteristics of soils with different suppressive capacities

Soil component	Suppressive capacity			
	High	Medium	Low	None
Clay (%)[x]	19.1 a[y]	17.7 a	14.3 a	13.2 a
Silt (%)	30.8 b	29.6 b	34.0 b	25.7 b
Organic matter (%)	3.7 c	3.4 c	2.7 c	3.6 c
Number of fields	15	41	33	7

[x] Percentage of dry volume.
[y] Data are the mean of all fields. Values followed by the same letter in the same line are not significantly different (*P* = 0.05) according to Duncan's multiple range test.

LITERATURE CITED

1. Anonymous. 1977. Düngungsrichtlinien für den Ackerbau. Sonderdruck der Mitteilungen für die Schweizerische Landwirtschaft 2/72:33-49.
2. Broadbent, P., and Baker, K. F. 1975. Soils suppressive to Phytophthora root rot in eastern Australia. Pages 152-157 in: Biology and Control of Soil-Borne Plant Pathogens. G. W. Bruel, ed. American Phytopathological Society, St. Paul, MN. 216 pp.
3. Gasser, R., and Défago, G. 1981. Mise en évidence de la résistance de certaines terres à la pourriture noire des racines du tabac causée par le *Thielaviopsis basicola*. Bot. Helv. 91:75-80.
4. Hantke, R. 1980. Eiszeitalter 2: Die jüngste Erdgeschichte der Schweiz und ihrer Nachbargebiete. Ott Verlag AG

Thun, Switzerland. Pp. 461-584.

5. Hsi, D. C. H. 1978. Effect of crop sequence, previous peanut blackhull severity, and time of sampling on soil populations of *Thielaviopsis basicola*. Phytopathology 68:1442-1445.

6. Louvet, J., Rouxel, F., and Alabouvette, C. 1976. Recherches sur la résistance des sols aux maladies. I. Mise en évidence de la nature microbiologique de la résistance d'un sol au développement de la fusariose vasculaire du melon. Ann. Phytopathol. 8:425-436.

7. Mornod, L. 1947. Sur les dépôts glaciaires de la vallée de la Sarine en Basse-Gruyère. Eclogae Geol. Helv. 40.

8. Parriaux, A. 1981. Contribution à l'étude des resources en eau du bassin de la Broye. Thesis 393. Université de Lausanne. 386 pp.

9. Rumeau, J. L. 1954. Géologie de la région de Payerne. Thesis. Université de Fribourg. 108 pp.

10. Scher, R. M., and Baker, R. 1980. Mechanism of biological control in a *Fusarium*-suppressive soil. Phytopathology 70:412-417.

11. Schuppli, H. M. 1950. Erdölgeologische Untersuchungen in der Schweiz. III. S.: Oelgeologische Untersuchungen im Schweizer. Mittelland zw. Solothurn u. Moudon. Mat. carte géol. Suisse. Série géotech. 26.

12. Smith, S. N., and Snyder, W. C. 1972. Germination of *Fusarium oxysporum* chlamydospores in soils favorable and unfavorable to wilt establishment. Phytopathology 62:273-277.

Effect of Varied NPK Nutrition and Inoculum Density on Yield Losses of Wheat Caused by Take-all

G. TROLLDENIER, Agricultural Research Station Büntehof, D-3000 Hannover 71, Federal Republic of Germany

Pot experiments artificially inoculated with *Gaeumannomyces graminis* var. *tritici* have proved highly suitable for studying the effect of soil factors and fertilization on take-all of wheat because they allow a direct comparison of yield responses between diseased and healthy plants. Moreover, the influence of environmental factors on the severity of the disease can be evaluated separately depending on the inoculum density (ID). The effect of soil moisture, soil acidity, and nitrogen source (6), as well as potassium nutrition (7), has been studied in this way.

This contribution reports the results of a polyfactorial pot experiment in which comparisons of yield responses were made between healthy plants and plants with varying severity of disease at optimal nutrition and at varied reduced rates of each of three major elements—nitrogen (N), phosphorus (P), and potassium (K).

MATERIALS AND METHODS

The pots contained 10.2 kg of acid humic sandy soil mixed with 700 g of peat. The experiments were carried out with one optimal NPK treatment and six treatments in which each of the three nutrients was reduced to one-half or one-quarter of the full rate (Table 1). All pots received one-fifth of the full rate of nitrogen and one-third of the full rates of phosphorus and potassium as basal dressing, the rest being top-dressed in split applications. Other nutrients were applied as basal dressing: 20 g of $CaCO_3$, 0.5 g of MgO as $MgSO_4$, and

Table 1. Quantities of NPK nutrients applied in pot experiments before inoculation with *Gaeumannomyces graminis* var. *tritici*

Nutrient	Treatments (rates in grams per pot)						
	$N_1P_1K_1$	$N_{0.5}P_1K_1$	$N_{0.25}P_1K_1$	$N_1P_{0.5}K_1$	$N_1P_{0.25}K_1$	$N_1P_1K_{0.5}$	$N_1P_1K_{0.25}$
N as NH_4NO_3	6	3	1.5	6	6	6	6
P_2O_5 as NaH_2PO_4	3	3	3	1.5	0.75	3	3
K_2O as KCl	3	3	3	3	3	1.5	0.75

Fig. 1. Effect of phosphorus application on growth of wheat. **Left:** infected with *Gaeumannomyces graminis* var. *tritici* (medium inoculum density); **right:** not infected.

minor elements as formerly reported (6). At the end of the experiment, the soil pH measured in 0.01 M $CaCl_2$ ranged between 4.6 and 4.8. Soil moisture was maintained at 60% water-holding capacity by weighing the pots daily and adding water when necessary. Each treatment comprised four replicate pots.

The inoculum of *G. graminis* var. *tritici* was prepared as earlier described (6). Depending on the inoculum density required, different numbers of infected kernels were placed 2.5 cm deep into the soil between the four seedling rows immediately before the sowing of spring wheat, variety Kolibri. The pots were supplied with the following numbers of kernels at different degrees of infection: 0 kernels per pot at no ID (0), 6 kernels at low ID (+), 15 kernels at medium ID (++), and 30 kernels at high ID (+++). Twenty-four seedlings were left per pot after germination. Yields and mineral contents of straw were measured and the results evaluated by analysis of variance and tested for significance using Tukey's test.

RESULTS

Starting from the phase of stem elongation, different fertilizer rates and different ID clearly affected plant growth. Lower N, P, or K rates led to a greater growth reduction in infected plants than in healthy plants at all three ID levels. Figure 1 compares the appearance of healthy and diseased plants (medium ID) at varied rates of phosphorus. As indicated by the number of grain-bearing ears per pot (Table 2), the effect of reduced fertilizer rates on tiller numbers increased with increasing ID. Whereas in healthy plants the reduction of P and K nutrition to one-half of the full rate did not decrease the number of ears, the greatest relative depression in the number of ears was observed at the highest ID.

Concomitant with the number of tillers, straw and grain yields in the various fertilizer treatments were also highest in healthy plants (Figs. 2 and 3). Increasing ID depressed straw weights more at insufficient than at complete nutrition, the depression being highest (up to 28%) at intermediate N and P rates. Grain yields were also remarkably reduced by infection (up to 50%). However, yield reduction was aggravated with increasing ID only at N reduction to one-half. At lower P and K nutrition, grain yield reduction did not change much with ID in spite of decreasing ear numbers. Obviously, the lower ear numbers were offset by slightly higher thousand-grain weights (Table 2) and numbers of grains per ear. Both parameters were highest in the healthy plants of the respective fertilizer treatments, but they were not negatively affected by increasing ID.

The mineral content of straw was considerably influenced by fertilization (Table 3). Generally, the content increased slightly with severity of disease.

Table 2. Yield components as influenced by NPK nutrition and inoculum density (ID)

Fertilizer treatment	Number of ears per pot at ID[a]				Thousand-grain weight (g) at ID			
	0	+	++	+++	0	+	++	+++
$N_1P_1K_1$	62	58	56	51	40.6	34.3	36.2	37.4
$N_{0.5}P_1K_1$	54	49	46	39	47.3	35.9	32.7	36.5
$N_{0.25}P_1K_1$	45	39	40	33	37.5	33.0	31.3	33.0
$N_1P_{0.5}K_1$	67	60	53	43	40.1	30.3	35.3	36.0
$N_1P_{0.25}K_1$	50	49	41	40	37.6	28.7	31.7	32.4
$N_1P_1K_{0.5}$	62	51	52	39	39.4	27.9	29.2	30.1
$N_1P_1K_{0.25}$	60	50	44	44	29.4	23.2	29.6	29.2
LSD at 5%: Fertilizer			5.2				2.7	
ID			2.9				1.9	

[a]+ = Low, ++ = medium, and +++ = high inoculum density.

Straw Yield
dependence on NPK nutrition and ID (g/pot)

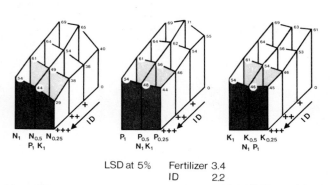

LSD at 5%   Fertilizer 3.4
            ID   2.2

Fig. 2. Straw yields of wheat as influenced by NPK nutrition and inoculum density (grams per pot). LSD at 5%: fertilizer, 3.4; inoculum density, 2.2.

Grain Yield
dependence on NPK nutrition and ID (g/pot)

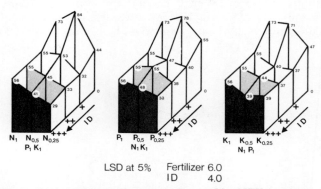

LSD at 5%   Fertilizer 6.0
            ID   4.0

Fig. 3. Grain yields of wheat as influenced by NPK nutrition and inoculum density (grams per pot). LSD at 5%: fertilizer, 6.0; inoculum density, 4.0.

Table 3. Mineral contents of straw (mg/g dry matter) as influenced by NPK nutrition and inoculum density (ID)

Fertilizer treatment	Nitrogen at ID[a]				Phosphorus at ID				Potassium at ID			
	0	+	++	+++	0	+	++	+++	0	+	++	+++
$N_1 P_1 K_1$	13.4	13.8	14.4	11.5	0.9	0.8	0.7	0.7	14.8	15.3	16.1	15.5
$N_{0.5} P_1 K_1$	5.3	6.6	6.6	7.7	0.7	0.9	0.9	0.7	16.5	17.3	17.5	18.0
$N_{0.25} P_1 K_1$	4.0	6.7	5.5	7.9	1.0	1.3	1.1	0.6	19.2	19.9	20.2	16.9
$N_1 P_{0.5} K_1$	11.6	13.5	12.7	15.8	0.3	0.5	0.6	0.5	12.6	16.1	15.6	17.3
$N_1 P_{0.25} K_1$	16.8	16.5	17.5	15.7	0.3	0.3	0.3	0.5	16.8	14.6	16.2	15.1
$N_1 P_1 K_{0.5}$	14.5	16.2	16.0	17.1	0.8	1.1	0.9	0.6	6.0	7.6	8.2	8.8
$N_1 P_1 K_{0.25}$	22.5	19.1	16.9	17.3	1.4	1.3	1.1	0.7	2.1	3.4	4.1	4.0
LSD at 5%: Fertilizer		0.9				0.1				1.4		
ID		0.6				0.6				0.9		

[a] + = Low, ++ = medium, and +++ = high inoculum density.

DISCUSSION

Confirming in part earlier findings (2), the results clearly show that wheat infected with *G. graminis* var. *tritici* responds to inadequate fertilization by greater yield depression than noninfected wheat. This applies equally to the three major elements nitrogen, phosphorus, and potassium. The significance of N source for severity of take-all (4) and severity of yield depressions (6) is known. The favorable effect of NH_4 nutrition may not be linked exclusively with a decrease of pH in the rhizosphere, although this is generally assumed. It might also be linked with the increase of the number of rhizosphere organisms and increased root respiration with NH_4 as compared with NO_3 nutrition (8). The ambivalent effect of nitrate nitrogen was already detected by Garrett (3). Although higher rates of N aggravate the severity of various diseases, they also enable the plant to escape the fungus by continuous production of new roots.

The causes of the favorable effect of higher P and K rates have not yet been investigated. It has been demonstrated in an earlier work (7) that the influence of K fertilization is most remarkable at medium ID. Recently it was suggested that chloride reduces the susceptibility of wheat to take-all and that this effect might be associated with the lowering of chemical potential of water in the plant (1). As water potential is similarly affected by potassium (5), the favorable effect of optimal K nutrition when applied as KCl is possibly caused by both of the components.

As already found with K, the present results show that straw yield depressions at increasing ID are higher with undersupply of N or P than with optimal nutrient supply. The grain yield response was less obvious, as the lower numbers of ears were offset by higher thousand-grain weights; this result was inconsistent with other observations. The unexpected slight increase in grain weight with higher ID probably results from the inability of the pathogen to spread in acid soil, producing a better supply of the remaining ears with photosynthates.

ACKNOWLEDGMENT

The author is indebted to Mrs. Labrenz for translation work.

LITERATURE CITED

1. Christensen, N. W., Taylor, R. G., Jackson, T. L., and Mitchell, B. L. 1981. Chloride effects on water potentials and yield of winter wheat infected with take-all root rot. Agron. J. 73:1053-1058.
2. Garrett, S. D. 1941. Soil conditions and the take-all disease of wheat. VI. The effect of plant nutrition upon disease resistance. Ann. Appl. Biol. 28:14-18.
3. Garrett, S. D. 1948. Soil conditions and the take-all disease of wheat. IX. Interaction between host plant nutrition, disease escape, and disease resistance. Ann. Appl. Biol. 35:14-17.
4. Huber, D. M., Painter, C. G., McKay, H. C., and Peterson, D. L. 1968. Effect of nitrogen fertilization on take-all of winter wheat. Phytopathology 58:1470-1472.
5. Mengel, K., and Arneke, W. W. 1982. Effect of potassium on the water potential, the pressure potential, the osmotic potential and cell elongation of leaves of *Phaseolus vulgaris*. Physiol. Plant. 54:402-408.
6. Trolldenier, G. 1981. Influence of soil moisture, soil acidity and nitrogen source on take-all of wheat. Phytopathol. Z. 102:163-177.
7. Trolldenier, G. 1982. Influence of potassium nutrition and take-all on wheat yield in dependence on inoculum density. Phytopathol. Z. 103:340-348.
8. Trolldenier, G., and von Rheinbaben, W. 1981. Root respiration and bacterial population of roots. I. Effect of nitrogen source, potassium nutrition and aeration of roots. Z. Pflanzenernaehr. Bodenkd. 144:366-377.

Influence of Environmental Factors and Sclerotial Origin on Parasitism of *Sclerotinia sclerotiorum* by *Coniothyrium minitans*

P. TRUTMANN, Plant Research Institute, Department of Agriculture, Burnley, Victoria, 3121, and Department of Botany, La Trobe University, Bundoora, Victoria, 3083; P. J. KEANE, Department of Botany, La Trobe University, Bundoora; and P. R. MERRIMAN, Plant Research Institute, Department of Agriculture, Burnley, Victoria, 3121, Australia

Inadequacies in fungicidal control of disease caused by *Sclerotinia sclerotiorum* (Lib.) de Bary prompted an investigation into additional methods of control. Previous studies had shown that postharvest plowing of soils carrying bean stubble and sclerotia reduced disease in subsequent crops in comparison to shallow disk harrowing, which is the normal cultivation practice (5). Plowing buried most debris to between 15 and 20 cm, whereas harrowing left most material at or near the soil surface. Buried sclerotia produce apothecia less frequently because most stipes cannot reach the soil surface (1) and because they are destroyed more rapidly (4).

Significant levels of disease still occur in crops grown after burial of previous crops debris, and this is thought to originate primarily from previously buried sclerotia that resist degradation and are returned to the surface by the plowing treatment. These "resistant" types of sclerotia are known to occur in bean straw (5). Their greater longevity is attributed not only to the protection offered by the straw, but also to the fact that the sclerotia are themselves more resistant to degradation. There is therefore a need to eradicate this inoculum.

Two hyperparasitic fungi, *Coniothyrium minitans* Campbell and *Sporidesmium sclerotivorum* Uecker have been reported to control disease caused by *Sclerotinia* species (2,3). *C. minitans* occurs naturally in Victorian soils and is currently under investigation as a treatment for *S. sclerotiorum*. Preliminary field studies with *C. minitans* have shown that although it can rapidly kill sclerotia in soil, its effectiveness depends on time of application (6). To maximize the effectiveness of hyperparasitism, information is required on factors that affect parasitism, especially under field conditions. This paper presents information on the effects of seasonal conditions, of sclerotial position, and of origin on sclerotial survival and parasitism by *C. minitans*.

MATERIALS AND METHODS

Experiment 1. This experiment studied the effect of season and burial on infection of sclerotia by *C. minitans*. Fiberglass mesh bags (7×7 cm^2, 2-mm^2 mesh size) each containing 20 sclerotia from agar culture were placed on fine sandy loam (pH 6.0) at an experimental site in southeast Victoria. Each bag was sprayed with 5 ml of a suspension of *C. minitans* (10^7 spores per milliliter). Bags were then either buried at 5–10 cm or left on the soil surface. The treatment was replicated three times. Bags were removed and replaced with freshly prepared ones at monthly intervals over a period of 14 months. Sclerotia recovered from bags were surface sterilized for 1 min in 5% sodium hypochlorite:95% ethanol (1:1, v:v) and placed in petri dishes on drops of potato-dextrose agar containing chloramphenicol at 50 mg/L. Plates were incubated at $20 \pm 2°$C, and sclerotia were examined for infection by *C. minitans* after 7–10 days.

Experiment 2. This experiment studied the effect of origin of sclerotia, burial, and *C. minitans* treatment on survival of sclerotia. Fiberglass mesh bags prepared as above and each containing 20 field sclerotia, formed either on the outside of plants or within stems of *Phaseolus vulgaris* L., were placed on 1-m^2 plots on fine sandy loam (pH 6.0) in southeast Victoria. Bags were then sprayed either with 5 ml of a suspension of *C. minitans* (10^7 spores per milliliter) or with 5 ml of water and then either left on the surface or buried 5 cm deep in the soil. Treatments were replicated five times. The number of viable sclerotia surviving after 8 months was assessed by recovering them from bags, then surface sterilizing and plating them onto potato-dextrose agar containing chloramphenicol as described above. The experiment was conducted from May (late autumn) until February (late summer).

Experiment 3. This experiment studied the effect of protection of sclerotia within bean straw on infection by *C. minitans*. In this laboratory study, sclerotia from agar culture were inserted into dry, hollow, 5-cm-long pieces of bean straw whose ends were then plugged with cotton wool. Five sclerotia were inserted in each straw piece. Four straw segments were placed on the surface of moist, autoclaved, acid-washed sand in 2-L containers and treated with 10 ml of a spore suspension of *C. minitans* (10^7 spores per milliliter) or with water. Treatments were replicated four times and incubated at $20 \pm 2°$C. After 1 month, sclerotia were assessed for infection by *C. minitans* by the method described above.

Experiment 4. This experiment studied the effect of protection of sclerotia within bean straw and of burial on hyperparasitism and survival of sclerotia. Polyethylene containers (50 cm in diameter $\times$ 50 cm deep) were filled to within 5 cm of their tops with fine sandy loam (pH 6.0) and buried in soil in the field to a depth of 45 cm. After 4 weeks, fiberglass mesh bags as described above were prepared so that each contained 20 sclerotia, obtained from agar cultures. Sclerotia were either free inside the bag or were contained inside pieces of hollow bean straw as described above. In half of the bags with bean straws, the straws were broken to simulate the effect of slashing of bean debris in field operations. Bags were placed on the surface of the soil in the plastic containers. Four days later, they were sprayed either with 5 ml of a spore suspension of *C. minitans* (10^7 spores per milliliter) or with water. Bags were then either left on the soil surface or buried 5 or 20 cm deep. Treatments were replicated four times. Italian ryegrass was sown in each container. Two bags from each treatment were recovered after 1 month, and sclerotia were assessed for infection by *C. minitans* as described above. The two remaining bags from each treatment were recovered after 8 months, and sclerotia were assessed for viability as above. The experiment was conducted from June (early winter) through February (late summer).

RESULTS

Experiment 1. High levels of parasitism of sclerotia occurred in the months of April through November (late autumn, winter, and spring) (Fig. 1). The proportion of sclerotia parasitized by *C. minitans* was low during the hotter summer and early autumn months of December to March. The proportion of sclerotia parasitized was, on average, similar in buried and in exposed sclerotia.

High temperatures (mean monthly maxima of 24–26°C), rather than availability of soil water, probably limited parasitism by *C. minitans* in summer because the field was regularly irrigated and adequate rainfall was recorded throughout the summer.

Experiment 2. Survival of sclerotia formed within bean straws was significantly reduced ($P = 0.05$) after 8 months by treatment with *C. minitans* (Table 1). However, survival of such sclerotia, both on the soil surface and at a depth of 5 cm, was significantly greater

($P = 0.05$) than that of sclerotia formed on the surface of plants.

Experiment 3. None of the sclerotia within bean straws was infected by *C. minitans* applied to the outside of the straws.

Experiment 4. Levels of infection of sclerotia by *C. minitans* were significantly greater ($P = 0.01$) in *C. minitans* treatments than in controls in all cases (Table 2). However, levels of infection were significantly lower ($P = 0.01$) when sclerotia were protected inside intact

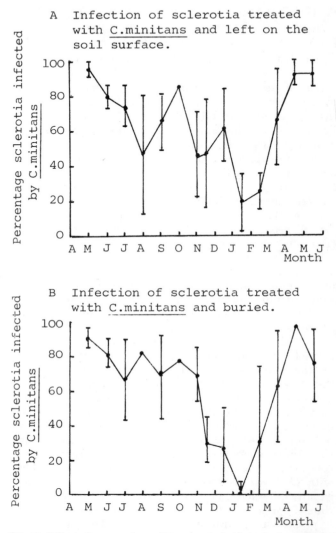

Fig. 1. Effect of seasonal conditions on infection of sclerotia by *Coniothyrium minitans* in the field (1979–80).

Table 1. Effects of burial, sclerotial origin, and *Coniothyrium minitans* treatment on survival of sclerotia

| Depth of burial (cm) | Biological treatment | Viable sclerotia recovered after 8 months (percentage of original inoculum) | |
		Sclerotia formed within stems	Sclerotia formed on outside of plants
0	Nil	24 (0.608)[a]	3 (0.120)
	C. minitans	3 (0.120)	0 (0.000)
5	Nil	13 (0.612)	5 (0.251)
	C. minitans	4 (0.276)	0 (0.000)

[a]Duncan's multiple range test values. Differences required for significance: $P (0.05) = 0.284$, $P (0.01) = 0.378$.

Table 2. Effect of burial and placement of sclerotia within intact or broken bean straws on hyperparasitism (after 1 month) and on sclerotial survival (after 8 months)

| Depth of burial (cm) | Biological treatment | Sclerotia infected after 1 month by *Coniothyrium minitans* (%)[a] | | | Viable sclerotia recovered after 8 months (percentage of original inoculum)[b] | | |
		Unprotected	Broken stems	Intact stems	Unprotected	Broken stems	Intact stems
0	Nil	0.0 (0.000)[c]	4.3 (0.226)	0.0 (0.000)	88.8 (1.270)	100.0 (1.322)	93.8 (1.293)
	C. minitans	86.3 (1.261)	93.2 (1.293)	46.1 (0.973)	0.0 (0.000)	0.0 (0.000)	71.0 (1.150)
5	Nil	0.0 (0.000)	0.0 (0.000)	0.0 (0.000)	85.0 (1.250)	97.5 (1.312)	88.3 (1.269)
	C. minitans	80.5 (1.223)	82.2 (1.238)	27.7 (0.778)	11.0 (0.443)	9.0 (0.369)	43.0 (0.943)
20	Nil	0.0 (0.000)	0.0 (0.000)	0.0 (0.000)	83.8 (1.237)	92.5 (1.288)	71.3 (1.180)
	C. minitans	71.6 (1.180)	83.6 (1.327)	10.8 (0.452)	12.5 (0.426)	0.0 (0.000)	40.0 (0.932)

[a]Differences required for significance: $P (0.05) = 0.143$, $P (0.01) = 0.190$.
[b]Differences required for significance: $P (0.05) = 0.302$, $P (0.01) = 0.400$.
[c]Duncan's multiple range test values.

straws. Parasitism of sclerotia inside intact straws decreased with increasing depth of burial.

Recovery of viable sclerotia after 8 months was significantly reduced ($P = 0.01$) in *C. minitans* treatments except when sclerotia were within intact bean straws. Mortality of sclerotia within intact straws increased with increasing depth of burial regardless of *C. minitans* treatment. This effect was probably caused by factors other than *C. minitans* because parasitism by *C. minitans* of protected sclerotia actually declined with increasing depth of burial. Temperatures were probably conducive to *C. minitans* activity for the first 5 months after the start of the experiment, with mean monthly maxima ranging from 13 to 33°C and minima from 5 to 12°C. Soils in the containers were moist at the beginning of the experiment, and regular rainfall was recorded in the first month. Soil water may have limited *C. minitans* activity in July and then from October onwards (late spring and summer).

DISCUSSION

The results of these experiments define conditions that limit parasitism of sclerotia by *C. minitans*. Under Victorian conditions, few sclerotia are parasitized during summer—especially during January and February, when most bean crops are at or near maturity and ambient temperatures often exceed 30°C. Such temperatures prevent parasitism of sclerotia by *C. minitans* (7). The hyperparasite is likely to be most effective when applied as a postharvest soil treatment of infected plant remains from April (mid autumn) onwards. There was no evidence that burial of sclerotia following *C. minitans* application adversely affected the parasitism of sclerotia by *C. minitans*. This allows the possibility of combining hyperparasite treatment with deep burial of crop debris in an integrated approach to control of *S. sclerotiorum* (1,4).

Sclerotia formed in the protected environment within bean straws survived somewhat better than sclerotia formed on the outside of bean plants. Survival of both lots of field-collected sclerotia was considerably less than that of agar-culture-grown sclerotia used in the subsequent field experiment (Table 2). Sclerotia formed under stress in the field seem to have fewer food reserves, less intact rinds, and a greater level of contamination by other microorganisms than culture-grown sclerotia (4).

Sclerotia protected within bean straws are clearly less likely to be parasitized by *C. minitans* than unprotected sclerotia or sclerotia in broken straws. In the field experiment, there was a moderate degree of parasitism of sclerotia within straws. This could have resulted from the activity of soil fauna in breaking into straws and transporting spores of *C. minitans* to the sclerotia within, explaining why parasitism of enclosed sclerotia was reduced at the greater depths where soil fauna are less prevalent. It is clear that *C. minitans* treatment of sclerotia within bean debris will be greatly enhanced by prior mechanical slashing of the debris to expose sclerotia.

C. minitans is currently being evaluated alone and as a component of an integrated program to control white mold of beans. The proposed strategy includes the use of fungicidal sprays during the growing season and the application of *C. minitans* spore suspensions to slashed crop debris containing sclerotia, which is subsequently buried by deep plowing. The goal is for burial and *C. minitans* treatments together to reduce sclerotial inoculum to negligible levels over a number of seasons, while the sprays are used as short-term protectants at least as an interim control measure.

LITERATURE CITED

1. Abawi, G. S., and Grogan, R. G. 1979. Epidemiology of diseases caused by *Sclerotinia* species. Phytopathology 69:899-904.
2. Adams, P. B., and Ayers, W. A. 1982. Biological control of Sclerotinia lettuce drop in the field by *Sporidesmium sclerotivorum*. Phytopathology 72:485-488.
3. Huang, H. C. 1980. Control of Sclerotinia wilt of sunflower by hyperparasites. Can. J. Plant Pathol. 2:26-32.
4. Merriman, P. R. 1976. Survival of sclerotia of *Sclerotinia sclerotiorum* in soil. Soil Biol. Biochem. 8:385-389.
5. Merriman, P. R., Pywell, M., Harrison, G., and Nancarrow, J. 1979. Survival of sclerotia of *Sclerotinia sclerotiorum* and effects of cultivation practices on disease. Soil Biol. Biochem. 11:567-570.
6. Trutmann, P., Keane, P. J., and Merriman, P. R. 1980. Reduction of sclerotial inoculum of *Sclerotinia sclerotiorum* with *Coniothyrium minitans*. Soil Biol. Biochem. 12:461-465.
7. Turner, G. J., and Tribe, H. T. 1976. On *Coniothyrium minitans* and its parasitism of *Sclerotinia* species. Trans. Br. Mycol. Soc. 66:97-105.

Part VI

Management Practices, Chemical Control, and Solarization

Impact of Herbicides on Plant Diseases

JACK ALTMAN, Botany and Plant Pathology Department, Colorado State University, Fort Collins, 80523, U.S.A.

The volume and use of herbicides and insecticides in the United States and the world have increased greatly in the past 30 years. They have revolutionized temperate fruit production and have provided growers with improved opportunities for low-cost production.

Although 400 chemicals involving about 40,000 commercial products are currently registered by the U.S. Department of Agriculture, as few as 25 basic chemicals constitute 90% of the pesticides used in the United States, which in 1978 comprised 11.6% fungicides, 49% herbicides, and 39% insecticides (9,25).

Between 1961 and 1981, insecticide use increased slightly to 212 million kilograms, whereas fungicide use during this time decreased from 146 million to 65 million kilograms per year. Herbicide use, however, increased from 78 million to 381 million kilograms. With the prospect of continued use of pesticides in agriculture, we need to develop a thorough understanding of the total effects of these important compounds on our environment and on our crops (28).

The influence of herbicides on plant diseases is not fully understood, and it is only from experience of their integration into commercial production that their full effects will emerge. That herbicides do affect plant diseases is beyond question, but so many factors are altered when herbicides are used that identifying specific effects on a particular pathogen or host-pathogen relationship is often very difficult (2).

In studying the influence of trifluralin (a,a,a-trifluoro-2,6-dinitro-N,N-dipropyl-p-toluidine) on clover root nodule symbiosis, DeRosa et al (7) proposed that trifluralin acts via an interaction with plant microtubule protein. In treated seedlings, root tips increased in diameter, decreased in length, decreased cell elongation, and had abnormal cell wall deposition; root hairs were deformed; and there was a marked reduction in number of infection threads (required for nodule development) induced by the bacterial symbionts. Ultrastructural studies indicated that there was a change in the nitrogen-fixing nodules of *Phaseolus vulgaris* as evidenced by structural alterations in the bacteroidal cells.

In addition, this host plant alteration due to trifluralin could also produce a less desirable infection court for soilborne plant pathogens in the root system of developing seedlings, accounting in part for the reduced phycomycetous disease complex reported by Jacobsen and Hopen (15) and Sacher et al (23).

Studies on the effect of herbicides on soil micro-organisms aim to determine their short- and long-term effects on beneficial as well as harmful soil organisms. To determine what influence any changes produced in soil microflora will have on diseases in commercial plantings, studies need to be made for prolonged periods under different soil environments and soil types (20,21). Interactions among herbicides, plant pathogens, and hosts must be thoroughly understood to minimize detrimental interaction and to maximize beneficial interactions. It is the purpose of this paper to review both the favorable and detrimental aspects of the herbicide-pathogen-host interaction.

EFFECT OF HERBICIDES ON DISEASES

Wheeler (29) has suggested that cell membranes and the cellulose microfibrils do not develop in the presence of a toxicant, compared with development without a toxicant, making them more prone to attack by attenuated low-grade pathogens. Heitefuss (13,14) and Paul (19) reported similar observations concerning altered cell wall development. Paul worked with a system involving corn, *Fusarium moniliforme*, and herbicide. Johnston et al (16), using a pea-*Fusarium* model, found that pea diseases were not reduced in Canada. This report contrasts with those by Jacobsen and Hopen (15) that pea root rot, based on a pea-*Aphanomyces* model, could be reduced by a herbicide (trifluralin) in Illinois. Duncan and Paxton (8), also working in Illinois in 1981, reported that trifluralin enhanced Phytophthora root rot of soybeans.

Pathogens such as *Rhizoctonia* and *Fusarium* are stimulated and can readily penetrate host tissues of beans, cotton, and sugar beets that have been exposed to dinitroanilines (18). Furthermore, we can speculate that a lack of secondary cell wall development because of herbicide use could make host plants more vulnerable to pathogen ingress. The use of conventional herbicides has produced an increase in *Rhizoctonia* incidence when this pathogen was indigenous to fields planted to sugar beets, soybeans, potatoes, and beans. In other herbicide-treated fields where sugar beet nematodes were present, an increase in the incidence of sugar beet nematode damage has been observed in 1970, 1979, and 1980; this increase was attributed to preplant treatment of beets with herbicides (1). Results of a greenhouse test using infested soil are summarized in Table 1. A similar observation of increased root knot nematode damage in alfalfa has been reported by Griffin and Anderson in Utah in 1979 and 1980 (11,12). Altman and Steele (4) and Riggs and Oliver (22) reported in 1982 on the effect of trifluralin on hatching and juvenile emergence from the sugar beet and soybean cyst nematodes, respectively. Data on sugar beet cyst nematode juvenile emergence influenced by herbicide are presented in Tables 2 and 3.

Table 1. Effects of cycloate on *Heterodera schachtii* and three *Beta* species growing under greenhouse conditions[x,y]

| | Beta vulgaris | | | | | | Beta procumbens | | | Beta patellaris | | |
| | Mono Hy A1 | | | Mono Hy D2 | | | | | | | | |
Treatment	Top (g/plant)	Root (g/plant)	Nema-todes[z] (no./g of root)	Top (g/plant)	Root (g/plant)	Nema-todes (no./g of root)	Top (g/plant)	Root (g/plant)	Nema-todes (no./g of root)	Top (g/plant)	Root (g/plant)	Nema-todes (no./g (of root)
Noninfested green-house soil (Unamended)	2.63 a	1.82 a	None	1.93 a	1.49 a	None	3.47 a	0.86 a	None	4.33 a	1.88 a	None
Infested field soil (Unamended)	1.05 b	0.56 b	126 a (70) a	0.58 b	0.30 b	277 a (83) a	2.15 ab	0.67 ab	149 a (100) a	1.68 b	0.69 b	162 a (112) a
Cycloate, 4 µg/g	0.24 c	0.13 c	186 b (24) bc	0.14 c	0.09 b	500 b (45) ab	1.80 ab	0.40 bc	168 a (67) ab	0.90 c	0.35 c	325 a (114) a
Cycloate, 8 µg/g	0.33 c	0.11 c	391 c (43) ab	0.07 c	0.03 c	1,100 c (33) b	1.45 bc	0.37 bc	211 ab (78) ab	0.62 cd	0.21 cd	167 a (35) ab
Cycloate, 16 µg/g	0.14 c	0.05 c	344 c (17) c	0.23 c	0.07 c	531 b (37) b	1.06 bc	0.19 c	248 b (47) bc	0.23 de	0.13 d	569 b (74) a
Cycloate, 32 µg/g	0.03 c	0.015 c	340 c (6) c	0.10 c	0.01 c	286 a (28) b	0.18 c	0.08 c	155 a (12) c	0.09 e	0.005 d	212 a (1) b

[x]Adapted, by permission, from Abivardi and Altman (1).
[y]Numbers are means of four replications. Column means followed by common letters are not different according to Duncan's multiple range test ($P = 0.05$).
[z]In columns reporting nematodes, the top number refers to mean number of nematodes per gram of roots and the number in parentheses refers to number of nematodes per root system in corresponding treatments.

Table 2. Numbers of larvae emerged from cysts of *Heterodera schachtii* treated 6 weeks with aqueous solutions of cycloate, EPTC, and zinc chloride

Treatment	Concentration (µg/ml)	Mean numbers of larvae
Tap water	...	122[x] a[y]
Cycloate	1.25	313 ab
Cycloate	2.50	315 ab
Cycloate	5.00	471 b
EPTC[z]	1.25	491 b
EPTC	2.50	363 ab
EPTC	5.00	459 b
Zinc chloride		6,834 c

[x]Figures given are the mean numbers of larvae emerged from six replications, each of 40 cysts.
[y]Means with unlike lowercase letters are significantly different according to Duncan's multiple range test ($P = 0.05$).
[z]EPTC = S-ethyl dipropylthiocarbamate.

Table 3. Numbers of larvae emerged from cysts of *Heterodera schachtii* treated 6 weeks with aqueous solutions of diallate, trifluralin, and zinc chloride

Treatment	Concentration (µg/ml)	Mean numbers of larvae
Tap water	...	242[y] a[z]
Diallate	1.25	630 b
Diallate	2.50	572 ab
Diallate	5.00	571 ab
Trifluralin	1.25	459 ab
Trifluralin	2.50	408 ab
Trifluralin	5.00	471 ab
Zinc chloride		6,075 c

[y]Figures given are the mean numbers of larvae emerged from six replications, each of 40 cysts.
[z]Means with unlike lowercase letters are significantly different according to Duncan's multiple range test ($P = 0.05$).

In a replicated field study of the effect of trifluralin on bean root and hypocotyl rot caused by *F. solani* f. sp. *phaseoli*, a disease severity index (DSI) developed by Bareta and Altman (6) showed that the field rate of trifluralin application (1.5 ppm a.i.) increased the DSI twofold or threefold. With twice the field rate, the DSI was increased fourfold or fivefold compared with inoculated replications that were not treated with herbicides (Table 4).

Table 4. Severity of bean root and hypocotyl rot caused by *Fusarium solani* f. sp. *phaseoli* with trifluralin use[z]

| | Disease severity index | | |
Replicate	No herbicide	Herbicide at field rate (1.5 ppm a.i.)	Herbicide at 2× field rate
R1	8.75	16.1	42.1
R2	19.9	29.9	37.2
R3	11.4	37.5	21.4
R4	11.8	33.4	43.0

[z]Preliminary field data (August 1983) based on Fusarium lesion measurement (length and depth) of 10 randomly selected plants from each treatment in each field replicate.

THE PESTICIDE-DISEASE SYNDROME

Traditional control of pests on crops was based on a foundation of biological principles that predicted the interaction between host and pest in the agroecosystem. After World War II, however, development of organic pesticides (e.g., 2,4-D and DDT) caused a marked shift in the philosophy of pest control. The broad foundation of control based on biological principles was essentially abandoned for a pesticide-oriented system in which a specific pest or group of pests was controlled by a pesticide with little regard to the total impact on the agroecosystem.

Increased reliance on pesticides for control produced a substantial change in the varieties of individual crops developed and on cultural conditions. Current crop varieties have been developed with the expectation of greater usage of pesticides, more fertilizer, and increased irrigation. This has led to an increasingly homogeneous cropping system in which the crop had a narrower gene pool and was more susceptible to attack from pests: for example, corn to southern corn leaf blight. This genetic vulnerability, in turn, created a higher demand for pesticides at higher concentrations and greater frequency of application. This practice has reduced the number of natural predators and parasites of pests normally found in the cropping system, thus allowing, in some instances, rather minor pests to assume a new, major importance. There is concern among crop protectionists that we cannot maintain our present level of crop protection capability without a concerted research and development effort to do so.

The basis for such pessimism stems from the recognition that modern intensive-production agriculture, which is needed to produce adequate food, unfortunately intensifies a number of crop protection problems. Many earlier practices, such as rotation and the use of diverse crop germ plasm, have largely been abandoned. Pest-resistant varieties are planted so extensively that the chances for the development of successful pest biotypes are increased. Monoculture, irrigation, and multiple cropping increase the intensity of many pest problems, and may even create new ones. The more recent innovation of "no-till" agriculture has also resulted in increased use of pesticides, providing a means for increasing selective pressures to produce resistant pests (insects, pathogens, and weeds).

Another basis for concern is the high attrition rate of existing pesticides resulting from increasingly restrictive regulations (banning) and loss of effectiveness as pests acquire resistance. Not only do insect and fungal pests acquire resistance, but so do weeds. Thus we cannot be certain that our present level of research effort will be adequate to maintain present crop protection status into the future.

PESTICIDE RESISTANCE IN WEEDS AND CROPS

LeBaron and Gressel (17), in referring to pesticide resistance, reported in 1982 that 428 species of insects and arthropods have become resistant to one or more insecticides that were once effective against them. In the past 12 years, 81 cases of pathogens resistant to benzimidazole fungicides and 8 cases of pathogens resistant to streptomycin have been reported. Recently, 30 weed species—including monocots and dicots—have been found resistant to triazines; in a few isolated recent instances, weed resistance to 2,4-D, trifluralin, paraquat, and diuron have also occurred.

Smith and Schweizer (24), recognizing the potential for the development of resistant weed species and the need for concomitant use of increasing herbicide rates to effect weed control, have demonstrated through research in 1983 the existence of higher tolerance to herbicides in *Beta vulgaris*. Selections of such tolerant *B. vulgaris* lines could then be grown in the presence of higher herbicide concentration without yield loss or adverse effect on the crop. Could such host selection be beneficial if a crop is exposed to a more "aggressive" root pathogen developing in the presence of higher herbicide concentrations?

Generally speaking, weed control researchers minimize the seriousness of the environmental impacts of herbicides, but they also conclude that their toxic effects will be more pronounced on lower rather than higher trophic levels of food-chain organisms. This is evident from reports by Quilt, Grossbard, and Wright in 1980 (20) regarding barban inhibition of *Rhizobium* nodule development; in reports by Griffin and Anderson

in 1980 (12) that nematode damage in alfalfa increased following herbicide use; and in the conflicting reports of Fusarium disease increase in processing peas following trifluralin use and the reports of Teasdale, Harvey, and Hagedorn (26,27), Sacher, Hopen, and Jacobsen (23), and Grau and Reiling (10) of an apparent inhibition of phycomycetous fungi such as *Pythium* sp. or *Aphanomyces* in peas using trifluralin and other nitroaniline herbicides, which resulted in less disease. However, such specific use of trifluralin as a fungicide does not merit a general recommendation for its use against other root diseases such as Rhizoctonia and Fusarium in peas and beans, which, according to Johnston, Ivany, and Cutcliffe (16), are more severe with trifluralin applications. Ashton and Crafts (5) have stated that dinitroaniline herbicides are known to inhibit normal meristem development and restrict lateral and secondary root development in legumes. Because these herbicides also restrict germ-tube development in zoospores of various phycomycetes, according to Grau and Reiling (10) and Jacobsen and Hopen (15), and result in a reduction in an infection court (by lack of root surface), perhaps these phenomena can account, in part, for the "phycomycetous" root rot reduction noted above.

CONCLUSION

In general, four major herbicide effects can lead to increased plant disease. First, the structural defenses of the host may be reduced. (Heitefuss [13,14] discussed induced morphological and physiological changes in host plants that may ultimately alter resistance and susceptibility to plant disease. These changes included reduction of wax formation on leaves; changes in carbohydrate, nitrogen, or glucoside metabolism; and retardation or stimulation of plant growth, which may alter the coincidence between pathogen presence and susceptible growth stages of the host.) Second, the herbicides may stimulate increased exudation from host plants. Third, pathogen growth may be stimulated. Fourth, microflora that compete with potential pathogens may be inhibited. Additional studies are needed to distinguish among these mechanisms as well as to show the overall effects of these interactions in plant protection programs. Studies on weed resistance and crop tolerance to herbicides are also needed.

In cooperative research in Germany in 1979, Altman, Dehne, and Shönbeck (3) have also shown that diallate used in beet culture in Europe resulted in an increase in root rot due to *Rhizoctonia* and an interference in sugar translocation and other photosynthetic assimilates from beet roots.

It is easy to accept the idea of some pesticides stimulating plant growth by eliminating weeds or by eliminating pathogens, but it is quite difficult to accept the concept that many pesticides are involved in predisposing hosts to disease. Yarwood (30), in defining predisposition, wrote that "if a chemical or physical treatment affects disease through its effect on the host, the condition is referred to as predisposition; if it acts directly on the pathogen the condition is called activation. Predisposition may be both detrimental when disease increases or beneficial when the chemical stimulates the host to produce a phytoalexin which in turn prevents disease."

Varieties of host plants that are resistant to disease would be ideally suitable for areas where such pesticide disease interactions do occur. Varieties of host plants that are tolerant of increasing levels of herbicides are also being developed. Such varieties would be suitable for "no-till" areas. Plants possessing both disease resistance and herbicide tolerance will be developed by breeders, thus overcoming two potential problems that would tend to reduce crop yield.

It should also be possible to develop appropriate management procedures by modifying the pesticides that are used. This can be accomplished by including combinations of fungicides with the required herbicides to reduce diseases or by developing agronomic programs and practices that reduce the need for these herbicides, therefore reducing the predisposing interactions.

Chemical weed control and insect pest control mark the continuation of a new era in crop production and bring their own problems and challenges. Altered management practices, increased plant density, and altered nutrient status and microclimate are some of the factors affected by herbicides that will influence disease incidence and severity in the future. A revision of fertilizer practices in the light of some of these factors is essential. Diseases that have been troublesome in the past may become less important, and others that have been slight or absent may become prominent. Disease management programs must be adjusted to meet these new situations, and the complexes resulting from plant herbicide resistance must also be considered; otherwise, we can only guess at the consequences of ever-increasing application rates of herbicides to combat these resistant weeds.

LaBaren and Gressel (17) summarize the pesticide-pest control dilemma best: "Our battle against pests is not inevitably one we are going to lose; it is dynamic and must be fought, as a complex war, with all available weapons. It is a constant interplay of measure and countermeasure by all organisms concerned, including man. Common sense and the laws of nature tell us that it is a game we can never entirely win. Yet there is no reason to believe that we cannot maintain a satisfactory level of crop protection indefinitely. We have simply learned again that we must never become complacent or assume that our present herbicides are the ultimate, the final solution. We must keep available all the tools we have ever had, including the hoe, while we continue searching for new and better answers."

What of the future development of resistant host crops and weeds? Should we breed plants resistant to diseases and tolerant of high doses of herbicides? Should we apply herbicides in mixtures containing softening agents, such as charcoal or antitoxicants or fungicides? Should we design herbicides with more specific target applications designed for specific weeds?

As scientists, it behooves us to take a better look at all the soil microflora exposed to herbicides: the mycorrhizae, the rhizobia, and the pathogens.

Perhaps the answer to the dilemma is to consider a better integrated approach to disease control—not just IPM (integrated pest management), but IHM (integrated host management).

LITERATURE CITED

1. Abivardi, C., and Altman, J. 1978. Effects of cycloate on development of Heterodera schachtii and growth of three Beta species. J. Nematol. 10:90-94.
2. Altman, J., and Campbell, C. L. 1977. The influences of herbicides on plant diseases. Annu. Rev. Phytopathol. 15:361-386.
3. Altman, J., Dehne, H., and Schönbeck, F. 1979. Effect of several carbamate herbicides on sugar beet. Pages 507-512 in: Soil-borne Plant Pathogens. B. Schippers and W. Gams, eds. Academic Press, London. 686 pp.
4. Altman, J., and Steele, A. E. 1982. Effect of herbicides on hatching and emergence of larvae from Heterodera schachtii cysts. (Abstr.) Phytopathology 71:985.
5. Ashton, F. M., and Crafts, A. S. 1981. Mode of Action of Herbicides. 2d ed. John Wiley and Sons, New York, p. 525.
6. Bareta, C., and Altman, J. 1983. Measurement of trifluralin effects on a pinto bean-Fusarium disease syndrome. (Abstr.) Phytopathology 73:811.
7. DeRosa, F., Haber, O., Williams, C., and Marqulis, L. 1978. Inhibitory effects of the herbicide trifluralin on the establishment of clover root nodule symbiosis. Cytobios 21:37-43.
8. Duncan, D. R., and Paxton, J. D. 1981. Trifluralin enhancement of Phytophthora root rot of soybean. Plant Dis. 65:435-436.
9. Glass, E. H. 1976. Research needs on pesticides and related problems for increased food supplies. NSF report to Science and Technology Policy Office. 63 pp.
10. Grau, C. R., and Reiling, T. P. 1977. Effect of trifluralin and dinitramine on Aphanomyces root rot of pea. Phytopathology 67:273-276.
11. Griffin, G. D., and Anderson, J. L. 1979. Effects of DCPA, EPTC and chloropropane on pathogenicity of Meloidogyne hapla to alfalfa. J. Nematol. 11:32-36.
12. Griffin, G. D., and Anderson, J. L. 1980. Herbicides increase alfalfa vulnerability to nematodes. (Abstr.) Agrichem. Age, May, p. 36C.
13. Heitefuss, R. 1972. Ursachen der Nebenwirkungen von Herbiziden auf Pflanzenkrankheiten. Z. Pflanzenkr. (Pflanzenpathol.) Pflanzenschutz. Sonderh. 6:79-87.
14. Heitefuss, R. 1973. Der Einfluss von Herbiziden auf Bodenbürtige Pflanzenkrankheiten. Proc. Eur. Weed Res. Conf. Symp. Herbicides-Soil. 99-128.
15. Jacobsen, B. J., and Hopen, H. J. 1975. Influence of herbicides on Aphanomyces root-rot of pea. (Abstr.) Proc. Am. Phytopathol. Soc. 2:26.
16. Johnston, H. W., Ivany, J. A., and Cutcliffe, J. A. 1980. Effects of herbicides applied to soil on Fusarium root rot of processing peas. Plant Dis. 64:942-943.
17. LeBaron, H. M., and Gressel, J. 1982. Herbicide Resistance in Plants. John Wiley & Sons, New York. 401 pp.
18. Neubauer, R., and Avizohar-Hershenson, Z. 1973. Effect of the herbicide, trifluralin, on Rhizoctonia disease in cotton. Phytopathology 63:651-652.
19. Paul, V. 1975. Untersuchugen über den Einfluss von Diallat auf den Wurzelbefall von Zea mays L. durch Fusarium moniliforme Sheldon. Dr. Agric. Diss. Rheinischen Friedrich-Wilhelms-Universitat. Bonn. 76 pp.
20. Quilt, P., Grossbard, E., and Wright, S. J. L. 1979. Effects of the herbicide barban and its commercial formulation 'Carbyne' on soil micro-organisms. J. Appl. Bacteriol. 46:431-442.
21. Quilt, P., Grossbard, E., and Wright, S. J. L. 1980. Effects of the herbicide barban and its commercial formulation 'Carbyne' on nitrification in soil. J. Appl. Bacteriol. 49:255-263.
22. Riggs, R. D., and Oliver, L. R. 1982. Effect of trifluralin on soybean cyst nematode. J. Nematol. 14:466.
23. Sacher, R. F., Hopen, H. J., and Jacobsen, D. J. 1978. Suppression of root diseases and weeds in peas treated with dinitrophenol and dinitroaniline herbicides. Weed Sci. 26:589-593.
24. Smith, G. A., and Schweizer, E. E. 1983. Cultivar × herbicide interaction in sugarbeet. Crop Sci. 23:325-328.
25. Strohbusch, D. F. 1974. Changes in agriculture and their influence on the use of agrochemicals. Adv. Dev. Pestic. Univ. Manchester, Sept. 1974. Bull. 8/74 BASF.
26. Teasdale, J. R., Harvey, R. G., and Hagedorn, D. J. 1979.

Mechanism for the suppression of pea (*Pisum sativum*) root rot by dinitroaniline herbicide. Weed Sci. 27:195-201.

27. Teasdale, J. R., Harvey, R. G., and Hagedorn, D. J. 1979. Factors affecting the suppression of pea (*Pisum sativum*) root rot (*Aphanomyces euteiches*) by dinitroaniline herbicides. Weed Sci. 27:467-472.

28. U.S. Department of Agriculture. 1982. The Pesticide Review. Washington, D.C.

29. Wheeler, H. 1975. Plant Pathogenesis. Springer-Verlag, New York. 106 pp.

30. Yarwood, C. E. 1976. Modification of the host-response predisposition. Pages 703-718 in: Physiological Plant Pathology. R. Heitefuss and P. H. Williams, eds. Springer-Verlag, New York.

Effects of Soil Application of Fungicides on Take-all in Winter Wheat

G. L. BATEMAN, Rothamsted Experimental Station, Harpenden, Herts., U.K.

Take-all of cereals caused by *Gaeumannomyces graminis* (Sacc.) von Arx & Olivier var. *tritici* Walker is a major disease for which no effective fungicide treatment is available. The pathogen is susceptible in vitro to many fungicides (2,6), but their application by conventional methods in the field to seed, soil surface, or aerial plant parts is usually ineffective. This is because of the well-distributed soil inoculum and the long period of root susceptibility. However, a recent report (4) showed that triadimenol seed treatment was effective where early infection in autumn affects yield. This may not apply where rapid disease development in early summer is more important; long-term protection of the crown would then, presumably, be required. Small-scale field trials have also shown that incorporating high concentrations of benomyl into the seedbed can partially control take-all (3). This paper gives a brief account of further fungicide trials to assess soil application methods, timing, fungicide concentrations, and residual effects.

MATERIALS AND METHODS

Three field experiments were done on silty clay loam sites naturally infested with take-all. In 1979–81 (experiment 1) and 1980–83 (experiment 2), wheat cultivar Flanders was drilled in October in successive years in 12-m × 2-m plots separated by 0.9-m paths and arranged in four randomized blocks. In 1982–83 (experiment 3), wheat cultivar Avalon was drilled in five randomized blocks of 13-m × 3-m plots separated by plots of the same width. Benomyl (50% wettable powder [w.p.]) or nuarimol (9% emulsifiable concentrate [e.c.] or, in experiment 3, 8% w.p.) were applied in the first year of each experiment by sprinkling suspensions in water at 5,000–9,000 L/ha, either to the seedbed in autumn followed by incorporation by rotary harrow just before drilling or over the crop in April.

A 3-m section at the end of each plot was sampled for disease assessment, the rest being harvested for yield determinations. Three or four samples were taken between April and July by digging up five 20-cm lengths of row chosen at random from each plot. The washed roots were then examined for take-all symptoms, and a disease rating for each plot was calculated as percentage of slight infection + (2 × percentage of moderate infection) + (3 × percentage of severe infection) (1,7).

RESULTS

Disease ratings were obtained from the final sample taken in each year of each experiment at Zadoks growth stages 55–77, and also grain yields where available (Fig. 1).

In experiment 1, year 1, autumn benomyl and autumn and spring nuarimol treatments reduced take-all. Nuarimol decreased yield slightly, but benomyl had no yield effect. In the second year, nuarimol continued to control the disease, and yields were increased in the plots with nuarimol and spring benomyl treatments.

In the first year of experiment 2, take-all was controlled significantly ($P = 0.05$) by benomyl applied in spring and nuarimol in autumn; the former treatment also increased yield. In the second year, take-all was significantly increased in the plots with autumn benomyl treatments, but there were no significant yield effects. There were no effects of treatment in the third year (when nuarimol spring treatments were not sampled).

In experiment 3, all concentrations of nuarimol applied in autumn significantly suppressed take-all.

DISCUSSION

Benomyl and nuarimol, two of the most promising fungicides from preliminary screens (2), were tested in field trials. Benomyl controlled take-all after autumn or spring treatments in different experiments, in one case producing a yield increase but at an uneconomic concentration (20 kg/ha). One benomyl treatment (experiment 2) caused an increase in disease in the second year. This can also occur in years following application of soil fumigants to control take-all (5), probably because of a reduction of the antagonistic microflora in the soil.

Nuarimol suppressed take-all after incorporation into the seedbed in autumn at much smaller concentrations than benomyl. However, yield increases were not achieved, probably because of a phytotoxic effect; an exception was the second year of experiment 1 after two applications of fungicide at 4.4 kg/ha, when the disease control also persisted. The value of the smallest concentrations in eliminating the phytotoxic effect will be known better after the harvest of experiment 3 in 1983. Nuarimol had no adverse residual effects on take-all.

Although these results show that fungicidal control of take-all by soil application is possible, the ideal fungicide has not yet been found. Further work is required to establish the properties of such a fungicide. An improved experimental design, to minimize the effects of variable inoculum distribution, is being developed. A possible role for fungicide application in an integrated control program, possibly taking advantage of subsequent natural take-all decline or treatment with microbiological control agents, is also being investigated.

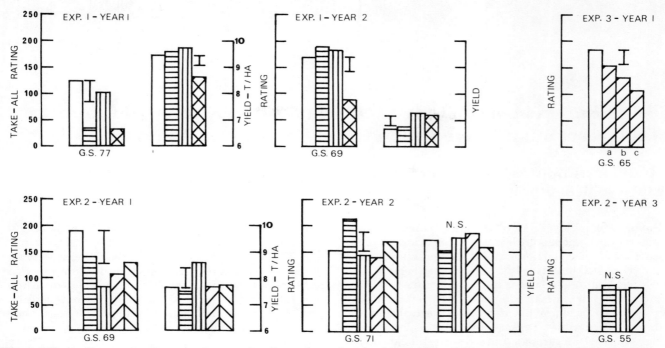

Fig. 1. Effects of benomyl and nuarimol upon take-all severity and grain yields in three field experiments. Vertical bars show least significant differences at $P = 0.05$. Treatments, applied in the first year only, were nil (□), benomyl applied in autumn (▤), benomyl in spring (▥), nuarimol in autumn (▨), nuarimol in spring (▧), and nuarimol in autumn and spring (▩). Benomyl was applied at 20 kg/ha. Nuarimol was applied at 4.4 kg/ha in experiment 1; 2.2 kg/ha in experiment 2; and 0.6 (a), 1.1 (b), and 2.2 (c) kg/ha in experiment 3.

SUMMARY

In field trials, benomyl or nuarimol suppressed take-all and sometimes increased yields when applied in drenches on naturally infested sites. A high concentration of benomyl (20 kg/ha) was effective in one experiment as an autumn presowing treatment, incorporated by rotary harrowing, and in another when applied in spring. The latter treatment increased yield by 12%. Disease increased slightly in some second crops after benomyl treatment. Nuarimol applied at concentrations down to 0.6 kg/ha in autumn suppressed disease and had no adverse residual effects in subsequent crops.

LITERATURE CITED

1. Anonymous. 1976. Manual of plant growth stages and disease assessment keys. Ministry of Agriculture, Fisheries and Food, Harpenden, Herts., U.K.
2. Bateman, G. L. 1980. Prospects for fungicidal control of take-all of wheat. Ann. Appl. Biol. 96:275-282.
3. Bateman, G. L. 1981. Effects of soil application of benomyl against take-all (*Gaeumannomyces graminis*) and footrot diseases of wheat. Z. Pflanzenkr. Pflanzenschutz 88:249-255.
4. Bockus, W. W. 1983. Effects of fall infection by *Gaeumannomyces graminis* var. *tritici* and triadimenol seed treatment on severity of take-all in winter wheat. Phytopathology 73:540-543.
5. Ebbels, D. L. 1969. Effects of soil fumigation on disease incidence, growth and yield of spring wheat. Ann. Appl. Biol. 63:81-93.
6. Gorska-Poczopko, J. 1971. Studies on chemical control of *Ophiobolus graminis* Sacc. I. Testing systemic fungicides against *Ophiobolus graminis* Sacc. Acta Phytopathol. Acad. Sci. Hung. 6:393-398.
7. Prew, R. D., and Dyke, G. V. 1979. Experiments comparing 'break crops' as a preparation for winter wheat followed by spring barley. J. Agric. Sci. Cambridge 92:189-201.

Use of Fungicides to Study Significance and Etiology of Root Rot of Subterranean Clover in Dryland Pastures of Victoria

F. C. GREENHALGH and R. G. CLARKE, Plant Research Institute, Department of Agriculture, Victoria, Burnley, 3121, Australia

Subterranean clover (*Trifolium subterraneum* L.) is the most important pasture legume in temperate regions of southern and eastern Australia (11,12). This annual plant originated in the Mediterranean region and parts of western Europe, and it was introduced into Australia last century. It was first exploited in improved pastures during the 1920s to provide protein for livestock and to improve soil fertility. Subsequently, the legume was widely adopted by farmers and is now established in more than 20×10^6 ha of dryland pasture (4).

Subterranean clover is grown in two types of dryland pastures in Australia: those that are permanent and remain undisturbed for many years, and those in the ley phase of crop rotations. In permanent pasture, subterranean clover seed germinates after adequate autumn rain, and the plant grows until late spring when it sets seed in burrs ("fruiting bodies" that bury in soil) and dies. In crop-pasture rotations, subterranean clover is normally established by sowing seed with a cereal. A pasture then establishes the following season from seed produced under the cover crop.

During the last 15–20 years, there has been a decline in productivity of subterranean clover in substantial areas of western and northeastern Victoria, and this has reduced animal and cereal production (B. Muir, *personal communication*; 13). The cause of the decline is unknown, but a number of factors including root rot have been associated with it. Root rots have also been associated with poor growth of the legume in other areas of Victoria (3,10) and in parts of Western Australia (1) and New South Wales (16). Tests showed that *Pythium irregulare* Buisman from rotted roots and *Fusarium avenaceum* (Corda ex. Fr.) Sacc. from roots and seed caused damping-off and root rot of subterranean clover in sand or pasteurized soils. These fungi were therefore implicated as causal agents of subterranean clover decline, but their effects on roots and plant productivity under natural conditions were not investigated in detail.

In preliminary surveys of proposed seed lines and experimental sites for this study, *F. avenaceum* was isolated frequently from burrs and some seed lines but infrequently from sieved soils. *P. irregulare* and *P. mamillatum* Meurs. were isolated commonly from burrs and soils, and *P. mamillatum* was shown to be pathogenic to subterranean clover roots in pasteurized soils. Other preliminary studies under controlled conditions showed that the selective fungicide metalaxyl (14) controlled disease caused by *P. irregulare* (6). In addition, previous work by Kellock (8) demonstrated that benomyl, which is not active against pythiaceous fungi (2), was fungicidal to *F. avenaceum* in laboratory tests. These fungicides were therefore used in this study to control root rots caused by *Pythium* spp. and *F. avenaceum* and to determine whether root disease causes the decline of subterranean clover. This paper reports on the effects of fungicide treatments on root rot and incidence of the fungi in roots and on production of subterranean clover in permanent pasture and in undersown crops.

MATERIALS AND METHODS

Permanent pasture. The decline of subterranean clover cultivar Mt. Barker was studied in a permanent pasture at Clunes in western Victoria. Metalaxyl and benomyl were applied alone and in combination as soil drenches to replicated plots (4 m²) on sandy loam of pH 5.4 (measured in 1:5 soil:water suspension). Drenches of metalaxyl (25 WP) at 0.125 g a.i./m² and benomyl (50 WP) at 2.5 g a.i./m² were applied 3 weeks before and during and 17 weeks after seedling emergence in April 1981. Preliminary tests under controlled conditions showed that drenches of the fungicides at these rates were not phytotoxic or stimulatory to seed germination and seedling growth.

Three sods each of 100 cm²× 15 cm deep were sampled at random from half of each plot at 6, 17, and 30 weeks after seedling emergence. Roots were washed free from soil and the number of subterranean clover plants in each recorded. Severity of root rot on each plant was rated on a 0–5 scale (7), and root nodulation was assessed (5). Isolations of fungi were made from necrotic and apparently healthy roots of plants collected for the first and final assessments. Lateral root tips and taproot segments of plants taken at random from each sod were dipped in 70% ethanol, washed in sterile water, and placed onto either P_{10}VP medium (17) or PCNB-achromycin medium (9). Root pieces on the media were incubated at 25°C; *Pythium* spp. were identified by examination of fungi on P_{10}VP medium and *F. avenaceum* by examination of colonies transferred from PCNB-achromycin medium to potato-sucrose agar.

Root and shoot dry weights of the total number of subterranean clover plants in each sod were determined, and weights for each plant were calculated. The herbage of half of each plot was mown and weighed at 17 and 30 weeks, and this together with assessments of the proportion of subterranean clover in the plots was used to estimate subterranean clover yield. At the final assessment, soil samples at depths of 0–5, 5–10, and 10–15 cm were collected from plots for pH determinations.

Establishment of subterranean clover. The influence of root rot caused by soilborne fungi on establishment of subterranean clover after cropping with wheat was studied at Boxwood in northeastern Victoria and at Joel South in western Victoria. Subterranean clover seed with low levels of contamination by *F. avenaceum* (<3% of seed) was sown with oats into cultivated sandy loam of pH 5.2 at Boxwood and of pH 5.7 at Joel South. Seed of cultivar Woogenellup was used at Boxwood and of cultivar Mt. Barker at Joel South. The fungicide treatments described for the permanent pasture experiment were applied to replicated plots (10 m²) at sowing in May 1981 and after 4 and 8 weeks.

The effect of sowing seed highly contaminated by *F. avenaceum* on root rot was also tested at the Boxwood and Joel South sites and at Rutherglen in northeastern Victoria. A commercial line of subterranean clover cultivar Mt. Barker with 22% of seed contaminated by *F. avenaceum* was either treated with benomyl (0.5 g a.i./kg of seed) to eliminate the fungus (8) or left untreated. Treated and untreated seeds were then sown in replicated plots (10 m²) at Boxwood and Joel South in May 1981. A second line of cultivar Woogenellup with 34% of seed contaminated by the fungus was treated in the same manner and sown (June) in clay loam (pH 4.9) at Rutherglen.

In studies on both the soilborne and seedborne fungi, subterranean clover plants were collected from five 1-m lengths of drill row in each plot at 4, 8, and 26 weeks after sowing. The exception to this was the final sampling at Rutherglen, which was made 18 weeks after sowing. The plants were assessed as described for the permanent pasture.

RESULTS

Permanent pasture. In the permanent pasture, the metalaxyl-benomyl treatment increased plant density and decreased root rot severity at 6, 17, and 30 weeks and increased yield in comparison to the control treatment (Table 1). Drenches of either metalaxyl or benomyl caused smaller increases in plant density at 30 weeks and reductions in disease severity at 6 and 30 weeks. They also increased yield, but not to the same extent as the metalaxyl-benomyl drenches. None of the treat-

ments affected root and shoot weights of each plant at 6 weeks, but metalaxyl-benomyl drenches significantly ($P = 0.01$) increased them at 17 weeks. This treatment also increased root but not shoot weights at 30 weeks.

The metalaxyl-benomyl treatment reduced the frequency of isolation of *P. irregulare*, *P. mamillatum*, and *F. avenaceum* from plants at 6 weeks (Table 2). The metalaxyl and benomyl treatments reduced the frequency of detection of *Pythium* spp. and *F. avenaceum*, respectively. At 30 weeks, *Pythium* spp. were not isolated from roots in soil treated with either metalaxyl or metalaxyl-benomyl, but they were isolated frequently from those in either untreated or benomyl-treated soil. At the same time, *F. avenaceum* was not detected in roots from soils treated with benomyl or metalaxyl-benomyl and was isolated infrequently from roots from soil that was either untreated or treated with metalaxyl.

Fungicide drenches did not affect soil pH or root nodulation, which was good at each assessment.

Establishment of subterranean clover. Experiments on the importance of soilborne fungi in subterranean clover establishment at Boxwood and Joel South showed that drenches of either metalaxyl, benomyl, or metalaxyl-benomyl did not significantly ($P = 0.05$) affect plant number at 4, 8, and 26 weeks. At each site and sampling time, the metalaxyl-benomyl drench was the most effective treatment in reducing the disease severity rating in comparison with that for the untreated control, which was low (1.0 or less). The fungicide treatments did not significantly ($P = 0.05$) affect root and shoot weights at each assessment, except at Joel South, where benomyl and metalaxyl-benomyl drenches decreased them in the final assessment.

At 4 and 26 weeks, *P. irregulare* and *P. mamillatum* were isolated consistently from rotted and apparently healthy roots in either untreated or benomyl-treated soil, but infrequently from roots in soil treated either with metalaxyl or metalaxyl-benomyl at both sites. During these assessments, *F. avenaceum* was detected infrequently in roots of plants from each treatment.

Assessments for each treatment showed that root nodulation was poor at Joel South and negligible at Boxwood.

Results of the three experiments on the significance of seedborne inoculum of *F. avenaceum* showed that at each sampling time the benomyl seed treatment did not significantly ($P = 0.05$) affect plant number, root rot severity, root and shoot weights, and frequency of

Table 1. Effects of metalaxyl and benomyl drenches on subterranean clover density and root rot severity at 6, 17, and 30 weeks after emergence and on yield in a permanent pasture at Clunes[a]

Treatment	Plant number/ 100 cm²			Root rot severity[b]			Estimated dry matter[c] (t/ha)
	6	17	30	6	17	30	
Metalaxyl + benomyl	57.3	62.1	55.3	1.2	1.3	0.5	5.00
Metalaxyl	46.4	55.0	46.3	1.4	2.2	0.8	3.37
Benomyl	46.2	46.4	39.8	1.8	2.4	0.8	3.19
Control (untreated)	42.9	45.6	30.7	2.2	2.3	1.4	2.57
LSD ($P = 0.05$)		8.4			0.2		0.55

[a] Values represent means of six replicates.
[b] Root rot severity on scale of 0–5: 0, roots healthy; 1, slight lateral root rot; 2, moderately severe lateral root rot or slight taproot rot; 3, severe lateral root rot or moderately severe taproot rot; 4, severe taproot rot; and 5, plant dead.
[c] Based on mowings at 17 and 30 weeks after seedling emergence.

Table 2. Effects of metalaxyl and benomyl drenches on frequency of isolation of *Pythium* spp. and *Fusarium avenaceum* from subterranean clover plants 6 weeks after emergence in a permanent pasture at Clunes

Treatment	Number of plants from which fungi isolated[a] (max. = 12)		
	P. irregulare	*P. mamillatum*	*F. avenaceum*
Metalaxyl + benomyl	0.7[b]	0.3	0.8
Metalaxyl	0.8	0.2	6.0
Benomyl	10.3	7.0	1.0
Control (untreated)	9.2	7.0	7.3
LSD ($P = 0.05$)	1.4	2.1	1.0

[a] Isolations were made from three lateral root tips and taproot segments of each plant.
[b] Values represent means of six replicates.

isolation of *F. avenaceum*. At each site and assessment, the disease severity rating for plants from either treated or untreated seed was <0.7. During the assessments, *F. avenaceum* was detected in the roots of <7, <17, and 0% of plants from either treated or untreated seed at Boxwood, Joel South, and Rutherglen, respectively.

DISCUSSION

This investigation demonstrates the benefits of using fungicides as experimental tools to study aspects of subterranean clover root rot in field situations. The approach provides estimates of yield loss that can be attributed to the disease and also information on relative effects of different groups of pathogenic fungi on plant growth.

It is not always possible to interpret responses to fungicide treatments accurately because of nonspecific effects of chemicals on plant growth (15). However, results of the study suggest that the disease is a major factor causing the substantial loss in subterranean clover yield in the permanent pasture at Clunes. Data in Table 1 and the increased productivity of plants in soil treated with metalaxyl-benomyl indicate that both damping-off and root rot contribute to yield loss and that the disease is caused by interactions between pathogens.

The metalaxyl and benomyl drenches reduced the frequency of isolation of their respective target organisms (Table 2), but they may have also affected other pathogens. For example, it is possible that metalaxyl affected a pathogenic *Phytophthora* sp. (P. Taylor, *unpublished*) detected recently on rotted roots of subterranean clover from the permanent pasture at Clunes and from other dryland pastures. This fungus was not isolated in the present study because it does not grow from rotted roots onto $P_{10}VP$ medium. In view of this finding, further studies are necessary to resolve the etiology of the disease in dryland areas.

In contrast to the situation in permanent pasture, experiments showed that root rot was of little significance during establishment of subterranean clover in cultivated soils, even though pathogens were present on seed and in the soils. Tests under controlled conditions have also shown that the disease does not develop in disturbed soil from the permanent pasture at Clunes (7). Therefore, soil disturbance is thought to be an important factor influencing disease development. The lack of cultivation in permanent pasture may favor disease outbreaks because it is conducive to soil compaction and it allows burrs and other organic material to accumulate at the soil surface. Soil compaction would stress plants by impeding root growth, and the formation of an organic mat would increase inocula of the fungi in direct contact with germinating seeds and with taproots of subterranean clover.

In conclusion, the study provides evidence that a complex of fungi causes root rot that can significantly reduce productivity of subterranean clover in permanent pasture under dryland conditions. It also indicates that cultivation of permanent pasture may reduce disease severity and therefore increase growth of the legume. The most appropriate cultivation treatment for disease control and the duration of its beneficial effects are the subjects of further investigations.

LITERATURE CITED

1. Barbetti, M. J., and MacNish, G. C. 1978. Root rot of subterranean clover in the irrigated areas of south-western Western Australia. Aust. J. Exp. Agric. Anim. Husb. 18:426-433.
2. Bollen, G. J., and Fuchs, A. 1970. On the specificity of the in vitro and in vivo antifungal activity of benomyl. Neth. J. Plant Pathol. 76:299-312.
3. Burgess, L. W., Ogle, H. J., Edgerton, J. P., Stubbs, L. L., and Nelson, P. E. 1973. The biology of fungi associated with root rot of subterranean clover in Victoria. Proc. R. Soc. Victoria 86:19-28.
4. Cocks, P. S., Mathison, M. J., and Crawford, E. J. 1978. From wild plants to pasture cultivars: Annual medics and subterranean clover in southern Australia. Pages 569-596 in: Advances in Legume Science, vol. 1. Proc. Int. Legume Conf., Kew, 1978. R. J. Summerfield and A. H. Bunting, eds. R. Bot. Gardens, Kew, England. 667 pp.
5. Corbin, E. J., Brockwell, J., and Gault, R. R. 1977. Nodulation studies on chickpea (*Cicer arietinum*). Aust. J. Exp. Agric. Anim. Husb. 17:126-134.
6. Greenhalgh, F. C. 1983. Growth cabinet evaluation of fungicides for control of Pythium damping-off and root rot of subterranean clover, 1982. Fungic. Nematic. Tests 38:47.
7. Greenhalgh, F. C., and Lucas, S. E. 1985. Effect of soil pasteurization on damping-off and root rot of subterranean clover caused by *Fusarium avenaceum* and *Pythium* spp. Soil Biol. Biochem. In press.
8. Kellock, A. W. 1975. A study of pathological and ecological factors causing root rot of subterranean clover (*T. subterraneum* L.). Ph.D. thesis, University of Melbourne. 202 pp.
9. Kellock, A. W., Stubbs, L. L., and Parberry, D. G. 1978. Seed-borne *Fusarium* species on subterranean clover and other pasture legumes. Aust. J. Agric. Res. 29:975-982.
10. McGee, D. C., and Kellock, A. W. 1974. *Fusarium avenaceum*, a seed-borne pathogen of subterranean clover roots. Aust. J. Agric. Res. 25:549-557.
11. Morley, F. H. W. 1961. Subterranean clover. Adv. Agron. 13:57-123.
12. Powell, S. C. 1970. Subterranean clover, our most important pasture legume. J. Agric. (Victoria, Aust.) 68:274-277.
13. Reeves, T. G., and Hirth, J. R. 1982. Survey of undersown and established subterranean clover pastures in the northeast of Victoria, 1980. Dep. Agric. (Victoria, Aust.) Res. Proj. Ser. 122. 23 pp.
14. Schwinn, F. J., Staub, T., and Urech, P. A. 1977. A new type of fungicide against diseases caused by oomycetes. Meded. Fac. Landbouwwet, Rijkisuniv. Gent 42:1181-1188.
15. Smiley, R. W. 1981. Nontarget effects of pesticides on turfgrasses. Plant Dis. 65:17-23.
16. Stovold, G. E. 1974. Root rot caused by *Pythium irregulare* Buisman, an important factor in the decline of established subterranean clover pastures. Aust. J. Agric. Res. 25:537-548.
17. Tsao, P. H., and Ocaca, G. 1969. Selective isolation of species of *Phytophthora* from natural soils on an improved antibiotic medium. Nature 223:636-638.

Suppression of Soilborne Diseases of Ornamental Plants by Tree Bark Composts

H. A. J. HOITINK and G. A. KUTER, Department of Plant Pathology, Ohio State University, Ohio Agricultural Research and Development Center, Wooster, 44691, U.S.A.

During the 1960s and 1970s, tree barks, sawdusts, and other organic wastes were introduced by growers to replace peats and reduce costs of container media. It was soon observed that losses caused by Phytophthora spp. were lower on plants produced in container media amended with barks than in the conventional peat media. Although this solved some disease problems, other difficulties were encountered. Some tree barks induced nitrogen (N) deficiency and had low water-holding and cation capacities. Some bark composts, if used in excessive quantities, also caused low oxygen tensions in the media because of their high biodegradability. This occurred particularly near the water-saturated base of the container medium. Finally, some container media prepared with "inadequately stabilized" composts fermented during transit and became toxic to plants.

At present, barks from most tree species are processed or composted so that plants can be produced without developing N deficiency (5). Barks also are used as soil amendments for suppression of soilborne diseases. For example, Fusarium wilt of Chinese yam is controlled by incorporation of composted larch bark (30 tons/ha) into field soil. Control was similar to that obtained with methyl bromide and better than that obtained with benomyl (12). This paper emphasizes research on suppression of soilborne plant pathogens in compost-amended container media.

It has been proposed to add specific pathogen suppressive field soils to container media to render them suppressive. However, soil-amended container media have disadvantages; for example, suppressive soils frequently affect only one pathogen and harbor others; herbicide residues frequently occur in soils; the bulk density of soil-amended media is frequently too high; clay and silt particles may migrate through media to the base of the container, resulting in inadequate drainage; and top soil is not a renewable resource. In this paper, effects of physical, chemical, and biological properties of bark-amended container media on suppression of soilborne plant pathogens are discussed.

EFFECTS OF PHYSICAL PROPERTIES

Observations made in nurseries indicate that Phytophthora root rots are most prevalent in media with air-filled pore space levels of less than 15% (after 2-hr drainage of saturated, 15-cm-tall columns to

Journal article 94-85 of the Ohio Agricultural Research and Development Center.

"container" capacity). Media amended with tree barks that suppress Phytophthora root rots typically have air-filled pore space levels of greater than 25% (7,15) and percolation rates greater than 2.5 cm/min. Addition of sand to these media reduces pore space levels and negates the suppressive effect (15), suggesting that physical properties have a significant effect, particularly on diseases caused by water molds. Differences in physical properties, however, do not explain all of the suppressive effects observed in bark-amended media. For example, composted hardwood bark (CHB)-sand medium with a soil moisture desorption curve similar to that of a peat-sand medium was suppressive to Phytophthora root rot in spite of similarities in drainage properties, suggesting that other factors played a role (7).

Precise methods have been proposed for determining percentages of air- and water-filled pore space levels and percentage of solid particle levels in container media. Idealized desorption curves have been described (3). Although substantial information is also available on the water relations of plant pathogens, particularly of water molds (4), the optimum physical properties described from the plant pathologist's and horti-culturalist's points of view, unfortunately, have not been integrated.

EFFECTS OF CHEMICAL PROPERTIES

Nutritional and chemical factors affect soilborne diseases. For example, low pH inhibits most Phytophthora root rots (1). The pH of Canadian sphagnum peat ranges from 3.1 to 4.0. Pine bark compost (PB) also may be a low pH (4.4), although the pH of most barks when first removed from trees ranges from 5.0 to 5.2. Not surprisingly, therefore, Phytophthora root rot of Aucuba japonica was suppressed in a PB medium (pH 4.4), even when saturated with water (14). However, the pH of the PB medium rose to 6.0 within a few weeks after planting and became conducive, so that even for Ericaceae, which tolerate acidity, low pH alone was not a practical control procedure for Phytophthora root rot (14,15).

Several reports indicate that tree barks release inhibitors of Phytophthora spp. into the water phase of container media. Filter-sterilized leachates prepared from PB (13,15), CHB (7,15), and eucalyptus barks and sawdusts (13) may have such inhibitory effects. Leachates from media amended with green CHB (<6 months old) are toxic. But leachates from 1- and 2-year-old (mature) CHB are not, even though these media are suppressive to disease (7). Extracts of fresh and composted PB also differ in levels of inhibition (13). Heating (60°C) negates the suppressive effect of 1-year-old CHB to lupine Phytophthora root rot, but not that of 6-month-old (green) CHB. It appears that the suppressive effect of mature CHB results from heat-labile factors.

Inhibitors of *Phytophthora* spp. in leachates from 6-month-old CHB, PB, and eucalyptus sawdusts reduce sporangium production and zoospore release (7,13,15). It is now known that the activity in leachates from green CHB is caused by many compounds. Toxic compounds present in highest concentrations were ethyl esters of hydroxy-oleic acids. These inhibitors do not affect *Rhizoctonia solani* Kuehn (H. A. J. Hoitink, *unpublished*).

EFFECTS OF BIOLOGICAL PROPERTIES

Suppression of Rhizoctonia damping-off (8,10,11), lupine Phytophthora root rot, and cucumber Pythium root rot in media amended with mature CHB is probably caused by biological factors. Several facts provide evidence for this: heat treatment (60°C) or radiation negates suppression; addition of a small volume of unheated suppressive CHB to heated media restores suppression; and individual antagonists that are isolated from CHB, then added to mature compost when temperatures decline to <50°C, restore suppression (10). Bacterial, fungal, and actinomycete antagonists of *R. solani* and *Pythium ultimum* Trow have been isolated, but only the role of fungal antagonists in suppression of *R. solani* in media has been examined in detail. Container media amended with mature composts prepared in windrows on field soil, or in mechanized reactor vessels, were suppressive, whereas those amended with green compost were conducive (8). *Trichoderma* spp. were characteristically abundant in the suppressive media and were effective fungal antagonists of *R. solani* in mature CHB (8,11). Although high population levels of *Trichoderma* antagonists could develop in media amended with green CHB, Rhizoctonia damping-off is not suppressed, suggesting that compost age affects activity of the antagonist (11).

The fate of plant pathogens in container media is a critical issue. Propagules, for example, could possibly remain dormant until after the suppressive effect disappears and cause disease after outplanting of nursery stock in landscapes. The effects of composts on plant pathogens were reviewed recently (5). Propagules of *R. solani* are killed rapidly in suppressive mature CHB (11), although the pathogen colonizes green CHB (6). Chlamydospores of *Phytophthora cinnamomi* Rands survive in media amended with various tree barks (7,15); however, mycelium, sporangia, and zoospores, which are produced from chlamydospores before infection, are lysed in media amended with green barks or composts (7,13,15). Suppressiveness to *R. solani* and *P. cinnamomi* persists for at least 2 years in CHB-amended media (5,7), which suggests that survival of propagules of plant pathogens in suppressive mature CHB media may be a lesser problem than in conducive peat media that are typically treated with fungicides.

Centers of compost piles often reach temperatures of 50–80°C, whereas the outer layers remain near 15–30°C. It was found recently that the addition of antagonists (*Trichoderma* spp. or *Pseudomonas* spp.) to the outer layer of mature compost, which is colonized by mesophilic microorganisms, did not significantly increase its suppressive effect to Rhizoctonia damping-off. However, addition of these same antagonists to compost removed from the center of a mature compost pile (40–55°C) increased suppressiveness to levels two to three times higher than that in the outer layer. Controlled production of antagonist-fortified compost, therefore, is feasible (OSU patent pending). This process also eliminates the variability in suppressiveness to damping-off typically encountered in "natural" bark composts (8,11). Composts prepared from municipal sludge also vary in their disease-suppressive effects (9).

The prevailing procedure in Ohio nurseries is to add 25–50% CHB to media on a volume basis. Pine bark, composted leaves, composted municipal sludge, and peat are other organic components used. Small volumetric amounts (<25%) of CHB do not consistently suppress the various soilborne plant pathogens of ornamentals. On the other hand, media prepared with 50% CHB may immobilize significant amounts of N because the compost frequently is not cured adequately before utilization. Addition of a lower volume of mature CHB levels with predictably high levels of suppression is therefore highly desirable.

In summary, suppressive composted media prepared from tree barks have been used successfully for control of diseases caused by a variety of soilborne plant pathogens. Techniques need to be developed that accurately assess compost maturity. This information needs to be paired against other factors, such as potential disease suppressive effects and nutrient release rates. Although some progress has been made in this area recently (2), much research remains to be done.

ACKNOWLEDGMENT

Salaries and research support have been provided by state and federal funds appropriated to the Ohio Agricultural Research and Development Center, Ohio State University.

LITERATURE CITED

1. Blaker, N. S., and MacDonald, J. D. 1983. Influence of container medium pH on sporangium formation, zoospore release, and infection of rhododendron by *Phytophthora cinnamomi*. Plant Dis. 67:259-263.
2. Chanyasak, V., and Kubota, H. 1981. Carbon/organic nitrogen ratio in water extracts as a measure of compost degradation. J. Ferment. Technol. 59:215-219.
3. DeBoodt, M., and DeWale, N. 1968. Study on the physical properties of artificial soils and the growth of ornamental plants. Pedologie 18:275-300.
4. Duniway, J. M. 1979. Water relations of water molds. Annu. Rev. Phytopathol. 17:431-460.
5. Hoitink, H. A. J. 1980. Composted bark, a lightweight growth medium with fungicidal properties. Plant Dis. 64:142-147.
6. Hoitink, H. A. J., Herr, L. J., and Schmitthenner, A. F. 1976. Survival of some plant pathogens during composting of hardwood tree bark. Phytopathology 66:1369-1372.
7. Hoitink, H. A. J., Van Doren, D. M., Jr., and Schmitthenner, A. F. 1977. Suppression of *Phytophthora cinnamomi* in a composted hardwood bark mix. Phytopathology 67:561-565.
8. Kuter, G. A., Nelson, E. B., Hoitink, H. A. J., and Madden, L. V. 1983. Fungal populations in container media amended with composted hardwood bark suppressive and conducive to Rhizoctonia damping-off. Phytopathology 73:1450-1456.
9. Lumsden, R. D., Lewis, J. A., and Millner, P. D. 1983.

Effects of composted sewage sludge on several soilborne plant pathogens and diseases. Phytopathology 73:1543-1548.

10. Nelson, E. B., and Hoitink, H. A. J. 1983. The role of microorganisms in the suppression of *Rhizoctonia solani* in container media amended with composted hardwood bark. Phytopathology 73:274-278.

11. Nelson, E. B., Kuter, G. A., and Hoitink, H. A. J. 1983. Effects of fungal antagonists and compost age on suppression of Rhizoctonia damping-off in container media amended with composted hardwood bark. Phytopathology 73:1457-1462.

12. Sekiguchi, A. 1977. Control of Fusarium wilt on Chinese yam. Annu. Rep. Dep. Plant Pathol. Entomol., Nagano Veg. Floriculture Exp. Stn. (Japan) 1:10-11.

13. Sivasithamparam, K. 1982. Some effects of extracts from tree barks and sawdusts on *Phytophthora cinnamomi* Rands. Aust. Plant Pathol. 10:18-20.

14. Spencer, S., and Benson, D. M. 1981. Root rot of *Aucuba japonica* caused by *Phytophthora cinnamomi* and *P. citricola* and suppressed with bark media. Plant Dis. 65:918-921.

15. Spencer, S., and Benson, D. M. 1982. Pine bark, hardwood bark compost, and peat amendment effects on development of *Phytophthora* spp. and lupine root rot. Phytopathology 72:346-351.

Effects of Cropping Sequences on Saprophytic Survival and Carry-over of *Gaeumannomyces graminis* var. *tritici*

J. F. KOLLMORGEN, Victorian Crops Research Institute, Horsham, Victoria, 3400; and J. B. GRIFFITHS and D. N. WALSGOTT, Mallee Research Station, Walpeup, Victoria, 3507, Australia

Take-all (*Gaeumannomyces graminis* (Sacc.) von Arx & Olivier var. *tritici* (Walker) is the most serious fungal disease of wheat in the Mallee region of Victoria, Australia. In years when climatic conditions favor the pathogen, wheat yields are lowered by an estimated 25%, which means a loss in yield of some 250,000 t. The disease can also cause appreciable losses in both yield and malting quality of barley. A 3-year rotation of wheat (*Triticum aestivum* L.), medic pasture (*Medicago* spp.), and fallow is commonly practiced in the Mallee, but the fallow has often been ineffective in breaking the disease cycle of *G. graminis* var. *tritici* possibly because of the presence of host grasses such as barley grass (*Hordeum leporinum* Link). Barley (*Hordeum vulgare* L.), oats (*Avena sativa* L.), and, more recently, triticale are sometimes grown as alternatives to, or in rotation with, wheat in the Mallee. Barley is reported to increase take-all in a following wheat crop, whereas oats are considered to reduce disease severity (9). There is no information on the effects of triticale on disease.

As neither effective fungicides nor resistant cultivars are available for the control of take-all, there is increasing interest in growing nongraminaceous "break" crops such as lupine (*Lupinus angustifolius* L.), field peas (*Pisum sativum* L.), and rapeseed (*Brassica napus* L.), especially now that cultivars with suitable agronomic characteristics are available. Selective herbicides can be used with these break crops to control grass weeds that are hosts of *G. graminis* var. *tritici*.

Crop rotation has been shown to control take-all elsewhere in Australia and overseas. Zogg (10) reported that it stimulated a microflora antagonistic to *G. graminis* var. *tritici*, and Chambers (5) showed that longevity of *G. graminis* var. *tritici* was reduced when various plant species were sown over infected straws.

We describe a series of glasshouse and field investigations to determine the effects of break crops and various cereals on saprophytic survival and carry-over of *G. graminis* var. *tritici* in Mallee soils. Some of the findings have been previously published (8).

MATERIALS AND METHODS

Soils. For glasshouse studies, a sandy loam (described by Chambers [4]) from the Mallee Research Station, Walpeup, Victoria, was amended with NPK fertilizer (7% nitrogen, 5% phosphorus, and 8% potassium) at the rate of 2.0 g/kg of oven-dry soil. Field studies were conducted in a similar soil at the Mallee Research Station that was amended with P at 13.1 kg/ha either before or at seeding.

Survival study. Segments of wheat straw were artificially inoculated with a single-spore isolate of *G. graminis* var. *tritici* using the method of Garrett (7) as modified by Butler (3). A mixture of 28 infected straws and 1.3 kg of soil (dry wt.) was placed on 460 g of soil (dry wt.) in a 17.5-cm-diameter plastic pot and covered by a further 736 g of soil (dry wt.). Germinated seeds of plant species listed in Table 1 were planted in the pots on 10 March 1981. The pots were watered to field capacity twice a week until 12 April, three times a week until 3 May, and daily thereafter. Mean maximum and minimum temperatures in the glasshouse were 25 and 11°C, respectively. The experimental design was a randomized block with six replications of each treatment. After 5, 10, and 23 weeks, buried straws from each treatment were recovered and tested for viability of *G. graminis* var. *tritici* by the wheat seedling test (7).

Table 1. Survival in the glasshouse of *Gaeumannomyces graminis* var. *tritici* in segments of wheat straw buried under various plant species

Species	Percentage of straws with viable pathogen after		
	5 weeks	10 weeks	23 weeks
Nil (unplanted)	99.5 a[z]	89.9 a	72.1 a
Wheat	98.3 a	83.0 ab	43.2 b
Oats	97.3 a	75.4 b	49.3 b
Rapeseed	98.6 a	72.3 b	49.8 b
Lupine	100.0 a	82.5 ab	42.4 b
Field peas	99.8 a	75.5 b	46.3 b
Medic	99.8 a	88.6 a	44.7 b

[z]Within each column, results with no letter in common are significantly different (*P* ≤ 0.05) as determined by Duncan's multiple range test.

Table 2. Effects of cropping sequence on take-all and wheat yield in microplots at Mallee Research Station

Preceding crop	Number of plants (of 15) with		Percentage with deadheads	Grain yield (kg/ha)
	Take-all	>5% of root affected		
Nil (fallow)	4.5 de[z]	1.3 cd	0.1 d	1,419 a
Rapeseed	3.4 ef	0.8 cd	0.2 cd	1,212 abc
Medic and grass weeds	5.2 cde	0.6 cd	0.2 cd	1,242 ab
Medic	3.0 ef	0.8 cd	0.7 bcd	1,346 a
Lupine	2.2 f	0.1 d	0.2 cd	1,301 a
Field peas	4.5 de	0.7 cd	0.1 d	1,214 abc
Wheat	9.6 ab	6.6 a	2.6 bc	754 de
Oats	6.6 bcd	2.2 bc	1.0 bcd	944 cd
Barley	8.2 abc	4.8 ab	3.2 b	985 bc
Triticale	12.3 a	8.7 a	10.7 a	679 e

[z]Within each column, results with no letter in common are significantly different (*P* ≤ 0.05) as determined by Duncan's multiple range test.

Carry-over studies. *Microplot trial.* Plots (4.6m²
with sixteen 2-m rows) were sown in May 1980 with the
crops listed in Table 2. Cropping history for the site was
wheat, medic pasture, medic pasture, and fallow
(August) in 1976, 1977, 1978, and 1979, respectively. The
medic pasture also contained barley grass and ryegrass
(*Lolium rigidum* Gaud.). There were few, if any, grass
weeds in the fallow. Initially, oats were sown in the
fallow plots and desiccated with a herbicide (glyphosate)
in August 1980 so that soil moisture was similar to that
on land fallowed in 1980. Two treatments were
incorporated in the medic rotation: in the first, weeds
were controlled by hand-weeding and herbicides; in the
second, no attempt was made to control weeds.
Experimental design was a randomized block with five
replications of each treatment. At maturity (November),
cereal plots were harvested and all others were mown. In
June 1981, the entire site was sown with wheat; at
flowering (October), 15 plants per plot were assessed for
take-all. At maturity (November), 15 plants per plot were
assessed for deadheads (containing little or no grain),
and grain was harvested in December.

Long-term rotation trials. The trial was established at
Mallee Research Station in May 1980 to study the effects
of 12 different cropping sequences on take-all. Cropping
history for the trial site was medic pasture (containing
some barley grass and ryegrass), fallow (containing
very few grass weeds), and wheat in 1977, 1978, and
1979, respectively. The trial consisted of three series of
experiments arranged sequentially so that different
phases of each cropping sequence were represented each
year. Those sequences within each series with a wheat
phase in 1981 or 1982 are shown in Tables 3 and 4. Plots

were 210 m² with sixteen 60-m rows, and the design for
each experiment was randomized block with three
replications of each cropping sequence. At flowering
(October), 10 samples of 30 cm of drill row were taken
from each wheat plot using a stratified sampling
procedure, and the roots were assessed for take-all. Plots
were harvested at maturity (November) and yields
were measured.

RESULTS

Survival study. At 5 weeks, there was no significant
difference in the survival of *G. graminis* var. *tritici* in the
various treatments (Table 1). At 10 weeks, however,
survival was greater ($P \leqslant 0.05$) in unplanted pots than in
pots with either oats, rapeseed, or field peas. This effect
increased with time so that at 23 weeks, survival was
greater in the absence of plants than under any of the
test crops. However, there was no significant difference
between survival under the various crops. Wheat plants
developed severe take-all symptoms, and lesions were
also present on oats and rapeseed. *G. graminis* var.
tritici was isolated from all three hosts.

Carry-over studies. In microplots, both the
incidence and severity of take-all were higher ($P \leqslant 0.05$)
when crops of wheat, barley, or triticale preceded wheat
than when the preceding phase was a fallow (Table 2).
Lupine reduced ($P \leqslant 0.05$) the incidence of disease more
than fallow. Neither disease incidence nor severity was
increased significantly by the presence of weeds in the
medic pasture, the most prevalent of which were
ryegrass and barley grass. Grain yields were largest
when either fallow or noncereals preceded wheat and
smallest when the preceding crops were triticale or
wheat. Deadheads were most prevalent following
triticale.

Results obtained in 1981 from the long-term rotation
trial are shown in Table 3. In series 1, fewer ($P \leqslant 0.05$)
wheat plants had take-all when grown after lupine than
after wheat, and there were more ($P \leqslant 0.05$) severely
affected plants after wheat than after either lupine or
medic pasture. In series 2, the incidence of take-all was
highest ($P \leqslant 0.05$) following a wheat phase and the
severity of infection was greater ($P \leqslant 0.05$) in wheat
after wheat than after fallow, rapeseed, lupine, or medic
pasture. In series 3, take-all was more severe ($P \leqslant 0.05$)

Table 3. Effects of cropping sequence on take-all and wheat yield in 1981
in a long-term rotation trial at Mallee Research Station

Preceding crop (1980)	Percentage of plants with take-all	Percentage of plants with affected root area of 51–100%	Grain yield (kg/ha)
	Series 1		
Wheat	66.4 a[y]	14.3 a	1,452 b
Lupine	31.5 b	0.6 b	2,571 a
Medic	39.4 ab	1.2 b	2,476 a
	Series 2		
Nil (fallow)	16.2 b	0.8 b	2,929 a
Wheat	82.2 a	33.7 a	1,500 c
Lupine	27.2 b	0.6 b	2,857 ab
Rapeseed	42.7 b	3.3 b	2,508 b
Medic	27.2 b	0.8 b	2,825 ab
	Series 3		
Wheat	95.5 a	41.6 a	675 b
Lupine (i)[z]	56.8 b	3.2 b	2,127 a
(ii)	57.8 b	2.1 b	2,198 a
(iii)	43.2 b	5.1 b	2,103 a
Rapeseed (i)[z]	66.2 ab	4.6 b	1,873 a
(ii)	57.4 b	4.3 b	1,976 a
Field peas	51.4 b	1.2 b	2,040 a
Medic	45.6 b	4.6 b	2,198 a

[y]Within each series, results in one column with no letter in common are
significantly different ($P \leqslant 0.05$) as determined by Duncan's multiple
range test.
[z]Each lupine sequence (i, ii, iii) and each rapeseed sequence (i, ii)
identical until 1982 phase.

Table 4. Effects of cropping sequence on take-all and wheat yield in 1982
in a long-term rotation trial at Mallee Research Station

Preceding crops 1980	1981	Percentage of plants with take-all	Grain yield (kg/ha)
	Series 2		
Wheat	Wheat	30.5 a[z]	55 a
Wheat	Medic	8.3 b	84 a
Wheat	Lupine	11.8 b	118 a
	Series 3		
Wheat	Wheat	30.0 a	51 a
Medic	Lupine	17.9 ab	106 a
Medic	Medic	16.3 ab	82 a
Medic	Nil (fallow)	20.9 ab	125 a
Medic	Rapeseed	11.0 b	60 a

[z]Within each series, results in one column with no letter in common are
significantly different ($P \leqslant 0.05$) as determined by Duncan's multiple
range test.

following wheat than after other crops. Where wheat followed wheat, most plants were moderately to severely affected by take-all and grain yields were correspondingly smaller.

In 1982 there were severe drought conditions in the Mallee, as evidenced by very low grain yields in the long-term rotation trial (Table 4). Although wheat roots were infected, the severity of attack was low compared with 1981. Thus, only the incidence of take-all is shown in Table 4. In series 2, break crops of medic pasture and lupine markedly reduced ($P \leqslant 0.05$) disease incidence. In series 3, the only break crop sequence that caused a significant reduction in take-all relative to the level in the continuous wheat rotation was medic pasture-rapeseed. The level of take-all after other sequences (medic pasture-lupine, medic pasture-medic pasture, and medic pasture-fallow) was not significantly different from that in continuous wheat, but this was caused in part by the wide variation between replications. There was little correlation between grain yield and take-all in series 2 and 3, possibly because of the drought.

DISCUSSION

The greater survival of *G. graminis* var. *tritici* in unplanted pots than in pots with various plant species substantiates the earlier results of Chambers (5) and suggests that crops of oats, rapeseed, lupine, field peas, and medic may be more effective than a fallow phase in reducing carry-over of *G. graminis* var. *tritici* in the Mallee.

Possible explanations for the reduced survival of *G. graminis* var. *tritici* under the various crops are that they stimulated a microflora unfavorable to the pathogen (10), exuded antifungal compounds from their roots (1), or depleted levels of soil nitrate (6). It is also possible that root exudates, by inducing activity in the fungus, may have decreased its survival where it was unable to infect a suitable host. Under wheat, and possibly to a lesser extent under rapeseed and oats, the new infections that were detected may have led to an increase in inoculum. Chambers (5) also isolated *G. graminis* var. *tritici* from supposedly nonsusceptible plant species, and this confirms the need for further studies on their role in perpetuating the disease.

Results from the field trials (Tables 2–4) clearly demonstrate the effectiveness of break crops and clean fallowing as control measures for take-all in the Mallee. Break crops are more attractive to farmers than a fallow phase because they generate income and, if legumes, also supply nitrogen to the soil. In some instances, farmers would prefer to grow successive crops of wheat or to alternate wheat with barley or triticale. However, until resistant cultivars or effective fungicides are

developed, such systems will not be feasible in areas prone to take-all.

The higher yields obtained when either noncereals or fallow preceded wheat can be attributed to control of take-all. Wheat yields following rapeseed were not significantly different from those after lupine, field peas, or medic, suggesting that the major effect of the legumes was to break the disease cycle of *G. graminis* var. *tritici* rather than to increase soil nitrogen. In addition, yields after the various break crops were generally similar to those after a fallow phase, indicating that the yield advantage over continuous wheat was the result of control of take-all rather than a buildup of soil moisture and mineral nitrogen during the fallow phase. The marked reduction in disease severity following noncereals or fallow is also strong evidence that these phases increased yields by controlling take-all. Because the break crops and fallow had more pronounced effects on disease severity than on disease incidence, it appears that they reduced the inoculum potential (2) of *G. graminis* var. *tritici*.

ACKNOWLEDGMENT

Funds for this investigation were provided by the Wheat Industry Research Committee of Victoria.

LITERATURE CITED

1. Baker, K. F., and Cook, R. J. 1974 (original ed.). Biological Control of Plant Pathogens. Reprint ed., 1982. American Phytopathological Society, St. Paul, MN. 433 pp.
2. Baker, R. 1978. Inoculum potential. Pages 137-157 in: Plant Disease, An Advanced Treatise. Vol. 2. J. G. Horsfall and E. B. Cowling, eds. Academic Press, New York.
3. Butler, F. C. 1953. Saprophytic behaviour of some cereal root-rot fungi. III. Saprophytic survival in wheat straw buried in soil. Ann. Appl. Biol. 40:305-311.
4. Chambers, S. C. 1970. Pathogenic variation in *Ophiobolus graminis*. Aust. J. Biol. Sci. 23:1099-1103.
5. Chambers, S. C. 1971. Some factors affecting the survival of the cereal root pathogen *Ophiobolus graminis* in wheat straw. Aust. J. Exp. Agric. Anim. Husb. 11:90-93.
6. Chambers, S. C., and Flentje, N. T. 1969. Relative effects of soil nitrogen and soil organisms on survival of *Ophiobolus graminis*. Aust. J. Biol. Sci. 22:275-278.
7. Garrett, S. D. 1938. Soil conditions and the take-all disease of wheat. III. Decomposition of the resting mycelium of *Ophiobolus graminis* in infected wheat stubble buried in the soil. Ann. Appl. Biol. 25:742-766.
8. Kollmorgen, J. F., Griffiths, J. B., and Walsgott, D. N. 1983. The effects of various crops on the survival and carry-over of the wheat take-all fungus *Gaeumannomyces graminis* var. *tritici*. Plant Pathol. 32:73-77.
9. Sims, H. J., Meagher, J. W., and Millikan, C. R. 1961. Deadheads in wheat—field studies in Victoria. Aust. J. Exp. Agric. Anim. Husb. 1:99-108.
10. Zogg, H. 1969. Crop rotation and biological soil disinfection. Qual. Plant Mater. Veg. 18:256-273.

Susceptibility of Apple Trees to *Phytophthora cactorum* and Effect of Systemic Fungicides

PETER G. LONG, Department of Horticulture and Plant Health, Massey University, Palmerston North, New Zealand

The 1974 Apple and Pear Disease Workers workshop on crown rot of apple trees outlined a number of areas in which further work on this disease was required (7). One was the elucidation of factors that predispose trees to infection and the effects of such factors on rootstock susceptibility. Sewell and Wilson (8) suggested, "Among the factors that could contribute to crown rot the influence of the scion variety on rootstock resistance should not be overlooked." Their suggestion does not appear to have been followed up, although the influence of the rootstock on scion susceptibility has been investigated (7,8). A second area was a means of protecting susceptible rootstocks. No mention was made of chemical protection, presumably because suitable systemic fungicides were not then available.

In the work described below, a laboratory technique was used to assess growth of *Phytophthora cactorum* in apple tissues. The method was used to investigate the effect of scion variety on rootstock resistance and was also a useful technique for testing the in vivo activity of systemic fungicides that had been applied topically to the intact bark.

MATERIALS AND METHODS

Tissue susceptibility was tested using Long's (5,6) modification of Borecki and Millikan's (2) technique. Pieces of stem or root 10 cm long were slit longitudinally down each side, and the outer layers were peeled off to expose the xylem on the core and the phloem/cambium tissue on the inside of the bark. The xylem and phloem tissues were inoculated with 5-mm-diameter plugs cut from cultures of *P. cactorum* (isolate P149) grown on Difco cornmeal agar and were incubated in humid chambers for 4 days at 20°C. The disease lesions were measured, and results are presented as lesion size less inoculum plug.

For an evaluation of the effect of scion variety on rootstock resistance, Malling-Merton 106 was used as rootstock for four scion varieties: Cox, Oratia Beauty, Splendour, and Granny Smith. The trial was a randomized block design with five trees per plot, four plots per scion variety. Trees were planted in July 1980 at a spacing of 2 m between trees and 3 m between rows. Half the trees in each treatment were assessed on 4 November 1982 (spring) and half on 3–4 February 1983 (late summer). Each tree was dug out of the ground, washed, stripped of leaves, and separated into previous season's growth, stem immediately above the graft union, stock, and roots. Four replicate pieces were taken from each of the four regions for susceptibility testing.

In the fungicide trials, fungicide was applied neat or diluted with water to a 10-cm length of previous season's shoot. After 2 days the shoot was cut from the tree, deleafed, and lightly rubbed under running water to remove fungicide remaining on the outer surface. Fungicides used were etridiazole (Terrazole, 25 EC), metalaxyl (Apron 35), and pyroxyfur (Dowco 444, 72EC). The trees were 6-year-old cultivar Oratia Beauty on Malling-Merton 106 stocks, except in the first trial, where the scion Gala was used for the full-strength fungicide treatments.

There were three sets of trials. In tests of fungicide penetration, the treated area and the next 10-cm piece above and below were prepared, inoculated, and tested for lesion development as described above. There were six replicates per treatment in each of seven trials (Table 1). In tests of translocation, four days after shoots were treated at the tips they were cut into the treated length plus the next four 10-cm lengths toward the base to test for basipetal translocation. There were six replicates per treatment. A corresponding set of shoots was treated at the base to test for acropetal translocation (Table 2). In tests of postinoculation treatments, 30-cm lengths of previous season's shoots were cut from the tree, deleafed, and inoculated by inserting a plug of cornmeal agar inoculum into a 5-mm-diameter hole cut with a cork borer. After 4 days of incubation at 20°C, the visible lesion length was measured and the fungicide applied to the intact bark. Lesion length was remeasured after a further 7 days of incubation at 20°C (Table 3).

RESULTS

There were no significant differences between the susceptibilities of the four scion varieties. Similarly, there were no differences between their rootstocks. However, roots were consistently less susceptible than stocks and stems (Table 4). Previous season's shoots were less susceptible in February than in November.

There was little or no lesion development in the xylem or phloem under the site of inoculation when fungicides were applied neat. A similar pattern of results was obtained from a series of trials in which two of the fungicides were tested at successively lower concentrations (Table 1). The etridiazole treatment caused extensive browning of tissues immediately under the area treated. Lesion development was inhibited in xylem and phloem above and below the treatment site, indicating both acropetal and basipetal movement of fungicides in stem tissues. Metalaxyl was the most effective of the fungicides.

Table 1. Effects of topical fungicides applied in a 10-cm band to previous season's shoots in tests of susceptibility of apple trees to *Phytophthora cactorum*

Variety Strength	Fungicide	Treatment[z] Above Xylem	Above Phloem	Site Xylem	Site Phloem	Below Xylem	Below Phloem
Gala	Untreated	26	25	30	25	37	25
35 EC (Full)	Metalaxyl	0	0	0	0	0	0
35 EC (Full)	Pyroxyfur	3	7	0	1	6	4
25 EC (Full)	Etridiazole	0	0	0	0	0	14
SD		5	7	7	7	6	8
Oratia Beauty	Untreated	28	24	14	18	5	23
1/2	Pyroxyfur	0	0	0	0	0	3
1/2	Metalaxyl	0	0	0	0	0	0
SD		5	6	5	5	0	7
	Untreated	19	32	3	5	8	37
1/4	Pyroxyfur	14	13	8	0	15	18
1/4	Metalaxyl	0	0	0	0	0	0
SD		7	9	3	2	5	9
	Untreated	10	36	10	49	19	52
1/8	Pyroxyfur	4	12	0	0	5	18
1/8	Metalaxyl	0	1	0	0	0	1
SD		6	9	7	4	9	7
	Untreated	29	22	26	21	25	28
1/16	Pyroxyfur	0	11	1	0	14	10
1/16	Metalaxyl	0	0	0	0	0	0
SD		8	7	4	13	11	12

[z]Treated section of the shoot plus the sections immediately above and below were cut from the tree after 4 days and tested for growth of *P. cactorum*. Mean of six replicates per treatment.

Table 2. Lesion size (mm) on twigs treated at base or tip with fungicide and separated into five pieces (base, 2, 3, 4, and tip) each 10 cm long 2 days after treatment[z]

Treatment	Base	2	3	4	Tip	SD
At base						
Metalaxyl	0	1	0	0	0	8
Pyroxyfur	0	0	0	10	14	
Control	10	18	8	12	14	
At tip						
Metalaxyl	1	3	1	2	0	8
Pyroxyfur	9	7	3	1	0	

[z]Measurements made after 4 days of incubation at 20°C. Mean of six replicates per treatment.

Table 3. Increase in lesion size 7 days after fungicide treatment[x]

Fungicide	Fungicide strength[y] Full	1/2	1/4	1/8	1/16
Metalaxyl	14	5	12	7	17
Pyroxyfur	133	111	155	178	167
Etridiazole[z]	6	6	53	112	82

[x]Results of four separate trials expressed as percentage of control treatments for comparison. Ten replicates per treatment.
[y]Metalaxyl and pyroxyfur, 35%; etridiazole, 25%.
[z]Some browning on most replicates, but no subsequent recovery of *Phytophthora cactorum*.

Similar evidence of acropetal and basipetal movement was obtained in a further trial where shoots were treated at the base or at the tip (Table 2). In both cases, movement of metalaxyl was more extensive than that of pyroxyfur. There was no unlignified xylem in the shoots when this trial was set up, so results are limited to phloem tissue only.

Metalaxyl was very effective in preventing further spread of established lesions. Etridiazole was effective at the higher concentrations but was frequently phytotoxic. Attempts to reisolate *P. cactorum* from brown areas were unsuccessful. Pyroxyfur had no controlling effect in this trial; indeed, lesions developed faster than in untreated controls.

DISCUSSION

Although scion varieties were chosen to represent both susceptible and resistant cultivars, there was no significant difference in susceptibility of the scions in this trial and consequently no detectable effect on rootstock resistance. Long (5) found that the most susceptible tissue in previous season's shoots in spring was nonlignified xylem and in late summer was phloem. This pattern was not as obvious in this trial partly because the spring assessment was later (end of bloom) and partly because there was still a thin layer of nonlignified xylem in February. Bielenin (1) found the greatest spread of varietal susceptibility between pink tip and end of bloom and considered this the best time for inoculations. However, sap movement in the xylem is acropetal and in the phloem is predominantly basipetal. If phloem is the major susceptible tissue at the time of the second peak of susceptibility, then that is when the scion could have most influence on rootstock susceptibility. Any future work on this topic should thus include both spring and late-summer inoculations. Roots were consistently less susceptible than stocks. McIntosh (7) noted that with the extensive use of Malling-Merton rootstocks, many disease problems now involve infections of rootstock tissue below soil level (crown rot); in the author's experience, crown rot is more common than root rot in New Zealand.

The use of the laboratory resistance testing method for evaluating movement and activity of fungicides in vivo gave consistent, clear-cut results. All the fungicides tested prevented lesion development under the site of application, although etridiazole was frequently phytotoxic at the concentrations used. There was evidence that all fungicides tested moved in both an acropetal and a basipetal direction. The detection of

Table 4. Susceptibility to *Phytophthora cactorum* of xylem and phloem of four scion varieties[x] on MM 106 rootstock in spring (November) and late summer (February) (lesion length in millimeters)

Site	Xylem					Phloem				
	OB	Cox	SPL	GS	Mean	OB	Cox	SPL	GS	Mean
				November						
PS[y]	59 c[z]	45 b	48 b	51 bc	51 c	49 b	46 b	42 b	44 b	45 b
Stem	51 c	43 b	37 b	56 c	47 bc	42 ab	47 c	35 ab	46 b	43 b
Stock	40 b	30 ab	41 b	46 b	39 b	32 a	33 ab	38 ab	43 b	37 ab
Roots	20 a	21 a	21 a	18 a	20 a	36 ab	29 a	29 a	31 a	31 a
				February						
PS	32 ab	29 a	35 a	39 b	34 ab	37 b	33 ab	29 a	29 a	32 a
Stem	42 b	42 ab	55 b	63 c	51 c	40 b	40 ab	37 ab	48 b	41 a
Stock	34 ab	52 b	56 b	59 c	50 c	19 a	42 b	45 b	52 b	40 a
Roots	23 a	25 a	29 a	21 a	25 a	34 ab	27 a	41 ab	35 a	34 a

[x]OB = Oratia Beauty, SPL = Splendour, and GS = Granny Smith.
[y]Previous season's shoots.
[z]Any two numbers in the same column followed by the same letter are not significantly different at the 5% level of Duncan's multiple range test.

basipetal movement was consistent but unexpected because (with the exception of ethyl hydrogen phosphonate) systemic fungicides are considered to show acropetal movement only (4). Metalaxyl was the most effective fungicide tested.

Brown and Hendrix (3) used fungicide trunk sprays and soil drenches but were unable to obtain field control of *P. cactorum*. These trials suggest that it may now be possible to use modern systemic fungicides as a prophylactic treatment during periods of favorable environmental conditions or as eradicants if the infected trees are identified at an early stage of disease development.

ACKNOWLEDGMENTS

The author thanks L. Davis, K. Kelliher, and H. Neilson for invaluable technical assistance.

LITERATURE CITED

1. Bielenin, A. 1977. Collar rot of apple trees. II. Seasonal fluctuations in susceptibility of apple trees to *Phytophthora cactorum* (Leb. et Cohn) Schroet. Fruit Sci. Rep. 4:27-39.
2. Borecki, Z., and Millikan, D. F. 1969. A rapid method for determining the pathogenicity and factors associated with the pathogenicity of *Phytophthora cactorum*. Phytopathology 59:247-248.
3. Brown, E. A., II, and Hendrix, F. F. 1980. Efficacy and in vitro activity of selected fungicides for control of *Phytophthora cactorum* collar rot of apple. Plant Dis. 64:310-312.
4. Erwin, D. C. 1977. Control of vascular pathogens. Pages 162-224 in: Antifungal Compounds. Vol. 1. M. R. Siegal and H. D. Sisler, eds. New York, Dekker.
5. Long, P. G. 1978. Apple tree resistance to collar rot disease. N. Z. J. Agric. Sci. 16:54-56.
6. Long, P. G. 1982. Testing of systemic fungicides against *Phytophthora cactorum*. Proc. N. Z. Weed Pest Control Conf., 35th, 265-268.
7. McIntosh, D. L., ed. 1975. Proceedings of the 1974 APDW workshop on crown rot of apple trees. Can. Plant Dis. Surv. 55:109-116.
8. Sewell, G. W. F., and Wilson, J. F. 1973. Phytophthora collar rot of apple: Influence of the rootstock on scion variety resistance. Ann. Appl. Biol. 74:159-169.

Enhanced Suppression of Take-all Root Rot of Wheat with Chloride Fertilizers

R. L. POWELSON, Department of Botany and Plant Pathology; and T. L. JACKSON and N. W. CHRISTENSEN, Department of Soil Science, Oregon State University, Corvallis, OR 97331, U.S.A.

Suppression of root attack by the wheat take-all fungus (*Gaeumannomyces graminis* var. *tritici*) with ammonium nitrogen (NH_4-N) fertilizers is well documented (28). Disease suppression with NH_4-N was first observed in field experiments conducted by Garrett (18,19) and has since been demonstrated in a wide variety of soils that differ in their chemical and physical properties (28,29,38,58,61,66,68). However, only recently has this knowledge been applied to commercial wheat production. Based on extensive field experiments conducted in Western Australia (36–38), the use of ammonium sulfate fertilizers drilled at planting is now recommended as a management option to reduce crop losses from take-all. Huber (27) and Huber et al (30) demonstrated that the nitrification inhibitor nitropyrin improved suppression of take-all with NH_4-N fertilizers. This practice is now used by growers in Indiana (D. M. Huber, *personal communication*) for management of take-all. Since 1979, use of NH_4-N with chloride (Cl) has increased in Oregon and is now an accepted practice in fields with a high risk of take-all (34).

Cook and Reis (16) discussed strategies involving water management, tillage, nitrogen, phosphate (P), and trace element nutrition to control take-all in the Pacific Northwest of the United States. The proposed practices of tillage to fragment and bury infected crowns, irrigation that allows the soil surface to dry between applications, use of NH_4-N, and use of P and trace element fertilizers are valid concepts. However, specific recommendations for take-all risk situations need to be developed for grower use.

CHLORIDE RESPONSES

Chloride has no effect on the growth of *G. graminis* var. *tritici* in vitro when used as an osmoticum (13,14) and does not affect take-all development in plants supplied with NO_3-N (25,46). However, nitrification rates are reduced when Cl is supplied with NH_4-N in soil (1,21,23,24). Chloride is also a competitive inhibitor of NO_3-N uptake (63). Both effects would enhance the intensity and duration of NH_4-N plant nutrition.

Our research on enhanced NH_4-N suppression of take-all by Cl began with a field experiment in the autumn of 1976. Different forms of nitrogen, including ammonium chloride (NH_4Cl), were banded with the seed (45). Potassium chloride (KCl) was included because Jackson et al (33) had observed reductions in disease caused by

barley yellow dwarf virus where KCl was added to N, P, S fertilizer banded with the seed at planting in western Oregon. The experiment was conducted on land that had previously been cropped to cereal and legumes. The field had been fallowed in 1974–75 and cropped to winter wheat in 1975–76. The autumn banded N rate was 22 kg of N per hectare, and the rate of KCl was 35 kg/ha (23-cm row spacing). An additional 135 kg of N per hectare was topdressed in early spring (late February or early March, prior to jointing). During the boot-dough growth stages, plots where Cl was used in combination with NH_4-N were easily identified. The plants were greener and taller where Cl was used, with less senescence of foliage and fewer whiteheads. Fresh weights of plants at this growth stage were greater where both NH_4-N and Cl were applied as opposed to NH_4-N alone (Table 1).

Since 1976, over 100 different field experiments have been conducted in seven western Oregon counties to determine under what cropping sequence and management practices a wheat grower could expect to benefit by delayed seeding; banding NH_4-N, P, and Cl fertilizer with the seed; and topdressing with Cl in the spring. A combination of the above practices has consistently improved grain yield and reduced severity of root attack in take-all risk fields (Tables 2, 3, and 4). Thirty experiments on growers' fields where wheat followed wheat during crop years 1979, 1980, 1981, and 1982 showed an average yield increase of ≈807 kg/ha where Cl was applied in both autumn and spring (Jackson and Powelson, 1979–82, *unpublished*). Results from these experiments have encouraged growers to apply Cl on a commercial basis. A survey conducted by

Table 1. Effect of different nitrogen sources and banded KCl on fresh weight of wheat infected with take-all[x]

Treatment[y]		Plant fresh weight
Autumn	Spring	(g/120 cm of row)
AmCl + KCl	AmCl	702 c[z]
AmCl	AmCl	470 bc
AmS + KCl	AmS	500 bc
AmS	AmS	260 ab
CaN + KCl	CaN	274 ab
CaN	CaN	160 a

[x] At soft-dough growth stage. North Willamette Experiment Station, 1977.

[y] AmCl = NH_4Cl, AmS = $(NH_4)_2SO_4$, and CaN = $Ca(NO_3)_2$. Autumn fertilizer banded with seed at planting in October at rate of 22 kg of N/ha, 22 kg of P/ha, and 35 kg of KCl/ha. Spring fertilizer topdressed in March at rate of 135 kg of N/ha.

[z] Means with same letter were not significantly different ($P = 0.05$) according to Duncan's multiple range test.

Oregon Agricultural Experiment Station Technical Paper 7153.

246

the Oregon State University Survey Research Center in 1982 has shown that Cl was applied to 55% of the fields in Polk and Yamhill counties where wheat was grown following wheat. For the 1982 crop year, growers who applied Cl at planting and in the spring reported yields that averaged 940 kg/ha more than those of growers not applying Cl.

In western Oregon, the greatest risk of crop losses from take-all usually occurs in the second, third, and fourth years of consecutive wheat production. In first-year fields, where wheat immediately follows a nonhost crop, the banding of NH₄-N with Cl and spring topdressing of Cl rarely gives a significant yield increase.

The practices of delayed seeding, banding fertilizers, and spring topdressing pose certain risks as well as benefits to wheat production (Table 5). A major benefit of delayed seeding is escape from autumn infection by the barley yellow dwarf virus because of a decline in aphid populations and delayed onset of infection by *G. graminis* var. *tritici, Pseudocercosporella herpotrichoides, Septoria* spp., and *Puccinia striiformis*. Late seeding, when the soil is likely to be cool and wet, also reduces the risk of delayed emergence or stand reduction from banded fertilizer salts. Other late-seeding risks are

a reduced yield potential and having to seed under unfavorable wet soil conditions (34).

RESPONSES IN THE RHIZOSPHERE

The importance of balanced plant nutrition to reduce crop losses from take-all has been reviewed by Huber (28). The need for adequate nitrogen has long been recognized (19), and Reis et al (47) have demonstrated that deficiencies of P as well as the micronutrients Zn,

Table 2. Yield of winter wheat (cv. Stephens) infected with *Gaeumannomyces graminis* var. *tritici* in response to seeding date, ammonium nitrogen, chloride, and phosphorus fertilizers[w]

Treatment[x]		Seeding date	Grain yield (kg/ha)
Autumn	Spring		
Hyslop Agronomy Farm			
AmS	AmS	15 Oct	3,898 a[y]
AmCl	AmCl	15 Oct	4,369 b
AmS	AmS	29 Oct	4,231 ab
AmCl	AmS	29 Oct	5,379 c
North Willamette Experiment Station			
AmS	AmS	10 Oct	3,160 a
AmS + P	AmS	10 Oct	3,631 a
AmCl + P	AmCl[z]	10 Oct	5,648 b

[w] During 1980–81 crop year at two locations in western Oregon.
[x] AmS = (NH₄)₂SO₄, AmCl = NH₄Cl, and P = Ca(H₂PO₄)₂. Autumn fertilizer banded with seed at planting at rate of 22 kg of N/ha and 22 kg of P/ha. Spring fertilizer topdressed in March at 146 kg of N/ha.
[y] Means within each location with the same letter were not significantly different (*P* = 0.05) according to Duncan's multiple range test.
[z] Split spring topdress application of AmCl on 1 February and 18 March.

Table 3. Yield of winter wheat (cv. Stephens) infected with *Gaeumannomyces graminis* var. *tritici* in response to ammonium chloride, ammonium sulfate, and phosphorus fertilizer[w]

Treatment[x]		Location[y] and crop year		Grain yield (kg/ha)	
Autumn	Spring	Hyslop 1979–80	Hyslop 1980–81	NWES 1980–81	NWES 1981–82
AmCl + P	AmCl	6,971 a[z]	...	5,648 a	6,253 a
AmCl	AmCl	6,186 ab	5,379 a	...	...
AmS + P	AmS	5,379 c	...	3,631 b	5,245 b
AmS	AmS	5,782 bc	4,236 b	3,160 b	...

[w] During three crop years (1979–80 to 1981–82) at two locations in western Oregon.
[x] AmCl = NH₄Cl, AmS = (NH₄)SO₄, and P = Ca(H₂PO₄)₂. Autumn fertilizer banded with seed at planting at rate of 22 kg of N/ha and 22 kg of P/ha. Spring fertilizer topdressed in March at 146 kg of N/ha.
[y] Hyslop = Hyslop Agronomy Farm and NWES = North Willamette Experiment Station.
[z] Means within columns with the same letter were not significantly different (*P* = 0.05) according to Duncan's multiple range test.

Table 4. Comparisons of take-all root attack of winter wheat (cv. Stephens) in response to different combinations of fertilizers banded with seed at sowing in autumn and topdressed in spring[x]

Treatment comparison[y]		Root attack (%)
Autumn	Spring	
AmCl + P	AmCl vs.	11***[z]
AmS + P	AmS	34
P	...	30***
Nil P	...	46
Cl	Cl vs.	28**
Cl	nil Cl	42
Cl	Cl vs.	28**
Nil Cl	nil Cl	40

[x] At North Willamette Experiment Station, 1981.
[y] AmCl = NH₄Cl, P = Ca(H₂PO₄)₂, AmS = (NH₄)₂SO₄, and Cl = AmCl, KCl, or CaCl₂ as sources of chloride banded with seed at planting at rate of 22 kg of N/ha. Spring fertilizer topdressed in March at 146 kg of N/ha and 185 kg of Cl/ha.
[z] Means of treatment comparisons were significantly different if followed by ** (*P* < 0.01) and *** (*P* < 0.001).

Table 5. Risks and benefits from delayed autumn seeding, banding NH₄-N, phosphorus, and chloride fertilizer salts, and spring-topdressed chloride fertilizers

Practice	Risks	Benefits
Delayed seeding	Lower yield potential Heavy, late-fall rains make seeding difficult Small seedlings more susceptible to winter damage and herbicide injury Increased soil erosion from winter rains	Escape fall infection by barley yellow dwarf virus Reduced severity of attack by *Gaeumannomyces graminis* var. *tritici* and *Pseudocercosporella herpotrichoides* Delay onset of Septoria leaf and glume blotch and stripe rust
Autumn-banded fertilizers	Delayed emergence or stand reduction if salt is placed with seed when soil is dry or warm Delayed emergence would increase delayed seeding risks Increased severity of root attack by *Fusarium* and *Rhizoctonia*	Reduced severity of root attack by *G. graminis* var. *tritici* P stimulates root and shoot growth in cool, wet soils Delayed emergence would give benefits similar to delayed seeding
Spring-topdressed chloride	Requires extra trip over field	Reduced severity of attack by *G. graminis* var. *tritici, Septoria* spp., and *Puccinia striiformis* Results in higher levels of NH₄-N in soil and may reduce leaching losses of NO₃-N

Cu, Mn, and Fe can increase the severity of root attack by *G. graminis* var. *tritici*. The hypothesis concerning form of nitrogen, as presented by Huber and Watson (31), certainly applied to take-all. Most workers testing this hypothesis on take-all (25,29,38,61,68) have reported therapeutic response from the uptake of NH4-N, especially in the presence of P (5,47,67) and Cl (9,25,45,69). However, as Huber (28) has pointed out, NH4-N is likely to be less effective in alkaline or warm soils and when there are high amounts of carry-over NO3-N from previous crops.

An adequate supply of NH4-N can increase P uptake in seedlings, which promotes root growth in cool, wet soils (43). However, when Cl is included, additional P may be required. An examination of data presented by Stynes (64) from an extensive synoptic study of wheat production in South Australia shows that low levels of take-all were associated with sites having high average salinity, high topsoil water-holding content, and increased incorporation of Cl in plants.

The rhizosphere pH (pHr) tends to be lowered when plants absorb NH4-N and raised when NO3-N is the dominant ion absorbed (44). This change in pHr has been clearly demonstrated by Marschner et al (40). It was this lowering of pHr with uptake of NH4-N that attracted the attention of plant pathologists working with take-all (25,56,57,61). The mechanisms proposed are a shift in pHr, producing a change in the composition or activity of the rhizosphere microflora suppressive to *G. graminis* var. *tritici*; and a lower pH decreasing the growth of runner hyphae along the root (12).

The mechanisms involved may be different for different pathogens; the contrast between *G. graminis* var. *tritici* and certain root-infecting *Fusarium* spp. is of particular interest. Although the severity of take-all is reduced by NH4-N and enhanced by NO3-N, the opposite is reported for Fusarium diseases (32,41,50,62). Recently a mechanism of competition for iron (Fe) induced by siderophore-producing bacteria has been proposed (35,53,55) to explain soil suppressiveness to certain *Fusarium* pathogens. The evidence suggests that siderophore-producing bacteria will complex Fe in the rhizosphere, thus limiting its availability to the pathogen. Competition for Fe would be reduced by lowering pHr through uptake of NH4-N (39) and could explain why NH4-N increases severity of attack by certain soilborne *Fusarium* pathogens. Because suppression of take-all is enhanced by NH4-N fertilizers, it seems unlikely that competition for Fe between suppressive organisms and the pathogen is the mechanism involved.

There are several similarities between the take-all decline phenomenon and suppression of take-all by NH4-N. Both occur only in natural soils (25,58) and are eliminated by soil treatments that destroy "general" or specific antagonism as defined by Rovira and Wildermuth (49). Halsey (25) found that suppression of take-all by NH4-N was reduced or eliminated by soil treatment with methyl bromide or moist heat at 60° C for 30 min. Both take-all decline and NH4-N suppressiveness are associated with biological activity. The group of organisms that have received the most attention are the fluorescent *Pseudomonas* spp. (25,45,56,57,59). Some of these bacteria are efficient soil (48,52) and root colonizers (65), are acid tolerant, and produce antibiotics

(7,55,59,60,71). Some of these have been tested for biological control of take-all by coating them on seed (15,25,59,72) and as plant-growth-promoting rhizobacteria (PGPR) (55).

Brown (3) made a general classification of the *Pseudomonas* spp. and other bacteria associated with take-all lesioned and nonlesioned wheat roots and concluded that the data "provided little support for a hypothesis involving these organisms in disease suppression." Schroth and Hancock (54) pointed out that in attempts to determine biologically suppressive associations of microorganisms, specific strains—rather than just the genus and species—need to be identified.

Gerlagh (20), Zogg and Jaggi (76), and Wildermuth (74) found that the biological suppressive factors presumably responsible for take-all decline could be induced by adding *G. graminis* var. *tritici* inoculum to soil but not by adding noninfected wheat roots. Rovira and Wildermuth (49) proposed that the suppression is caused by buildup of antagonistic bacteria associated with the infection propagules and that their activity reduced the number of *G. graminis* var. *tritici* hyphae reaching the root. They proposed that "some of the action is beyond the rhizosphere" because there was reduced colonization of the root by *G. graminis* var. *tritici* hyphae when suppressive soil and inoculum were 5–8 mm from the root. The activity could still be within the rhizosphere, as Gilligan (22) has calculated a sphere of influence of wheat roots with respect to *G. graminis* var. *tritici* of about 12 mm. We suggest that suppression or take-all decline is under the influence of the rhizosphere and can be enhanced by plant uptake of NH4-N. Although previously infected parts of the wheat plant may serve as a source of both *G. graminis* var. *tritici* and suppressive bacteria, wheat root exudates would provide the substrates and environment necessary to stimulate both the pathogen and suppressive organisms. The rhizosphere, not the inoculum source (previously colonized plant parts), would be the site of active suppression of *G. graminis* var. *tritici*. Recent evidence supporting the concept that the rhizosphere is the site of suppression is the work with PGPR (55). A characteristic of these bacteria is that they can be used for bacterization and will aggressively colonize the plant roots (4,55,65). Weller and Cook (72), using antibiotic markers, have demonstrated that strains of bacteria, effective as a seed coating for suppression of take-all, were good root colonizers.

PLANT WATER RELATIONSHIPS

Plants fertilized with Cl in the spring have more erect flag leaves during the middle of a warm day than control plants not receiving Cl (9). These plants also have less take-all root rot (9), and the foliage is less severely attacked by stripe rust (*Puccinia striiformis*) and *Septoria* spp. (*S. tritici* and *S. nodorum*) (8). Chloride uptake may influence plant water relationships and reduce the ability of the pathogen to attack the wheat plant.

Chloride is an inert anion that plants normally accumulate, and it affects plant and perhaps parasite water relationships (11). Water potential has a profound influence on fungal development; its influence on the take-all fungus has been recently reviewed by Cook (12).

Both stripe rust and Septoria diseases are also favored by wet environmental conditions that would increase leaf osmotic potential (π). The severity of these diseases can be influenced by Cl salts. Russell (51) reported that the severity of attack by *P. striiformis* and recently *P. recondita* (26) was decreased with soil applications of KCl and NaCl. Cunfer et al (17) reported that although high rates of P increased severity of Septoria glume blotch, this detrimental affect could be reduced by an increased rate of applied KCl.

Plant uptake of Cl reduced leaf π and increased leaf turgor potential (P) without markedly altering leaf xylem potential (ψ) (9). An increase in the Cl solute content of plant cells may result in slower growth and activity by the pathogen through a reduced turgor gradient. It is possible that the π and P component of the pathogen may be changed through osmoregulation (42,75), but this would require a metabolic energy input. Thus, under wet environmental conditions that are favorable for disease development, Cl uptake may produce changes in cell physiology that would limit pathogen activity similar to that caused by moisture stress (10,75).

WATER STRESS

The best responses from the use of NH_4-N and Cl have occurred in seasons or under conditions where moisture has not been limiting. The 1980–81 and 1981–82 growing seasons in western Oregon clearly demonstrated this moisture relationship. The entire 1980–81 season, from planting through grain filling, was wet. Good suppression of take-all, stripe rust, and Septoria leaf and glume blotch (Tables 3 and 6) was obtained with NH_4-N and Cl fertilizers. The 1981–82 season was similar to the previous year, except that during the boot-dough stages the weather was dry. There was good suppression of take-all, but there were no significant reductions in the severity of attack by *P. striiformis* and *Septoria* spp. where Cl had been topdressed (Tables 3 and 6). Apparently the moisture stress conditions in late spring were already limiting pathogen activity (6,70) beyond a point that could not be further modified by uptake of Cl.

Moisture stress also creates other problems that would reduce the efficacy of NH_4-N and Cl fertilizers. A major risk is the banding of fertilizer salts with the seed under low soil-moisture conditions. This situation can produce a salt toxicity that will delay emergence and may reduce stands. Low soil-moisture conditions decrease uptake of NH_4-N (10) and anions such as Cl (2). Root-colonizing epiphytic bacteria do not tolerate moisture stress situations and are less likely to be effective biological suppressors under low soil moisture, as has been pointed out in reviews by Cook (12), Rovira and Wildermuth (49), and Schroth and Hancock (55). Thus, NH_4-N and Cl should be less effective in wheat-growing areas or during seasons with prolonged periods of moisture stress.

CONCLUSION

A combination of NH_4-N, P, and Cl, banded with the seed at planting, should enhance biological suppression of take-all. A major role of Cl is to enhance plant uptake of NH_4-N, which favors the activity of epiphytic bacteria suppressive to *G. graminis* var. *tritici*. The mechanism of suppression may be similar to that described for take-all decline. Thus, the use of fertilizer should facilitate the establishment of an effective take-all decline situation in a shorter period of time. Early spring applications of Cl as a topdressing change plant water relationships and give additional suppression of *G. graminis* var. *tritici*. The applications may also reduce severity of attack by *Septoria* spp. and *P. striiformis*.

ACKNOWLEDGMENTS

This research was supported in part by grants from the Oregon Wheat Commission, Japanese Ammonium Chloride Institute, and the Potash/Phosphate Institute. We are indebted to the following graduate students for their research contributions; Ron Taylor, Mark Halsey, Don Sullivan, Gary Kiemnel, Becky Mitchell, Joyce Mack, and Marcia Brett.

LITERATURE CITED

1. Agarwal, A. S., Singh, B. R., and Kanehiro, Y. 1971. Ionic effect of salts on mineral nitrogen release in an allophanic soil. Soil Sci. Soc. Am. Proc. 35:454-457.
2. Alston, A. M., and Miller, M. H. 1978. Effect of water stress on subsequent uptake of chloride by wheat plants. Plant Soil 49:305-315.
3. Brown, M. E. 1981. Microbiology of roots infected with the take-all fungus (*Gaeumannomyces graminis* var. *tritici*) in phased sequences of winter wheat. Soil Biol. Biochem. 13:285-291.
4. Burr, T. J., Schroth, M. N., and Suslow, T. 1978. Increased potato yields by treatment of seedpieces with specific strains of *Pseudomonas fluorescens* and *P. putida*. Phytopathology 68:1377-1383.
5. Butler, F. C. 1961. Root and foot rot diseases of wheat. N.S.W. Dep. Agric. Sci. Bull. 77. Agric. Res. Inst., Wagga Wagga.
6. Cartwright, D. W., and Russell, G. E. 1980. Histological and biochemical nature of 'durable' resistance to yellow rust in wheat. Proc. Eur. Mediterr. Cereal Rust Conf., 5th, Bari and Rome, Italy, 28 May-4 June.
7. Chan, E. C. S., and Katznelson, H. 1961. Growth interactions of *Arthrobacter globiformis* and *Pseudomonas* spp. in relation to the rhizosphere effect. Can. J. Microbiol. 7:759-767.
8. Christensen, N. W., Jackson, T. L., and Powelson, R. L. 1982. Suppression of take-all root rot and stripe rust diseases of wheat with chloride fertilizers. Plant Nutrition, 1982. Proc. Int. Plant Nutr. Colloq., 9th, 1:111-116. Warwick University, England. 22–27 August.
9. Christensen, N. W., Taylor, R. G., Jackson, T. L., and Mitchell, B. L. 1981. Chloride effects on water potentials

Table 6. Influence of spring-topdressed ammonium chloride and ammonium sulfate fertilizers on severity of attack of winter wheat by *Puccinia striiformis*, *Septoria tritici*, and *S. nodorum* during 1981 and 1982

	Disease severity[x]			
	Stripe rust		Septoria leaf and glume blotch	
Treatment[w]	1981[y]	1982	1981	1982
AmCl	5.8***[z]	38.2 n.s.	8.5***	5.0 n.s.
AmS	9.0	40.0	75.0	5.2

[w] AmCl = NH_4Cl and AmS = $(NH_4)_2SO_4$. Fertilizers topdressed in March at the rate of 146 kg of N/ha.

[x] Given as percentage of attack. Visual estimate of disease severity for stripe rust was made in May and for Septoria leaf and glume blotch in June.

[y] Mean percentage of attack for 381 cultivars. After Christensen et al (8).

[z] Means of comparisons within columns were significantly different if followed by *** ($P<0.001$) and not significantly different (n.s.) ($P<0.10$).

and yield of winter wheat infected with take-all root rot. Agron. J. 73:1053-1058.

10. Clark, R. B. 1981. Effect of light and water stress on mineral element composition of plants. J. Plant Nutr. 3:853-885.

11. Clarkson, D. T., and Hanson, J. B. 1980. The mineral nutrition of higher plants. Annu. Rev. Plant Physiol. 31:239-298.

12. Cook, R. J. 1981. Effect of soil reaction and physical conditions on take-all. Pages 343-352 in: Biology and Control of Take-all. P. J. Shipton and M. J. C. Asher, eds. Academic Press, New York. 538 pp.

13. Cook, R. J., and Christen, A. A. 1976. Growth of cereal root-rot fungi as affected by temperature-water potential interactions. Phytopathology 66:193-197.

14. Cook, R. J., Papendick, R. I., and Griffin, D. M. 1972. Growth of two root-rot fungi as affected by osmotic and matric water potentials. Soil Sci. Soc. Am. Proc. 36:78-82.

15. Cook, R. J., and Rovira, A. D. 1976. The role of bacteria in the biological control of *Gaeumannomyces graminis* by suppressive soils. Soil Biol. Biochem. 8:269-273.

16. Cook, R. J., and Reis, E. 1981. Cultural control of soilborne pathogens of wheat in the Pacific North-West of the U.S.A. Pages 164-177 in: Strategies for the Control of Cereal Disease. J. F. Jenkyn and R. T. Plumb, eds. Blackwell, Oxford. 219 pp.

17. Cunfer, B. M., Touchton, J. T., and Johnson, J. W. 1980. Effects of phosphorus and potassium fertilization on Septoria glume blotch of wheat. Phytopathology 70:1196-1199.

18. Garrett, S. D. 1941. Soil conditions and the take-all disease of wheat. VI. The effect of plant nutrition upon disease resistance. Ann. Appl. Biol. 28:14-18.

19. Garrett, S. D. 1948. Soil conditions and the take-all disease of wheat. IX. Interactions between host plant nutrition, disease escape and disease resistance. Ann. Appl. Biol. 35:14-17.

20. Gerlagh, M. 1968. Introduction of *Ophiobolus graminis* into new polders and its decline. Neth. J. Plant Pathol. 74: (Suppl.2) 1-97.

21. Ghosh, A. B., Raychaudhari, S. P., and Bhur, N. D. 1956. Ammonium chloride and ammonium sulphate-nitrate as sources of nitrogen to paddy and wheat. J. Indian Soc. Soil Sci. 4:23-30.

22. Gilligan, C. A. 1980. Zone of potential infection between host roots and inoculum units of *Gaeumannomyces graminis*. Soil Biol. Biochem. 12:513-514.

23. Golden, D. C., Sivasubramanian, S., Sandanam, S., and Wijedas, M. A. 1981. Inhibitory effects of commercial potassium chloride on the nitrification rates of added ammonium sulphate in an acid red-yellow podzolic soil. Plant Soil 59:147-151.

24. Hahn, B. E., Olson, F. R., and Roberts, J. L. 1942. Influence of potassium chloride on nitrification in Bedford silt loam. Soil Sci. 55:113-121.

25. Halsey, M. E. 1982. Suppression of take-all (*Gaeumannomyces graminis* var. *tritici*) of wheat by banded ammonium and chloride fertilizers. Ph.D. thesis, Oregon State University, Corvallis. 104 pp.

26. Hashim, L. O., and Russell, G. E. 1982. Possible control of brown (leaf) rust on winter wheat by applying chlorides to the soil. Cereal Rusts Bull. 10:29-34.

27. Huber, D. M. 1976. Nitrogen management—a key to control of soilborne diseases of wheat. Proc. Am. Phytopathol. Soc. 3:286.

28. Huber, D. M. 1981. The role of mineral nutrients and agricultural chemicals in the incidence and severity of take-all. Pages 317-341 in: Biology and Control of Take-all. M. J. C. Asher and P. J. Shipton, eds. Academic Press, New York. 538 pp.

29. Huber, D. M., Painter, C. G., McKay, H. C., and Peterson, D. L. 1968. Effect of nitrogen fertilization on take-all of winter wheat. Phytopathology 58:1470-1472.

30. Huber, D. M., Warren, H. L., Nelson, D. W., Tsai, C. Y., and Shaner, G. E. 1980. Response of winter wheat to inhibiting nitrification of fall-applied nitrogen. Agron. J. 72:632-638.

31. Huber, D. M., and Watson, R. D. 1974. Nitrogen form and plant disease. Annu. Rev. Phytopathol. 12:139-165.

32. Huber, D. M., Watson, R. D., and Steiner, G. W. 1965. Crop residues, nitrogen and plant disease. Soil Sci. 100:302-308.

33. Jackson, T. L., Foote, W. E., and Dickason, E. A. 1962. Effect of fertilizer treatments and planting dates on yield and quality of barley. Oreg. Agric. Exp. Stn. Tech. Bull. 65. 20 pp.

34. Jackson, T. L., Powelson, R. L., and Christensen, N. W. 1982. Combating take-all root rot of winter wheat in western Oregon. Oregon State University, Corvallis, Ext. Serv. FS 250.

35. Kloepper, J. W., Leong, J., Teinzte, M., and Schroth, M. N. 1980. *Pseudomonas* siderophores: A mechanism explaining disease-suppressive soil. Curr. Microbiol. 4:317-320.

36. MacNish, G. C. 1979. Take-all disease of cereals. West. Aust. Dep. Agric. Farmnote 148/79.

37. MacNish, G. C. 1980. Management of cereals for control of take-all, 1980. J. Agric. West. Aust. 21:48-51.

38. MacNish, G. C., and Speijers, J. 1982. The use of ammonium fertilizers to reduce the severity of take-all (*Gaeumannomyces graminis* var. *tritici*) on wheat in Western Australia. Ann. Appl. Biol. 100:83-90.

39. Marschner, H. 1978. Role of the rhizosphere in iron nutrition on plants. Iran J. Agric. Res. 6:69-80.

40. Marschner, H., Romheld, V., and Ossenberg-Neuhaus, H. 1982. Rapid method for measuring changes in pH and reducing processes along roots of intact plants. Z. Pflanzenphysiol. 105:407-416.

41. Maurer, C. L., and Baker, R. 1965. Ecology of plant pathogens in soil. II. Influence of glucose, cellulose, and inorganic nitrogen amendments on development of bean root rot. Phytopathology 55:69-72.

42. Mengel, K., and Kirkby, E. A. 1978. Principles of plant nutrition. Int. Potash Inst. Worblaufen-Bern. 593 pp.

43. Miller, M. H. 1974. Effects of nitrogen on phosphorus absorption by plants. Pages 310-329 in: The Plant Root and Its Environment. E. W. Carson, ed. University of Virginia, Charlottesville. 691 pp.

44. Nye, P. H. 1981. Changes of pH across the rhizosphere induced by roots. Plant Soil 61:7-26.

45. Powelson, R. L., and Jackson, T. L. 1978. Suppression of take-all (*Gaeumannomyces graminis*) root-rot of wheat with fall applied chloride fertilizers. Pages 175-182 in: Proc. Annu. North-West Fert. Conf., 29th, Beaverton, Oregon, July 11-13.

46. Reis, E. M. 1980. Effect of mineral nutrition on take-all of wheat. Ph.D. thesis, Washington State University, Pullman. 117 pp.

47. Reis, E. M., Cook, R. J., and McNeal, B. L. 1982. Effect of mineral nutrition on take-all of wheat. Phytopathology 72:224-229.

48. Ridge, E. H. 1976. Studies on soil fumigation. II. Effects of bacteria. Soil Biol. Biochem. 8:249-253.

49. Rovira, A. D., and Wildermuth, G. B. 1981. The nature and mechanisms of suppression. Pages 385-415 in: Biology and Control of Take-all. M. J. C. Asher and P. J. Shipton, eds. Academic Press, New York. 538 pp.

50. Rowaished, A. K. 1981. The influence of different forms of nitrogen on Fusarium root-rot diseases of winter wheat. Phytopathol Z. 100:331-339.

51. Russell, G. E. 1978. Some effects of applied sodium and potassium chloride on yellow rust in winter wheat. Ann. Appl. Biol. 90:163-168.

52. Sands, D. C., and Rovira, A. D. 1971. *Pseudomonas fluorescens* biotype G, the dominant fluorescent pseudomonad in South Australian soils and wheat rhizospheres. J. Appl. Bacteriol. 34:261-275.

53. Scher, F. M., and Baker, R. 1982. Effect of *Pseudomonas putida* and a synthetic iron chelator on induction of soil suppressiveness to Fusarium wilt pathogens. Phytopathology 72:1567-1573.

54. Schroth, M. N., and Hancock, J. G. 1981. Selected topics in biological control. Annu. Rev. Microbiol. 35:453-476.

55. Schroth, M. N., and Hancock, J. G. 1982. Disease-suppressive soil and root colonising bacteria. Science 216:1376-1381.

56. Sivasithamparam, K., Parker, C. A., and Edwards, C. S. 1979. Bacterial antagonists to the take-all fungus and fluorescent pseudomonads in the rhizosphere of wheat. Soil Biol. Biochem. 11:161-165.

57. Smiley, R. W. 1978. Antagonists of *Gaeumannomyces graminis* from the rhizoplane of wheat in soils fertilized with ammonium or nitrate nitrogen. Soil Biol. Biochem. 10:169-174.

58. Smiley, R. W. 1978. Colonization of wheat roots by *Gaeumannomyces graminis* inhibited by specific soils, micro-organisms and ammonium nitrogen. Soil Biol. Biochem. 10:175-179.

59. Smiley, R. W. 1979. Wheat-rhizoplane pseudomonads as antagonists of *Gaeumannomyces graminis*. Soil Biol. Biochem. 11:371-376.

60. Smiley, R. W. 1979. Wheat rhizosphere pH and the biological control of take-all. Pages 329-338 in: The Soil-Root Interface. J. L. Harley and R. S. Russell, eds. Academic Press, New York.

61. Smiley, R. W. and Cook, R. J. 1973. Relationship between take-all of wheat and rhizosphere pH in soils fertilized with ammonium vs. nitrate-nitrogen. Phytopathology 63:882-890.

62. Smiley, R. W., Cook, R. J., and Papendick, R. I. 1972. Fusarium root rot of wheat and peas as influenced by soil applications of anhydrous ammonia and ammonia-potassium azide solutions. Phytopathology 62:86-91.

63. Street, H. E., and Sheat, D. E. G. 1958. The absorption and availability of nitrate and ammonium. Pages 150-165 in: Encyclopedia of Plant Physiology. Vol. VIII. W. Ruhland, ed. Springer-Verlag, Berlin.

64. Stynes, B. A. 1975. A synoptic study of wheat. Ph.D. thesis, University of Adelaide. 291 pp.

65. Suslow, T. E., and Schroth, M. N. 1982. Rhizobacteria of sugar beets: Effects of seed application and root colonization on yield. Phytopathology 72:199-206.

66. Syme, J. R. 1966. Fertilizer and varietal effects on take-all in irrigated wheat. Aust. J. Exp. Agric. Anim. Husb. 6:246-249.

67. Taylor, R. G., Jackson, T. L., Powelson, R. L., and Christensen, N. W. 1983. Chloride, nitrogen form, lime, and planting date effects on take-all root rot of winter wheat. Plant Dis. 67:1116-1120.

68. Trolldenier, G. 1981. Influence of soil moisture, soil acidity and nitrogen source on take-all of wheat. Phytopathol. Z. 102:163-177.

69. Trolldenier, G. 1982. Influence of potassium nutrition and take-all on wheat yield in dependence on inoculum density. Phytopathol. Z. 103:340-348.

70. Van der Wal, A. F., Smettink, H., and Mann, G. C. 1975. An ecophysiological approach to crop losses exemplified in the system wheat, leaf rust and glume blotch. III. Effects of soil water potential on development, growth, transpiration, symptoms and spore production of leaf rust-infected wheat. Neth. J. Plant Pathol. 81:1-13.

71. Vojinovic, Z. 1973. The influence of micro-organisms following *Ophiobolus graminis* Sacc. on its further pathogenicity. EPPO Bull. 9:91-101.

72. Weller, D. M., and Cook, R. J. 1983. Suppression of take-all of wheat by seed treatments with fluorescent pseudo-monads. Phytopathology 73:463-469.

73. Wiese, M. V. 1977. Compendium of Wheat Diseases. American Phytopathological Society, St. Paul, MN. 106 pp.

74. Wildermuth, G. B. 1980. Suppression of take-all by some Australian soils. Aust. J. Agric. Res. 31:251-258.

75. Yancey, P. H., Clark, M. E., Hand, S. C., Bowlus, R. D., and Somero, G. N. 1982. Living with water stress: Evolution of osmolyte systems. Science 217:1214-1222.

76. Zogg, H., and Jaggi, W. 1974. Studies on the biological soil disinfection. VII. Contribution to the take-all decline (*Gaeumannomyces graminis*) imitated by means of laboratory trials and some of its possible mechanisms. Phytopathol. Z. 81:160-169.

Effect of Tillage on *Heterodera avenae* in Wheat

D. K. ROGET and A. D. ROVIRA, CSIRO Division of Soils, Glen Osmond, S.A., 5064, Australia

Heterodera avenae (cereal cyst nematode or cereal eelworm) causes widespread losses in cereals in Victoria and South Australia (1,4,7). Following the prediction (3) that tillage practices could affect *H. avenae*, the study described in this paper was undertaken to assess the effects of tillage on the damage to wheat roots caused by *H. avenae*, the populations of females on roots, and the yields of grain in fields infested with *H. avenae*.

MATERIALS AND METHODS

The field experiments were conducted in 1980 and 1981 at Calomba, S. A. (34°23′S, 138°24′E), which has a winter dominant rainfall with an annual average of 407 mm (316 mm in 1980, 377 mm in 1981). The soil type at the 1980 site was a calcareous sandy loam and at the 1981 site a calcareous clay loam, classified as Gc1.12 and Gc2.12 respectively (2); both soils are classified alternatively as solonized brown soil or calcic xerosol. The 1980 experiment followed volunteer grass-*Medicago* spp. pasture, and the 1981 experiment followed a barley crop. The population of *H. avenae* in 1980 was 8.3 eggs per gram of soil, giving a root damage rating of 3.6; in 1981 it was 4.2 eggs per gram of soil, giving a rating of 2.7 (6).

In 1980, two forms of tillage were used: "conventional cultivation," in which the wheat (cv. Condor) was sown after three cultivations with a 10-row conventional seed drill fitted with 15-cm shares (a combined cultivator drill) (8); and "direct drilling," in which the volunteer pasture was killed by spraying with a mixture of paraquat (0.125 kg/L) and diquat (0.075 kg/L) (Spray Seed; ICI Australia) at 4 L/ha, with 2,4-D ester at 0.5 L/ha and diuron at 0.7 kg/ha. One day after spraying, plots were sown with the SIRODRILL (9), a 10-row seed drill designed for minimum soil disturbance. Each row has a fluted colter wheel cutting through surface plant residues and 7 cm into the soil, followed by both a 2-cm pointed tine penetrating 5 cm and delivering the seed and a press wheel (5 cm wide) to consolidate the soil around the seed.

The 1981 experiment included four tillage treatments: conventional and seeding with the combined cultivator seed drill; direct drilling with the SIRODRILL, as in the 1980 experiment; and two other direct-drill treatments using the combined cultivator seed drill with either 15-cm or 3-cm shares. One day before sowing, all plots including cultivated plots were sprayed with the paraquat-diquat mixture at 3 L/ha, diuron at 0.5 kg/ha, 2,4-D ester at 0.15 L/ha, and dimethoate at 0.01 L/ha.

All tillage treatments were sown with and without the nematicide aldicarb (Temik 10G) applied at 2 kg/ha in the drill row via a small seed box attachment on each seed drill. Aldicarb at this rate is not economic and was used to provide an almost nematode-free control.

Early root damage by *H. avenae* was assessed on plants from each plot 6–8 weeks after sowing using a 0–5 scale (6). The number of females of *H. avenae* per root system was determined at 12 weeks after seeding. Both assessments were made on an average of 25 plants per plot. Grain yields were derived from machine harvests of the whole plots.

The experimental design consisted of a split plot with four replications in 1980 and a randomized complete block with three replications in 1981. Plots in both years were 50 m long by 1.5 m (10 rows) wide.

RESULTS

Early root damage by *H. avenae* in 1980 and 1981 was significantly lower when the wheat was sown by direct drilling than when sown following cultivation (Table 1). In 1981, all three direct-drill treatments reduced root damage from *H. avenae*, but the effect with 3-cm shares was not significant

The application of aldicarb in both 1980 and 1981 almost eliminated invasion of wheat roots by *H. avenae* (Table 1) and the consequent severe stunting and branching of roots. The dry weight of plant tops for the different treatments 6 weeks after seeding in 1980 largely reflected the level of *H. avenae* damage; without aldicarb application, direct-drilled plots had significantly higher plant weights than cultivated plots. In 1981, tillage had no effect on plant dry weight in the absence of aldicarb; in both years, application of aldicarb significantly increased plant weight (Table 2).

Table 1. Effect of tillage and nematicide on the early damage to wheat roots by *Heterodera avenae*[a]

Tillage	Aldicarb[b]	1980	1981
Cultivated and sown with combined cultivator drill	−	4.6	2.7
	+	1.0	0.4
Direct drill: SIRODRILL	−	3.2	1.6
	+	0.8	0.2
Direct drill: combined cultivator drill with 15-cm shares	−		1.4
	+		0.2
Direct drill: combined cultivator drill with 3-cm shares	−		2.3
	+		0.4
LSD (*P* = 0.05)	Tillage	0.8	0.6
	Nematicide	0.8	0.6

[a] Rating of damage on 0–5 scale (6).
[b] Aldicarb at 2 kg/ha, in-furrow application.

The number of *H. avenae* females per plant was reduced by 61% in 1980 and by 46% in 1981 when wheat was direct drilled (without aldicarb); this effect was significant ($P = 0.05$) in 1980. Aldicarb significantly reduced female numbers in both years (Table 3).

In 1980, grain yield of wheat was significantly increased by 0.38 t/ha by direct drilling with the SIRODRILL, which gave minimum soil disturbance. The difference in yield between tillage treatments was eliminated with application of aldicarb, indicating that the beneficial effect of direct drilling results from reduced *H. avenae* damage to roots. In 1980, aldicarb significantly increased yield in cultivated plots by 1.01 t/ha (119%) and in direct-drilled plots by 0.43 t/ha (35%) (Table 4). Tillage or aldicarb had no effect on yield in 1981 (Table 4).

DISCUSSION

This study has shown that there was less early damage to roots by *H. avenae* and fewer females on roots of wheat sown by direct drilling than on wheat sown after cultivation. The increase in grain yield in 1980 of 0.38 t/ha (45%) with direct-drilled wheat compared with wheat sown following cultivation is attributed largely to the reduction in *H. avenae* damage; this conclusion is supported by the result that there was no effect of tillage when *H. avenae* was controlled with aldicarb.

The 1980 experiment indicated that this tillage effect was the result of the minimal soil disturbance characteristics of the SIRODRILL. In 1981, however, when the trial was extended to direct drilling with a combined cultivator seed drill with 15-cm shares (as used by farmers), it showed that the degree of soil disturbance with direct drilling did not affect root damage by *H. avenae*. The failure of direct drilling with 3-cm shares to reduce root damage by *H. avenae* is attributed to the delayed emergence and impaired early growth with this treatment, as caused by clods of soil above many seeds. These adverse conditions accentuate the effects of *H. avenae*.

The effects of tillage on early root damage and female numbers were consistent in both years. Yield differences occurred in 1980 but not in 1981 when neither direct drilling nor aldicarb gave the yield increases that could be expected from the reduction in early root damage. Such a result is attributed to the dry seasonal conditions being exacerbated by the high clay content of the 1981 experimental site: 30% clay, compared with 15% for the 1980 site. Another factor that influenced the 1981 result was the lower level of *H. avenae* at the site: 4.2 eggs per gram of soil, compared with 8.3 in 1980. At this level, the effects of tillage and aldicarb would not be as marked.

Direct drilling is a management strategy available when *H. avenae* is the predominant root disease. The reduction of root damage and female numbers and the increase in yield with direct drilling at the 1980 site were comparable to that obtained by commercial nematicides in an adjacent trial at Calomba in the same year (5).

The mechanisms involved in the reduction of damage by *H. avenae* with direct drilling have yet to be resolved. Several mechanisms that may be involved are more even distribution of nematodes in the soil with cultivation; lower bulk density of cultivated soil, favoring movement of nematodes in soil; changed aeration affecting nematode activity; and roots of volunteer annual grasses (growing before spraying and

Table 2. Effect of tillage and nematicide on the top dry weight (mg/plant) of wheat plants 6 weeks after seeding

Tillage	Aldicarb[a]	1980	1981
Cultivated and sown with combined cultivator drill	−	26	11
	+	52	16
Direct drill: SIRODRILL	−	38	13
	+	49	16
Direct drill: combined cultivator drill with 15-cm shares	−		12
	+		14
Direct drill: combined cultivator drill with 3-cm shares	−		12
	+		12
LSD ($P = 0.05$)	Tillage	12	2.6
	Nematicide	12	2.6

[a] Aldicarb at 2 kg/ha, in-furrow application.

Table 3. Effect of tillage and nematicide on the numbers of females of *Heterodera avenae* on roots of wheat at anthesis

Tillage	Aldicarb[a]	1980	1981
Cultivated and sown with combined cultivator drill	−	59	71
	+	8	8
Direct drill: SIRODRILL	−	23	32
	+	8	10
Direct drill: combined cultivator drill with 15-cm shares	−		48
	+		10
Direct drill: combined cultivator drill with 3-cm shares	−		36
	+		10
LSD ($P = 0.05$)	Tillage	29	NSD[b]
	Nematicide	29	39

[a] Aldicarb at 2 kg/ha, in-furrow application.
[b] No significant difference.

Table 4. Effect of tillage and nematicide on the grain yield of wheat (t/ha)

Tillage	Aldicarb[a]	1980	1981
Cultivated and sown with combined cultivator drill	−	0.85	0.85
	+	1.86	1.03
Direct drill: SIRODRILL	−	1.23	0.85
	+	1.66	0.96
Direct drill: combined cultivator drill with 15-cm shares	−		0.99
	+		0.95
Direct drill: combined cultivator drill with 3-cm shares	−		0.87
	+		0.94
LSD ($P = 0.05$)	Tillage	0.27	NSD[b]
	Nematicide	0.27	NSD

[a] Aldicarb at 2 kg/ha, in-furrow application.
[b] No significant difference.

seeding) being invaded by nematodes, which are then unable to attack the roots of wheat plants.

In conclusion, we have shown that wheat sown by direct drilling had significantly less root damage due to *H. avenae* at the five-leaf stage and significantly fewer females ("white cysts") per plant at flowering than wheat sown following cultivation. Grain yield increases of up to 0.38 t/ha occurred with direct drilling as the result of reduced root damage by *H. avenae*. Our results have important implications in relation to the buildup of *H. avenae* in soil following wheat crops, as the production of cysts that carry the eggs was reduced by 50% following direct drilling.

ACKNOWLEDGMENTS

We acknowledge the assistance of N. R. Venn, A. Simon, A. P. Neate, and P. I. Forrester. Financial contributions from the Wheat Industry Research Council, Union Carbide (Aust.), Cyanamid (Aust.) Ltd, and ICI (Aust.) Ltd assisted this project. The cooperation of Mr. J. McEvoy throughout our research on his farm is most gratefully acknowledged.

LITERATURE CITED

1. Meagher, J. W. 1968. The distribution of the cereal cyst nematode (*Heterodera avenae*) in Victoria and its relation to soil type. Aust. J. Exp. Agric. Anim. Husb. 8:637-640.

2. Northcote, K. H., Hubble, G. D., Isbell, R. F., Thompson, C. H., and Bettenay, E. 1975. A Description of Australian Soils. CSIRO Publ., Aust. 170 pp.

3. Rovira, A. D. 1979. Direct drilling—friend or foe to soil-borne root diseases. Proc. Natl. Direct Drilling Conf. ICI Aust. Publ. 79-93.

4. Rovira, A. D., Brisbane, P. G., Simon, A., Whitehead, D. G., and Correll, R. L. 1981. Influence of cereal cyst nematode (*Heterodera avenae*) on wheat yields in South Australia. Aust. J. Exp. Agric. Anim. Husb. 21:516-523.

5. Rovira, A. D., and Simon, A. 1982. Integrated control of cereal cyst nematode (*Heterodera avenae*). EPPO Bull. 12:517-523.

6. Simon, A. 1980. A plant assay of soil to assess potential damage to wheat by *Heterodera avenae*. Plant Dis. 64:917-919.

7. Simon, A., and Rovira, A. D. 1982. The relation between wheat yield and early damage of roots by cereal cyst nematode. Aust. J. Exp. Agric. Anim. Husb. 22:201-208.

8. Venn, N. R., and Gilbert, R. J. 1982. Description of the CSIRO-Bagshaw Trash Clearance Seeder. CSIRO Div. Soils Tech. Mem. 7/1982. 5 pp. (Available upon request.)

9. Venn, N. R., Whitehead, D. A. R. G., and Swaby, B. A. 1982. Description of the SIRODRILL direct drill seeder. CSIRO Div. Soils Tech. Mem. 5/1982. 8 pp. (Available upon request.)

Effect of Rotation and Tillage on Take-all and Rhizoctonia Root Rot in Wheat

A. D. ROVIRA and N. R. VENN, CSIRO Division of Soils, Glen Osmond, S.A., 5064, Australia

With the development of desiccant herbicides in the 1970s, the technique of conservation tillage (reduced tillage or direct drilling) became an alternative to cultivation in Australian wheat farming. At that time, no work had been undertaken in Australia on the effects of tillage on soilborne root diseases of cereals. Studies outside Australia have reported take-all (caused by *Gaeumannomyces graminis* (Sacc.) von Arx & Olivier var. *tritici* Walker) to be lower (1,3,5,12), higher (3,8), or no different (15,16) in wheat sown by direct drilling compared with wheat sown following cultivation.

This study was undertaken to investigate the effect of tillage and crop rotation on the incidence and severity of root diseases of wheat in a soil and climate representative of a large proportion of the dryland wheat-farming area of southeastern Australia.

MATERIALS AND METHODS

Experimental site. The experiments described in this paper were conducted at Avon, 200 km north of Adelaide, South Australia, located at 34°14′S, 138°18′E. The climate is Mediterranean with hot, dry summers and a winter dominant rainfall. The average annual rainfall is 350 mm. In the year when the experiments described in the paper were conducted, only 164 mm fell, resulting in severe water shortage toward the end of the growing season. The rainfall (mm) throughout 1982 was 7 in January, 1 in February, 18 in March, 34 in April, 25 in May, 25 in June, 10 in July, 12 in August, 10 in September, 20 in October, 0 in November, and 2 in December.

The soil is an alkaline calcareous sandy loam, pH 8.4, classification Gc1.12, solonized brown soil or calcic xerosol (9). The site had moderate levels of disease caused by *G. graminis* var. *tritici* and Rhizoctonia root rot caused by *Rhizoctonia solani* Kuehn and low levels of *Fusarium, Helminthosporium, Pythium*, and *Heterodera avenae*.

Experimental design. Six replicated plots, 100 m long and 1.6 m wide, were established for each of the rotations—self-sown annual pasture/wheat, medic pasture/wheat, peas/wheat, and wheat/wheat—for each of two tillage systems—direct drill (no tillage before seeding) or cultivation (short fallow). Annual pasture is a self-seeded pasture composed of 20% medic (*Medicago truncatula* var. Harbinger), 70% barley grass (*Hordeum glaucum*), and 10% ryegrass (*Lolium rigidum*), which germinate after autumn rains. Medic pasture was *M. scutellata* sown at 12 kg/ha. Peas were *Pisum sativum* sown at 90 kg/ha. These rotations were commenced in 1979 following wheat; the results reported in this paper for 1982 represent the end of the second cycle of the 2-year rotation. Results over the 4 years will be presented in detail elsewhere.

Cultivation and seeding. The soil was cultivated to a depth of 7 cm with a tine cultivator on three occasions in April and May. Seeding was done with a 10-row combined cultivator seed drill with a 15-cm share on each tine delivery tube. Direct-drilled plots were sprayed with a mixture of paraquat (0.125 kg/L) and diquat (0.075 kg/L) (Spray Seed; ICI Australia) 2 days before seeding with the SIRODRILL, an implement designed to seed with minimum soil disturbance (13).

Wheat, cv. Condor, was sown when soil conditions were suitable in early June at 50 kg/ha with superphosphate at 120 kg/ha.

Disease assessment. Ten weeks after seeding (early tillering), plants were taken at 10-m intervals over the 100-m length of plots—10 to 15 from each of the 10 sampling points. The roots were washed free of soil and examined for symptoms of take-all and Rhizoctonia root rot.

The symptom for take-all was well-defined, black lesions in the stele of the main axes of the seminal roots; isolations from such roots consistently produced *G. graminis*, and these isolates produced identical symptoms on roots of wheat grown in the greenhouse. The level of take-all infection or incidence has been expressed as the percentage of plants with one or more stelar lesions.

Severity of disease caused by *Rhizoctonia* spp. was assessed using a 0–5 rating (7). McDonald (6) showed that of 13 potential pathogens isolated from roots of wheat showing symptoms of Rhizoctonia root rot, only *R. solani* caused the severe cortical rot and brown "spear tip" symptoms seen on plants from the field. Of 53 *Rhizoctonia*-like isolates from wheat roots from patches of poor growth, 25 were identified as *R. solani* or *Thanatephorus cucumeris* on the basis of colony characteristics and perfect stage fructification (10); 13 were multinucleate "*Rhizoctonia*-like" fungi, whereas 7 were binucleate "*Rhizoctonia*-like," probably *Ceratobasidium*. The remaining 8 isolates did not fit the above three categories.

Plant growth and grain yields. The top weight at tillering was determined on plants used for root disease assessment. Grain yield was obtained by machine harvest over the 100-m-long plots at maturity in December.

RESULTS

Take-all. The incidence of take-all was higher in direct-drilled wheat following all rotations, the highest being in wheat following wheat and wheat following annual pasture (Fig. 1). With conventional cultivation, wheat following sown medic and peas, and continuous wheat, had low levels of take-all.

The incidence of take-all was lower in 1982 than in 1980, probably because of the higher rainfall between April and July in 1980 (Fig. 2). There was a greater decline of take-all in continuous wheat with cultivation in 1982 than in continuous wheat sown by direct drilling or in wheat sown by both tillage methods following pasture.

Rhizoctonia root rot. There was consistently more disease caused by *Rhizoctonia* on roots of wheat sown by direct drilling than when sown following cultivation (Fig. 3). Although there were significant differences in disease following different rotations, these effects were small compared with the large effect of rotation on take-all.

Plant growth and grain yield. There were large effects of tillage and rotation on early plant growth (Fig. 4), and statistical analysis of the results showed that Rhizoctonia root rot accounted for 70% of the variability in early growth.

Grain yield varied considerably with the various treatments, even though yields were low because of the drought (Fig. 5). The district average yield was 0.4 t/ha; in wheat following peas and medic with cultivation in our experiments, however, yields up to 0.85 t/ha were obtained, indicating that in the absence of root disease more effective use was made of this limited rainfall.

A multiple regression analysis demonstrated that 69% of the variation in grain yield could be accounted for by three factors: the incidence of take-all lesions accounted for 43%, nitrate at seeding for 21%, and Rhizoctonia root rot for 5%.

DISCUSSION

This study confirms earlier observations (2,4,15) that rotation plays a major part in the incidence of take-all in wheat (Figs. 1 and 5). The host range of *G. graminis* var.

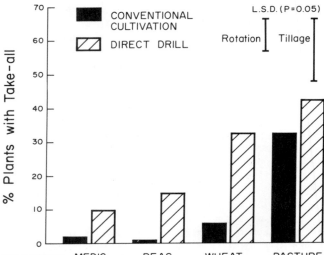

Fig. 1. Effect of tillage and rotation on the incidence of take-all on wheat in 1982.

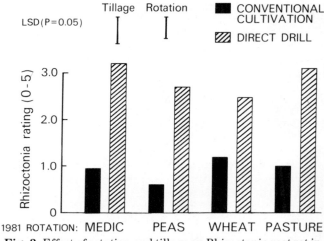

Fig. 3. Effect of rotation and tillage on Rhizoctonia root rot in 1982.

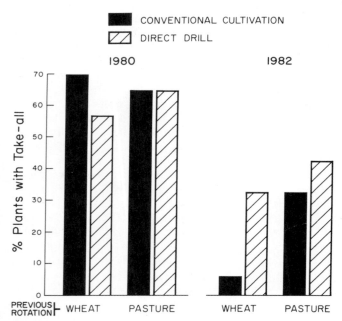

Fig. 2. Effect of tillage and rotation on the incidence of take-all in continuous wheat and pasture-wheat rotations in 1980 and 1982.

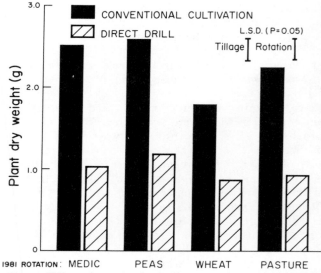

Fig. 4. Effect of rotation and tillage on top weight of wheat plants at tillering (11 weeks after seeding).

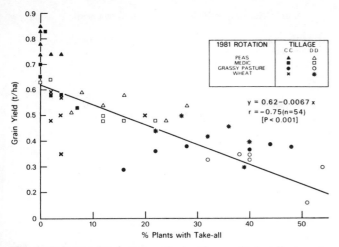

Fig. 5. Relationship between the incidence of take-all on roots at tillering and grain yield.

tritici is such that disease was confined to the roots of grasses, wheat, and barley; this explains the low levels of root disease after medic and peas compared with grass-medic pasture—especially with cultivation that gave control of annual grasses.

The results illustrate for the first time in Australia the consistently higher levels of take-all and Rhizoctonia root rot in direct-drilled wheat (Fig. 3). Although the higher incidence of take-all in direct-drilled wheat may be reduced by rotation, such management methods may not be effective for the control of Rhizoctonia root rot because of the wide host range of *R. solani*.

The higher level of take-all in direct-drilled wheat compared with wheat sown into cultivated soil may be attributed to several factors: cultivation breaks up the crowns and roots of wheat, giving smaller propagules that are more vulnerable to attack by microorganisms and have a lower inoculum potential (8); cultivation stimulates microbial activity, which results in more rapid decomposition of crop residues; the annual pasture in direct-drilling treatment grows until the procedure of spraying and seeding, so that the roots of the grasses act as a highly nutritious "bridge" for the take-all fungus between old roots and crowns (in which the pathogen oversummered) and the new wheat roots; and a higher level of mineral nitrogen is present at seeding because of cultivation (Rovira, *unpublished*).

The greater decline in take-all in continuous wheat sown after cultivation than when sown by direct drilling may result from fragmentation of infected crowns and roots (which increases decomposition and death of the pathogen) and the spreading of suppressive organisms associated with take-all lesions (11). However, a further factor could be the high population of barley grass that grows in the annual pasture and is killed with herbicide 2–3 days before seeding the wheat and the higher incidence of barley grass survival within the direct-drilled wheat crop (Rovira and Correll, *unpublished*). *G. graminis* var. *tritici* was isolated from the roots of barley grass at the time of seeding and also during the growth of the crop.

Direct drilling leads to an accumulation on and near the surface of particulate plant residues that *Rhizoctonia* can colonize because of its strong competitive saprophytic ability (14). This colonized organic matter forms infective propagules, and *R. solani*

can be isolated from plant residues on and near the surface of the soil in patches of poor growth in the wheat crop. The direct effect of soil disturbance on *Rhizoctonia* competitiveness or infectivity could also be important. Laboratory experiments with soil to which *R. solani* propagules were added and allowed to grow through the soil have demonstrated that root infection is reduced when the soil is disturbed before seeding (7). The roots of the annual grasses that germinate after the autumn rains and explore the soil in direct-drilled plots until sprayed with herbicide also act as major carriers of *Rhizoctonia*; experiments in 1983 demonstrated that killing grasses 3 weeks before seeding reduced *Rhizoctonia* damage to roots and increased grain yields of wheat.

ACKNOWLEDGMENTS

We acknowledge the help of H. McDonald and D. Roget in the field and laboratory studies. The generous help and support of Mr. and Mrs. R. Manley, upon whose farm the trial was conducted, are gratefully acknowledged. This project was supported by the Australian Wheat Industry Research Council.

LITERATURE CITED

1. Brooks, D. H., and Dawson, M. G. 1968. Influence of direct-drilling of winter wheat on incidence of take-all and eyespot. Ann. Appl. Biol. 81:57-64.
2. Butler, F. C. 1961. Root and foot rot diseases of wheat. N. S. W. Dep. Agric. Sci. Bull. 77.
3. Hornby, D. 1975. Inoculum of take-all fungus: Nature, measurement, distribution and survival. EPPO Bull. 5:319-333.
4. Kollmorgen, J. F., Griffith, J. B., and Walsgott, D. M. 1983. The effects of various crops on the survival and carry-over of the wheat take-all fungus *Gaeumannomyces graminis* var. *tritici.* Plant Pathol. 32:73-77.
5. Lockhart, D. A., Heppel, V. A., and Holmes, J. C. 1975. Take-all (*Gaeumannomyces graminis* (Sacch.) Arx and Olivier) incidence in continuous barley growing and effect of tillage method. EPPO Bull. 5:375-383.
6. McDonald, H. 1982. Pathogenicity of *Rhizoctonia solani* and *Rhizoctonia*-like fungi isolated from direct drilled wheat grown at Avon, S.A. in 1981. CSIRO Div. Soils Tech. Mem. 24/1982. (Available upon request.)
7. McDonald, H. J., and Rovira, A. D. 1985. Development of inoculation technique for *Rhizoctonia solani* and its application to screening cereal cultivars for resistance. Pages 174-176 in: Ecology and Management of Soilborne Plant Pathogens. C. A. Parker, A. D. Rovira, K. J. Moore, P. T. W. Wong, and J. F. Kollmorgen, eds. American Phytopathological Society, St. Paul, MN.
8. Moore, K. J. 1978. The influence of no tillage on take-all of wheat. Ph.D. thesis, Washington State University.
9. Northcote, K. H., Hubble, G. D., Isbell, R. F., Thompson, D. H., and Bettenay, E. 1975. A Description of Australian Soils. CSIRO Publ., Aust. 170 pp.
10. Parmeter, J. R., Jr., and Whitney, H. A. 1970. Taxonomy and nomenclature of the imperfect state. Pages 7-19 in: *Rhizoctonia solani*, Biology and Pathology. John R. Parmeter, Jr., ed. University of California Press, Berkeley.
11. Rovira, A. D., and Wildermuth, G. B. 1981. The nature and mechanisms of suppression. Pages 385-416 in: Biology and Control of Take-all. M. J. C. Asher and P. J. Shipton, eds. Academic Press, New York.
12. Shipton, P. J. 1969. Take-all decline with particular reference to spring wheat and barley. Ph.D. thesis, University of Reading.
13. Venn, N. R., Whitehead, D. A. R. G., and Swaby, B. A. 1982. Description of the SIRODRILL direct drill seeder. CSIRO Div. Soils Tech. Mem. 5/1982. 8 pp. (Available upon request.)

14. Weinhold, A. R. 1977. Population of *Rhizoctonia solani* in agricultural soils determined by a screening procedure. Phytopathology 67:566-569.

15. Yarham, D. J. 1981. Practical aspects of epidemiology and control. Pages 353-384 in: Biology and Control of Take-all. M. J. C. Asher and P. J. Shipton, eds. Academic Press, New York.

16. Yarham, D. J., and Hirst, J. M. 1975. Diseases in reduced cultivation and direct-drilling systems. EPPO Bull. 5:287-296.

Activity of Fungicides in Soil Against Infection of Wheat Roots by *Gaeumannomyces graminis* var. *tritici*

A. D. ROVIRA and D. G. WHITEHEAD, CSIRO Division of Soils, Glen Osmond, S.A., 5064, Australia

Take-all caused by *Gaeumannomyces graminis* (Sacc.) von Arx & Olivier var. *tritici* Walker is a disease of worldwide significance. However, it is difficult to assess crop losses from this disease, especially at the chronic or subclinical level when the incidence and severity of infection can be seen on the roots but the classical take-all symptoms in seedlings or dead-heads or whiteheads at flowering do not occur.

Fumigation of soil in the field has given increases of 50 to 400% in grain yield (11); but because of the release of ammonium-N and the control of all root pathogens (including *G. graminis* var. *tritici*) in these situations, the increases cannot be attributed to the control of any one pathogen. The use of chemicals with action against specific wheat root pathogens has demonstrated the magnitude of losses caused by *Heterodera avenae* (9,12) and *Pythium* spp. (6).

Recent studies have shown that benomyl, iprodione, nuarimol, triadimenol (1–4), and triadimefon (7) reduce damage by *G. graminis* var. *tritici*. Our pot experiments were undertaken in 1974 to assess the activity of a range of fungicides against the infection of wheat roots by *G. graminis* var. *tritici*. Effective compounds found in these pot experiments may be of value in field studies to assess crop losses caused by take-all.

MATERIALS AND METHODS

For the first experiment, the chemicals in powder form were added to soils previously wet to field capacity and mixed through the soil at 10, 25, and 50 mg/kg of dry soil (Table 1). In the second experiment, five fungicides that showed promise in the first test—quintozene, mancozeb, benomyl, thiram, and zineb—were used at 50 mg/kg of soil with and without aldicarb at 10 mg a.i./kg of soil. Aldicarb was included in this trial because it has been used extensively to assess yield losses caused by the nematode *H. avenae* (9,12,13), often in fields in which take-all occurs. Hence we consider it desirable to know the effect of aldicarb on the infection of wheat roots by *G. graminis* var. *tritici*.

Two calcareous loamy sands classified as Ucl.12 calcic regasols (10) from western Eyre Peninsula in South Australia were used; both soils had low levels of *G. graminis* var. *tritici*, and one had a moderate level of *H. avenae*. The soils, taken from under wheat following pasture, had the properties shown in Table 2.

For preparation of take-all inoculum, *G. graminis* var. *tritici* was grown on sterilized oat kernels for 4 weeks, then dried and ground before mixing through the soils at 0.2% in experiment 1 and 0.1% in experiment 2.

Soils wet to −0.1 bar and with the appropriate chemical were dispensed (300 g of wet soil per pot) into 300-ml tapered plastic drink containers. In experiment 1, with 0.2% inoculum, there were duplicate pots of each treatment; in experiment 2, with 0.1% inoculum, five replicates were used. Seven wheat seeds (cv. Condor) were planted in each pot; the seedlings were thinned to five per pot. Pots were watered to weight twice each week. After 4 weeks in a water bath (root temperature tank) at 15°C in a glasshouse, soil was washed from roots and the amount of disease on the roots assessed. The assessment included number of roots with lesions, total length of lesions per plant, and a disease rating of 0–5 (5).

RESULTS

Of the 10 fungicides tested in the initial study, only quintozene and benomyl applied at 25 and 50 mg/kg of soil reduced disease when 0.2% oat inoculum was added to the Streaky Bay soil. In Nunjikompita soil, eight of

Table 1. Chemicals used in pot experiments assessing activity of fungicides against *Gaeumannomyces graminis* var. *tritici*

Common name	Chemical name[a]
Benomyl	1-Butylcarbomoyl-2-benzimidazole carbamic acid (methyl ester)
Captafol	*N*-(1,1,2,2-Tetrachloroethylthio)-cyclohex-4-ene-1,2-dicarboximide
Captan	*N*-(Trichloromethylthio)cyclohex-4-ene-1,2-dicarboximide
Carboxin	5,6-Dihydro-2-methyl-1,4 oxathiin-3-carboxanilide
Chlorothalonil	Tetrachloroisophthalonitrile
Fenaminosulf	*p*-Dimethylamino benzenediazo sodium sulphonate
Mancozeb	Complex of zinc and manganese: ethylene-1,2 bisdithiocarbamate (maneb) with 20% manganese and 2.5% zinc
Quintozene (PCNB)	Pentachloronitrobenzene
Thiabendazole	2(Thiazol-4-yl)benzimidazole
Thiram (TMTD)	Tetramethylthiuram disulphide
Zineb	Zinc ethylene-1,2 bisdithiocarbamate

[a] As given in "Materials registered under the Agricultural Chemicals Act, 1955, 1978." South Australian Department of Agriculture, 31 January 1981.

Table 2. Properties of calcareous loamy sand soils used in pot experiments

Soil	pH	CaCO₃ (%)	Organic carbon (%)	Total nitrogen (%)	Total phosphorus (%)	Water at −0.1 bar (%)
Streaky Bay	8.6	84	1.3	0.13	0.05	24
Nunji-kompita	8.6	57	1.3	0.097	0.03	15

the fungicides—quintozene, benomyl, mancozeb, chlorothalonil, fenaminosulf, thiram, thiabendazole, and carboxin—were active at 25 or 50 mg/kg of soil (Tables 3 and 4). In Streaky Bay soil, the total length of lesions on roots caused by *G. graminis* var. *tritici* was increased by chlorothalonil, fenaminosulf, thiram, and captan.

When inoculum was applied at a lower rate (0.1% ground oat inoculum) and the selected fungicides were tested, quintozene and benomyl gave the greatest control, whereas mancozeb and zineb gave some control in both soils. Thiram significantly reduced the incidence and length of lesions in the Nunjikompita soil. Aldicarb had no effect on take-all (Tables 5 and 6).

Table 5 shows an inverse relationship between the incidence of lesions caused by *G. graminis* var. *tritici* and *H. avenae*. Roots with heavy *G. graminis* var. *tritici* infection had few *H. avenae* "knots," possibly because the nematode invades only root tips well supplied with assimilate—a situation that does not occur in roots with stelar lesions of *G. graminis* var. *tritici*. Benomyl and thiram reduced the damage caused to roots by *H. avenae*.

DISCUSSION

In these glasshouse experiments, both quintozene and benomyl applied to soil gave the greatest reduction of infection of roots by *G. graminis* var. *tritici*, whereas thiram, mancozeb, and zineb reduced infection to a lesser extent. Although Bateman (3) reported that coating seeds with benomyl gave effective control of take-all in pots, neither Dolezal and Jones (7) nor A. Simon (*personal communication*) obtained control by coating seeds with benomyl. Subsequently, it has been reported that seed treatment with triadimefon (Bayleton) (7) and triadimenol (Baytan 150FS) (4) reduced take-all on roots and increased yields.

A problem with the use of chemicals such as quintozene and benomyl is that they are active against many pathogens other than *G. graminis* var. *tritici*, and altering the balance of fungi in the rhizosphere may exacerbate the effects of other pathogens. The increased infection of roots by *G. graminis* var. *tritici* following the use of several fungicides suggests that such chemicals modify the soil and/or rhizosphere microflora in such a way that the *G. graminis* var. *tritici* is more damaging.

Table 3. Effect of fungicides on the incidence and length of take-all lesions on wheat roots after 4 weeks at 15°C in Streaky Bay soil with 0.2% ground oat inoculum

Fungicide	No. of roots with lesions[a] at rate of			Total length of lesions[b] at rate of		
	10 mg[c]	25 mg	50 mg	10 mg	25 mg	50 mg
Benomyl	23	13	6	550	270	141
Captan	21	17	30	588	600	840
Carboxin	22	19	25	567	530	570
Chlorothalonil	19	23	24	450	640	720
Fenaminosulf	22	25	29	562	662	790
Mancozeb	18	17	22	475	425	560
Quintozene	18	12	6	453	300	108
Thiabendazole	19	...[d]	...	510	...	...
Thiram	27	27	30	780	779	766
Zineb	22	24	28	570	682	731
LSD (5%)		7			211	

[a]Number of seminal roots with lesions out of 30 examined in each replicate. Control (no fungicide) = 23.
[b]Total length (cm) of take-all lesions on 30 roots. Control (no fungicide) = 510.
[c]Per kilogram of soil.
[d]Phytotoxic to the point of killing the plants at 25 and 50 mg/kg soil.

Table 4. Effect of fungicides on the incidence of take-all lesions and disease rating on wheat roots after 4 weeks at 15°C in Nunjikompita soil with 0.2% ground oat inoculum

Fungicide	No. of roots with lesions[a] at rate of			Disease rating[b] at rate of		
	10 mg[c]	25 mg	50 mg	10 mg	25 mg	50 mg
Benomyl	30	9	0	3.4	1.0	0
Captafol	NR[d]	NR	30	NR	NR	3.2
Captan	NR	NR	30	NR	NR	4.9
Carboxin	30	30	21	3.4	3.0	3.0
Chlorothalonil	30	30	24	4.1	3.3	2.8
Fenaminosulf	NR	NR	30	NR	NR	5.0
Mancozeb	30	30	30	4.2	3.1	2.8
Quintozene	30	24	0	3.7	2.1	0
Thiabendazole	NR	30[e]	...[e]	NR	2.7[e]	...[e]
Thiram	30	30	30	3.3	3.5	3.1
Zineb	30	30	30	3.2	3.7	3.0
LSD (5%)		10			1.0	

[a]Number of seminal roots with lesions out of 30 roots from five plants in each replicate pot; values are the means of two replicate pots. Control (no fungicide) = 30.
[b]Rating of 0–5 (maximum disease) (5). Control (no fungicide) = 4.5.
[c]Per kilogram of soil.
[d]Not recorded.
[e]Phytotoxic to the point of killing the plants at 50 mg/kg of soil; at 25 mg/kg of soil, plants were severely stunted.

Table 5. Effect of fungicides and aldicarb on take-all on wheat roots after 4 weeks at 15°C in Streaky Bay soil with 0.1% ground oat inoculum

Treatment Inoculum[a]	Chemical[b]	No. of roots with lesions[c]	Total lesion length[d]	Disease rating[e]	*Heterodera avenae* knots[f]
Nil	Nil	6.4	74	0.8	40
0.1% dead	Nil	6.6	122	1.0	40
0.1% live	Nil	25	690	2.6	1.5
	Benomyl	0.4	3	0.1	1
	Mancozeb	20.8	379	2.1	70
	Quintozene	1.8	12	0.5	70
	Thiram	25	675	2.5	1.5
	Zineb	21	413	2.1	40
	Aldicarb	25	703	2.2	0
	Benomyl + aldicarb	0.0	0	0	0
	Mancozeb + aldicarb	18.6	377	2.0	1.5
	Quintozene + aldicarb	1.0	8	0.2	0
	Thiram + aldicarb	25	703	2.3	0
	Zineb + aldicarb	25	700	2.1	0
LSD (5%)		3.0	102	0.3	NC[g]

[a] *Gaeumannomyces graminis* var. *tritici*.
[b] Fungicides applied at 50 mg/kg of soil, aldicarb at 10 mg/kg of soil.
[c] Number of seminal roots out of 25 examined in each replicate.
[d] Total length (cm) of take-all lesions on 25 roots from five plants from each replicate pot.
[e] Rated on 0–5 scale (5).
[f] Per root system. Rated according to Simon (13).
[g] Not calculated.

Table 6. Effect of fungicides and aldicarb on take-all on wheat roots after 4 weeks at 15°C in Nunjikompita soil with 0.1% ground oat inoculum

Treatment Inoculum[a]	Chemical[b]	No. of roots with lesions[c]	Total lesion length[d]	Disease rating[e]
Nil	Nil	0	0	0
0.1% dead	Nil	0	0	0
0.1% live	Nil	23.4	700	2.2
	Benomyl	0	0	0
	Mancozeb	20.5	435	2.1
	Quintozene	2.2	9	0.5
	Thiram	21.0	509	2.0
	Zineb	18.8	472	2.1
	Aldicarb	25.0	687	2.2
	Benomyl + aldicarb	0.4	4	0.2
	Mancozeb + aldicarb	17.0	417	2.0
	Quintozene + aldicarb	1.8	14	0
	Thiram + aldicarb	19.0	508	2.0
	Zineb + aldicarb	14.6	336	2.1
LSD (5%)		3.0	88	0.5

[a] *Gaeumannomyces graminis* var. *tritici*.
[b] Fungicides applied at 50 mg/kg of soil, aldicarb at 10 mg/kg of soil.
[c] Number of seminal roots out of 25 examined in each replicate.
[d] Total length (cm) of take-all lesions on 25 roots from five plants from each replicate pot.
[e] Rated on 0–5 scale (5).

There is a need for further studies in the area of microbial interactions in the rhizoplane.

The finding that aldicarb neither increased nor decreased infection by *G. graminis* var. *tritici* is reassuring considering that it has been used in soils with both take-all and *H. avenae*. However, the report that aldicarb applied to potatoes increased damage by *Rhizoctonia solani* (8) needs to be considered when undertaking trials on *H. avenae* in southern Australia because *H. avenae* and *R. solani* often occur together (9).

Field tests are now required to assess the effectiveness of the fungicides under the difficult growing conditions often experienced in Australian dryland wheat farming. In the initial field trials, the levels of fungicides need not be limited for economic reasons—their primary use at first could be as a tool to assess crop losses caused by cereal root diseases. Then, with more reliable crop loss assessment data for root diseases than we have now, farmers and chemical companies would be in a position to assess the feasibility of chemical control. Such an approach with cereal cyst nematode (*H. avenae*) in southern Australia has led to the development of management strategies for the integrated control of this disease (12).

LITERATURE CITED

1. Bateman, G. L. 1980. Prospects for fungicidal control of take-all of wheat. Ann. Appl. Biol. 96:275-282.
2. Bateman, G. L. 1981. Effects of soil application of benomyl against take-all (*Gaeumannomyces graminis*) and footrot diseases of wheat. Z. Pflanzenkr. Pflanzenschutz 88:249-255.
3. Bateman, G. L. 1982. Formation of soil-applied fungicides for controlling take-all (*Gaeumannomyces graminis* var. *tritici*) in experiments with pot-grown wheat. Z. Pflanzenkr. Pflanzenschutz 89:480-486.
4. Bockus, W. W. 1983. Effects of fall infection by *Gaeumannomyces graminis* var. *tritici* and triadimenol seed treatment on severity of take-all in winter wheat. Phytopathology 73:540-543.
5. Cook, R. J., and Rovira, A. D. 1976. The role of bacteria in

the biological control of *Gaeumannomyces graminis* by suppressive soils. Soil Biol. Biochem. 8:269-273.

6. Cook, R. J., Sitton, J. W., and Waldher, J. T. 1980. Evidence for *Pythium* as a pathogen of direct-drilled wheat in the Pacific Northwest. Plant Dis. 64:102-103.

7. Dolezal, W. E., and Jones, J. P. 1980. A systemic seed control treatment of take-all disease of wheat. Arkansas Farm Res. 29:10.

8. Leach, S. S., and Frank, J. A. 1982. Influence of three systemic insecticides on Verticillium wilt and Rhizoctonia disease complex of potato. Plant Dis. 66:1180-1182.

9. Meagher, J. W., Brown, R. H., and Rovira, A. D. 1978. The effect of cereal cyst nematode (*Heterodera avenae*) and *Rhizoctonia solani* on the growth of wheat. Aust. J. Agric. Res. 29:1127-1137.

10. Northcote, K. H., Hubble, G. D., Isbell, R. F., Thompson, C. H., and Bettenay, E. 1975. A Description of Australian Soils. CSIRO Publ., Aust. 170 pp.

11. Rovira, A. D., and Ridge, E. H. 1979. The effect of methyl bromide and chloropicrin on some chemical and biological properties of soil and on the growth and nutrition of wheat. Pages 231-250 in: Soil Disinfestation. D. Mulder, ed. Elsevier, Amsterdam.

12. Rovira, A. D., and Simon, A. 1982. Integrated control of *Heterodera avenae*. EPPO Bull. 12:517-523.

13. Simon, A. 1980. A plant assay of soil to assess potential damage to wheat by *Heterodera avenae*. Plant Dis. 64:917-919.

Integrated Control of Root Rot of Soybean Caused by *Phytophthora megasperma* f. sp. *glycinea*

A. F. SCHMITTHENNER and D. M. VAN DOREN, JR., Departments of Plant Pathology and Agronomy, Ohio Agricultural Research and Development Center, Ohio State University, Wooster, 44691, U.S.A.

Phytophthora root rot caused by *Phytophthora megasperma* f. sp. *glycinea* is the most destructive disease of soybeans in the upper midwest region of the United States. Development of resistance to this disease has been complicated by the many different races of the fungus, now a total of 22 (13). Genes for resistance have been identified at six different loci (2). However, isolates have been found in soil in Ohio that are virulent to most of these resistant genes (9). If these virulent races build up, root rot may no longer be controlled by resistance. Integrated control is being investigated as an alternative.

The key to integrated control is cultivars that are not severely damaged by *P. megasperma* f. sp. *glycinea*, even though susceptible and infected. This disease-limiting phenomenon has been referred to as field resistance (10), rate-reducing resistance (19), field tolerance (5,11,14,16,17), or tolerance (4,15). Other components of integrated control being studied are drainage, tillage, *Phytophthora*-selective fungicides, and crop rotation. This paper summarizes 2-year results of the effects of these components and some of their interactions upon soybean yield in *Phytophthora*-infested soil.

MATERIALS AND METHODS

Experiments were conducted at two sites in Ohio. The first was at the Northwest Branch of the Ohio Agricultural Research and Development Center, Custar, in Hoytville silty clay loam, a Mollic Ochraqualf (16.7% sand, 42.1% silt, and 41.2% clay). The second was at the North Central Branch, Vickery, in a Fulton silty clay, a Mollic Haplaquept (9.0% sand, 50.3% silt, and 40.7% clay). The predominant *P. megasperma* f. sp. *glycinea* races were 1, 3, and 7 at both sites. Cultural factors were evaluated at the Custar site only. The fungicide metalaxyl—DL-methyl *N*-[2,6-dimethyl-phenyl]-*N*-[2-methoxyacetyl]alaninate—was evaluated at both sites. Cultivars were Beeson 80 (resistant to races 1–3, 6–11, 13, 15, 17, and 21) used as the resistant treatment, Voris 295 and Goldtag 1250 (resistant to race 1 and with high tolerance to other races), and Beeson (resistant to race 1 or Sloan with no resistance, both with low tolerance to races 1, 3, and 7). The fungicide metalaxyl was used as a soil treatment (1.12 kg/ha) with Beeson in 1980 and as a seed treatment (0.31 g/kg) with Goldtag 1250 in 1982.

Journal article 112-83 of the Ohio Agricultural Research and Development Center.

Cultural factors evaluated were drainage (tile or no tile), tillage (zero or complete), seedbed surface (flat or ridged), and rotation (continuous soybeans, or corn and soybeans, in alternate years). Alternating blocks of land measuring 90 m × 60 m either had no tile drainage or had tiles 15 m apart perpendicular to the crop rows at 1-m depth. All blocks were land leveled and shaped to provide surface water flow to surface drains at the ends of the plot rows as described by Triplett and VanDoren (20). Tillage and rotation variables were established in 1978 and 1979, respectively, in preparation for the

Table 1. Effect of selected cultivar, drainage, tillage, ridging, rotation, and metalaxyl soil or seed treatments on yield of soybean in *Phytophthora*-infested soil

Integrated control factor combinations[w]	Yield (kg/ha)[x] 1980	Yield (kg/ha)[x] 1982
Low-tolerant cultivar		
Flat, no tile, no till	1,510 a	1,772 a
+ Tile drainage	2,120 b	2,223 b
+ Complete tillage	1,523 a	NT[y]
+ Tile + tillage	2,483 c	NT
+ Tile + ridging	2,705 d	2,034 b
+ Tile + tillage + ridging	2,880 ef	NT
+ Tile + ridging + rotation	NT[z]	2,213 b
+ Metalaxyl soil treatment	3,330 ij	NT
+ Metalaxyl + tile + tillage + ridging	3,715 k	NT
High-tolerant cultivar		
Flat, no tile, no till	2,533 cd	2,680 cd
+ Tile drainage	2,475 c	2,645 c
+ Complete tillage	3,160 ghi	NT
+ Tile + tillage	3,260 ij	NT
+ Tile + ridging	3,130 fghi	2,689 cd
+ Tile + tillage + ridging	2,945 efgh	NT
+ Tile + ridging + rotation	NT	2,805 cd
+ Metalaxyl seed treatment	NT	2,898 de
+ Metalaxyl + tile + ridging + rotation	NT	3,085 ef
Resistant cultivar		
Flat, no tile, no till	3,408 ij	3,210 f
+ Tile drainage	3,310 ij	2,858 cde
+ Complete tillage	3,205 hi	NT
+ Tile + tillage	3,540 jk	NT
+ Tile + ridging	2,815 e	2,841 cde
+ Tile + tillage + ridging	2,905 efg	NT
+ Tile + ridging + rotation	NT	2,707 cd

[w] All possible combinations of cultivar (low tolerant = Beeson in 1980 or Sloan in 1982, high tolerant = VS295 in 1980 or Goldtag 1250 in 1982, resistant = Beeson 80); drainage (tile vs. no tile); tillage (zero vs. complete tillage); ridging (flat vs. ridged seedbed); rotation (continuous soybean vs. corn and soybean in alternate years); metalaxyl soil treatment (1.12 kg/ha) of Beeson in 1980; metalaxyl seed treatment (0.31 g/kg of seed) of Goldtag 1250 in 1982.

[x] Numbers followed by the same letter are not significantly different based on LSD for *P* = 0.05.

[y] Not tested; tillage factor missing.

[z] Not tested; rotation factor not analyzed.

experiment. Complete tillage was fall moldboard plowing to 20 cm + spring preplant preparation with a field cultivator to 10 cm for a flat seedbed, or fall chisel plowing to 20 cm + fall ridging with disk-hillers. Ridges were spaced at a distance of 152 cm, 30 cm high and 80 cm wide at the top for two rows spaced 76 cm apart. Zero-tillage ridges were permanent beds spaced at a distance of 152 cm, 30 cm high and 80 cm wide for two rows spaced 76 cm apart. Some soil was moved from furrow to ridge top each spring to maintain ridge height. Most combinations of the above factors were tested in 2 years. Rotation was not evaluated in 1980. The nontiled plots were destroyed by flooding in 1981. Wet weather prevented fall tillage in 1981, so the tillage factor was missing in 1982.

Metalaxyl was evaluated at both sites under no tile, complete tillage, and flat seedbed cultural conditions. In eight tests over a 3-year period, seed treatment rates of 0.31 or 0.62 g/kg of seed and a soil treatment rate of 1.12 kg/ha were compared on Beeson 80, Voris 295, and either Beeson or Sloan. Several lower soil fungicide rates were also tested. Soil treatments were applied as an 18-cm spray or granule band to the soil surface over the seed furrow. Metalaxyl was applied during planting after the seed had dropped into the furrow and before the furrow closed; thus, it was partially incorporated with the press wheel.

Plot size varied in different tests from one to four rows 10–30 m long and spaced 76 cm apart. For the cultural factor tests, the main plot was drainage; the split plots were tillage, seedbed surface, and rotation; and the split-split plots were the cultivars and single metalaxyl treatment. In the metalaxyl tests, the main plot was the cultivar and the split plot the fungicide treatments. Stand and yield data were taken. All data were subjected to a factorial analysis of variance. LSD values (Prob. $F = 0.05$) for the highest order significant interaction were used for means comparisons.

RESULTS AND DISCUSSION

The first cycle of testing multiple cultural factors utilized Beeson 80, Voris 295, and Beeson as the resistant, high-tolerant, and low-tolerant cultivars, respectively. Rainfall was higher than normal with 7.4 cm in May, 9.7 cm in June, 15.6 cm in July, and 18.8 cm in August. Plots were planted on 9 May 1980 and received the first significant rainfall 2 to 4 days after planting. The rotation component was not evaluated but was used as replication for other factor combinations. Stands were adequate (nine plants per meter) in all treatments except the nontiled, low-tolerant cultivar combination. In general, the lowest yield was obtained with the low-tolerant cultivar in nontiled soil with zero tillage and a flat seedbed. Cultivar, tiling, tillage, and fungicide had positive effects on yield. Ridging reduced yield except in certain combinations. Some of the more interesting positive effects of cultivar combinations are summarized in Table 1. Tile drainage, tile + tillage, tile + ridging, tile + tillage + ridging, and application of metalaxyl to soil all significantly increased yield of the low-tolerant cultivar. Yield of the high-tolerant cultivar was higher than of the low-tolerant cultivar in all comparisons; in contrast to the low-tolerant cultivar, tillage had a large effect on yield, whereas tiling had

none. The high-tolerant cultivar under the best cultural conditions was as good as the resistant cultivar.

In the second testing cycle (1982), Beeson 80, Goldtag 1250, and Beeson were the resistant, high-, and low-tolerant cultivars, respectively. The planting season was very dry, with no significant rain between 4 April and 22 May. Plots were planted on 27 April. There was no tillage factor because wet weather precluded fall tillage operations in 1981. Metalaxyl seed treatment of the high-tolerant cultivar was substituted for metalaxyl soil treatment of the low-tolerant cultivar because the former is more economically acceptable ($5 vs. $130/ha). Again, the poorest stands (fewer than nine plants per meter) and lowest yields were in the low-tolerant cultivar, poorly drained, flat-surface treatment combinations. In general, only cultivar and drainage increased yield. Ridging and rotation reduced yield except in certain combinations. Metalaxyl seed treatment was effective only in certain combinations (Table 1). Tiling significantly increased the yield of the low-tolerant cultivar only. The combination of metalaxyl seed treatment + tile + ridging + rotation significantly increased yield of the high-tolerant cultivar. The highest yield of the resistant cultivar was under cultural conditions most conducive for Phytophthora root rot, probably because these conditions were most favorable for water conservation.

Metalaxyl seed and soil treatments were compared in separate tests because space limitations precluded imposing fungicide treatments on all other cultural combinations. Data for yields from eight tests are summarized in Table 2. Lowest yields were obtained with the low-tolerant, untreated cultivar. Seed treatments increased yields of the low-tolerant cultivar significantly, and soil treatments increased yields even more. Yields of the high-tolerant cultivar were significantly higher than the low-tolerant cultivar for untreated and metalaxyl-treated seed ($P = 0.05$). The highest yield was obtained with metalaxyl soil treatment combined with high tolerance ($P = 0.05$). Neither metalaxyl seed or soil treatment increased the yield of the resistant cultivar.

It was assumed in these tests that most of the poor yields were the result of Phytophthora root rot damage even though plant loss did not occur extensively. This assumption is based on three observations. First, the yield potential of the cultivars used was approximately the same. Beeson 80, Voris 295, and Goldtag 1250 yielded 94.5, 98.7, and 103.5 as much as Sloan, respectively, in several tests in areas without Phytophthora root rot. Second, metalaxyl soil treatment

Table 2. Effect of metalaxyl on yield of soybean cultivars with different disease reactions in *Phytophthora*-infested soil (summary of eight tests)

Cultivar disease reaction[x]	Yield (kg/ha)		
	No metalaxyl	Metalaxyl seed treatment[y]	Metalaxyl soil treatment[z]
Low tolerance	1,256	1,928	2,678
High tolerance	2,340	2,707	2,966
Resistance	2,532	2,427	2,602
LSD (0.05) = 331 kg			

[x] Low tolerance is cultivar Sloan or Beeson, high tolerance is VS295, and resistance is Beeson 80.
[y] 0.31 g/kg of seed.
[z] 1.12 kg/ha, applied as an 18-cm band over the row at planting time.

completely eliminated low yields of the low-tolerant cultivar. Third, the performance of the multirace-resistant Beeson 80 was relatively uniform across all cultural and metalaxyl treatments.

Metalaxyl soil treatment was the most effective factor for control of Phytophthora root rot of soybeans. However, the rate used was too high to be economically acceptable. Anderson and Buzzell (1) have also concluded that high rates of soil-applied metalaxyl would be necessary for root rot control on low-tolerant cultivars. Seed treatment was considerably less effective and would provide satisfactory control only with certain cultivars. Cultivar-metalaxyl seed treatment interactions have been reported elsewhere (6). In preliminary tests not reported here, soil application rates of 280 g/ha were almost as effective as those of 1.12 kg/ha. More work is needed to verify that more economic soil rates are effective, especially with low-tolerant cultivars.

Cultivar tolerance was the next most significant factor for Phytophthora control in these tests. However, under Phytophthora-conducive cultural conditions (no tile, reduced tillage, flat seedbed), cultivar tolerance by itself was not as good as resistance. Combination of several disease-suppressive cultural factors + metalaxyl may be necessary for high-tolerant cultivars to provide acceptable Phytophthora control. There are many high-tolerant cultivars available to growers in the north central United States (3). Tolerant cultivars should be the starting point for integrated control, provided they can be identified.

Drainage was the third most significant factor in these tests. The importance of drainage to reduce Phytophthora damage in soybean has been reported elsewhere (7,12). Drainage was more important in the wet year, 1980, than in the dry year, 1982. Drainage had a greater impact on the yield of the low- than on the high-tolerant or the resistant cultivar, but did not completely reverse Phytophthora damage.

Complete tillage resulted in a higher yield in the high-tolerant cultivars. Phytophthora-conducive effects of conservation tillage have recently been reported (18). The explanation of these tillage effects is not available. Zero tillage might reduce drainage or soil bulk density, both of which affect Phytophthora severity (8,12). Also, inherent in both zero and conservation tillage is an accumulation of plant residue on the soil surface and generally close to the plant row. Complete tillage distributes residues throughout the plow zone, which might effectively dilute inoculum adjacent to the germinating seedling. Tillage effects need further study.

The effects of ridging and rotation were least pronounced. Ridging was effective with the low-tolerant cultivar and only in combination with tile and complete tillage. It is interesting to note that some of the practices designed to reduce excess water actually reduced the yield of the resistant cultivar in the dry year. Rotation was significant only with the high-tolerant cultivar in combination with metalaxyl seed treatment, tile, and ridging. It would appear from these 2 years of results that ridging and rotation might not be essential for integrated control of Phytophthora root rot of soybean. Additional information is needed to verify this conclusion.

Optimizing cultural conditions did not completely reverse Phytophthora damage in the high-tolerant cultivar. However, high tolerance + optimum cultural conditions + metalaxyl seed treatment were as effective as resistance. High-tolerant cultivars + metalaxyl soil treatment might even be superior, but work is needed to find an economic rate. It was concluded that a combination of high tolerance, tile drainage, complete tillage, and metalaxyl seed treatment would be as effective as multirace resistance for Phytophthora control. Integrated control is now feasible and will provide the soybean grower with a second option to control Phytophthora root rot.

ACKNOWLEDGMENTS

Salaries and research support were provided by state and federal funds appropriated to the Ohio Agricultural Research and Development Center. The research was partially supported by gifts from the Agricultural Division, Ciba-Geigy Corporation, and Gustafson Incorporated.

LITERATURE CITED

1. Anderson, T. R., and Buzzell, R. I. 1982. Efficacy of metalaxyl in controlling Phytophthora root and stalk rot of soybean cultivars differing in field tolerance. Plant Dis. 66:1144-1145.
2. Athow, K. L., and Laviolette, F. A. 1982. Rps_6, a major gene for resistance to Phytophthora megasperma f. sp. glycinea in soybean. Phytopathology 72:1564-1567.
3. Beuerlein, J. E., Ryder, G. J., Walker, A., and Schmitthenner, A. F. 1981. Ohio soybean performance trials—1981. Ohio State Univ. Agron. Dep. Ser. 212. 8 pp.
4. Buzzell, R. I., and Anderson, T. R. 1982. Plant loss response of soybean cultivars to Phytophthora megasperma f. sp. glycinea under field conditions. Plant Dis. 66:1146-1148.
5. Cartter, J. L., and Hartwig, E. E. 1963. The management of soybeans. Pages 161-239 in: The Soybean: Genetics, Breeding, Physiology, Nutrition, Management. A. J. Norman, ed. Academic Press, New York. 239 pp.,
6. Diatloff, A., Irwin, J. A. G., and Rose, J. L. 1983. Effects of systemic fungicidal seed dressings on the incidence of Phytophthora megasperma, nodulation, and nitrogen-fixation in two soybean cultivars. Aust. J. Exp. Agric. Anim. Husb. 23:89-90.
7. Dirks, V. A., Anderson, T. R., and Bolton, E. F. 1980. Effect of fertilizer and drain location on incidence of Phytophthora root rot of soybeans. Can. J. Plant Pathol. 2:179-183.
8. Fulton, J. M., Mortimore, C. G., and Hildebrand, A. A. 1961. Note on the relation of soil bulk density to the incidence of Phytophthora root and stalk rot of soybeans. Can. J. Soil Sci. 41:247.
9. Hobe, M. A., and Schmitthenner, A. F. 1981. Direct isolation of new races of Phytophthora megasperma var. sojae from NW Ohio soils. (Abstr.) Phytopathology 71:226.
10. Irwin, J. A. G., and Langdon, P. W. 1982. A laboratory procedure for determining relative levels of field resistance in soybeans to Phytophthora megasperma f. sp. glycinea. Aust. J. Agric. Res. 33:33-39.
11. Jimenez, B., and Lockwood, J. L. 1980. Laboratory method for assessing field tolerance of soybean seedlings to Phytophthora megasperma var. sojae. Plant Dis. 64:775-778.
12. Kittle, D. R., and Gray, L. E. 1979. The influence of soil temperature, moisture, porosity and bulk density on the pathogenicity of Phytophthora megasperma var. sojae. Plant Dis. Rep. 63:231-234.
13. Laviolette, F. A., and Athow, K. L. 1983. Two new physiologic races of Phytophthora megasperma f. sp. glycinea. Plant Dis. 67:497-498.
14. Probst, A. H., Athow, K. L., and Laviolette, F. A. 1966. Wayne soybean in Indiana. Indiana Agric. Stn. Res. Prog. Rep. 221. 2 pp.

15. Schmitthenner, A. F., and Walker, A. K. 1979. Tolerance versus resistance for control of Phytophthora root rot of soybeans. Pages 35-44 in: Soybean Seed Res. Conf., 9th. Am. Seed Trade Assoc. Publ. 9. 91 pp.

16. Slusher, R. L., and Sinclair, J. B. 1973. Development of *Phytophthora megasperma* var. *sojae* in soybean roots. Phytopathology 63:1168-1171.

17. Tachibana, H., Epstein, A. H., Nyvall, R. F., and Musselman, A. F. 1975. Phytophthora root rot of soybean in Iowa: Observations, trends, control. Plant Dis. Rep. 59:994-998.

18. Tachibana, H., and Van Diest, A. 1983. Association of Phytophthora root rot of soybean with conservation tillage. (Abstr.) Phytopathology 73:844.

19. Tooley, P. W., and Grau, C. R. 1982. Identification and quantitative characterization of rate-reducing resistance to *Phytophthora megasperma* f. sp. *glycinea* in soybean seedlings. Phytopathology 72:727-733.

20. Triplett, G. B., Jr., and VanDoren, D. M., Jr. 1963. Development of a drainage variable facility for soil and crop management studies on a lakebed clay soil. Ohio Agric. Exp. Stn. Res. Circ. 117. 25 pp.

Cropping Practices and Root Diseases

DONALD R. SUMNER, Department of Plant Pathology, University of Georgia, Coastal Plain Experiment Station, Tifton, 31793, U.S.A.

Research on the influence of cropping practices on root diseases was initiated long before fungicides and fumigants were commonly available to control root diseases. Excellent reviews have been published on earlier research on the effects of crop rotations, herbicides, soil fertility, irrigation, and other aspects of cropping on root diseases. This paper will focus primarily on recent research concerning the influence of cropping practices on root diseases.

TILLAGE

Tillage practices directly influence physical and chemical properties of the soil, such as soil moisture, aeration, temperature, and therefore root growth and nutrient uptake. The physical disruption of microhabitats and the movement of both soilborne plant pathogens and beneficial soil microorganisms by tillage may directly influence root disease severity. Conservation tillage and minimum tillage practices may increase, decrease, or have no effect on root diseases (37).

Tillage practices used in applying pesticides or during crop production may have an unintended effect on root and crown diseases. Root rot of wheat was increased following preemergence incorporation of triallate and trifluralin in wheat in Canada, but the increase was apparently related to the method of incorporation rather than to the herbicides (44). In multiple-cropping systems in the Georgia coastal plain, incorporation of nematicides with a rotary cultivator immediately following a peanut crop apparently increased the inoculum potential of *Pythium aphanidermatum* and subsequent severe damping-off in cucumber (39).

Deposition of soil on sugar beet crowns (hilling) from cultivation or preparing irrigation furrows by operating equipment at speeds of 6.4–12.8 km/hr increased incidence and severity of Rhizoctonia root rot in susceptible cultivars in Colorado (29). Turning soil 15–25 cm with a moldboard plow to bury organic debris has frequently been effective in reducing crown rot diseases caused by *Rhizoctonia* in vegetables, peanut, and other crops in Georgia (37). However, tillage practices have less influence on root diseases caused by *Pythium* spp. and *Fusarium* spp., and little is known about the influence of tillage practices on biological control of soilborne pathogens.

IRRIGATION

The influences of irrigation on root disease severity may vary among crops. Frequent light irrigations increased peanut pod rot in Israel (6) and Phytophthora root rot of avocado in California (47), but they reduced potato scab in Denmark (24) and storage rot in onions in India (32) when compared with infrequent, heavy irrigations. Flooding had a limited effect on the production of microsclerotia of *Verticillium dahliae* in tomato tissues (19), but flooding of paddy rice for 17 weeks reduced populations of *V. dahliae* below detectable levels and controlled Verticillium wilt of cotton for 2–3 years (26). Ponding of irrigation water increased potato tuber rot in silt loam and sandy loam soils (5). The longer the period of soil saturation, the more Phytophthora root rot occurred in citrus; and low oxygen reduced growth and regeneration of roots (33). There was more Phytophthora root rot in alfalfa in irrigated than in nonirrigated loam and silt loam soils in New York (46).

Irrigation reduced colonization of sound, mature peanut kernels by *Aspergillus flavus*, and therefore aflatoxin contamination, compared with drought stress. Irrigation caused a higher incidence of *A. niger* than drought. This may have prevented aflatoxin contamination of undamaged peanuts (9).

Plant pathogens have been found in canal water (17), surface irrigation pond water (31), and recycled irrigation water (43). Contaminated irrigation water may disseminate plant pathogens and could be a primary source of reinfestation of fumigated soil (31). In Israel, furrow irrigation disseminated the cotton wilt pathogen *Fusarium oxysporum* f. sp. *vasinfectum* (7).

Irrigation increased corn yields but also increased corn stalk rot (*F. moniliforme*) in Nebraska (34). In California, pretassel stress increased corn stalk rot but not postpollination and grain fill stress as compared with corn grown without water stress (30).

In the Georgia coastal plain, methods of irrigation scheduling having little influence on grain yield of corn had a significant effect on stalk rot of corn grown in Bonifay sand. When the average suction on tensiometers in the 15- to 30-cm zone was maintained below 30 centibar (cb), corn stalk rot was greater than when the average suction was maintained below 30 cb in the 15- to 60- or the 45- to 60-cm zone. The latter two treatments had fewer irrigations. Stalk rot percentage was always lower in nonirrigated than in irrigated corn, but grain yield was also less (D. R. Sumner and J. Hook, *unpublished*).

The increased use of overhead sprinkler irrigation has stimulated a great deal of interest in applying pesticides through irrigation water. Metham sodium has been applied through sprinkler irrigation water to control a Pythium pod rot disease complex and Verticillium wilt in peanut in Israel (18). The same method has been used to control root diseases of vegetables caused by *Pythium* spp. and *Rhizoctonia solani* in the Georgia coastal plain

(35). The fumigant has been applied successfully through both solid-set and center-pivot irrigation systems. Application through irrigation water has given more consistent control of *Pythium* spp. and *R. solani* than conventional soil injection with chisels. However, the chemical has low efficacy against *Fusarium* spp., and it probably could not be used in crops susceptible to Fusarium wilts and crown and root rots.

HERBICIDES

Herbicides may have the desirable consequence of reducing root disease severity by affecting the pathogen directly. Trifluralin and dinoseb interfere with zoospore production in *Aphanomyces euteiches* and reduce root rot in fields of pea where the fungus is the primary pathogen (12). However, in pea fields where root rot is caused primarily by a *Fusarium* complex, herbicides have not had a beneficial affect on root diseases (15). Dinoseb decreased root rot in bean (8); and terbacil, simazine, and linuron increased plant survival and reduced crown rot in asparagus (20).

In contrast, other herbicides may enhance the growth or the virulence of soilborne pathogens or predispose crop roots to invasion by root pathogens (1). In spring and fall crops of turnip in multiple-cropping systems in the Georgia coastal plain, herbicides contributed to increased root disease severity and decreased yields compared with cultivation without herbicides (40). In other multiple-cropping systems, chloramben caused root enlargements and apparently increased root disease severity in cucumber (39). A foliage application of mecoprop increased take-all disease and crown root malformations in barley in Sweden (23). In corn, pendimethalin increased crown and brace root rot induced by *R. solani* anastomosis group (AG) 2 type 2 and reduced root growth and grain yield on a sand soil in an early planting when soil temperatures were 5–18°C as compared with no herbicide treatment. The herbicide did not influence root disease severity in later plantings on sand, nor in any plantings on a loamy sand soil (38). Cycloate inhibits growth of sugar beet and increases root disease caused by *R. solani* in the western United States. The increased root disease and damping-off may be related to increased exudates from seedling hypocotyls (1).

Indirect effects of herbicides on soil microorganisms antagonistic to pathogens and on the soil microclimate may also influence root diseases (1). The influence of herbicides on the host-parasite relationship interacting with other cropping practices is difficult to evaluate. In many crops, the possible effects on root diseases should be considered when selecting herbicides.

OTHER PESTICIDES

Much research has been published on interactions of herbicides with root diseases, but nematicides and soil insecticides may influence root diseases in ways similar to those described for herbicides. One of the earliest discoveries of an interaction of a nematicide with root diseases was with 1,2-dibromo-3-chloropropane (DBCP). This nematicide reduced the inoculum potential of *R. solani* for 106 days in a field of Spanish peanut (2). In contrast, ethoprop and fensulfothion decreased whereas DBCP increased the incidence of *Sclerotium rolfsii* in peanut (27).

Ethoprop increased root disease severity in cowpea and spring crops of snapbean but not in fall crops of snapbean in multiple-cropping systems. Increased disease was associated with increased infection by *Pythium* spp. (41). This finding contrasts with decreased disease in peanut attributed to the toxic effects of ethoprop on *S. rolfsii* and *R. solani* (27).

Systemic insecticides and nematicides have had varying effects on root diseases in different crops. Aldicarb and phorate as side-dressings 1 month after planting increased Rhizoctonia root rot in sugar beet, but carbofuran did not (28). In potato, aldicarb increased Rhizoctonia disease and reduced marketable yields of tubers, but disulfoton and carbofuran did not (21). None of the pesticides influenced wilt severity caused by *V. albo-atrum*. Carbofuran has been associated with symptoms of seedling decline in sweet corn but not with increased root diseases (42).

MULTIPLE CROPPING

It has been known for several decades that planting into decomposing plant residues could increase root diseases and result in poor stands of stunted, deformed plants. In Connecticut, head lettuce grown after sweet corn and vetch and leaf lettuce grown after clover were severely injured (3). In California, spinach and lettuce planted into green residues of barley, rye, wheat, bean, Sudan grass, or broccoli germinated poorly and were usually stunted (25). The possibility of phytotoxins formed in decomposing residues should be considered in multiple-cropping systems, especially with green residues in cool, wet soils.

Because the relative susceptibility of crops to root pathogens is different, care should be taken in selecting crop sequences in multiple-cropping systems. In particular, continuous cropping of legumes, crucifers, cucurbits, and other vegetables should usually be avoided. Alternating grasses, wheat, rye, corn, or sorghum with legumes and vegetables in combination with moldboard plowing has contributed to reduced root disease severity in vegetables in the Georgia coastal plain. Peanut seed can serve as a reservoir for *R. solani* AG 2 type 2, which causes crown and brace root rot of corn; the disease appears to be unique to irrigated corn rotated with peanut (36).

OTHER CROPPING PRACTICES

Growing high-value crops with film mulches, soil fumigation, and trickle irrigation is becoming more common in many areas of the world. Soil fumigation with DD-MENCS (20% methylisothiocyanate + 80% chlorinated hydrocarbons [C_3]) allowed the production of three successive crops of cucumber or cucumber-squash-cucumber without additional fumigation. Populations of soilborne pathogens and root-knot nematodes were suppressed for 17 months (13). Adequate nutrition and watering can compensate for injury by root pathogens in some crops grown under film mulch, but in okra where a Fusarium wilt and root-knot nematode complex limits production, soil fumigation is essential for full-season multiharvests of plentiful, tender pods (14).

Solar heating with transparent polyethylene during a hot season before planting is a recent practice that has been useful in Israel, California, Jordan, Greece, and Italy for soil disinfestation and control of root diseases in vegetables, peanut, and cotton (16).

Soil fertility may influence the incidence and severity of root diseases, and recent studies have shown that changes in fertilization practices can substantially reduce injury by soilborne pathogens. Equipment to simultaneously add nitrification inhibitors, trace minerals, or soil sterilants to soil with anhydrous ammonia may facilitate field application while reducing take-all and root rot in wheat and corn stalk rot in the midwestern United States (10,11). Calcium plays an important role in peanut nutrition, and peanut cultivars having a high calcium requirement may be more susceptible to pod rot (45). Recent research indicates that sources of Ca may affect the disease complex causing pods. Applications of gypsum to peanut increased Ca in pegs and pods at mid-season and in hulls and kernels at harvest, and they reduced peg and pod rot compared with calcite lime at equivalent rates of Ca (4).

Other cropping practices that may influence root diseases are fluid drilling and fertilization with wastewater and sludge (22).

LITERATURE CITED

1. Altman, J. and Campbell, C. L. 1977. Effect of herbicides on plant diseases. Annu. Rev. Phytopathology 15:361-385.
2. Ashworth, L. J., Jr., Langley, B. C., and Thames, W. H, Jr. 1964. Long-term inhibition of *Rhizoctonia solani* by a nematicide, 1,2-dibromo-3-chloropropane. Phytopathology 54:187-191.
3. Cochran, V. W. 1949. Crop residues as causative agents of root rots of vegetables. Conn. Agric. Exp. Stn. New Haven Bull. 526. 34 pp.
4. Csinos, A. S., Gaines, T. P., and Walker, M. E. 1982. Influence of calcium source on peanut peg and pod rot complex. (Abstr.) Proc. Am. Peanut Res. Educ. Soc. 14:105.
5. Easton, G. D., and Nagle, M. E. 1977. Tuber rot—the result of ponding of irrigation water. Plant Dis. Rep. 61:1064-1066.
6. Frank, Z. R. 1967. Effect of irrigation procedure on Pythium rot of ground-nut pods. Plant Dis. Rep. 51:414-416.
7. Grinstein, A., Fishler, G., Katan, J., and Hakohen, D. 1983. Dispersal of the Fusarium wilt pathogen in furrow-irrigated cotton in Israel. Plant Dis. 67:742-743.
8. Hagedorn, D. J., and Binning, L. K. 1982. Herbicide suppression of bean root and hypocotyl rot in Wisconsin. Plant Dis. 66:1187-1188.
9. Hill, R. A., Blankenship, P. D., Cole, R. J., and Sanders, T. H. 1981. Effects of soil moisture and temperature on preharvest invasion of peanuts by the *Aspergillus flavus* group and subsequent aflatoxin development. Appl. Environ. Microbiol. 45:628-633.
10. Huber, D. M., Karamesines, P. D., Warren, H. L., and Nelson, D. W. 1981. Equipment to mix chemical additives with anhydrous ammonia during application to soil. Agron. J. 73:1046-1048.
11. Huber, D. M., Warren, H. L., Nelson, D. W., and Tsai, C. Y. 1977. Nitrification inhibitors—new tools for food production. Biol. Sci. 27:523-528.
12. Jacobsen, B. J., and Hopen, H. J. 1981. Influence of herbicides on Aphanomyces root rot of peas. Plant Dis. 65:11-16.
13. Johnson, A. W., Sumner, D. R., and Jaworski, C. A. 1979. Effect of film mulch, trickle irrigation, and DD-MENCS on nematodes, fungi, and vegetable yields in a multicrop production system. Phytopathology 69:1172-1175.
14. Johnson, A. W., Sumner, D. R., Jaworski, C. A., and Chalfant, R. B. 1977. Effects of management practices on nematode and fungi populations and okra yield. J. Nematol. 9:136-142.
15. Johnston, H. W., Ivany, J. A., and Cutcliffe, J. A. 1980. Effects of herbicides applied to soil on Fusarium root rot of processing peas. Plant Dis. 64:942-943.
16. Katan, J. 1981. Solar heating (solarization) of soil for control of soilborne pests. Annu. Rev. Phytopathol. 19:211-236.
17. Klotz, L. J., Wong, P.-P., and DeWolfe, T. A. 1959. Survey of irrigation water for the presence of *Phytophthora* spp. pathogenic to citrus. Plant Dis. Rep. 43:830-832.
18. Krikun, J., and Frank, Z. R. 1982. Metham sodium applied by sprinkler irrigation to control pod rot and Verticillium wilt of peanut. Plant Dis. 66:128-130.
19. Ioannou, N., Schneider, R. W, and Grogan, R. G. 1977. Effect of flooding on the soil gas composition and the production of microsclerotia by *Verticillium dahliae* in the field. Phytopathology 67:651-656.
20. Lacy, M. L. 1979. Effects of chemicals on stand establishment and yields of asparagus. Plant Dis. Rep. 63:612-616.
21. Leach, S. S., and Frank, J. A. 1982. Influence of three systemic insecticides on Verticillium wilt and Rhizoctonia disease complex of potato. Plant Dis. 66:1180-1182.
22. Lumsden, R. D., Lewis, J. A. and Millner, P. D. 1983. Effect of composted sewage sludge on several soilborne pathogens and diseases. Phytopathology 73:1543-1548.
23. Nilsson, H. E., and Larsson, B. 1978. Effect of mecoprop and bentazon on take-all disease and plant growth of barley. Swed. J. Agric. Res. 8:203-207.
24. Ostergaard, S. P., and Nielson, S. 1979. Control of potato scab (*Streptomyces scabies*) by irrigation. Tidsskr. Planteavl 83:201-204.
25. Patrick, Z. A., Toussoun, T. A., and Snyder, W. C. 1963. Phytotoxic substances in arable soils associated with decomposition of plant residues. Phytopathology 53:152-161.
26. Pullman, G. S., and DeVay, J. E. 1981. Effect of soil flooding and paddy rice culture on the survival of *Verticillium dahliae* and incidence of Verticillium wilt in cotton. Phytopathology 71:1285-1289.
27. Rodriguez-Kabana, R., and Curl, E. A. 1980. Nontarget effects of pesticides on soilborne pathogens and disease. Annu. Rev. Phytopathol. 18:311-332.
28. Ruppel, E. G., and Hecker, R. J. 1982. Increased severity of Rhizoctonia root rot in sugar beet treated with systemic insecticides. Crop Prot. 1:75-81.
29. Schneider, C. L., Ruppel, E. G., Hecker, R. J., and Hogaboam, G. J. 1982. Effect of soil deposition in crowns on development of Rhizoctonia root rot in sugar beet. Plant Dis. 66:408-410.
30. Schneider, R. W., and Pendery, W. E. 1983. Stalk rot of corn: Mechanism of predisposition by an early season water stress. Phytopathology 73:863-871.
31. Shokes, F. M., and McCarter, S. M. 1979. Occurrence, dissemination, and survival of plant pathogens in surface irrigation ponds in southern Georgia. Phytopathology 69:510-516.
32. Singh, J. P., and Bhatnagar, D. K. 1981. Note on the irrigation of onion as a factor predisposing its bulbs to infection by *Fusarium solani* (Martius) Appel & Wollenweber in storage. Indian J. Agric. Sci. 51:686-687.
33. Stolzy, L. H., Letey, J., Klotz, L. J., and Labanauskas, C. K. 1965. Water and aeration as factors in root decay of *Citrus sinensis*. Phytopathology 55:270-275.
34. Sumner, D. R. 1968. The effect of soil moisture on corn stalk rot. Phytopathology 58:761-765.
35. Sumner, D. R. 1981. Application of foliar and soil fungicides and a soil fumigant through overhead irrigation water. Pages 82-88 in: Proc. Natl. Symp. Chemigation. Rural Development Center, Tifton, GA, 20-21 August 1981.
36. Sumner, D. R., and Bell, D. K. 1982. Root diseases induced in corn by *Rhizoctonia solani* and *Rhizoctonia zeae*. Phytopathology 72:86-91.

37. Sumner, D. R., Doupnik, B., Jr., and Boosalis, M. G. 1981. Effects of reduced tillage and multiple cropping on plant diseases. Annu. Rev. Phytopathol. 19:167-187.

38. Sumner, D. R., and Dowler, C. C. 1983. Herbicide, planting date, and root disease interactions in corn. Plant Dis. 67:513-517.

39. Sumner, D. R., Dowler, C. C., Johnson, A. W., Glaze, N. C., Phatak, S. C., Chalfant, R. B., and Epperson, J. E. 1983. Root diseases of cucumber in irrigated multiple-cropping system with pest management. Plant Dis. 67:1071-1075.

40. Sumner, D. R., Glaze, N. C., Dowler, C. C., and Johnson, A. W. 1979. Herbicide treatments and root diseases of turnip in intensive cropping systems. Plant Dis. Rep. 63:801-805.

41. Sumner, D. R., Johnson, A. W., Glaze, N. C., and Dowler, C. C. 1978. Root diseases of snapbean and southern pea in intensive cropping systems. Phytopathology 68:955-961.

42. Sumner, D. R., Young, J. R., Johnson, A. W., and Bell, D. K. 1982. Seedling decline and root diseases in sweet corn. Prot.

Ecol. 4:115-125.

43. Thomson, S. V., and Allen, R. M. 1974. Occurrence of *Phytophthora* species and other potential plant pathogens in recycled irrigation water. Plant Dis. Rep. 58:945-949.

44. Tinline, R. D., and Hunter, J. H. 1982. Herbicides and common root rot of wheat in Saskatchewan. Can. J. Plant Pathol. 4:341-348.

45. Walker, M. E., and Csinos, A. S. 1980. Effect of gypsum on yield, grade and incidence of pod rot in five peanut cultivars. Peanut Sci. 7:109-113.

46. Wilkinson, H. T., and Millar, R. L. 1982. Effects of soil temperature and moisture on activity of *Phytophthora megasperma* f. sp. *medicaginis* and alfalfa root rot in the field. Phytopathology 72:790-793.

47. Zentmeyer, G. A., and Richards, S. J. 1952. Pathogenicity of *Phytophthora cinnamomi* to avocado trees, and the effect of irrigation on disease development. Phytopathology 42:35-37.

Root Rot of Irrigated Subterranean Clover in Northern Victoria: Significance and Prospects for Control

PETER A. TAYLOR, R. G. CLARKE, and K. KELLY, Department of Agriculture, Victoria, Australia; and R. SMILEY, Department of Plant Pathology, Cornell University, Ithaca, NY 14853, U.S.A.

Subterranean clover (*Trifolium subterraneum* L.) is valued as a source of protein for livestock and is an important component of irrigated pasture in northern Victoria. A recent decline in its productivity is reducing stocking rates and causing considerable economic losses to farmers. Under natural conditions in southern Australia, seed of subterranean clover germinates after adequate autumn rain. The plants grow during autumn, winter, and spring and then die after setting seed. Frequently, shortages of feed occur in early winter, creating a need for early establishment of subterranean clover. This is achieved by irrigation before the onset of autumn rains, and it is under these conditions that clover decline problems become most apparent. Root disease has been associated with the decline in northern Victoria (5), but its effect on growth of subterranean clover in the field is not known. Therefore, the objectives of this study were to determine whether the decline is caused by root rot and to assess crop losses.

MATERIALS AND METHODS

Field study. A randomized block experiment was established in 1982 on an irrigated subterranean clover (cv. Yarloop) pasture in northern Victoria. The soil type was a Wana loam (8) of pH 5.9 (1:5 soil:water), comprising a shallow silty clay loam topsoil overlying deep, heavy clay. The following soil treatments were applied 2 days before the first irrigation: control, drenched with water at 5 L/m²; and fungicides, a soil application of metalaxyl (Ridomil 5G) at 0.5 g/m² a.i. and benomyl (Benlate 50 WP) at 2.5 g/m² a.i. The metalaxyl granules were scattered on the soil surface and then watered in with benomyl in suspension at 5 L/m².

Plant density and incidence and severity of root rot were assessed six times during the growing season (Fig. 1). At each date, three sods (400 cm² × 10 cm deep) were sampled from one-half of each 20-m² plot. The other half of the plot was mown in May and October for measurement of production of dry matter. Plants were washed free of soil and scored as follows: 0 = roots healthy; 1 = few rotted lateral roots; 2 = taproot lesions, but not girdling; 3 = girdling taproot lesion within 5 cm of crown; 4 = root system entirely rotted; 5 = plant dead. Root rot severity was expressed as the average rating of plants examined, and incidence of disease was recorded as the percentage of plants with ratings of one or more.

Greenhouse experiments. Twenty clover seeds (cv. Yarloop) were sown into pots (12 cm in diameter by 13 cm deep) containing pasteurized potting mix (1:1:1, peat:loam:coarse sand, by volume) and grown to the two to three trifoliate leaf stage. Air-dried, sieved (10-mm mesh) topsoil from the field trial site was added to the surface of the mix to form a layer about 10 mm deep. Benomyl and triadimefon were then applied as drenches

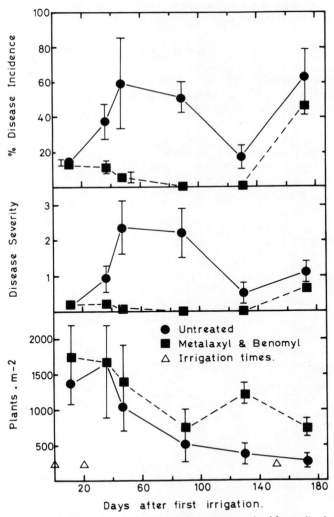

Fig. 1. Changes in subterranean clover root rot incidence (top), severity (middle), and plant population density (lower) with time in untreated and fungicide-treated plots in field trial. First irrigation (day 0) was in early autumn (3 March). Disease severity was averaged from individual ratings of plants on a scale from 0 (roots healthy) to 5 (plant dead). Bars represent standard deviations.

Postal address of first author: Irrigation Research Institute, Private Bag, Tatura, Victoria, 3616.

(Table 1) and metalaxyl and fenamiphos as dry granules to the surface soil. Phosethyl Al (Aliette 80 WP) was applied as a foliar spray when plants were at the first true leaf stage. The pots were saturated with water for 24 hr each week, and after 3–4 weeks the root systems were assessed for root rot as described for the field experiment. Temperatures in the greenhouse were 15–22° C. There were nine replicates per treatment in each experiment.

The procedure was modified for a test of the efficacy of metalaxyl as a seed dressing. The fungicide-treated seed (Apron 35 SD at 1.75 g a. i./kg of seed) was sown on the surface of the potting mix and covered immediately with field soil as described above. The pots were sprinkled with water daily until seedlings emerged, then saturated with water for 24 hr once each week.

Evaluation of cultivars for resistance. The cultivars Larisa, Trikkala, Enfield, and Woogenellup were screened for resistance to root rot in greenhouse and field experiments. Ten plants of each cultivar were grown in 10 replicate pots until 8 weeks old and then assessed for root disease. Field assessments were made in 1983 on replicated plots in a trial established in 1980. Sampling and assessment procedures were as described for the field study.

RESULTS

Field trial. After the first irrigation there was no evidence of preemergence damping-off, and the incidence and severity of disease were low (Fig. 1).

Table 1. Relative efficacy of selective chemicals against root rot of subterranean clover in greenhouse tests

Chemical and application rate[a]	Root rot severity[b]
Experiment 1	
Triadimefon, 0.3 g/m^2	3.7 (0.2)[c]
Benomyl, 3.0 g/m^2	3.5 (0.4)
Fenamiphos, 5.0 g/m^2	2.9 (0.4)
Metalaxyl, 0.6 g/m^2	0.1 (0.1)
Metalaxyl, 0.6 g/m^2, + benomyl, 3.0 g/m^2	0.1 (0.0)
Metalaxyl, 0.6 g/m^2, + fenamiphos, 5.0 g/m^2	0.4 (0.4)
Control (untreated)	3.4 (0.4)
Experiment 2	
Phosethyl Al, 1.25 g/m^2 (foliar spray)	0.1 (0.2)
Metalaxyl, 0.9 g/m^2	0
Metalaxyl, 0.3 g/m^2	0
Metalaxyl, 0.09 g/m^2	0
Metalaxyl, 0.03 g/m^2	0.1 (0.1)
Control (untreated)	3.6 (0.3)

[a] Application rates calculated on the basis of surface area of soil in pots.
[b] Average of ratings on a scale of 0 (roots healthy) to 5 (plant dead).
[c] Standard error in parentheses.

Table 2. Effect of pregermination soil treatments on total and subterranean clover dry matter at May and October harvests in a pasture affected by root rot

Treatment	Subterranean clover (t/ha)		Total (t/ha)	
	May	October	May	October
Untreated	0.10 (0.07)[a]	0.89 (0.51)	1.00 (0.24)	1.60 (0.31)
Metalaxyl + benomyl	0.40 (0.15)	2.60 (0.40)	1.59 (0.25)	3.13 (0.15)

[a] Standard deviation in parentheses; n = 3.

Incidence and severity of root rot in untreated plots increased after the second irrigation, and this was accompanied by severe foliar symptoms including wilting, chlorosis of cotyledons, and red discoloration of trifoliate leaves. The taproots of diseased plants were girdled by brown necrotic lesions 1–4 cm below soil level. Most lateral roots appeared healthy.

The metalaxyl/benomyl treatment reduced the incidence and severity of root rot in comparison with the control, and these treatment effects persisted for the duration of the experiment. In addition, the fungicide treatment maintained higher clover population densities than the controls at 130 and 180 days (Fig. 1) and increased total and clover dry matter production in both autumn and spring harvests (Table 2).

Activity of selective fungicides. In the greenhouse experiments, benomyl, triadimefon, and fenamiphos had no effect on the disease, whereas metalaxyl alone or in combination with benomyl or fenamiphos controlled root rot (Table 1). In the second experiment, the disease was controlled by metalaxyl at rates as low as 0.03 g/m^2 and by foliar sprays of phosethyl Al.

Seed treatment with metalaxyl protected the plants from root rot for 8 weeks. Root rot severity ratings were 0.1 ± 0.1 and 2.7 ± 0.6 for treated and untreated plants, respectively.

Resistance of cultivars. Greenhouse and field assessments of root rot severity and incidence showed that Larisa and Trikkala were resistant whereas Woogenellup and Enfield were susceptible (Table 3).

DISCUSSION

The positive responses with fungicides from the preliminary field and greenhouse tests indicate that root disease caused by pathogenic fungi is a major cause of the clover decline in irrigated pasture. Root rots of subterranean clover have been attributed to a complex of pathogens in the genera *Fusarium*, *Pythium*, and *Rhizoctonia* (1–5,7,9). In this study, the response to treatments with the selective fungicide metalaxyl (6) suggests that root rot in irrigated pastures is caused by a pythiaceous fungus.

Recently, a species of *Phytophthora* was associated with root rot of subterranean clover at widely scattered locations in northern Victoria and was highly pathogenic to established plants in pot tests (Taylor, *unpublished data*). The fungus rotted the taproot and produced symptoms similar to those observed on roots in the field. It is therefore more likely to be the causal

Table 3. Relative susceptibility to root rot of cultivars of subterranean clover in greenhouse and field experiments

Cultivar	Greenhouse assessment		Field assessment	
	Root rot severity[a]	Plants diseased (%)[b]	Root rot severity	Plants diseased (%)[c]
Larisa	0.07 (0.08)[d]	6	0.08 (0.16)	3
Trikkala	0.47 (0.39)	27	0.10 (0.12)	4
Enfield	2.64 (0.58)	90	2.84 (0.09)	97
Woogenellup	2.80 (0.27)	94	2.61 (0.46)	94

[a] Average of ratings on a scale of 0 (roots healthy) to 5 (plant dead).
[b] N = 100 plants.
[c] N = 120, with 30 plants subsampled at random from each of four plots.
[d] Standard error in parentheses.

272

organism of root rot in the field than *Pythium* spp., which only rot the lateral roots of established plants (9). Further research is needed to determine the precise etiology of the disease.

The results of the greenhouse tests with phosethyl Al and reduced application rates of metalaxyl present a possibility for control of root rot with these fungicides in the field, and this aspect is currently being investigated. The seed dressing of metalaxyl, although effective, would have only limited practical application because subterranean clover is self-seeding once established. The cultivar susceptibility studies provide encouraging prospects for control of root rot using the resistant cultivars Larisa and Trikkala. Further research to evaluate the susceptibility to root rot of a wide range of cultivars and breeding lines is in progress.

ACKNOWLEDGMENTS

We acknowledge financial support from the Australian Dairy Research Committee and thank Mr. R. Smith for the use of his land, Dr. P. R. Merriman for constructive advice, and Mr. J. Barrie for management of the cultivar field experiment.

LITERATURE CITED

1. Barbetti, M. J., and MacNish, G. C. 1978. Root rot of subterranean clover in the irrigation areas of south-west Western Australia. Aust. J. Exp. Agric. Anim. Husb. 18:426-433.
2. Burgess, L. W., Ogle, H. J., Edgerton, J. P., Stubbs, L. L., and Nelson, P. E. 1973. The biology of fungi associated with root rot of subterranean clover in Victoria. Proc. R. Soc. Victoria 86:19-28.
3. Kellock, A. W. 1972. A fungus that rots the roots of subterranean clover. J. Agric. (Victoria, Aust.) 70:112-113.
4. Ludbrook, W. V., Brockwell, J., and Riceman, D. S. 1953. Barepatch disease and associated problems in subterranean clover pastures in South Australia. Aust. J. Agric. Res. 4:403-414.
5. McGee, D. C., and Kellock, A. W. 1974. *Fusarium avenaceum*, a seed-borne pathogen of subterranean clover roots. Aust. J. Agric. Res. 25:549-557.
6. Schwinn, F. J., Staub, T., and Urech, P. A. 1977. A new type of fungicide against diseases caused by oomycetes. Meded. Fac. Landbouwwet. Rijksuniv. Gent 42:1181-1188.
7. Shipton, W. A. 1967. Fungi associated with "purple patch" of subterranean clover in Western Australia. Aust. J. Sci. 30:65-66.
8. Skene, J. K. M. 1963. Soils and land use in the Deakin Irrigation Area, Victoria. Dep. Agric. Victoria Tech. Bull. 16.
9. Stovold, G. E. 1974. Root rot caused by *Pythium irregulare* Buisman, an important factor in the decline of established subterranean clover pastures. Aust. J. Agric. Res. 25:537-548.

Solar Disinfestation of Soils

J. KATAN, Department of Plant Pathology and Microbiology, Hebrew University of Jerusalem, Faculty of Agriculture, Rehovot 76-100, Israel

Soil disinfestation is an effective but drastic means for controlling soilborne pathogens, insects, mites, weeds, and other soil pests by applying broad-spectrum physical or chemical killing agents to the soil, thus reducing populations of soilborne pests to a given soil depth. Because these killing agents are harmful to plants, soil disinfestation is usually carried out before planting.

Two basic approaches to soil disinfestation were developed in the early days of modern plant pathology: a physical one, such as artificially heating the soil by steam; and a chemical one, such as fumigation with CS_2. In the following decades, new technologies and fumigants were developed. Soil disinfestation is a complicated and often expensive and hazardous technique. It is used for pest control (mainly pathogens) as well as for special purposes—replanting, treating soil sickness, and improving plant health in soils with a long history of cropping.

Soil solarization is a third and new approach to soil disinfestation that is based on solar heating of the soil by covering (mulching, tarping) it with transparent polyethylene during the hot season, thereby increasing the temperatures and killing the pests. The polyethylene sheets laid on the soil serve as traps for capturing solar energy. Therefore, the concept of soil solarization combines the heating principle as a killing agent for pest control with that of soil heating by polyethylene mulching. The concept differs in principle from previous uses of polyethylene mulching in that the soil is heated during the hottest months (rather than the coldest) to increase the *maximal* temperatures to a level lethal to pests. This idea, based on observations of extension workers and growers in the Jordan Valley, who noticed that intensive heating occurs in mulched soils, was developed to a control method by a team including A. Greenberger, A. Grinstein, and H. Alon (21). Since that first publication in 1976, the many studies in this area, carried out in Israel and at least 13 additional countries (including the United States, Greece, Jordan, Italy, Australia, Egypt, Iraq, Portugal, England, South Africa, Japan, Morocco, and Pakistan [5,6,19,21,24, 28,29,31,36,37, *unpublished data*], deal mostly with pathogen control. However, the control of weeds and other pests has also been studied. No attempt will be made here to cover all studies or publications. Although this method is effective mainly in regions with climatic conditions that allow adequate heating of the soil, attempts have been made to adapt it to cooler regions (see below).

Soil solarization (also referred to as "solar heating," "solar pasteurization," or "soil tarping") was described in earlier works and reviews (18,19,21). In this article, although the principles and mechanisms involved will be described briefly, the emphasis will be placed on recent and future developments.

PRINCIPLES

Soil solarization is a technique for heating the soil by means of solar energy. Mulching the soil with transparent (not black) polyethylene sheets is, at present, the only means for achieving this purpose. Effective control by solar heating, providing climatic conditions are adequate, can be achieved under the following conditions: (a) Soil mulching should be carried out during the period of high temperatures and intense solar irradiation. (b) Soil should be kept moist to increase thermal sensitivity of resting structures and to improve heat conduction. (c) The thinnest polyethylene tarp possible (25–30 mm) is recommended because it is both cheaper and somewhat more effective. (d) Because the upper layers are more quickly and intensively heated than the lower ones, the mulching period should be long enough, usually 4 weeks or more, to achieve pest control at all desired depths. The longer the mulching period, the higher the pathogen-killing rates and the deeper its effectiveness (20,21,31). Four experiments carried out in Israel (*unpublished data*) showed the following duration of solarization required for killing 90–100% of *Verticillium dahliae* sclerotia: 3–6 days at 10 cm of soil depth, 14–20 days at 30 cm, 20–30 days at 40 cm, and 30–42 days at 50 cm.

Typical maximal temperatures in the solarized plots where effective disease and weed control were obtained were within the range of 45 to 50 and 38 to 45° C at depths of 10 and 20 cm, respectively, although temperatures that were 5–10° C higher were also recorded (12–17,20,21, 23,31,33,37).

There are important biological and technological differences between soil solarization and artificial soil heating by steam or other means, which is usually carried out at 60–100° C (3,4). Because there is no need to transport the heat from its source to the field, soil solarization can be carried out directly in the open field. Solar heating is carried out at relatively low temperatures, compared with artificial heating; therefore, its effect on living and nonliving soil components is likely to be less drastic. Thus, microbial activities occurring in solarized soils may lead to biological control of pathogens in addition to the thermal effect. Negative side effects, such as the phytotoxicity caused by release of manganese or other toxic products and a rapid soil reinfestation resulting from the creation of a biological vacuum (4,11) sometimes observed with soil steaming, have not yet been reported with solar heating; however, these possibilities should not be excluded a priori. Artificial soil heating is carried out as a single, short, and drastic treatment, whereas soil solarization can be regarded as an extended, mild pasteurization process that might reduce the chances for creating a biological vacuum in the soil. However, during solarization the soil is unavailable for crop growing, thus limiting its use.

CONTROL OF PESTS
AND INCREASES IN YIELD

Many studies have shown that solarization effectively reduces pathogen populations and disease incidence in various regions. In many cases these effects lasted for the whole season or even longer. The list of the controlled pathogens and diseases is long, with the following given as examples: Verticillium diseases (tomato, potato, eggplant, cotton, peanuts, pistachio, olive), *Rhizoctonia solani* (potato, onion, iris), *Sclerotium rolfsii* (peanuts), *Pyrenochaeta lycopersici* (tomato), *P. terrestris* (onion), Fusarium diseases (cotton, melon, tomato, onion, strawberry), the free nematode *Pratylenchus thornei* (potato), *Orobanche* (carrot, eggplant, broad bean), pod rot (peanuts), and *Ditylenchus dipsaci* (garlic) (2,9,10,12,14,15,17-24, 31,32,34,36, *unpublished information*). Disease control was usually accompanied by an increase in yield and quality. Yield increase is a function of disease reduction as well as the level of soil infestation and the damage caused to the crop by the disease. For example, controlling Verticillium and Fusarium diseases in cotton increased the yield by 60% (31) and 40–120% (20), respectively, whereas controlling corky root of tomatoes resulted in an increase of 140–350% (19,36). With *D. dipsaci* in garlic and *Orobanche* in carrots, yields were entirely destroyed in the nontreated plots, whereas those in the solarized plots were normal (17,34). In one experiment, storage quality of onions grown in solarized soil was improved (*unpublished data*). As expected, some pathogens are not controlled by solarization, e.g., *Macrophomina phaseolina* and *F. oxysporum* f. sp. *pini* (28). It is also ineffective in controlling melon and pepper collapse in Israel, and its effectiveness against *Meloidogyne* spp. is variable (19).

Reproducibility of results is a crucial parameter for evaluating any method for controlling soilborne pests, especially when dealing with one that is dependent on climatic conditions. Repeated experiments conducted over some years and in various locations have shown that, at least under Israeli and California conditions, this method is indeed reproducible with several diseases and crops. Reproducibility studies should, however, be carried out in every climatic region.

Weed control, an evident phenomenon in solarized soils already discovered in the earlier experiments (21), is considered an additional benefit to disease control. An increasing number of studies of weed control by solarization have recently been carried out (e.g., 16,33). In addition, solarization drastically reduces populations of the soil mite *Rhizoglyphus robini* (13).

Long-term effectiveness. A long-term effect, for more than one season after soil disinfestation, is a very desirable phenomenon that reduces cost per season and also indicates that the treatment did not create a biological vacuum. Such a long-term effect was found in solarized soils with Fusarium wilts of cotton (20) and

melon and four other diseases (22, *unpublished data*) and with Verticillium wilt of cotton (31) and potatoes (10). It might well be that combining solarization with other control methods will not only increase but also prolong its effectiveness. A long-term effect may result from drastic reduction in inoculum or from a shift in favor of the antagonists in the biological equilibrium in the soil, thus retarding soil reinfestation and pathogen population buildup (see below).

MECHANISMS OF PEST CONTROL
AND YIELD INCREASE

Thermal inactivation. Thermal death or inactivation of a population of an organism depends on both the temperature and exposure time, which are inversely related. In many cases, heat mortality curves have an exponential nature. Some studies investigating the time-temperature relationship of thermal killing have shown that straight lines are obtained by plotting the logarithms of the number of survivors against the exposure time units at a given temperature. Pullman et al (30) obtained a linear relationship by plotting the logarithm of the time required to kill 90% of the propagules of various soilborne fungal pathogens against the temperature. Other methods for such computations are also used (19).

Studying the heat sensitivity of a pathogen in heated baths or incubators may indicate the feasibility of its control by solarization; *V. dahliae* sclerotia, for example, being more sensitive to elevated temperatures than those of *S. rolfsii*, are also more easily controlled by solarization. However, we should be cautious when interpreting the significance of heat mortality curves obtained under laboratory conditions. The choice of an adequate inoculum type of a natural origin is crucial for meaningful results, and environmental conditions (e.g., moisture level) have to be adjusted. Heat sensitivity of resting structures also might differ when embedded in soil or heated in a water suspension. The course of heating in a bath also differs from the solarizing of soils. In the latter, it is necessary to account for the cumulative effect of the fluctuations in temperature during the day. Propagules might be weakened by sublethal heating, which may not be reflected in a reduction in their germinability. However, pathogenicity and survival might be adversely affected. Finally, during the extended process of solarization, biological control activities might operate. Thus, the rate of pathogen killing might be higher than expected from mere physical killing.

Biological control. Microbial processes in the soil have a pronounced impact on soilborne pathogens at all stages of their life cycle, in both the presence and absence of the host plant; consequently, these processes influence pathogen survival and disease incidence. Inoculum density and inoculum potential of the pathogens might be affected. Soil disinfestation (or any agricultural practice) affects soil microbial activities and may, finally, lead to a positive or negative effect on the pathogens depending on the shift created in the antagonistic population in the soil. The more drastic the effect of disinfestation on soil microorganisms, the more likely is the creation of a biological vacuum that brings quick reinfestation. Therefore, when artificial heating is used, relatively lower temperatures (60°C) are preferable to higher ones (80–100°C) because of the lowered chances of reinfestation and phytotoxicity (3,4,11). Soil solarization is carried out at even lower temperatures. The mechanisms of biological control, which may be created or stimulated by solarization (or any disinfestation method) (19), include effects on inoculum existing in or introduced to the soil after solarization.

Data from various experiments show that one or more biological control mechanisms might be induced in solarized soils. This may also explain the long-term effect of solarization in disease control mentioned earlier. Propagules of pathogens weakened by sublethal heating may be more easily attacked and killed by soil microorganisms, as shown with *S. rolfsii* (25). These mechanisms may contribute to pathogen control beyond the physical effect of heating. However, the possibility that solarization may have a detrimental effect on antagonists should not be excluded. Volatiles are involved in key processes in the soil. The permeability of polyethylene to many gases is not high; for example, CO_2 accumulates under plastic mulch up to 35 times more than in nonmulched soil (33). Thus, the possibility that heated volatiles may accumulate under the polyethylene mulch and play a role in pathogen control should be considered.

Increased growth response. The phenomenon of increased growth response (IGR) denotes the improvement in plant growth when disinfestation is carried out in soils free of known pathogens. This phenomenon was discovered in both artificially heated and in fumigated soils several decades ago. It was also found in solarized soils (8,35). The mechanisms for explaining IGR are either chemical (release of mineral nutrients or growth factors; nullification of toxins) or biological (elimination of minor or unknown pathogens; stimulation of beneficial microorganisms). Higher concentrations of mineral and organic substances were found in solutions of solarized soils. The growth of tomato seedlings in these extracts was enhanced (8).

The increase in yields resulting from IGR, in certain cases more than 50%, may be of great economic importance. Methods for IGR prediction in various soils should be developed. Improved plant growth in disinfested soils without a previous history of pathogens raises the question of whether this is related to plant health improvement beyond disease control. Plant health describes the relative freedom from biotic and abiotic stresses of the green plant and its ecosystem that limit their production to the maximum of their greatest potential over time (7).

PROCESSES OF SOIL HEATING BY SOLARIZATION

Solar heating of the soil depends on climatic factors (solar radiation and air temperature, humidity, velocity), soil characteristics, and polyethylene properties (e.g., transmissivity to short- and long-wave atmospheric infrared radiation) (26,38). Because the rate of heat flow through the soil depends on its thermal properties and the temperature gradient to which it is subjected, the temperature regime at a given depth can be predicted by equations for energy involved in soil heating. The temperatures in mulched soils were very similar (0.6–1.2°C differences) to those predicted using Mahrer's one-dimensional numerical model (26). This model enables us to choose the suitable climatic regions and the time of year most adequate for soil solarization, providing data on the heat sensitivity and population density of the pathogen at various depths are available and adequately interpreted. Moreover, the model enables the evaluation of the relative importance of each factor—type of mulching material, soil type, moisture, and climate. Analysis of the spatial soil temperature regime in mulched soil, using a more complicated numerical model, shows that the heating at the edges of the mulch is lower than at the center and that a narrow mulch strip is less efficient than a wider one (27).

In earlier experiments, soil was kept moist under the polyethylene tarp by means of a drip system enabling additional light irrigation during mulching. Drip irrigation is, however, not always available. Later studies show that satisfactory results can be obtained in many soils by a simpler method: soil is sprinkler irrigated 1–4 days before mulching, with no additional irrigations.

SOLARIZATION IN COOLER REGIONS

Various approaches might be followed in regions where climatic conditions are less favorable for solarization. First, the soil can be mulched inside a closed glasshouse or plastic house, thus improving the heating. This was successfully tried in northern Italy for controlling corky root of tomatoes (36) and in Japan for controlling Fusarium wilt of strawberries (24). Second, solarization can be combined with other methods of control (pesticides, biological control agents) either at full or reduced rates. This approach might also be considered in the hot regions for improving control of pests (especially those more tolerant to heat) and for extending the effectiveness of solarization. Whenever a pathogen or a pest is weakened by heating, a synergistic effect is expected; reduced dosages might thus be sufficient for improved control.

Third, certain pathogens might be controlled even in cooler regions. For example, solarization in Idaho, in the United States, successfully controlled Verticillium wilt and increased potato yield; the effect was still evident in the second year after solarization (10).

IMPROVEMENTS AND FUTURE RESEARCH

Where a long-term effect is achieved, costs are reduced by the less frequent need for solarization The use of thinner or used polyethylene also reduces costs. There are various research possibilities that may lead to future improvements: developing improved types of plastic materials with better heating efficiency for use in cooler regions; examining the effectiveness of solarization against additional pests; combining solarization with other methods of control; developing mulching machines for continuous soil coverage with polyethylene; using solarization for special purposes (e.g., with trees), as was successfully developed for controlling *V. dahliae* in pistachio (2) and for sterilizing contaminated tomato supports (5); developing techniques for plastic disposal or using degradable plastic materials; and carrying out long-term experiments in which solarization is repeated in some plots for detecting possible negative side effects.

Recent studies indicate that the fate of pesticides in solarized (or fumigated) soils is different from that in untreated ones (1). This may enable a reduction in soil pesticide application in solarized soils and the use of solarization as a tool for decontaminating soils from pesticides.

CONCLUSIONS

In an era when important pesticides and fumigants are banned from time to time, any additional potential method is very desirable. Soil solarization is a new method of control that has advantages and limitations. It is a simple, nonchemical, nonhazardous method that does not involve toxic materials and is, in the right instances, economical. It is, however, limited to certain regions and can be used only if the soil is free of crops for about a month. It is too expensive for certain crops and is ineffective for controlling some diseases. Solarization should not be regarded as a universal method, but rather as an additional option for pest control to be used alone or in combination with other methods. Should any negative side effects be detected, its use, in those specific situations, should be either avoided or combined with suitable preventive measures. Solarization is simple, easy to teach to less trained farmers, and can be used in small plots. It has, therefore, a great potential for developing countries. It is a pest management method that should be integrated in pest control plans. The extension service should therefore play a major role in translating the results from the research data to large commercial use.

ACKNOWLEDGMENTS

I wish to express my deep appreciation and thanks to A. Greenberger, A. Grinstein, and H. Alon for cooperating in a joint effort to develop the solarization method. Part of our studies was supported by grants provided by the Israel Ministry of Agriculture, the United States-Israel Binational Agricultural Research and Development Fund, and the United States-Israel Binational Science Foundation, Jerusalem, Israel.

LITERATURE CITED

1. Ahronson, N., Rubin, B., Katan, J., and Benjamin, A. 1982. Effect of methyl bromide or solar heating treatment on the persistence of pesticides in the soil. Pages 180-194 in: Pesticide Chemistry. Vol. 3. J. Miyamoto and P. C. Kearny, eds. Pergamon Press, Oxford.
2. Ashworth, L. J., Jr., and Gaona, S. A. 1982. Evaluation of clear polyethylene mulch for controlling Verticillium wilt in established pistachio nut groves. Phytopathology 72:243-246.
3. Baker, K. F. 1962. Principle of heat treatment of soil and planting material. J. Aust. Inst. Agric. Sci. 28:118-126.
4. Baker, K. F., and Cook, R. J. 1974 (original ed.). Biological Control of Plant Pathogens. Reprint ed., 1982. American Phytopathological Society, St. Paul, MN. 433 pp.
5. Besri, M. 1982. Solar heating (solarization) of tomato supports for control of Didymella lycopersici Kleb. stem canker. (Abstr.) Phytopathology 72:939.
6. Borges, L. V. 1982. Solarizacon do solo novo metodo de pasteuriacao do solo. Rev. Cienc. Agrar. 5:1-15.
7. Browning, J. A. 1983. Whither plant pathology? Whither plant health? Plant Dis. 67:575-577.
8. Chen, Y., and Katan, J. 1980. Effect of solar heating of soils by transparent polyethylene mulching on their chemical properties. Soil Sci. 130:271-277.
9. Chet, I., Elad, Y., Kalfon, A., Hadar, Y., and Katan, J. 1982. Integrated control of soilborne and bulbborne pathogens in iris. Phytoparasitica 10:229-236.
10. Davis, J. R., and Sorensen, L. H. 1983. Carry-over effects of plastic tarping on Verticillium wilt of potato. Abstr. Pap. Int. Congr. Plant Pathol. 4th, Melbourne, Australia. p. 157.
11. Dawson, J. R., Johnson, R. A., Adams, P., and Last, F. T. 1965. Influence of steam/air mixtures, when used for heating soil, on biological and chemical properties that affect seedling growth. Ann. Appl. Biol. 56:243-251.
12. Elad, Y., Katan, J., and Chet, I. 1980. Physical, biological, and chemical control integrated for soilborne diseases in potatoes. Phytopathology 70:418-422.
13. Gerson, U., Yathom, S., and Katan, J. 1981. A demonstration of bulb mite control by solar heating of the soil. Phytoparasitica 9:153-155.
14. Grinstein, A., Katan, J., Abdul-Razik, A., Zeidan, O., and Elad, Y. 1979. Control of Sclerotium rolfsii and weeds in peanuts by solar heating of soil. Plant Dis. Rep. 63:1056-1059.
15. Grinstein, A., Orion, D., Greenberger, A., and Katan, J. 1979. Solar heating of the soil for the control of Verticillium dahliae and Pratylenchus thornei in potatoes. Pages 431-438 in: Soilborne Plant Pathogens. B. Schippers and W. Gams, eds. Academic Press, London. 686 pp.
16. Horowitz, M., Regev, Y., and Herzlinger, G. 1983. Solarization for weed control. Weed Sci. 31:170-179.
17. Jacobsohn, R., Greenberger, A., Katan, J., Levi, M., and Alon, H. 1980. Control of Egyptian broomrape (Orobanche aegyptiaca) and other weeds by means of solar heating of the soil by polyethylene mulching. Weed Sci. 28:312-316.
18. Katan, J. 1980. Solar pasteurization of soils for disease control: Status and prospects. Plant Dis. 64:450-454.
19. Katan, J. 1981. Solar heating (solarization) of soil for control of soilborne pests. Annu. Rev. Phytopathol. 19:211-236.
20. Katan, J., Fishler, G., and Grinstein, A. 1983. Short- and long-term effects of soil solarization and crop sequence on Fusarium wilt and yield of cotton in Israel. Phytopathology 73:1215-1219.
21. Katan, J., Greenberger, A., Alon, H., and Grinstein, A. 1976. Solar heating by polyethylene mulching for the control of diseases caused by soilborne pathogens. Phytopathology 66:683-688.
22. Katan, J., Grinstein, A., Fishler, G., Frand, A. Z., Rabinowitch, H. D., Greenberger, A., Alon, H., and Zig, U. 1981. Long-term effects of solar heating of the soil. (Abstr.) Phytoparasitica 9:236.
23. Katan, J., Rotem, I., Finkel, Y., and Daniel, J. 1980. Solar heating of the soil for the control of pink root and other soilborne diseases in onions. Phytoparasitica 8:39-50.
24. Kodama, T., and Fukai, T. 1982. Solar heating in closed plastic house for control of soilborne diseases. V. Application for control of Fusarium wilt of strawberry. Ann. Phytopathol. Soc. Jpn. 48:570-577.
25. Lifshitz, R., Tabachnik, M., Katan, J., and Chet, I. 1983. The effect of sublethal heating on sclerotia of Sclerotium rolfsii. Can. J. Microbiol. 29:1607-1610.
26. Mahrer, Y. 1979. Prediction of soil temperature of a soil mulched with transparent polyethylene. J. Appl. Meteorol. 18:1263-1267.
27. Mahrer, Y., and Katan, J. 1981. Spatial soil temperatures regime under transparent polyethylene mulch: Numerical and experimental studies. Soil Sci. 131:82-87.
28. Old, K. M. 1981. Solar heating of soil for the control of nursery pathogens of Pinus radiata. Aust. For. Res. 11:141-147.
29. Porter, I. J., and Merriman, L. R. 1983. Effects of solarization of soil on nematode and fungal pathogens at two sites in Victoria. Soil Biol. Biochem. 15:39-44.
30. Pullman, G. S., DeVay, J. E., and Garber, R. H. 1981. Soil solarization and thermal death: A logarithmic relationship between time and temperature for four soilborne plant pathogens. Phytopathology 71:959-964.
31. Pullman, G. S., DeVay, J. E., Garber, R. H., and Weinhold,

A. R. 1981. Soil solarization: Effects on Verticillium wilt of cotton and soilborne populations of *Verticillium dahliae, Pythium* spp., *Rhizoctonia solani*, and *Thielaviopsis basicola*. Phytopathology 71:954-959.

32. Rabinowitch, H. D., Katan, J., and Rotem, I. 1981. The response of onion to solar heating, agricultural practices and pink root disease. Sci. Hortic. 15:331-340.

33. Rubin, B., and Benjamin, A. 1983. Solar heating of the soil: Effect on soil incorporated herbicides and weed control. Weed Sci. 31:819-825.

34. Sitti, E., Cohen, E., Katan, J., and Mordechai, M. 1982. Control of *Ditylenchus dipsaci* in garlic by bulb and soil treatments. Phytoparasitica 10:93-100.

35. Stapleton, J. J., and DeVay, J. E. 1982. Effect of soil solarization on populations of selected soilborne microorganisms and growth of deciduous fruit tree seedlings. Phytopathology 72:323-326.

36. Tamietti, G., and Garibaldi, A. 1980. Control of corky root in tomato by solar heating of the soil in greenhouse in Riviera Ligure. (In Italian) La Difesa delle Piente 3:143-150.

37. Usmani, S. M. H., and Ghaffar, A. 1982. Polyethylene mulching of soil to reduce viability of sclerotia of *Sclerotium oryzae*. Soil Biol. Biochem. 14:203-206.

38. Waggoner, P. E., Miller, P. M., and DeRoo, H. C. 1960. Plastic mulching principles and benefits. Conn. Agric. Exp. Stn. Bull. 623. 44 pp.

Soil Solarization: Effects on Fusarium Wilt of Carnation and Verticillium Wilt of Eggplant

G. E. ST. J. HARDY and K. SIVASITHAMPARAM, Soil Science and Plant Nutrition Group, School of Agriculture, University of Western Australia, Nedlands, 6009

Serious loss of production of eggplant (*Solanum melongena* L.) caused by *Verticillium dahliae* and of carnation (*Dianthus caryophyllus* L.) by *Fusarium oxysporum* f. sp. *dianthi* is a major concern for the small but significant horticultural industry in Western Australia. The Verticillium wilt of eggplant appears to be severe even after three times larger than recommended rates (1 kg/10 m^2) of soil fumigation. Soil solarization (1) was tested as a control method for both these diseases in soils with a history of severe disease.

MATERIALS AND METHODS

The trials were conducted at five sites in the southwest of Australia, which enjoys a Mediterranean type of climate; with the exception of one sandy site that had been amended with loam, all the sites were on plots on sand of the Karrakatta series (4). Clear, transparent polyethylene (0.1 mm thick) was used to tarp the soil, which was kept moist with trickle-drip irrigation. Treatments consisted of solarization for 14 days (S$_{14}$), 21 days (S$_{21}$), 30 days (S$_{30}$), methyl bromide fumigation, and untreated controls, all of which were replicated at least three times and laid in rows 1.8 m wide with lengths ranging from 3 to 10 m. A field site in Wanneroo was used for eggplants; another site in the same area was used for tunnel-house production of Sims carnations, and a site in Spearwood under cropping with field carnations was used for the carnation trials. Intact cores of soil (12.5 cm in diameter × 15 cm deep) from the affected eggplant and field carnation sites were buried

in the university experimental plots before treatments. Temperatures were measured by introducing the bulb of a field thermometer into the soil at required depths.

RESULTS

A significant rise in soil temperature following tarping was evident at all sites (Table 1), especially in the top layers of the soil. This effect was observed even at depths of 30 cm. A significant reduction in the "most probable number" of *Verticillium* and *Fusarium* propagules was observed as a result of mulching, the effect being most pronounced after solarization for 30 days (Table 2) (3). Disease reduction was evident in visual ratings (Fig. 1) made through time following tarping. The rapid reinvasion of the pathogen that occurred in the fumigation treatment was hardly noticeable in the eggplants and was totally absent in the carnations growing in solarized soils. Yield increases as a result of disease control were evident with both eggplants and carnations (Table 3). There was a significant increase in yield following 30 days of solarization for both crops. All weeds were controlled by mulching with the exception of *Portulaca oleracea*, which appeared to survive the treatment. Involvement of nutrients may be minimal in the increased productivity of mulched soils, as all treatments received standard fertilizer amendments.

Observations made during the most-probable-number assay (Table 2), indicating that soil solarization might be introducing suppressiveness to *F. oxysporum* f. sp.

Table 1. Soil temperatures (degrees centigrade) at five sites used to evaluate solarization as control for Fusarium wilt of carnation and Verticillium wilt of eggplant

Depth (cm)	University site 1[u] (17 Dec 1982–26 Jan 1983)		University site 2[u] (25 Feb–24 Mar 1983)		Paulick's site[v] (6 Dec 1982–5 Jan 1983)		Lishman's site[w] (5 Dec 1982–7 Jan 1983)		Boroni's site[x] (1 Dec 1982–3 Jan 1983)	
	Mean solarized	Mean bare[y]	Mean solarized	Mean bare	Mean solarized	Mean bare	Mean solarized	Mean bare	Mean solarized	Mean bare
5	53.6 (58.5)[z]	35.9 (39)	46.5 (51)	31.0 (36)	42.0 (49.5)	30.7 (39)	43.5 (53)	34.5 (43.5)	41.3 (48.5)	31.4 (37.5)
10	48.6 (54)	34.9 (39.5)	44.0 (47.5)	29.0 (32)	38.8 (46)	28.9 (35)	40.6 (57)	32.3 (38.5)	39.5 (41)	28.0 (36)
15	46.1 (51)	33.6 (38.5)	40.5 (43.5)	27.5 (29.5)	36.0 (44)	27.1 (34)	37.6 (48)	30.5 (34.5)	31.5 (38.5)	25.6 (32.5)
20	43.0 (47)	32.1 (37.5)	38.5 (40.5)	26.5 (28.5)	32.8 (39.5)	25.5 (33)	34.9 (44)	28.5 (33)	29.2 (34)	24.0 (30.5)
25	40.3 (45)	30.9 (36.5)	37.0 (38.5)	25.5 (27.5)	30.8 (36)	25.5 (32)	33.2 (39)	28.1 (30.5)	28.1 (31.5)	23.2 (29)
30	38.7 (42)	30.5 (35)	35.5 (38.5)	25.0 (26.5)	29.75 (35)	24.7 (30.5)	32.2 (37.5)	27.7 (30)	27.3 (30.5)	22.7 (28)
Average air temperature	21.9		31		22.1		22.1		22.2	

[u] Soil is a loamy sand.
[v] Field carnation site.
[w] Sims carnation in tunnel house.
[x] Field with eggplants.
[y] Not tarped.
[z] Figures within parentheses are absolute maximum temperatures.

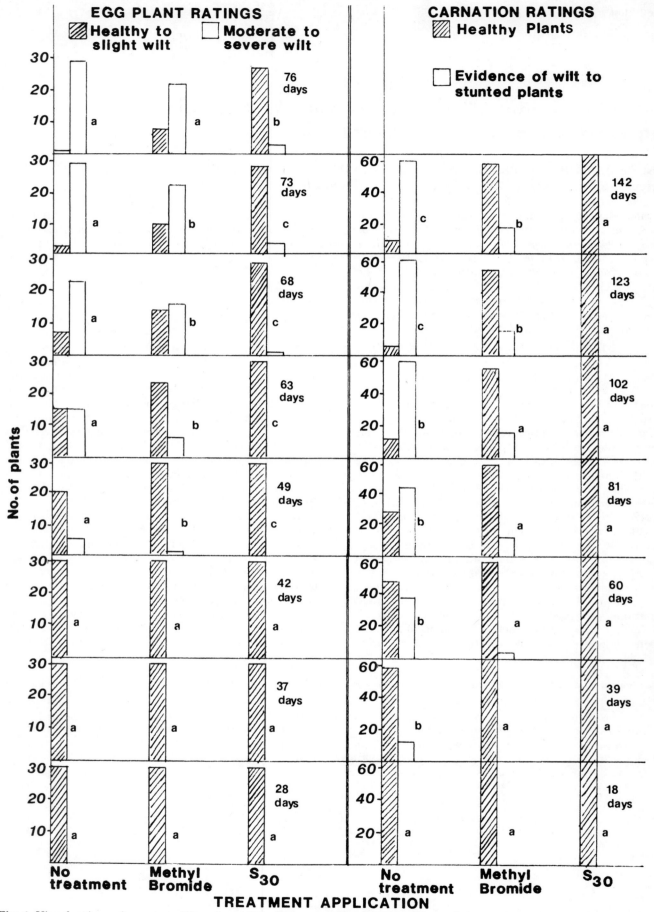

Fig. 1. Visual ratings of presence of Fusarium wilt in Sims carnation and Verticillium wilt in eggplant after no treatment, methyl bromide fumigation, and solarization.

Table 2. Most probable number of propagules of *Verticillium dahliae* and *Fusarium oxysporum* f. sp. *dianthi* at different depths following treatments with solarization and methyl bromide at three sites

Host Treatments[v]	Depth (cm) 0-15	15-30	30-45	45-60
Eggplant[w]				
NT	36.83a[x]	136.66 a	310.0 a	9.23 a
S₂₁	0 b	0 b	0 b	0 b
S₃₀	0 b	0 b	0 b	10.0 a
MB	0 b	0 b	2.2 b	22.3 a
Field carnations[y]				
NT	1,533.3 a	1,300.0 a	1,743.3 a	223.3 a
S₂₁	0 b	790.0 a	...[z]	...
S₃₀	0 b	0 b	38.5 b	17.6 a
MB	0 b	0 b	15.4 b	616.6 a
Tunnel house (Sims) carnations[y]				
NT	4,266.6 a	1,879.66 a	506.0 a	51.0 a
S₂₁	83.6 b	102.6 b	...	...
S₃₀	0 b	0.22 b	126.6 a	167 ab
MB	0 b	1.72 b	21.16 a	64.9 b

[v] NT = no treatment, S₂₁ = solarization for 21 days, S₃₀ = solarization for 30 days, and MB = methyl bromide (1 kg/10 m²).

[w] Assay of *Verticillium dahliae* propagules per gram of soil.

[x] Figures followed by the same letter are not significantly different from one another (*P* = 0.05).

[y] Assay of *Fusarium oxysporum* f. sp. *dianthi* propagules per gram of soil.

[z] Assay data not available.

Table 3. Yield responses to solarization in eggplant and Sims carnation

Treatment[y]	Weight of eggplant fruits (%)	Sims carnation flowers (mean no.)
S₂₁	448.0 a[z]	7.4 b
S₃₀	425.6 a	18.8 a
MB	314.21 b	12.4 bc
S₁₄	286.6 b	1.4 c
NT	141.66 c	0 c

[y] S₂₁, S₃₀, and S₁₄ = solarization at 21, 30, and 14 days, respectively; MB = methyl bromide (1 kg/10 m²); and NT = no treatment.

[z] Figures followed by the same letter are not significantly different (*P* = 0.05) according to Duncan's multiple range test.

Table 4. Number of days required for symptoms to appear on rooted carnations dip-inoculated in spore suspensions of *Fusarium oxysporum* f. sp. *dianthi* and planted in soil cores from treated soils within the tunnel house

Treatment[w]	Number of days Soil not steamed	Soil steamed
S₃₀		
A[x]	57 a[y]	31 b
B	NI c[z]	37 ab
C	NI c	NI c
MB		
A	34 ab	40 a
B	34 ab	41 a
C	NI c	NI c
NT		
A	46 a	30 b
B	53 a	34 ab
C	53 a	NI c

[w] S₃₀ = solarization for 30 days, MB = methyl bromide (1 kg/10m²), and NT = no treatment.

[x] A = 10⁶–10⁷ and B = 10³–10⁴ spores per milliliter; C = sterile water.

[y] Figures followed by the same letter are not significantly different (*P* = 0.05) according to Duncan's multiple range test.

[z] No infection.

---

dianthi, were tested. Inoculation with 10^3–10^4 spores per milliliter produced no infection in unsteamed S_{30} soils, in contrast to steamed S_{30} soils and steamed and unsteamed soils treated with methyl bromide (Table 4). In unsteamed solarized soil, however, the addition of a higher level of inoculum (10^6–10^7 spores per milliliter) overcame this suppressive effect, but with a delay in the expression of the symptoms. Thus steaming reduced the suppressive effect of solarized soil.

DISCUSSION

Solar heating of the soil may involve both a physical and biological control as a result of various mechanisms (1,2). Both such phenomena probably occurred in the present study. The addition of heavy inoculum (10^6–10^7 spores per milliliter) of *F. oxysporum* f. sp. *dianthi*, which overcame the suppressive effect of the solarized soil, may be considered excessive because the inoculation technique involved dipping the bare roots into the spore suspension. This places the heavy load of infective propagules directly in the infection court, reducing the area and time in which potential competitors or antagonists could interfere.

The rapid reinfestation of fumigated soil (Fig. 1) indicates that methyl bromide application creates suitable biological conditions for reinvasion by the pathogen. This is in contrast to the more specific disinfestation effect of soil solarization. Soil solarization aims at reducing pathogens that are less resistant to the temperatures produced by solar heating (40–60° C) than many saprophytes and antagonists (1).

The duration of solarization appears to be important, disease control being greater after 30 than 21 days of solarization. Solarization, as a method of control of soil-borne plant pathogens of horticulturally important crops in Western Australia, appears to hold great potential.

LITERATURE CITED

1. Katan, J. 1981. Solar heating (solarization) of soil for control of soilborne pests. Annu. Rev. Phytopathol. 19:211-236.
2. Katan, J., Greenberger, A., Alon, H., and Grinstein, A. 1976. Solar heating by polyethylene mulching for the control of diseases caused by soil-borne pathogens. Phytopathology 66:683-688.
3. Maloy, O. C., and Alexander, M. 1958. The "most probable number" method for estimating populations of plant pathogenic organisms in the soil. Phytopathology 48:126-128.
4. Northcote, K. H., Bettenay, K. M., Churchward, H. M., and McArthur, W. A. 1967. Atlas of Australian Soils. CSIRO and Melbourne University Press.

Evaluation of Soil Solarization for Control of Clubroot of Crucifers and White Rot of Onions in Southeastern Australia

I. J. PORTER and P. R. MERRIMAN, Plant Research Institute, Department of Agriculture, Burnley, Victoria, 3121, Australia

Midsummer temperatures in southeastern Australia are highest in January and February, and soil temperatures (Table 1) are similar to those in Israel and California where solarization is being successfully developed for control of root diseases (2,7,8,10). This region of Australia, particularly Victoria, produces a range of temperate vegetables such as crucifers, lettuce, potatoes, tomatoes, and onions. However, the soils are often infested by pathogens, and root diseases regularly reduce productivity and quality. Disease control practices include applications of fungicides or fumigants to soil and management practices such as deep plowing. As these methods are frequently expensive and results unreliable, an evaluation of soil solarization as an alternative method to control these pathogens was undertaken in Victoria in 1980. The first objectives were to determine the temperatures required (thermal sensitivity) to kill pathogens isolated from Victorian soils and the effects of solarization on the viability and distribution of pathogens in the soil profile. The soil pathogens under test were those that commonly cause disease in vegetable crops in Victoria, including *Fusarium oxysporum, Plasmodiophora brassicae, Pythium irregulare, Sclerotinia minor, S. sclerotiorum, Sclerotium cepivorum, S. rolfsii, Verticillium dahliae,* and the nematodes *Meloidogyne javanica* and *Pratylenchus penetrans.*

Preliminary experiments (9) showed that the thermal sensitivity of pathogens varied. For example, *Sclerotium cepivorum* was killed between 40 and 45°C, whereas temperatures in excess of 50°C were required to kill *Plasmodiophora brassicae.* In these experiments (9), solarization eliminated most pathogens from the surface layers (0–10 cm) of the soil and reduced their numbers to a depth of 32 cm.

This paper presents results from experiments on the control of clubroot (*P. brassicae*) of crucifers and of white rot (*S. cepivorum*) of onions by soil solarization and also discusses some problems that require resolution before the treatment can be recommended to the industry.

MATERIALS AND METHODS

Soil artificially inoculated. The effects of soil solarization on white rot and clubroot were first tested at a Vegetable Research Station in southern Victoria. A gray sand (pH 6.5, determined from a 1:5 dilution of soil to water) was inoculated with either sclerotia from a maize-meal sand culture of *S. cepivorum* (300 sclerotia per kilogram of soil) or a suspension of spores of *P. brassicae* in water (10⁷ spores per gram of soil). In the latter case, spores were derived from the maceration of cabbage roots naturally affected with clubroot (3). The inoculated soil was mixed uniformly and used to fill small (10 × 32 cm) and large (30 × 32 cm) polyvinyl chloride cylinders, which were buried in field plots so that the tops of the cylinders were level with the soil surface. Soil temperatures were monitored continuously by thermocouples placed at 5-cm intervals from 0 to 30 cm. The plots containing the cylinders were irrigated and then covered with 50-μm polyethylene for 4 weeks from 25 January to 23 February 1982. Cylinders of inoculated soil in control plots were not covered with polyethylene. After treatment, the small cylinders were recovered to determine the effect of solarization on the viability of each pathogen with increasing depth. Soil in the large cylinders was sown with either 5 broccoli seeds, cv. Green Duke, or 20 onion seeds, cv. Savages Flat White. The plants were then monitored over intervals of 12 and 23 weeks for the development of clubroot or white rot respectively.

Soil naturally infested. Solarization was evaluated against clubroot of Chinese cabbage (*Brassica pekinensis*) and white rot of dry bulb onions at sites on growers' properties in southern Victoria where soil was naturally infested with *P. brassicae* or *S. cepivorum,* respectively.

The treatment was tested against clubroot in a gray clay loam (pH 5.5) where previous crops of Chinese cabbage had been severely affected with disease. Soil was prepared as a seedbed and irrigated, and plots (45 × 1.5 m) were treated with 50-μm polyethylene from 31 December 1982 to 3 February 1983. After treatment, soil samples from depths of 0–10 cm and 10–20 cm were tested by bioassay for the presence of *P. brassicae.* Plots were sown on 4 February 1983 and disease assessed on plants sampled from plots either 6 or 12 weeks after sowing. Marketable yield was determined by weighing

Table 1. Effects of solarization on soil temperatures (°C) in gray sand in southern Victoria (23 January–18 February 1982)

Depth (cm)	Solarized		No treatment	
	Maximum (average)	Maximum (absolute)	Maximum (average)	Maximum (absolute)
5	46.7	54.4	33.4	41.7
15	37.0	42.2	27.5	31.7
25	31.0	35.0	26.4	27.8
Air temperature			28.5	42.2

the remaining plants from each plot after 12 weeks.

The effect of solarization on white rot was tested in a black clay loam (pH 6.0) with a resident population of sclerotia of between 50 and 150 per kilogram of soil. Plots (5 × 1.2 m) were solarized from 19 December 1980 to 16 January 1981; after treatment, the number of viable sclerotia was determined in cores of soil sampled from each plot to a depth of 20 cm. Control plots remained untreated. Plots were kept weed free by controlled applications of herbicides. Onion seeds, cv. Pukekohe, were sown at the usual sowing time 26 weeks after treatment on 16 July 1981. Plant counts were recorded at monthly intervals and expressed as a percentage of those that emerged. Samples were also removed to check for the presence of white rot. At maturity (8 February 1982), the incidence of white rot and the total weight of healthy bulbs were determined.

RESULTS

In soils artificially inoculated with the pathogens, the effects of solarization on clubroot and white rot varied. The results of a bioassay using cabbage seedlings, cv. Golden Acre (3), indicated that solarization reduced numbers but did not eliminate spores of *P. brassicae* from the surface layers of soil (Tables 2 and 3). Although most plants were affected by disease at harvest, yields from treated plots were considered marketable whereas those from untreated plots were not.

Solarization eliminated to a depth of 5 cm and reduced to 15 cm the number of viable sclerotia of *S. cepivorum* in

soil. The rate at which plants were killed by white rot (Fig. 1) shows that the main effect of solarization was to delay the onset of disease by 50 days. However, although yields were greater in solarized plots at harvest, most of the plants were affected by white rot.

In soils that were naturally infested with the pathogen, solarization reduced inoculum levels of both *P. brassicae* and *S. cepivorum* in the surface layers of soil (Tables 4 and 5). Solarization delayed the onset of clubbing on roots of Chinese cabbages and also increased yields. The treatment reduced the number of plants killed by white rot (Fig. 2), but the yield of onions was not significantly increased.

DISCUSSION

The results are significant for the vegetable industry because they indicate that solarization can be used by farmers for the control of clubroot. Our experiments showed that although *P. brassicae* was present in the roots of all plants from treated soil, disease was not severe enough to restrict growth and reduce yields to levels that were commercially unacceptable. In contrast, the prospects for using solarization for control of white rot in dry bulb onions on soils with high levels of *S. cepivorum* are less promising because of greater

Table 2. Effects of solarization on distribution of *Plasmodiophora brassicae* in soil artificially inoculated and on disease and yield of broccoli

Treatment	Bioassay rating[a] of seedlings grown in soil from				Disease rating[a] at harvest	Yield of tops (kg/m²)
	3–5 cm	8–10 cm	13–15 cm	23–25 cm		
Solarized	1.5**[b]	1.1**	3.0	3.0	2.0**	2.3**
Nil	3.0	3.0	3.0	3.0	3.0	0.2

[a]0 = No infection in root systems; 1 = microscopic identification of spores; 2 = mild infection, less than half of root system clubbed; 3 = severe infection, more than half of root system clubbed.
[b]** = Significantly different from controls at *P* = 0.01.

Table 3. Effects of solarization on distribution of *Sclerotium cepivorum* in soil artificially inoculated and on disease and yield of onions

Treatment	Percentage of sclerotia viable at				Percentage of dead and diseased plants at harvest	Yield of healthy bulbs (kg/m²)
	3–5 cm	8–10 cm	13–15 cm	23–25 cm		
Solarized	0**[a]	3.6**	72**	98	76	1.8*
Nil	98	98	100	100	92	0.3

[a]* = Significantly different from controls at *P* = 0.05, ** = significantly different at *P* = 0.01.

Table 4. Effects of solarization on the distribution of *Plasmodiophora brassicae* in soil naturally infested with the pathogen and on disease and yield of Chinese cabbage

Treatment	Bioassay rating[a] of seedlings grown in soil from		Disease rating[a]		Yield (t/ha)
	0–10 cm	10–20 cm	6 weeks	12 weeks	
Solarized	2.1**[b]	3.0	0.8**	2.9	14**
Nil	3.0	3.0	2.8	3.0	0

[a]0 = No infection in root systems; 1 = microscopic identification of spores; 2 = mild infection, less than half of root system clubbed; 3 = severe infection, more than half of root system clubbed.
[b]** = Significantly different from controls at *P* = 0.01.

Table 5. Effects of solarization on the number of sclerotia, incidence of white rot, and yield of dry bulb onions in soil naturally infested with *Sclerotium cepivorum*

Treatment	No. of viable sclerotia/kg soil from 0–20 cm		Percentage of plants killed by white rot	Yield (t/ha)
	Before	After treatment		
Solarized	128	41**[a]	35*	6.2
Nil	127	87	53	5.2

[a]* = Significantly different from controls at *P* = 0.05, ** = significantly different at *P* = 0.01.

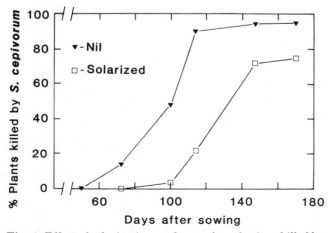

Fig. 1. Effect of solarization on the number of onions killed by *Sclerotium cepivorum* in soil artificially inoculated with the pathogen (southern Victoria).

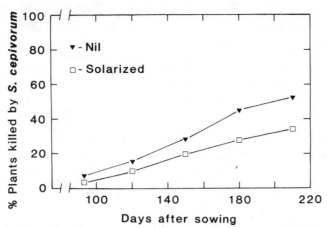

Fig. 2. Effect of solarization on the number of onions killed by *Sclerotium cepivorum* in soil naturally infested with the pathogen (southern Victoria).

losses from disease.

The different effects of solarization against clubroot and white rot are somewhat unexpected because the data on thermal sensitivity and on the viability and distribution of inoculum in soil suggest that solarization would be more effective against *S. cepivorum* than *P. brassicae*. For a better understanding of differences in the rate of development of these diseases and others in solarized soil, variables influencing the host-pathogen-environment interaction must be evaluated. We consider the more important of these variables to be the nature and distribution of the host root system and the duration of the cropping period; the inoculum level after treatment and the rate of saprophytic and/or parasitic recolonization by the pathogen; the influence on the pathogen of the microflora that recolonize soil during and after solarization; and the effects of soil moisture, temperature, and soil type on disease (1,4,6).

One main reason for the effectiveness of solarization is that the root system of the plant is able to exploit the surface layers of soil that are relatively free from infestation by pathogens. Ultimately, disease occurs when the roots contact inoculum. The data from these experiments show that, for long-term crops, there is substantial recolonization of roots by the pathogens.

The results for white rot indicate that there is a need for better control either by improvements to the solarization treatment (7) or by integration with additional treatments, such as chemicals, antagonists, or competitors that restrict the spread of the pathogen in the upper layers of soil or in the host tissue (5). Notwithstanding these problems, the development of solarization represents a major advance in the control of soilborne diseases, and there is little doubt that it will have application as a treatment for root diseases of horticultural crops in Victoria.

LITERATURE CITED

1. Adams, P. B., and Papavizas, G. C. 1971. Effect of inoculum density of *Sclerotium cepivorum* and some soil environmental factors on disease severity. Phytopathology 61:1253-1256.
2. Ashworth, L. J., Jr., and Gaona, S. A. 1982. Evaluation of clear polyethylene mulch for controlling Verticillium wilt in established pistachio nut groves. Phytopathology 72:243-246.
3. Buczacki, S. T. 1973. Glasshouse evaluation of some systemic fungicides for control of clubroot of brassicae. Ann. Appl. Biol. 74:85-90.
4. Crowe, F. J., and Hall, D. H. 1980. Soil temperature and moisture effects on sclerotium germination and infection of onion seedlings by *Sclerotium cepivorum*. Phytopathology 70:74-78.
5. Elad, Y., Katan, J., and Chet, I. 1980. Physical, biological, and chemical control integrated for soilborne diseases in potatoes. Phytopathology 70:418-422.
6. Karling, J. S. 1968. The Plasmodiophorales. Hafner Publishing Co., New York. 256 pp.
7. Katan, J. 1981. Solar heating (solarization) of soil for control of soilborne pests. Annu. Rev. Phytopathol. 19:211-236.
8. Katan, J., Rotem, I., Finkel, Y., and Daniel, J. 1980. Solar heating of the soil for the control of pink root and other soilborne diseases in onions. Phytoparasitica 8:39-50.
9. Porter, I. J., and Merriman, P. R. 1983. Effects of solarization of soil on nematode and fungal pathogens at two sites in Victoria. Soil Biol. Biochem. 15:39-44.
10. Pullman, G. S., Devay, J. E., Garber, R. H., and Weinhold, A. R. 1981. Soil solarization: Effects on Verticillium wilt of cotton and soilborne populations of *Verticillium dahliae, Pythium* spp., *Rhizoctonia solani*, and *Thielaviopsis basicola*. Phytopathology 71:954-959.

Relative Efficiency of Polyethylene Mulching in Reducing Viability of Sclerotia of *Sclerotium oryzae* in Soil

S. M. HAROON USMANI and A. GHAFFAR, Department of Botany, University of Karachi, Karachi-32, Pakistan

Sclerotium oryzae Catt. (= *Magnaporthe salvinii* (Catt.) Krause and Webster), the cause of stem rot of rice, causes considerable damage to the paddy crop. Yield losses as high as 72–75% have been recorded in certain rice-growing areas of the Punjab in Pakistan and Arkansas, 30–80% in the Philippines, 18–56% in India, and 5–10% for the rest of the world (2,8,10). The disease causes increased tillering, unfilled panicles, chalky grain, and widespread lodging of the plants (2,7,13). The pathogen survives as sclerotia in soil or on plant debris, and this propagule serves as the primary source of inoculum for the subsequent crop. Attempts to control stem rot disease by fungicides have been limited and are not recommended because of pollution effects in fish cultures (5,12). Resistant varieties are also not available (8).

In recent years, solar heating with polyethylene mulching of soil has been used in Israel and elsewhere for the control of soilborne pathogens (4,6,9,11,15,16). In Pakistan, mulching of soil with transparent polyethylene sheets was successfully used to reduce the viability of sclerotia of *S. oryzae* (15). The present paper reports the results of experiments on the efficiency of a mulching treatment using transparent compared with black plastic mulch under wet vs. dry soils during winter and summer to reduce the viability of sclerotia of *S. oryzae* in soil.

MATERIALS AND METHODS

Experiments were performed at the university experimental farm in Karachi. The soil is a sandy loam of pH 7.2. One-month-old sclerotia of *S. oryzae* (culture 158, University of Karachi Botany Dept.) grown on sterile rice culms were separated from the culms and mixed with soil at the rate of 10 sclerotia per gram of soil. Ten-gram portions of infested soil were placed in nylon net bags (15 × 15 cm, 150 μm pore) and buried in soil at depths of 5 or 20 cm. Plots measuring 3 × 1 m were mulched with transparent or black polyethylene sheets that were 40 μm thick (Sehgal Plastic Industry, Karachi). A comparable nonmulched control was used. Soil was maintained at field capacity (40% water content) by drip irrigation using a system of plastic tubes laid beneath the plastic mulch. Both mulched and nonmulched plots were irrigated daily. In addition, nylon bags containing *S. oryzae*-infested soils were buried in mulched and nonmulched plots in which soil was not irrigated but kept dry throughout the experiment. There were four replicates of each treatment, and all the plots were randomized.

At intervals of 3, 7, and 15 days, the polyethylene sheets were removed and four bags of each treatment containing sclerotia were recovered from the surface (0–5 cm) and depth (15–20 cm). The populations of sclerotia were determined by a wet sieving and flotation technique (14) and their viability tested by transferring sclerotia onto water agar containing penicillin G and streptomycin sulphate at 3 mg/ml each. Germination was assayed after 3 days at 30°C.

Experiments were performed during the winter (December 1981) and summer (June 1982). Soil temperature and moisture were recorded every 2 hr from 0800 to 1800 hr for 2 weeks using thermometers and gypsum blocks buried 5 and 20 cm deep.

RESULTS

A 3-day mulch treatment of wet soil in June reduced the viability of sclerotia of *S. oryzae* to zero in the surface 5 cm of soil compared with 94% viability in nonmulched treatments (Fig. 1). In mulched soils, sclerotia at depths of 15–20 cm were 80% viable. Mulching wet soil for 15 days during winter gave no significant reduction in viability of sclerotia compared with nonmulched treatments at 5- and 20-cm depths. In dry soil, a gradual decline in the viability of sclerotia was observed in mulched and nonmulched treatments at the surface and at 20-cm depth regardless of whether mulching was performed in summer or winter. There was no significant difference in the use of transparent or black plastic mulch in wet soil at the surface or at 20-cm depth during summer or winter. In dry soil, however, there was greater reduction in viability of sclerotia both at the soil surface and at depth under transparent plastic as compared with black plastic mulch. A 15-day transparent mulch treatment of dry soil during summer reduced the viability of sclerotia at 5 cm to zero as compared with 79% loss in viability under black mulch (Fig. 1).

The reduction in viability of sclerotia was closely related to soil temperature measurements. During the summer (Fig. 2, Table 1), highest soil temperatures (47–53°C) were attained daily between 1400 and 1600 hr in the upper 5 cm of soil in mulched plots (53°C in dry and 47°C in wet soil). At this period, air temperatures near the ground surface ranged between 27 and 36°C, maximum temperature being attained between 1200 and 1400 hr. The greatest temperature differences, ranging from 12°C in wet soil and 8°C in dry soil under transparent plastic and 10°C in wet soil and 0°C in dry soil under black plastic, were noticed in the mulched and nonmulched treatments at the peak temperatures of surface soil.

During winter (Fig. 3, Table 1), the air temperature at the soil surface ranged between 16 and 25°C, and highest soil temperatures did not exceed 35°C in the upper layer of 5-cm depth in nonmulched treatments. During the winter, where the soil was kept mulched, the temperatures were 34°C in dry and 33°C in wet soil under transparent and 29°C in dry and 31°C in wet soil under black plastic. The temperature attained in winter at the soil surface after tarping was 14–19°C less than that of summer months (Figs. 2 and 3, Table 1).

DISCUSSION

Mulching of soil with polyethylene sheets has been reported to bring a marked reduction in the population of and diseases caused by *Verticillium dahliae* and *Fusarium oxysporum* f. sp. *lycopersici* (6,9), *S. rolfsii* (4),

S. oryzae (15), *Macrophomina phaseolina* (11), *Plasmodiophora brassicae* (16), *Rhizoctonia solani, Thielaviopsis basicola,* and *Pythium* spp. (9). An increase in yield of paddy by 23% has been observed after mulching treatment (S. M. H. Usmani and A. Ghaffar, *unpublished results*). A soil mulching practice can be considered superior to fumigation because it is safer with no pesticidal residues. In addition, soluble mineral nutrients in soil are also reported to increase following solarization (1).

In the present investigation, mulching of soil with polyethylene increased soil temperature and reduced the viability of sclerotia of *S. oryzae*. The high temperatures reached in the upper layers of mulched soil were lethal to *S. oryzae*. A high temperature of 47°C was not attained in nonmulched wet soil and did not reduce the viability of sclerotia. Reduction in the viability of sclerotia in mulched and nonmulched dry soil series where the soil temperature did not exceed 35°C during winter could

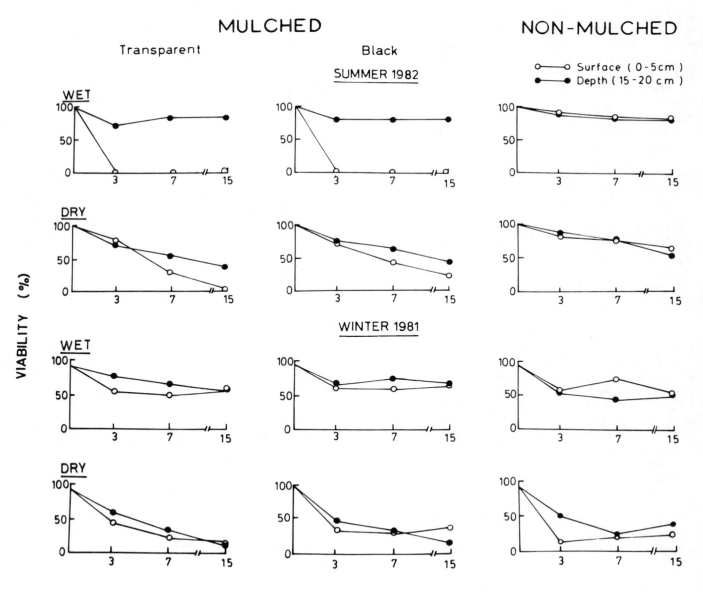

Fig. 1. Effect of soil solarization by transparent or black plastic mulching during summer or winter months on the viability of sclerotia of *Sclerotium oryzae*. Data are for periods of 5–19 June 1982 and 2–17 December 1981. Moisture content of wet soil, 96–100%; of dry soil, 2–3%. Standard errors are small and are covered by the points.

therefore not be attributed to temperature effects alone. Continued desiccation effects or soil radiation presumably contributes to the inactivation of sclerotia in dry soil. A comparison between mulching of wet soil vs. dry soil showed a more rapid decline in viability of sclerotia in mulched wet soil. Maintenance of high soil moisture is thus necessary for increasing sensitivity of sclerotia to high temperatures. The mechanisms of heat inactivation in fungi are not clearly understood, but they may involve enzyme inactivation, phase changes

in fatty acids, and membrane components (3). Although direct thermal inactivation effects are probably the major factors involved in loss in viability of sclerotia following soil solarization, the possible role of biological control mechanisms cannot be eliminated. For example, greater numbers of bacteria and actinomycetes inhibitory to *S. oryzae* were found in soils incubated in an oven at higher rather than lower temperatures (S. M. H. Usmani and A. Ghaffar, *unpublished results*).

Detrimental effects of high soil temperature on soilborne plant pathogens by polyethylene mulching

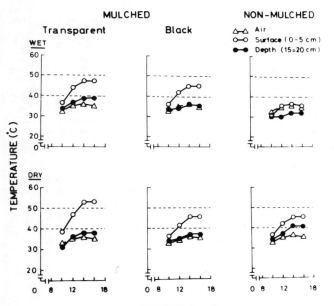

Fig. 2. Effect of soil solarization in the summer by transparent or black plastic sheets on the temperatures of wet (96–100% moisture) and dry (2–3% moisture) soil at depths of 5 and 20 cm. Data are average temperatures recorded at each interval from 5–19 June 1982 at University of Karachi experimental farm. Standard errors are small and are covered by the points.

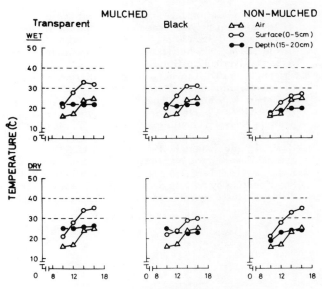

Fig. 3. Effect of soil solarization in the winter by transparent or black plastic sheets on the temperatures of wet (96–100% moisture) and dry (2–3% moisture) soil at depths of 5 and 20 cm. Data are average temperatures recorded at each interval from 2–17 December 1981 at University of Karachi experimental farm. Standard errors are small and are covered by the points.

Table 1. Effect of polyethylene mulching on increases in soil temperature (degrees Centigrade)

| | Maximum temperatures | | | | | | Differences between winter and summer |
| | Winter 1981 | | | Summer 1982 | | | |
Treatment	Air = 25°[a]	Transparent vs. black	Mulched vs. nonmulched	Air = 36°	Transparent vs. black	Mulched vs. nonmulched	Air = 11°
Transparent							
Wet							
Surface[b]	33	+2	+6	47	+2	+12	14
Depth	22	0	+2	39	+3	+7	17
Dry							
Surface	34	+5	−1	53	+8	+8	19
Depth	26	+3	+2	38	+1	−2	12
Black							
Wet							
Surface	31	−2	+4	45	−2	+10	14
Depth	22	0	+2	36	−3	+4	14
Dry							
Surface	29	−5	−6	45	−8	0	16
Depth	23	−3	−1	37	−1	−3	14
Nonmulched							
Wet							
Surface	27			35			8
Depth	20			32			12
Dry							
Surface	35			45			10
Depth	24			40			16

[a] Measured at the soil surface.

[b] Surface = upper 5 cm of soil; depth = 20 cm below soil surface.

suggest that the method can be applied to the field to eliminate stem rot disease of rice. Mulching with transparent or black plastic in wet soil should be done during hot summer when it is more efficacious.

ACKNOWLEDGMENTS

We are grateful for the support of this project by funds from the PL-480 program of the U.S. Department of Agriculture and the Pakistan Agriculture Research Council.

LITERATURE CITED

1. Chen, Y., and Katan, J. 1980. Effect of solar heating of soils by transparent polyethylene mulching on their chemical properties. Soil Sci. 130:271-277.
2. Cralley, E. M. 1936. Resistance of rice varieties to stem rot. Arkansas Agric. Exp. Stn. Bull. 329-331.
3. Crisan, E. V. 1973. Current concepts of thermophilism and the thermophilic fungi. Mycologia 65:1171-1198.
4. Grinstein, A., Katan, J., Abdul Razik, A., Zeydan, O., and Elad, Y. 1979. Control of Sclerotium rolfsii and weeds in peanuts by solar heating of the soil. Plant Dis. Rep. 63:1056-1059.
5. Hori, M., and Izuka, K. 1951. Control effect of ceresan against rice stem rot. Nogyo Gijitsu 6:35-37.
6. Katan, J., Greenberger, A., Alon, H., and Grinstein, A. 1976. Solar heating by polyethylene mulching for the control of diseases caused by soilborne pathogens. Phytopathology 66:683-688.
7. Misawa, T., and Kato, S. 1962. On the lodging of the rice plant caused by stem rot. Ann. Phytopathol. Soc. Jpn. 27:102-108.
8. Ou, S. H. 1972. Rice Diseases. Commonwealth Mycological Institute, Kew, Surrey, England.
9. Pullman, G. S., DeVay, J. E., Garber, R. H., and Weinhold, A. R. 1981. Soil solarization: Effects on Verticillium wilt on cotton and soilborne populations of Verticillium dahliae, Pythium spp., Rhizoctonia solani, and Thielaviopsis basicola. Phytopathology 71:954-959.
10. Shafi, M. 1970. Ten years of rice findings. Kalashah Kaku Rice Research Station, Lahore. Pp. 2-4.
11. Sheikh, A. H., and Ghaffar, A. 1982. Reduction in population of sclerotia of Macrophomina phaseolina by solar heating of soil through polyethylene mulching. Pak. J. Bot. 14:29-30.
12. Shioyama, O., Kurono, H., Murata, K., and Matsumoto, S. 1964. Fungicidal effect of organoarsenic compounds against rice stem rot fungus. Noyaku Seisan Gijutsu 11:8-12.
13. Tisdale, W. H. 1921. Two sclerotial disease of rice. J. Agric. Res. 21:649-658.
14. Usmani, S. M. H., and Ghaffar, A. 1974. Biological control of Sclerotium oryzae Catt., the cause of stem rot of rice. I. Population and viability of Sclerotium oryzae. Pak. J. Bot. 6:157-162.
15. Usmani, S. M. H., and Ghaffar, A. 1982. Polyethylene mulching of soil to reduce viability of sclerotia of Sclerotium oryzae. Soil Biol. Biochem. 14:203-206.
16. White, J. G., and Buczacki, S. T. 1979. Observations on suppression of club root by artificial or natural heating of soil. Trans. Br. Mycol. Soc. 73:271-275.

Proceedings of the
First International Workshop
on Take-all of Cereals

Victorian Crops Research Institute, Horsham, Victoria, Australia
10–11 August 1983

Edited by
J. F. Kollmorgen

Preface to the Take-all Workshop

Take-all has long been recognized as a major disease problem in most cereal-producing regions of the world. It has been a topic of scientific papers for more than 100 years, and in 1981 the monograph *Biology and Control of Take-all* (Asher and Shipton, eds.) summarized the research that had been done worldwide. However, although scientists have discussed take-all at various national and international meetings over the years, no formal international workshop has ever been devoted entirely to the disease. This situation was remedied when the scheduling of the Fourth International Congress of Plant Pathology in Melbourne, Australia, in August 1983 provided an opportunity for many of the leading take-all workers to participate in such a workshop. The First International Workshop on Take-All of Cereals was held at Horsham on 10–11 August and was hosted by the Victorian Department of Agriculture's Crops Research Institute. The 63 participants represented nine countries and included four Australian cereal growers.

It was most appropriate that this historic workshop was held in southern Australia, because it was here that take-all was first recognized as a serious problem in cereals. Daniel McAlpine of the Victorian Department of Agriculture was a pioneer researcher on the disease; as early as 1902 he made a major contribution toward its understanding by demonstrating that "take-all" and "deadheads" were symptoms of the same fungus.

The key objective of the workshop was to enable scientists and growers to discuss in detail aspects of take-all that were of concern to them, with a view to helping in the development of control strategies. Participants (and several key workers who were unable to attend) were invited to submit questions for discussion. These questions were then allocated to sessions according to subject area. The session chairperson not only introduced the questions and stimulated discussion but was also responsible for organizing and editing stenographer's transcripts to provide a record of the session. The edited transcripts are presented here, with the name of the participant who submitted the question appearing in parentheses just before the text of the question.

I express my sincere appreciation to the chairpersons for their help with these Proceedings and to all participants for their enthusiastic contributions. Thanks are also due to the Director (David Barber) and staff of the Victorian Crops Research Institute for their help and encouragement in organizing and running the workshop. Many individuals were involved in the final preparation of the Proceedings, but I especially thank my technicians, Fiona May and Wendy Lehmann, for their help.

The Proceedings provide a unique account of the views of a group of international scientists on take-all and its control. As such, they will be a valuable reference for students of this important disease and for researchers seeking fertile ground for future studies.

James F. Kollmorgen
Editor

Gaeumannomyces spp. Referred to in Text

Gaeumannomyces graminis (Sacc.) von Arx & Olivier (*G. graminis*)

Gaeumannomyces graminis (Sacc.) von Arx & Olivier var. *avenae* (Turner) Dennis (*G. graminis* var. *avenae*)

Gaeumannomyces graminis (Sacc.) von Arx & Olivier var. *graminis* (*G. graminis* var. *graminis*)

Gaeumannomyces graminis (Sacc.) von Arx & Olivier var. *tritici* Walker (*G. graminis* var. *tritici*)

Participants

D. Ballinger
Victorian Crops Research Institute
Horsham, Victoria, 3400
Australia

D. Barber
Victorian Crops Research Institute
Horsham, Victoria, 3400
Australia

G. Bateman
Rothamsted Experimental Station
Harpenden, Hertfordshire AL5 2JQ
England

T. Birley
Bayer Australia Ltd.
Botany, New South Wales, 2019
Australia

T. Bretag
Victorian Crops Research Institute
Horsham, Victoria, 3400
Australia

P. Brisbane
CSIRO Division of Soils
Glen Osmond, South Australia, 5064
Australia

N. Charigkapakorn
Department of Soil Science and Plant Nutrition
University of Western Australia
Nedlands, Western Australia, 6009
Australia

P. Collingwood
ICI Australia Operations Pty. Ltd.
Merrindane Research Centre
Croydon, Victoria, 3136
Australia

R. Cook
U.S. Department of Agriculture
Agricultural Research Service
Washington State University
Pullman, Washington 99164
U.S.A.

P. Cotterill
Department of Soil Science and Plant Nutrition
University of Western Australia
Nedlands, Western Australia, 6009
Australia

C. Cunningham
Oak Park Research Centre
Carlow
Irish Republic

A. Dubé
Department of Agriculture
Waite Agricultural Research Institute
Glen Osmond, South Australia, 5064
Australia

N. Eshed
Faculty of Agriculture
The Hebrew University of Jerusalem
Rehovot 76100
Israel

S. Garrett
179 Hills Road
Cambridge CB2 2RN
England

O. Glenn
Department of Soil Science and Plant Nutrition
University of Western Australia
Nedlands, Western Australia, 6009
Australia

R. Graham
Department of Agronomy
Waite Agricultural Research Institute
Glen Osmond, South Australia, 5064
Australia

J. Griffiths
Mallee Research Station
Walpeup, Victoria, 3507
Australia

P. Hardy
Alf Hannaford and Co. Pty. Ltd.
Welland, South Australia, 5007
Australia

G. Hayward
c/o Post Office
Piangil, Victoria, 3597
Australia

G. Hollamby
Roseworthy Agricultural College
Roseworthy, South Australia, 5371
Australia

D. Hornby
Rothamsted Experimental Station
Harpenden, Hertfordshire AL5 2JQ
England

D. Huber
Department of Botany and Plant Pathology
Purdue University
West Lafayette, Indiana 47907
U.S.A.

B. Ivers
Crossburn Grazing Co.
Kojonup, Western Australia, 6395
Australia

J. Kollmorgen
Victorian Crops Research Institute
Horsham, Victoria, 3400
Australia

J. LeRoux
Department of Agriculture
Grain Crops Research Institute
Small Grain Centre
Bethlehem, 9700
Republic of South Africa

G. MacNish
Department of Agriculture
South Perth, Western Australia, 6151
Australia

R. Manley
c/o Post Office
Avon, South Australia, 5501
Australia

M. Mebalds
Department of Agriculture
Seed Testing Station
Richmond, Victoria, 3121
Australia

K. Moore
New South Wales Department of Agriculture
Agriculture Research Centre
Tamworth, New South Wales, 2340
Australia

S. Neate
Department of Plant Pathology
Waite Agricultural Research Institute
Glen Osmond, South Australia, 5064
Australia

D. Nicoll
Bayer Australia Ltd.
Hindmarsh, South Australia, 5007
Australia

J. Parke
Department of Botany and Plant Pathology
Oregon State University
Corvallis, Oregon 97331
U.S.A.

C. Parker
Department of Soil Science and Plant Nutrition
University of Western Australia
Nedlands, Western Australia, 6009
Australia

L. Penrose
Department of Plant Pathology
Waite Agricultural Research Institute
Glen Osmond, South Australia, 5064
Australia

D. Pitt
Biotechnology Australia Pty. Ltd.
Roseville, New South Wales, 2069
Australia

P. Pittaway
Department of Agronomy
Waite Agricultural Research Institute
Glen Osmond, South Australia, 5064
Australia

R. Powelson
Department of Botany and Plant Pathology
Oregon State University
Corvallis, Oregon 97331
U.S.A.

G. Purss
Division of Plant Industry
Brisbane, Queensland, 4001
Australia

A. Rathjen
Waite Agricultural Research Institute
Glen Osmond, South Australia, 5064
Australia

I. Rodgers
c/o Post Office
Piangil, Victoria, 3597
Australia

A. Rovira
CSIRO Division of Soils
Glen Osmond, South Australia, 5064
Australia

P. Scott
Plant Breeding Institute
Trumpington, Cambridge CB2 2LQ
England

R. Sigvald
Department of Plant and Forest Protection
Swedish University of Agricultural Sciences
S-750 07 Uppsala
Sweden

A. Simon
CSIRO Division of Soils
Glen Osmond, South Australia, 5064
Australia

K. Sivasithamparam
Department of Soil Science and Plant Nutrition
University of Western Australia
Nedlands, Western Australia, 6009
Australia

J. Skou
Agricultural Research Department
Risoe National Laboratory
DK-4000 Roskilde
Denmark

R. Smiley
Department of Plant Pathology
Cornell University
Ithaca, New York 14853
U.S.A.

Z. Solel
Agricultural Research Organization
The Volcani Centre
Institute of Plant Protection
Bet-Dagan, 50-250
Israel

S. Svensson
Department of Plant and Forest Protection
Swedish University of Agricultural Sciences
S-750 07 Uppsala
Sweden

B. Swaby
CSIRO Division of Soils
Glen Osmond, South Australia, 5064
Australia

G. Trolldenier
Landwirtschafliche Forschungsanstalt Bütehof
Bünteweg 8
D-3000 Hannover 71
Federal Republic of Germany

P. Trutmann
Plant Research Institute
Burnley, Victoria, 3121
Australia

N. Venn
CSIRO Division of Soils
Glen Osmond, South Australia, 5064
Australia

M. von Wechmar
Department of Microbiology
University of Cape Town
Rondebosch, 7700
South Africa

J. Walker
Biological and Chemical Research Institute
New South Wales Department of Agriculture
Rydalmere, New South Wales, 2116
Australia

H. Wallwork
Agronomy Department
Waite Agricultural Research Institute
Glen Osmond, South Australia, 5064
Australia

D. Walsgott
Mallee Research Station
Walpeup, Victoria, 3507
Australia

D. Weller
U.S. Department of Agriculture
Agricultural Research Service
Washington State University
Pullman, Washington 99164
U.S.A.

G. Wildermuth
Queensland Wheat Research Institute
Toowoomba, Queensland, 4350
Australia

W. Williams
Victorian Crops Research Institute
Horsham, Victoria, 3400
Australia

J. Wilson
Department of Soil Science and Plant Nutrition
University of Western Australia
Nedlands, Western Australia, 6009
Australia

P. Wong
New South Wales Department of Agriculture
Agricultural Research Centre
Tamworth, New South Wales, 2340
Australia

D. Yarham
Agricultural Development and Advisory Service
Block C, Government Buildings
Cambridge CB2 2DR
England

S. Garrett, K. Moore, J. Walker, and D. Yarham were unable to attend the workshop but submitted questions. J. Walker also prepared written replies to some of the questions in Session 1.

Session 1: Culture and Taxonomy

CHAIRPERSON: M. MEBALDS

Question 1 (COOK):
Why name the wheat take-all fungus "var. *tritici*" rather than "f. sp. *tritici*"? In China the strain pathogenic to wheat has lobed hyphopodia. Length of ascospores and lobed versus simple hyphopodia are variable traits. Pathogenicity to wheat, oats, rye, rice, or other grass hosts, however, is a reliable trait and the primary reason for naming members of *Gaeumannomyces graminis* at the subspecific level. Naming according to pathogenicity rather than morphology would therefore seem more logical and useful.

COOK:
There are some complications as to what constitutes var. *tritici*. Some of you are aware that the fungi observed by Nilsson in Sweden had the characteristics of var. *graminis* but were pathogenic on wheat. In China, the take-all fungus has lobed hyphopodia, and this has been observed elsewhere as well. I believe these anomalies will be observed more often in the future. Whatever the morphology of *G. graminis* in a given area, if it grows continually on wheat, eventually there will be enrichment for a strain virulent on wheat. The same would happen on oats or rye. Perhaps naming the fungus according to forma specialis would be more practical for plant pathologists.

WALKER:
The term "variety" is used for distinguishing, within a species, entities that show minor morphological differences (and perhaps other differences as well). It is a category covered by the International Code of Botanical Nomenclature. The category forma specialis is used to distinguish subspecific entities that do not differ morphologically but that show differences in physiologic characters (pigment production, pathogenicity, etc.). The category forma specialis is not currently covered by the International Code of Botanical Nomenclature. Three varieties of *G. graminis* can be distinguished using minor morphological characters, and these have enabled many hosts for these varieties in many countries to be distinguished from hundreds of collections. To reject this varietal classification would severely reduce the information content of the classification of the take-all fungi.

Within each variety, it may be possible to distinguish formae speciales pathogenic on particular cereal and grass hosts. Morphology (to distinguish varieties) and pathogenicity (to distinguish formae speciales and races) should be used together in classifying these fungi.

SCOTT:
Classification is difficult whether it is based on morphology or pathogenicity. Furthermore, to expect that the morphological characters will always be associated with pathogenic characters would be unrealistic.

COOK:
Most plant pathologists do a pathogenicity test before naming an isolate. But what about dead specimens in a herbarium? The host may be known, but how does one identify to a special form? Nevertheless, morphological characteristics have provided a very practical method of identification, at least to species level.

SCOTT:
You do not know that the isolate is specialized to attack the host in question.

PARKER:
Yeates made an exhaustive study of oat-attacking take-all isolates in Western Australia. He discovered oat-attacking strains of the take-all fungus that, although not var. *avenae*, were serious pathogens of wheat. There may be very good reasons for not naming them too precipitantly because he found *G. graminis* var. *avenae* only in lawns and not on oats in the field. His oat-attacking strains appeared to be variants of *G. graminis* var. *tritici*.

Question 2 (HORNBY):
Do other workers have difficulty distinguishing the simple hyphopodium of *G. graminis* var. *tritici*, and how frequently is it encountered? The simple hyphopodium distinguishes *G. graminis* var. *tritici* and var. *avenae* from var. *graminis*. I find it hard to identify and wonder how frequently the impression of hyphal swelling caused by an underlying lignituber when a hypha penetrates the epidermis is interpreted as a simple hyphopodium. The use of the term hyphopodia for the large, dark, lobed bodies of var. *graminis* leads one to expect something a little more distinctive than the absence of color and swelling when it is applied to the hyphae of *G. graminis* var. *tritici* above points of initial penetration. Do isolates of this fungus vary in producing more or less conspicuous simple hyphopodia?

WALKER:
Simple hyphopodia of *G. graminis* var. *tritici* are common on lower stems and leaf sheaths of infected hosts and often produce a refractive lignituber in the underlying host cell (at a lower level of focus of the microscope). Simple hyphopodia are variable in color, degree of swelling, position of the hyphae, and arrangement; but they are usually readily found in most

specimens. The term *hyphopodium* is used for a hyphal organ of attachment and penetration in various fungi, and its appropriateness for use for the structures seen in *G. graminis* var. *tritici* was first pointed out over 60 years ago by Arnaud. The fungi in this group that have brown, lobed hyphopodia also have simple hyphopodia.

HORNBY:

My experience has been that most penetrations arise from hyphae that do not differ in appearance from surrounding hyphae. Dr. Sivasithamparam, do you see the hyphopodium simply as any lateral outgrowth of a hypha that has no other distinguishing feature except that it is producing an infection?

SIVASITHAMPARAM:

I appreciate your point. I have the same problem you have and the way I have overcome it is to look for lobed hyphopodia on the coleoptile, absence of which is taken as indication of the presence of simple hyphopodia.

Question 3 (HORNBY):

What fungi related to the take-all fungus are found on cereal root systems, and what is their distribution? In the United Kingdom, *Phialophora graminicola*, *Phialophora* sp. (lobed hyphopodia), and *G. graminis* var. *graminis* are being found with increasing frequency on root systems with take-all. A fungus from Brazil, with a *Phialophora* state like that of *G. graminis*, has been isolated from wheat roots with symptoms resembling take-all; but its cultural characteristics are quite distinct from those of *G. graminis* var. *tritici*. Another fungus that resembles *P. graminicola* but is culturally distinct has been isolated from roots with take-all in spring barley monoculture in England. Do other workers look for and find similar fungi? Is it possible to take this opportunity to compare regional isolates of *G. graminis* and related fungi?

WONG:

I have looked at many wheat root systems as well as other grass root systems and have found *Phialophora* spp. different from the *Phialophora* sp. with lobed hyphopodia. At this time it is difficult to put them into different species. More work needs to be done to collect data on these different *Phialophora* spp., and perhaps different workers in the world will have to come together and decide on the best way to describe them.

WALKER:

The main relatives of the take-all fungus found on cereal roots are varieties of *G. graminis*, *G. cylindrosporus*, and some *Phialophora* spp. As further collections are made, previously unknown taxa will be discovered and knowledge of the range of variation of the known taxa will be extended. Fungi from other orders such as the *Pleosporales* are also common on roots and crowns of Poaceae. Careful study of a wide range of isolates from different hosts in different geographic areas is needed, before new taxa are described, to avoid undue proliferation of names and literature confusion.

WONG:

Dr. Hornby, is the Brazilian isolate a species of *Phialophora* or *Gaeumannomyces*?

HORNBY:

It is a *Phialophora* species that may have been grouped with the take-all fungus in Brazil.

WONG:

Maybe we will have to devise physiologic tests to try to differentiate between them. Do you find that most of them produce hyphopodia and look identical?

HORNBY:

Yes. In England we have a lobed hyphopodial *Phialophora* that has been called *Prr* in British publications; and very recently we have found what we believe to be *G. graminis* var. *graminis* in certain fields. This forms an imperfect state that looks very like *Prr*. Neither fungus causes typical take-all, and any lesioning is very mild. What I wish to draw attention to is a small amount of literature from the United Kingdom on the concurrence of these organisms. For instance, several people have produced correlation coefficients to show a negative relationship between *P. graminicola* and the take-all fungus in natural situations. I have published records of three related fungi, *G. graminis* var. *tritici*, *Prr*, and a *P. graminicola*-like fungus, occurring on the same root system. Are such occurrences widespread and commonplace, or are they peculiar to the United Kingdom?

WONG:

I think our situation (in Australia) is different from that in the United Kingdom, in that many of these *Phialophora* spp. occur on grass roots. In Australia, we don't have many grasses that favor *Phialophora* spp., so we tend not to see those species on cereal roots.

Question 4 (COOK):

What is the frequency of heterokaryosis in natural populations of *G. graminis* var. *tritici*? Circumstantial evidence of Naiki and myself indicates that heterokaryosis is the normal condition for mycelium of this fungus and that changes in nuclear condition rather than expressions of hypovirulence can account for the variable cultural appearance and tendency of this fungus to become nonpathogenic after a period on agar media.

COOK:

We were able to get three isolates of the take-all fungus to produce perithecia before and after they lost pathogenicity. A very large number of the ascosporic progeny of those isolates that had lost pathogenicity were weakly pathogenic, but a few were highly pathogenic. The simplest explanation would be in terms of heterokaryosis within the parent mycelium, where a certain frequency of nuclei-carrying genes existed that were associated with weak pathogenicity or a suppressed expression of virulence. Their frequency was such that, overall, the parent was pathogenic. We separated these nuclear types into their homotypes, based on the fact that we took single spores that presumably were homokaryotic; except for mutations, we recovered about a 10% frequency of weakly pathogenic types. We interpreted the selection pressure on agar as being in favor of a shift toward a higher frequency of those nuclear types associated with a lack of or a suppression of virulence. This could then explain why, when we isolated the homokaryotic types from these cultures in the form of ascosporic progeny, we found a much higher frequency of the nuclear type associated with the inability to cause disease. This idea goes back 50 years to Hansen's dual phenomenon, which we decided would be the logical interpretation for our results. Our data provided very strong evidence against the mycovirus explanation because a high percentage of ascosporic progeny were weakly virulent when they came from a weakly virulent parent, and also because we were able to recover weakly virulent types from a virulent parent. However, we agreed that our evidence was strongly against the mycovirus explanation and did not believe that heterokaryosis explained our results. Anagnostakis suggested instead that there was very strong evidence for a cytoplasmic factor transmitted to the ascospore progeny, a phenomenon similar to what Jinks reported some 20 years ago as so-called vegetative death in *Aspergillus*.

SCOTT:

It would seem that heterokaryosis on the one hand and a cytoplasmic factor on the other could explain what you observed. Can you rule out a third possibility, i.e., meiotic recombination?

COOK:

We were dealing with ascosporic progeny in this case, perithecia from selfed cultures.

SCOTT:

What did you start from?

COOK:

We started from hyphal tips.

SCOTT:

Which you thought contained more than one nucleus per cell?

COOK:

We did not think that at the beginning, but in tracking back we concluded that there may have been more than one nuclear type per cell. That is just one possible explanation. We also took mycelium of a culture that had lost pathogenicity. We allowed it to grow over cellophane, then lifted the cellophane and floated it in water so we had free mycelium, which was then treated with a homogenizer to fragment it. Some cells were destroyed but others were intact, and we could then pick out individual cells that were beginning to grow. Fifty of these were cut out from each of our three cultures that had lost pathogenicity. We recovered 2 normal pathogenic types out of the 50 in one test, and 3 or 4 out of 50 in another test. We were recovering primarily what we considered to be types similar to the parent that were very weakly pathogenic, and some of these were absolutely unrecognizable as *Gaeumannomyces* spp. It was parallel to the ascosporic analyses that we had done, but here we were using single-cell cultures.

Question 5 (SIVASITHAMPARAM):

Can we recommend a standard preservation technique for the take-all fungus? Has liquid nitrogen been tried yet?

WALKER:

Lyophilization (L-drying), used and described by Fang and Parker, has been tried at Rydalmere, Australia, and works very well, as long as the straws are properly colonized by the fungus and given an adequate incubation in the vials before L-drying. We have found that 14 days incubation gives good survival. Keeping isolates on potato-Vegemite-dextrose-agar (PVDA; 10 g of dextrose and 1 g of Vegemite) under mineral oil or water has also been successful. However, with this method some isolates are lost over a long period (15 years), even with subculturing every 2 to 3 years. Isolates seem to vary a lot in their ability to survive and maintain their original characteristics when this method is used.

SIVASITHAMPARAM:

Many laboratories now use liquid nitrogen as a standard technique. They are confident that the technique will maintain the original characteristics of the culture.

PARKER:

I support the idea of using liquid nitrogen because some of the L-dried cultures that we made 4 or 5 years ago in Western Australia are still losing pathogenicity. It seems that loss of pathogenicity is often a matter of aging or transfer frequency and not necessarily nutrition.

KOLLMORGEN:

What is the experience with keeping cultures under oil; do they lose their pathogenicity too?

PARKER:

Yes, they can lose many characteristics.

KOLLMORGEN:

I read an article by you and one of your students that indicated that the part of the culture that was under oil maintained its pathogenicity for years, whereas the part of the culture that was above the surface lost it.

PARKER:

This is correct, but not a general thing. I have one culture, for instance, that has been a virulent pathogen for 13 years; but all the other cultures of this particular batch of isolates gradually lost virulence.

SKOU:

The fungus will remain viable for a longer period under oil only if it is not transferred as often as it would be on conventional media.

Question 6 (WONG):

What is the best way to maintain pathogenicity in cultures of *G. graminis* var. *tritici*?

SIVASITHAMPARAM:

I would like to know more from the Rothamsted group who have been working in this area. One of the techniques that showed promise involved silica gel, which not only maintained viability of cultures but pathogenicity as well.

HORNBY:

I have no experience of this silica-gel method, but

freeze drying certainly preserved the fungus, although there was some unexpected loss of pathogenicity. I now keep my stock collection under water, having shown that test isolates were alive and easily grown after 6 or 7 years of storage. It is a very simple method, described by Boesewinkel in *Transactions of the British Mycological Society* 60:183-185. Squares cut from the edge of fungal cultures on agar are placed in sterile water in MacCartney bottles and stored at room temperature. So far, I have had no losses and there is some evidence that pathogenicity is maintained.

WALKER:

L-drying is probably the best method. In my experience with PVDA cultures kept under oil, some isolates have maintained pathogenicity for several years (14 years), whereas others that have not been in culture as long have lost it.

PITT:

Very often L-drying requires a carrier medium to maintain the organism. Does anybody have any experience in L-drying the take-all fungus when it is in host tissue?

PARKER:

We have L-dried isolates that were in host tissue and in infected straw. It would be valuable if we could learn more about the liquid nitrogen method because bacteriologists who have been using it for some time claim that the technique will retain all the characteristics of the organism.

COOK:

We compared different methods of maintaining pathogenicity: we let the cultures mature and stored them in a refrigerator, kept them growing continuously by transferring every 10 days, passed them through the host, obtained ascosporic cultures and refrigerated them. In all cases, except when passed through the host, cultures lost pathogenicity. Even the ascosporic cultures lost pathogenicity, which was Anagnostakis's reason for suggesting a cytoplasmic determinant as responsible in some way for loss of pathogenicity. But why preserve cultures? I realize that there are good reasons to go back to cultures, but most of our work only requires a highly pathogenic culture, which is best obtained fresh from nature.

MacNISH:

That is my comment too. Really, why keep them for so long? If they are really so precious, passage them through plants to maintain pathogenicity.

ROVIRA:

I think that we are trying to preserve every culture in its virgin field form. Within *Gaeumannomyces* spp., there is a lot of variability between isolates in stability and retention of pathogenicity. Professor Parker had one isolate for 10 or 12 years that retained its pathogenicity even in storage. We have one that Dr. Smiley isolated 10 years ago that has remained stable. Each year, however, we take the culture out of distilled water and passage it through a plant. Maybe we are too concerned with preserving *G. graminis* var. *tritici* in culture.

HORNBY:

I endorse that. We have cultures that retained their pathogenicity and ability to form perithecia for 10 or more years. It is quite possible, however, to make several subcultures from one isolate and find some that maintain pathogenicity and others that do not.

KOLLMORGEN:

If you can induce those cultures that lose their pathogenicity to produce perithecia, are the ascosporic progeny also pathogenic?

HORNBY:

I have not been able to do this. Usually when they lose their pathogenicity, they lose their ability to sporulate.

PARKER:

I agree with Dr. Rovira about putting isolates through plants. There is just one point I would make. I think this should be done using natural soil, preferably of course without *G. graminis* var. *tritici* so that both pathogenicity and competitiveness are preserved.

ROVIRA:

I agree. When we pass an isolate through a plant it is always done in natural soil, but it is a soil with a low level of *Gaeumannomyces* spp.

SKOU:

I received a culture from the United States isolated by Kirby in 1922. When I tested it, it was 45 years old and was quite pathogenic. I have tested several cultures, however, that have lost much of their pathogenicity within 3 months.

SCOTT:

If isolates lose their pathogenicity in culture, and if that pathogenicity can sometimes be revived, for example, by passing the isolate through a host, it raises an important question about the genetics of this character. Are we dealing with genetic change when isolates lose pathogenicity in culture or are we dealing with some change in physiological state that can be reversed by passing the fungus through a host?

WONG:

To understand the physiological aspects of the fungus it must be preserved such that it doesn't change.

SCOTT:

It is important that we preserve the individual cultures, especially when studying interactions between isolates and host species.

COOK:

I agree; in that instance, one would want to maintain an original culture and hope that it would not change. Much of our routine work involves experiments on disease control. When we want a pathogenic isolate we isolate one from host tissue, verify its pathogenicity, and then use it.

ROVIRA:

If new cultures are used each year, there is likely to be variability. If you were screening for host resistance, what technique would you use?

COOK:

I would use a highly pathogenic isolate. If you identified resistance, you may want to keep the isolates that you started with to confirm your results.

Session 2: Inoculum

CHAIRPERSON: A. SIMON

Question 1a (WALLWORK, RATHJEN, and BALLINGER):

What is the best method for applying artificial inoculum in field trials? The uneven distribution of *G. graminis* var. *tritici* over naturally affected areas creates problems in obtaining reliable field data. Artificial inoculum is commonly used to overcome this, but how and in what quantity should this inoculum be applied to give a fair test? There are problems with mixing inoculum and seed together at sowing, in that the emerging seedlings can be killed following early infection. Similarly, random distribution of inoculum in the soil profile can lead to chance and uneven attack of young seedlings especially where small plots are used. Is it perhaps better to add the inoculum as a layer below the seed so that the roots grow through the infection zone at a later stage?

Question 1b (PENROSE):

Is it correct to say that current methods of artificially infecting field plots with *G. graminis* var. *tritici* weight induced disease toward seedling attack?

BALLINGER:

In my work on testing of fungicides for control of take-all and screening cultivars for resistance, the use of artificial inoculum is necessary to achieve uniform distribution of the pathogen over a large area.

ROVIRA:

The most common form of inoculum is oat kernels colonized by *G. graminis* var. *tritici*; however, I see three problems associated with this technique. First, the resultant propagule is larger than most natural propagules in soil. Second, the propagules have unnaturally high substrate reserves available to other soil organisms. Third, if the oat kernels are drilled in with the seed, as is traditional, the seed and the propagules will be together in the drill row and the seedlings will be attacked at a very early growth stage. By distributing smaller propagules, for example, ryegrass or millet seed, throughout a known weight or volume of soil, we can overcome these problems, regulate the severity of attack, and achieve uniform distribution of the pathogen over the test area.

MacNISH:

Dr. Rovira, I was very impressed with the evenness of disease in the field obtained with your inoculation technique. However, I am worried that you are using only one isolate of the fungus.

SCOTT:

Could I ask why you are worried?

MacNISH:

There is a range of pathogenicity in isolates from the field. Therefore, I do not think we should screen cultivars against a "super-strain," discarding those cultivars that do not survive.

SCOTT:

I prefer to use several isolates only in case a single isolate loses its pathogenicity. However, there is very little evidence of either variation in host specificity (allowing a single isolate to overcome some forms of resistance and not others) or even of variation in resistance between cultivars.

ROVIRA:

I believe one isolate is sufficient if the original level of virulence is maintained.

CUNNINGHAM:

I am not sure that a lower rate of inoculum will give a later attack but rather a milder form of attack. I put the inoculum below the seed, thereby delaying the onset of the disease.

ROVIRA:

In the natural situation, there would be propagules from previous seasons above the seed so that occasionally the pathogen would infect the crown or the stem. That is possibly one difference between your technique, Dr. Cunningham, and what would occur in the natural situation.

HORNBY:

In one season when we used oat inoculum (approximately two colonized oat grains drilled with every wheat seed), the crop was almost devastated. In the following season we rotavated colonized grains into the soil to a depth of 15 cm, and there was a considerable reduction in disease produced with artificial inoculum. You are in danger, therefore, of producing epidemics of disease with artificial inoculum that bear very little relationship to what occurs naturally. In this example, the year when the crop was seriously affected artificially was one when natural infection was minimal.

HUBER:

We have established the disease by applying inoculum to the soil one year and then sowing cultivars to be screened for resistance the next year. In this way, we produced uniform disease but avoided problems associated with high rates of artificial inoculum.

PARKER:

Professor Huber's approach interests me because our work shows that black runner hyphae can grow out 4–5 cm from inoculum. These hyphae can survive several months in dry soil and then act as infective propagules; however, they have little stored energy and the attack is weak and late. On the other hand, large propagules give a strong, early attack.

TROLLDENIER:

In pot experiments, we position inoculum between rows of the host plant to avoid early infection. By varying inoculum density we are able to obtain an excellent dose-response curve.

CUNNINGHAM:

In Ireland, we use sugar beet seed clusters and vary the rate by varying the proportions of noninfested sugar beet clusters in mixtures for different infestation treatments. Of all forms of inoculum that we have tried, this is the easiest to manage. However, the amount of disease still varies at different sites in the same season, and even more so from season to season.

SMILEY:

We were concerned about the large food base that is available for the fungus in oat kernels, and so we tried to reduce propagule size by grinding the inoculum. We have also tried fescue seeds to avoid grinding. I feel a better alternative would be millet seeds because they are spherical propagules and would have better flow characteristics than most other small seeds.

SIMON:

The different-sized fractions of ground oat kernels vary considerably in the level of disease they produce. For that reason we now use small seeds—for example, ryegrass and millet.

WELLER:

When testing biological control agents, either in the greenhouse or field, the concentration and size of the G. graminis var. tritici inoculum will influence the effectiveness of the agent.

HUBER:

In addition to the technique I discussed earlier, wheat cultivars are also screened for resistance in the field with the highest natural level of take-all that we can find. They are screened with and without added nitrogen, and we select those that show a response to nitrogen and resistance and/or tolerance. We have found tolerance even under severe nitrogen stress.

WILDERMUTH:

In our work with *Bipolaris sorokiniana* we have the same differential response to artificial inoculum as to natural inoculum.

HORNBY:

We seem to have talked mainly about testing cultivars, but we may also wish to test putative

biological or chemical controls. Here I think it is important that some effort be made with artificial inoculum to try to simulate the vertical distribution of natural inoculum in the soil, because that may play a significant role in natural epidemics.

PARKER:

We examined the distribution of the fungus in a soil profile where the farmer had grown three consecutive wheat crops. Most of the propagules were in the top 7 cm, and at 15 cm there were none. The soil had never been plowed more than 8–10 cm deep.

HORNBY:

I think there is a lot of evidence to show that distribution of inoculum in soil is related to the cultural practice. If only the top few centimeters are cultivated, then the greatest concentrations of plant residues and inoculum occur in that region. For many years we have studied sites that are cultivated to a greater depth by moldboard plowing, and this has resulted in bands of infectious residues at about 15–20 cm.

COOK:

Each researcher should use the method that provides the best level of disease for the particular situation. For research on winter wheat, which can sometimes be covered with snow, we do not mix the oat kernels with the seed because of difficulties in controlling seeding rate. Rather, we seed and then introduce the inoculum into the same drill rows in a second operation. The onset of disease is very slow, and only the whitehead phase is evident after heading. However, this method is too severe for spring wheat, and I believe that the method of Mr. Simon and Dr. Rovira would be more suitable for spring wheat in our area.

RODGERS:

As a grower, I would comment that after we cultivate the soil, the natural inoculum must be spread through the surface layer to a depth of 7–10 cm. Therefore, I feel that any method of applying inoculum should simulate what occurs in the natural situation.

Question 2 (POWELSON):

What is the relationship between kind and size of artificial inoculum units? Artificial inoculum is prepared by growing *G. graminis* var. *tritici* on various sterilized substrates, such as oats, wheat, millet, ryegrass, and mixtures with sand. After colonization of the substrate, the inoculum is dried and used in various ways. Because there is only partial utilization of the substrate by *G. graminis* var. *tritici*, it becomes available as a food base for other soil microorganisms. How does this influence the performance of *G. graminis* var. *tritici*?

POWELSON:

The use of different concentrations of inoculum enables us to formulate response curves of inoculum density and disease incidence. If the aim is to generate subtle differences, then the amount of disease should be directly proportional to the amount of inoculum.

COOK:

In Washington State, U.S.A., Wilkinson worked with

three kinds of inoculum: colonized oat kernels, infested crowns (with roots removed) from a wheat field, and infected roots produced by growing wheat in a vermiculite rooting medium containing *G. graminis* var. *tritici*. Each form of inoculum was milled and sieved to different sizes—90 µm to larger than 1 mm. He used a soil from a field after take-all decline and a virgin soil. The soils were pasteurized, fumigated, or left untreated. The inocula were added to the soils at different concentrations on a weight basis. The bioassay ensured that the roots had to grow some distance before hitting the layer of inoculum. The disease was assessed on the wheat roots by counting lesions. He found that for a given particle size, there were slight but not significant differences between forms of inoculum; the trend in levels of disease produced was oat kernels greater than infected crowns greater than infected roots. This confirmed an observation by Hornby and Brown that root inoculum is not as effective as crown inoculum. The particle size that produced most disease was 250–500 µm, but 90-µm particles produced disease. There was an increase in number of lesions per unit mass in soil that had been fumigated or pasteurized. Once particle size exceeded the threshold size, particle mass was the only important factor in disease production.

KOLLMORGEN:

I believe the type of inoculum and method of inoculation depends on the aim of the work. If cultivars are to be screened for resistance or fungicides for efficacy, then, ideally, every plant should be infected. We should be very cautious of reducing inoculum density to a point where plants escape infection. In some of our work, we inoculate wheat with oat-kernel inoculum one year and carry out our trials on the same area the next year.

WELLER:

In greenhouse and field research to select biological control agents, sometimes the oat grains colonized or not colonized with *G. graminis* var. *tritici* provide a food source for other pathogens. *Pythium* is an example; under our conditions, it can cause a reduction in stand and a stunting of plants that make assessment of take-all suppression difficult.

POWELSON:

We have just begun to use a mixture of colonized millet seed and sand inoculum. We added prescreened sand to keep the millet seeds separate during incubation. The sand grains were extensively colonized by the fungus, which could be separated from the millet and used as a nil food-base inoculum unit.

Question 3 (COTTERILL):

Can a system of inoculum preparation be recommended to create a range of infections on wheat plants?

COTTERILL:

I have tried to obtain a graded response to inoculation based on a rating scale. The concentrations of various inocula ranged from 5 to 500 mg/kg of soil; however, all the disease levels obtained were at the top of the response curve and I was unable to obtain a graded response.

HUBER:

I believe you could obtain a graded response more effectively by changing the environment—the rate, ratio, or form of nitrogen—than by changing the inoculum density.

WELLER:

In reply to Professor Huber's comment, I should say that at the CSIRO experimental site at Avon in South Australia, we have produced varying levels of take-all with different amounts of *G. graminis* var. *tritici* inoculum.

HUBER:

I would not imply that you cannot achieve a curve for inoculum density and disease incidence in sterile soil, or even in natural soil; but the greatest factor determining disease incidence in our work is the environment, not the inoculum density.

PITTAWAY:

I think that both inoculum density and environment are important factors in determining disease incidence. However, the best way to discern what environmental factors are influencing disease incidence is to establish a dose-response curve. Regression analysis can then be used to determine how season or trace elements of phosphorus, to name a few, influence disease incidence.

MacNISH:

Methods of inoculation, such as the one used by Mr. Simon and Dr. Rovira, do allow you to obtain a dose-response curve in different seasons with associated changes in soil suppressiveness and climatic conditions.

PARKER:

It is necessary to obtain dose-response curves because seasons and soils vary greatly. Some of my experiments have failed because the inoculum was either too weak or too strong. Soil pH and soil temperature are just two factors that vary widely between cereal-growing areas. All the isolates of *G. graminis* var. *tritici* tested in Western Australia have a pH optimum below 7 (one had a pH optimum of 5). Cold soils favor the growth of the fungus. It competes well with the soil microflora between 10 and 15° C; but it becomes less competitive as the temperature rises until, at 20–25° C, growth is almost totally suppressed.

MacNISH:

For Mr. Cotterill to achieve the range in disease he is seeking, he must use bigger pots or inoculum with a much smaller particle size.

COOK:

In our work on mass screening of bacteria for biocontrol, we use the threshold size of 250–500 µm and adjust the inoculum density to operate at the low end of the dose-response curve—high enough to prevent the bacterial seed treatment from having a major effect, but not high enough to swamp potentially good treatments. In your case, I suggest sieving your inoculum to a particle size of less than 125 µm.

MacNISH:

I can produce a range of disease levels in the field by

using different sources of nitrogen. However, addition of nitrogen would influence other factors as well.

SIMON:

In 1982 we screened wheat cultivars for resistance or tolerance to *G. graminis* var. *tritici* in irrigated microplots in the field, using two rates of colonized ryegrass-seed inoculum—16 and 64 propagules per kilogram of soil. Each propagule was a single ryegrass seed. At the low rate, none of the cultivars tested showed any disease symptoms at tillering and grain yields were not reduced. At the high rate, there was top-growth reduction at tillering and reduction in grain yield with most cultivars.

ROVIRA:

I believe that in small field plots losses below 25% are barely visible but would be detectable in large plots. The plots Mr. Simon has just referred to were too small to detect a reduction in yield caused by the rate of 16 propagules per kilogram. This rate of inoculum is close to a realistic field inoculum density, but the plot size needs to be at least 2 m².

BRISBANE:

I think it would be irresponsible to add take-all inoculum to a field where it did not occur.

PARKER:

I know of no such wheat field.

BRISBANE:

In that case, you cannot achieve a zero disease level for a dose-response curve. If inoculum is applied in a nonconducive season, then inoculation will not be very successful. In a conducive year the uninoculated treatment will contain the pathogen, reducing the possible range for disease expression. I think, therefore, that it is difficult to achieve a dose-response curve in a field soil. A better technique would be to apply different levels of a fungicide if one was available.

Question 4 (HORNBY):

What would be an acceptable test for putative control agents? The problem of getting convincing evidence of the control of take-all in the field is complex. Opinions differ as to the value of artificial inoculum, which is often applied to try to introduce some degree of disease homogeneity into a site or as an insurance against a low disease incidence from natural inoculum. What would be a fair test, and how many sites and how many seasons should it involve? What methods of application are currently in use, and what are their relative merits?

HORNBY:

We decided some years ago to use natural inoculum, and I originally posed this question from that viewpoint. Climate often dictates the amount of disease obtained, whether artificial inoculum has been used or not.

HOLLAMBY:

We are seeking resistance in wheat to soilborne diseases but not necessarily take-all. The screening is done, as much as possible, in fields without added inoculum so that we can screen a large number of cultivars. If natural infection fails, the material is not promoted into the next generation but is held back until it has gone through a successful screening.

SCOTT:

I would like to extend Dr. Hornby's question to the screening of cultivars for resistance to take-all. Consistent differences between cultivars should be obtained in at least three seasons and under different environments before they are accepted as real. I am doubtful whether some of the varietal differences discussed here by various researchers really exist.

HORNBY:

We deliberately chose to work at two sites when testing bacterial and chemical control agents. At both sites, which had different soil types and cropping histories, we used natural inoculum. The site at Rothamsted (U.K.), on a heavy clay loam, had serious take-all in the season before we began these experiments but is now in the take-all decline phase, giving low levels of disease and not providing a good test of the control agents. However, the site at Woburn (U.K.), on a sandy loam and in only its second year of cropping at the time the experiments began, is building up severe take-all and is providing a good test. Because of the unpredictability of take-all, the precaution of working at more than one site is a sensible one.

WELLER:

I believe a series of tests, rather than a single test, is necessary. In my work in the Pacific Northwest (U.S.A.), it is necessary to use introduced *G. graminis* var. *tritici* inoculum to produce uniform disease in small field plots. Having identified isolates that provide control in the small plots, the next step is to test those isolates in fields where only natural inoculum exists.

COOK:

I maintain that any method of inoculation is justified as long as you can produce the desired level of disease. Dr. Moore and I developed a method of critically assessing both the effects and the possible mechanisms for those effects of specific cultural practices, such as tillage and fertility treatments. At one end of each plot, an area was fumigated with methyl bromide; an adjacent area of equal size was left untreated. The two areas were then split longitudinally: one side was seeded to wheat alone and the other side to wheat and oat-kernel inoculum. The growth response of wheat in the fumigated soil gave an idea of the level of natural inoculum, although we realized there were nontarget effects on fumigation. In some cases, the fumigation response was small; however, that may have resulted from lack of inoculum or soil suppressiveness. Hence, the treatment where inoculum was applied indicated whether disease could be produced in the soil; if not, could it be produced in fumigated soil? Dr. Weller's biocontrol agents might be superimposed on this system.

Question 5 (SIVASITHAMPARAM):

Does inoculum of the take-all fungus break down quickly in warm, moist soils?

SIVASITHAMPARAM:

South Australian experiments in the 1930s proved that take-all could be reduced by late sowing. This may be caused by breakdown of the inoculum; however, in some areas of Western Australia, occasional summer rains produce ideal conditions for breakdown of inoculum and yet wheat cannot be grown in these southern areas because of severe take-all.

HUBER:
Several other factors could be operating in this situation. One may be loss of nitrogen through denitrification in the warm, moist soils because nitrogen loss in desert soils in the United States is primarily from denitrification. Other microbial interactions could cause plant stress by altering trace element availability, and this stress may impose changes in host physiology.

PARKER:
The soils under question are very sandy, and under our dry, hot summer conditions there is no host. Denitrification is unlikely because it has two requirements—an anaerobic situation and an organic substrate to provide the denitrifying bacteria with electrons.

MacNISH:
My work in South Australia showed that the fungus disappeared very quickly from hot, wet (35°C, −0.01 bars) soils. However, I would describe the Western Australian soils in question as warm and moist, and my work showed that under similar conditions (15°C, −4 bars or 15°C, −0.1 bars) the fungus survived in soil for at least 45 weeks.

ROVIRA:
Temperatures at depths of 0–5 cm following summer rains in southern Australia range between 35 and 50°C. Microbial activity in the ensuing 2–3 days would be high, and I would expect G. graminis var. tritici to fare rather poorly in that situation. Certainly, in the glasshouse, natural biological suppression eliminates disease symptoms at temperatures exceeding 25°C.

SCOTT:
Earlier discussion has dealt with the poor relationship between inoculum density and disease incidence. I believe, therefore, that we should consider the effect of late planting on the parasitic phase of the fungus and not just concentrate on reduction of inoculum density by extending the saprophytic phase with late planting. Could late planting also be relevant by reducing the period of time in which the fungus can infect the root and produce disease?

COOK:
In the Pacific Northwest, I think both mechanisms operate. On the one hand, warm conditions, adequate soil moisture from irrigation, and soil disturbance before late planting in a system of wheat monoculture reduce survival of the fungus. On the other hand, late planting reduces the time the fungus has to become established on the roots before the cold winter.

Question 6 (HUBER):
What is the method of survival of the take-all fungus in soils cropped continuously to corn or soybeans for over 20 years? Are our techniques for determining populations and soil activity sensitive enough to provide needed answers?

HUBER:
Corn and soybean crops were alternated during this period; grass control was excellent, and no small grains were sown. However, severe take-all occurred on wheat grown in these soils in the glasshouse. This raises questions of parasitic versus saprophytic survival and the importance of inoculum density. Our belief is that G. graminis var. tritici and G. graminis var. graminis are ubiquitous soil organisms and that disease severity is regulated by interactions other than inoculum density.

PARKER:
Can corn host G. graminis var. tritici?

HUBER:
With the intensity of take-all observed in second and third consecutive wheat crops following corn, I am convinced that G. graminis var. tritici survives very well with corn.

KOLLMORGEN:
Does G. graminis var. tritici infect soybean roots?

HUBER:
In the midwestern areas of the United States, G. graminis var. tritici and G. graminis var. graminis may both be epiphytic on soybean roots. G. graminis var. graminis also has extensive epiphytic activity on foliage and pods of soybean.

PARKER:
There are reports of certain legumes being mildly infected by G. graminis var. tritici. In Western Australia, we cannot grow soybeans because our winters are too cold. Lupines, however, are grown extensively and appear to be an excellent cleaning crop, giving big yield increases in the following wheat crop.

HORNBY:
I wonder whether some of these observations could be glasshouse artifacts because we have also found take-all in field soils where we would not have expected to find it. It is not known whether plants grown in a glasshouse can be infected via aerial or water contamination, especially if infested soil is used nearby.

WONG:
G. graminis var. graminis is pathogenic on wheat in Indiana, in the United States. If hyphopodia can be found on wheat grown in this soil in the glasshouse, then it is likely that G. graminis var. graminis is surviving on the roots of corn and soybean, even though tritici is the dominant variety of G. graminis.

SCOTT:
Wheat-soybean rotation has been practiced in many parts of the world. Has Gaeumannomyces been reported on soybeans from such rotations elsewhere?

HUBER:
Roy has reported G. graminis var. graminis to be a common epiphyte throughout the midwestern and southern United States.

SCOTT:

Do these isolates of *G. graminis* var. *graminis* cause take-all symptoms? If so, do you actually account for the occurrence of take-all in wheat after long-term soybean cropping in terms of these lobed hyphopodial isolates, or are they rather incidental?

HUBER:

G. graminis var. *graminis* infects roots, but the infection is weaker than with *G. graminis* var. *tritici*. Most of the take-all we see is caused by *G. graminis* var. *tritici*.

COOK:

Is *G. graminis* var. *tritici* being carried over by the corn and soybean crops?

HUBER:

Because the baiting technique selects only for pathogenic isolates, avirulent strains of *G. graminis* var. *tritici* are seldom isolated from roots when the soil is suppressive. Corn and soybeans are more likely affecting situations in the soil that influence the pathogenic ability of *G. graminis* var. *tritici* rather than its survival.

Question 7 (MOORE):

Are current techniques of quantifying levels of take-all in the soil adequate? It would be desirable to have a selection of standard procedures for estimating "populations" of *G. graminis* var. *tritici* in soil. I appreciate that *G. graminis* var. *tritici* does not occur in soil as discrete propagules such as microsclerotia, chlamydospores, etc.; but there must be some way of preparing soils for processing in dilution series or bioassays, similar to methods used for detecting numbers of *Rhizobia*.

HORNBY:

I have used the most probable number (MPN) technique for many years and am reasonably satisfied with it. In the presence of biocontrol organisms, the host-infection technique records only that part of the inoculum able to infect under the conditions of the test. Is there any value in trying to obtain an absolute measure of inoculum density, or is it preferable—for practical experimentation—to measure soil infectivity? I believe the MPN technique does the latter very well.

WILSON:

The MPN technique measures infective propagules, and the result is dependent on the spatial distribution of roots in the test system. How do you reliably translate that data back to the field situation?

HORNBY:

I do not visualize inoculum in terms of uniform propagules. I estimate what I call "infectious fragments" that are discrete particles of different size and capacity, but able to cause infection. I have tried, with some success, to relate this to the condition and distribution of inoculum in the field by extracting different sizes of plant residues from soil and using these to produce a dose response.

WILSON:

The estimate is very dependent on root density.

HORNBY:

The assumption is made that the root system samples the entire volume of the pot. I have published work showing the relationship between numbers of plants per pot and the MPN estimate of infectious fragments in a field soil; the MPNs differed little after two plants per pot, which seems to be the optimum for my system.

WILSON:

I am still unsure how the MPN estimate is translated back to the field situation where the spatial distribution of roots and of disease propagules is different from the test system. The population of infective propagules in the field may therefore be different from that in the pot system.

HORNBY:

The MPN technique assesses how much inoculum there is, not where it is. Other techniques are used to establish the vertical distribution of infectious units in undisturbed cores. A combination of these techniques will provide a good impression of the inoculum situation in the soil. However, I do not think an absolute measure of the actual amount of soil inoculum will ever be achieved with host infection techniques.

PARKER:

We examined the growth of *G. graminis* var. *tritici* in soil and looked for chlamydospores, but we did not see them; perhaps they would form in the correct environment. It appears that the thick-walled, resistant, black runner hyphae act as a resting phase because they can survive for months in dry soil. Microsclerotia have been reported by Nilsson and also observed by Dr. Sivasithamparam and myself.

Question 8 (KOLLMORGEN):

Are segments of wheat straw artificially inoculated with *G. graminis* var. *tritici* useful in studies on survival of the fungus?

KOLLMORGEN:

Professor Garrett used segments of wheat straw colonized by *G. graminis* var. *tritici* in survival studies. The straws are precise units that are convenient to use and give an unequivocal answer to whether or not the fungus has survived burial in a particular soil. Are these units acceptable to other researchers, or should we be using naturally infected material?

WONG:

Because the infectious unit is uniform, I believe that the method can be used to compare the effects of physical and chemical factors on the survival of *G. graminis* var. *tritici* as well as the antagonists of take-all—*G. graminis* var. *graminis*, *Phialophora graminicola*, and *Phialophora* sp. with lobed hyphopodia—under controlled laboratory conditions. We will not know the absolute period of survival, but the survival periods of these fungi may be compared.

HORNBY:

I have artificially inoculated soil with colonized straw

segments and then tested the MPN technique by using it to estimate the number of segments. The relationship between the MPN estimate and the actual number was good.

SCOTT:

Professor Garrett proposed this method to compare the effects of experimental treatments on survival. One should not expect to model the absolute period of survival of the fungus under natural conditions using this approach. It may be a very useful technique for comparing, for example, the effects of certain nutrients on survival.

HUBER:

What alternative techniques are there for this type of study?

KOLLMORGEN:

Naturally infected wheat crowns from the field could be used as the test unit, but they are more difficult to work with and other organisms are often present.

PARKER:

A point in favor of colonized straw segments is that they would probably keep the pathogen in a fit physiological state to invade living cellulosic and pectic tissues.

HORNBY:

There would be some years in the United Kingdom when the development of an epidemic of take-all would not involve, to a large extent, the colonization of the stem base. Therefore, to consider colonized straws as a model for survival of the fungus in soil may be somewhat limited.

ROVIRA:

I see it as a useful method in some areas such as the colonization of a host root from a propagule. However, I see danger in studying the survival of the fungus on straw segments in the field and then extrapolating to the real situation. Rather, I would favor the use of infected crowns, however variable that method may be.

MacNISH:

The fungus would probably survive for longer on natural inoculum than on artificially colonized straw segments.

KOLLMORGEN:

I believe that segments of straw would be useful in determining relative and not absolute effects: for example, the survival of the fungus under different crop rotations.

MacNISH:

I agree, but I think it would be important that the straw be colonized by dark runner hyphae and not hyaline hyphae.

COOK:

The natural colonization of straw in the field occurs as the result of the parasitic phase of infection, whereas artificial colonization of straw segments would occur via the saprophytic phase. The fungus uses its parasitism to overcome competition from secondary organisms. The extent of colonization by secondary organisms and the consequent survival of the take-all fungus could therefore be different in naturally and artificially colonized straw.

PARKER:

Lucas's early work showed that some isolates of *G. graminis* var. *tritici* colonize straw in nonsterile soil under optimum temperature and moisture conditions. We have found that the fungus can colonize straw placed some centimeters away from the original source; this might happen in the field when straw is turned under.

COOK:

Straw other than that which is clean and bright would be so thoroughly colonized by other saprophytes that take-all could not compete.

HORNBY:

There seems to be too little known about the saprophytic ability of *G. graminis* var. *tritici* in soil. In my own work, the fungus, on occasions, has appeared to increase by saprophytic growth, and there are one or two reports in the literature to support this finding.

Session 3: Pathogenic Variation

CHAIRPERSON: D. PITT

Question 1 (POWELSON):

How should one select isolates for pathogenicity tests? There seems to be a wide range of pathogenicity among isolates, and pathogenicity may be influenced by host species, geographic origin, host tissue from which isolations were made, or age of culture. How consistent and stable is this difference? Does cross-protection occur if isolates are combined to obtain a wider range of pathogenicity?

POWELSON:

In our attempts to obtain pathogenic isolates from the field, we are concerned that we may not be obtaining true values of their pathogenicity in subsequent tests. From the studies of Dr. Wong with *G. graminis* var. *graminis* and *Phialophora* and of the French group on hypovirulent isolates, we may be underestimating the field importance of antagonistic fungi.

SKOU:

This is a very important question, but as yet we do not know what the field levels of pathogenicity are. Isolates subcultured frequently on rich media without the host often adapt to the medium. In some cases, isolates may become more adapted to growth on the artificial medium than on the host. Conversely, the pathogenicity of saprophytically adapted isolates can be regained by passage through the wheat plant. Therefore, the only true measure of pathogenicity is that obtained by transferring the fungus from wheat plant to wheat plant to minimize artificial adaptation. When you transfer an isolate to rye, you use an organism that is more adapted to wheat than to rye, and so pathogenicity is higher on wheat than on rye. Therefore, a true pathogenicity measure is only obtained when an isolate is subcultured from one plant directly onto another of the same species.

PITT:

This suggests that the form in which you present your inoculum in field studies may affect the expression of pathogenicity. For example, perhaps the use of wheat roots or crowns is preferable to oat-grain inoculum because adaptation will be partly dependent upon selection pressures imposed by the substrate.

WONG:

With regard to cross-protection, the important factors in determining the abilities of *G. graminis* var. *tritici* isolates to attack wheat are cortical colonization and invasion of the stele. Weakly pathogenic isolates that are more invasive than highly pathogenic isolates may cross-protect against the disease. Even though the weakly pathogenic isolate may cause some disease, the overall disease expression would be reduced in economic terms.

POWELSON:

We tested three isolates of *G. graminis* var. *tritici* for pathogenicity and then made equal admixtures of these isolates and repeated our tests. We found one isolate that, on its own, caused severe damage; but on mixing with other isolates, a considerable reduction in disease severity resulted.

HORNBY:

In my opinion, the question should read, "How should one use pathogenicity tests to select isolates?" We have a standard pathogenicity test in pots that establishes whether an isolate is useful for our studies. Performance of isolates in the test can be compared against a scale derived from isolates of known pathogenicity in the field.

POWELSON:

I agree. We also use pathogenicity tests to select isolates. Isolate responses varied considerably in our tests depending on the isolate source, species of plant obtained from, etc.

KOLLMORGEN:

Does greater variation occur with mycelial cultures than with cultures from single ascospores?

POWELSON:

We use mycelial tip cultures.

KOLLMORGEN:

Surely these are more variable than cultures from single spores.

POWELSON:

Yes.

PARKER:

I don't agree. Studies on the rice blast fungus, for example, have shown that considerable variability can result even from spores with a single haploid nucleus. I have always isolated from a series of tips from outgrowing cells rather than from a single tip. I feel this approach gives me a better representation of the isolate's potential.

CUNNINGHAM:

The literature indicates that high levels of pathogenic

variation occur with both hyphal tip and single ascospore isolates of the take-all fungus.

SCOTT:

Does one ascospore originate from one nucleus in *G. graminis* var. *tritici*? It is often assumed that a multinucleate spore has a mononuclear origin. If so, then all hyphae growing from this ascospore would have genetic information derived from one nucleus only. In contrast, hyphal tips usually contain more than one nucleus and so must constitute a source of greater genetic variation.

PARKER:

Chambers published work in which he isolated all the spores from one ascus and tested them independently for pathogenicity. Considerable variation was observed between the ascospores.

Question 2 (WALLWORK and RATHJEN):

Which are the best isolates of *G. graminis* var. *tritici* to use in trials? Either more or less pathogenic isolates of the fungus can be used to assess varietal differences in resistance. The former may provide a more rigorous test of differences, whereas use of the latter may have an advantage in that slight differences in resistance may be easier to detect. What factors influence the choice of other workers?

WALLWORK:

In Session 2, it was suggested that the most virulent isolates should be used. However, this will cause problems if plants are killed prematurely. An alternative to varying inoculum levels may be to use less pathogenic isolates.

COOK:

I prefer to use a very highly pathogenic isolate, which is stable in culture, and to regulate disease severity by varying inoculum density or some environmental factor. With less pathogenic isolates, a greater variability in pathogenicity often occurs and disease expression may fail altogether.

WONG:

I would like to ask Dr. Cook what criteria he uses to determine a stable isolate.

COOK:

We observe growth of isolates on agar plates and look for colonies without abnormalities such as sectoring. We make sure the isolate exhibits typical growth rate and good pigmentation and, when tested for pathogenicity, that it consistently produces severe root and crown rot. The isolate should also produce perithecia on plants.

WONG:

Some of my isolates have maintained a stable condition for up to 2 years, but abnormalities such as sectoring occurred on recovering them from oil storage.

COOK:

We started with several hundred isolates obtained from randomly selected plants and maintained these in various ways. There was a gradual decline in the stability of isolates regardless of how they were

maintained. I feel that, although storage method may affect the rate of decline, isolate degeneration is, in the long term, inevitable.

ROVIRA:

Our experience differs slightly from this. We subcultured over 40 *G. graminis* var. *tritici* isolates about 20 times and found that the pathogenicity of some isolates was reduced, whereas for others it remained unchanged. We also introduced our virulent strain "500" into the field along with these attenuated isolates, but we found no evidence for cross-protection. The reasons behind strain degeneration should be examined by fungal geneticists.

Question 3 (GLENN):

How variable is the response of different isolates of *G. graminis* var. *tritici* to changes in environmental factors? Even in pot experiments on take-all, the fungus does not always behave with the predictability one would like. This implies that variables of significance to the fungus are not being recognized as such in the design of experiments. Is there any work in progress on natural variation within and between strains of *G. graminis* var. *tritici* in response to environmental variables?

GLENN:

Earlier it was mentioned that perhaps most of the observed pathogenic variation arises from variation in climatic or soil conditions during the course of an experiment and not inherent factors within the fungus.

SKOU:

It is very difficult to identify the factors governing variation. In one experiment, various isolates may be graded for pathogenicity based on a standard pathogenicity test. A repeat of this same experiment 1 month later would give a different range of pathogenicities for the same isolates.

GLENN:

Have we really covered all the options of what causes variability in pathogenicity of an isolate?

SKOU:

No. I believe that the environment is responsible for much more variation than we presently understand.

PITT:

I feel that there is a disproportionate amount of effort directed toward genetic factors controlling pathogenic variation relative to environmental factors.

HORNBY:

I don't think there is a great deal of genetic analysis in progress. Both these areas are rather neglected.

PARKER:

Funding is not available for these types of studies, although it should be. Awarding bodies are more interested in funding applied, field-oriented research.

HORNBY:

Very few such studies have been made with ascomycetous fungi in general. Much of our understanding of the take-all fungus comes from analogies with such organisms as *Neurospora crassa*.

Question 4 (CUNNINGHAM):

What qualitative changes occur in the pathogen population with different cropping sequences?

CUNNINGHAM:

Our earlier studies of collections of isolates from different length sequences (1–8 years) have indicated definite trends in the different populations for various cultural characteristics and virulence. Isolate virulence increased in populations from the first to the third or fourth crops and declined in longer sequences. This trend mirrored disease severity observed in the field, suggesting selection pressure on the parasitic phase of the pathogen until the take-all peak was reached. Because of erosion of saprophytic traits during this period, it was postulated that from the take-all peak onwards, there was selection pressure for saprophytic as well as parasitic characters; this ultimately ensured a population with a less extreme blend of saprophytic and virulence traits. Asher in Scotland and Cook in Washington State (U.S.A.) neither support nor refute these results, as they found limited evidence only of population shifts. Currently, we are conducting similar studies of isolates from continuous winter wheat, winter wheat alternating with spring barley and with potatoes, and also continuous spring barley to corroborate selection pressure for virulence in populations. Our preliminary results suggest that the longer the pathogen is in contact with its host, the more virulent is the resulting population. Have other workers detected shifts in pathogen populations from different-length cereal sequences and different crop rotations?

HUBER:

We examined this during our studies on the relationships between take-all decline and loss of virulence. We found that, during a given cropping sequence, the frequency of pathogenic isolates obtained from a population increased; whereas with continuous monocropping, take-all decline appeared to be associated with loss of virulence in the pathogen population.

SCOTT:

There are two separate points here. One is the overall pathogenicity of isolates, independent of host species, which may be relevant to the take-all decline phenomenon. The other point relates to host specificity of isolates, and Dr. Cunningham's results provide some evidence of specificity for wheat in isolates obtained from continuous wheat.

CUNNINGHAM:

Yes. We also have evidence of specificity toward barley.

SCOTT:

In regard to this point, have any other workers obtained similar evidence on the effects of cropping sequence on pathogenicity? Many scientists assume, for example, that cropping with oats selects *G. graminis* var. *avenae*.

SKOU:

I don't think it's a question of host specificity but rather one of adaptation to the host. We have obtained similar results to Dr. Cunningham's in Denmark and Sweden.

SCOTT:

Yes. I do not mean to imply absolute specificity to a single host. In England, we find rye-adapted isolates. These are not particularly frequent in the small areas where rye is grown, however, so in this instance the cropping sequence has not affected host adaptation.

HORNBY:

Slope and Gutteridge published a short item in the Rothamsted Report for 1981 about *G. graminis* var. *tritici* populations from monoculture wheat and wheat in rotation with lucerne. They found that larger proportions of isolates from monoculture wheat sectored on potato-dextrose agar and that isolates that were floccose or grew slowly at pH 4 were obtained from monoculture wheat only. This is perhaps some of the strongest evidence yet for qualitative differences in the pathogen population in different cropping sequences, and, presumably, between take-all decline and nondecline situations.

CUNNINGHAM:

I believe that host-crop selective effects have a greater and more direct influence on the take-all peak than on take-all decline. Although the magnitude of the take-all peak seems to be the product of selected parasitic traits, take-all decline would appear to be largely determined by stabilizing constraints on the saprophytic phase.

ROVIRA:

Dr. Cunningham, do you find that during continuous wheat cropping your isolates become more pathogenic to wheat as time increases?

CUNNINGHAM:

Our recent results from different crop sequences and rotations would indicate that. However, about 10 years ago in our spring-sequence experiments we found an increase in virulence up to the third or fourth crop only, following which a reduction in pathogen virulence occurred.

ROVIRA:

Did this reduction in virulence occur when the disease was actually in the decline situation?

CUNNINGHAM:

Yes.

ROVIRA:

Then this does not conflict with Dr. Hornby's last comments.

CUNNINGHAM:

I agree.

COOK:

Recently, we published work that compared the pathogenicity of isolates from fields in monoculture wheat with isolates from fields in rotation sequences. From several hundred isolates obtained at random from randomly selected plants in wheat monoculture, we found a slightly higher frequency of weakly pathogenic isolates. In fact, we did not recover any such isolates from fields that had just been sown to wheat after a break and in which take-all was just starting to appear. Nevertheless, the frequency of weakly pathogenic types isolated from monoculture fields could not account for take-all decline. Some 90–95% of isolates obtained from our monoculture fields were still highly pathogenic. One explanation for this may be the proposal of Chambers, namely, that the weakly pathogenic types also have poor survival characteristics. In other words, weakly pathogenic isolates are least likely to survive during a break crop, leaving only the strong pathogens to reestablish the disease when wheat is again sown. Weakly pathogenic types perhaps only survive during wheat monoculture, where they can persist on a steady supply of susceptible wheat roots. However, this does not explain take-all decline.

ROVIRA:
Five percent is not a high proportion.

COOK:
It was not statistically significant, but it did occur.

GLENN:
Why would you assume that the weakly pathogenic types were less likely to survive in the saprophytic phase than the stronger pathogenic types?

COOK:
We have found a high correlation (0.90–0.95) between those isolates that are weakly pathogenic to wheat in a vermiculite rooting assay and those that grow poorly on agar at pH 4. In studies reported by Romanos and others from Imperial College in cooperation with workers at Rothamsted, it was reported that certain isolates of *Gaeumannomyces* produced an inhibitory compound when grown at pH 4. If our recent results on heterokaryosis or cytoplasmic determinants of variation are correct, then perhaps one or the other of these genetic mechanisms is controlling an inhibitor produced at pH 4. Isolates with such determinants grow abnormally and poorly, sector frequently, and are poor survivors. This may explain the observation of Chambers 16 years ago.

PARKER:
Slow growth can sometimes have a protective function with *Rhizobia*. The slow growers are better colonists of infertile soils.

COOK:
Slow growth of these isolates is thought to be due to

the activity of this inhibitory factor.

Question 5 (PENROSE):
What characteristics of *G. graminis* var. *tritici* isolates used to inoculate field trials are necessary to permit progressive disease development—in other words, where disease is more severe in adult plants than in seedlings?

PITT:
During the preparation of bulk inoculum for field trials, researchers presumably start with a single isolate grown up on a saprophytic medium as a preinoculum for bulk inoculum preparation. What is the frequency and degree of pathogenic variation in the final prepared bulk material, and how much does this vary from the pathogenicity of the original culture isolated?

ROVIRA:
Provided a stable, strongly virulent isolate is used, the method of inoculum preparation standardized and contamination avoided, a fairly reproducible inoculum preparation of consistently high virulence can be obtained. Each batch should be started from the original culture and not by subculturing from one batch to the next. On some occasions, however, a preparation may produce fluffy, white mycelial growth on the grain or seed instead of the usual uniform gray growth. The reason for this is unknown, but such abnormal preparations are generally less virulent and are not used in our experiments.

COTTERILL:
During our preparation of a *G. graminis* var. *tritici* isolate last year, we found that one preparation used at high inoculum doses proved less virulent than another used at lower doses. The same procedure for bulk inoculum was used throughout.

CUNNINGHAM:
We have observed some saprophytic and parasitic traits that change with frequency of cereal cropping following the plowing under of old grassland. The first saprophytic trait considered was isolate pigmentation; whereas isolates from the first and second cereals darkened quite rapidly in culture, those from the third and sometimes the fourth crops remained hyaline and those from longer sequences consisted of mixtures of both as well as intermediate types. Similarly, the linear growth rate of cultures diminished significantly from the first to the third and fourth crops (as virulence increased) and increased subsequently to a level intermediate between that of the first and third cereal. However, such cultural traits may be of little value in predicting isolates that will give progressive disease development.

Session 4: Growth Regulators, Pesticides, and Herbicides

CHAIRPERSON: D. BALLINGER

Question 1 (NICOLL):

Have any estimates been made of losses due to take-all?

NICOLL:

My company has several fungicides that look promising for take-all control, but before funds are committed to develop this approach, we need information on losses caused by the disease and an acceptable cost for the fungicides.

BRISBANE:

We cannot adequately answer this question until a suitable fungicide is available. Initially, the cost is not important provided the fungicide is effective.

ROVIRA:

That is a good point. When chemical control of cereal cyst nematode (CCN) was first researched, the most effective chemicals were uneconomic; but when it was shown that CCN was costing growers millions of dollars, this encouraged chemical companies to test chemicals for CCN control. In this way, chemicals and application techniques were developed, and economic control of CCN has eventuated. A similar situation may apply with take-all.

NICOLL:

I suspect that the use of sterol-inhibiting fungicides may affect other soilborne pathogens as well as take-all. It may be difficult to isolate these differences.

SMILEY:

The sterol-inhibiting fungicides seem to have the greatest promise in this area. However, they are likely to cause physiological and morphological responses in the plant other than disease control that may also affect ultimate yields.

HUBER:

I have attempted to estimate yield losses of plants in the glasshouse. I have done this work with a series of disease levels to predict the yield reductions for different levels of disease and different environmental conditions. The same approach has also been tried in the field. By combining all the information obtained, we have published some estimates of yield loss. I have surveyed wheat fields in Indiana in the United States for 7 years and estimated yield losses due to take-all, taking into account cropping history, fertilizer program, and potential yield for all cultivars grown under those conditions. Estimates of losses for each of the years varied from 2 to 27% of potential yield. This estimate was

probably conservative and, although the method was not perfect, it was the best we could manage.

BALLINGER:

Have any estimates of yield loss due to take-all been made in other countries?

HORNBY:

There have been a number of experiments at Rothamsted and elsewhere in Britain that have attempted to establish a relationship between disease and yield. From memory, for every 1% of tillers with take-all in the crop there was about a 0.4–0.6% reduction in yield. I recall making some estimates in 1975 when an estimated 40% of the U.K. wheat crop was at risk. Assuming this, 40% sustained losses of 15–20%; in other words, about 8% of the wheat crop was lost, for a loss in monetary terms of about 20 million pounds. So these kinds of exercises have been done, but how precise they are is another question. There is still much work needed in this area.

Question 2 (PITT):

What is the current state of fungicide research on control of take-all? Few fungicides are available for the effective control of soilborne cereal pathogens, and most success appears to have been made with products containing benomyl and triadimefon in various seed treatments. Clearly, a systemic fungicide such as these is preferred over a protectant (contact) chemical because application must be made at the time of sowing, but protection of a ramifying root system must be afforded. Aerial spraying is likely to be too expensive for farmers to contemplate in broad acre, low-intensive cropping regions such as those typifying Australian cereal agriculture. What then is known about translocation of systemic fungicides within the root system of cereal plants, and which fungicides currently available or under test have shown the greatest potential for systemic control? Do any protectant fungicides exist that have control possibilities for take-all or other soilborne plant pathogens? Which methods are available to measure the efficacy of protectant and systemic fungicides in laboratory studies, and can these yield minimum inhibitory concentration (MIC) and minimum fungicidal concentration (MFC) values?

PITT:

I am particularly interested in any work done with systemic fungicides that remain active within the roots rather than being translocated up through the stem of the plant.

BATEMAN:

My experience is that systemic control will be too short-lived to protect roots.

HUBER:

We have looked at the sterol inhibitors triadimenol and triadimefon; along with others working in this area, we observed a 6- to 8-week delay in fall infection. Under our conditions, we obtain a yield response one year in three from this delayed infection. In other years, spring infections are more than adequate to offset the yield potential from a delay in infection achieved with chemicals.

BALLINGER:

Most of the systemic chemicals we are currently working with are mainly translocated upwards. Ideally, we need a fungicide with more systemic activity in the root area.

PARKER:

I think that if you can protect the seminal roots for the first few weeks of their life it would be a powerful protective mechanism. Seminal roots usually get infected first and provide a new powerful source of inoculum for subsequent nodal roots. Therefore, fungicidal control of initial infections would be quite valuable, and I would like to see more effort in this area.

Question 3a (WONG):

Can the use of systemic seed dressings be useful in delaying take-all infection and reducing disease?

Question 3b (BALLINGER):

What effect does control of early seminal root infection have on severity of take-all?

WONG:

This has been discussed somewhat by Professor Parker. We also have information to suggest that a delay in infection of seminal roots will result in less disease on the crown-root system. Some field observations have suggested that late sowing, resulting in a shorter growing season and a shorter season where roots are exposed to the pathogen, would lead to less severe infection. I wonder whether we can get an economic return if we protect the seminal root system for 6–8 weeks by using a fungicide.

SOLEL:

Most of the fungicides being tested move upward in the plant and are only in the root tissues for a very short time and at low concentrations. I think that systemic fungicides are affecting the fungus in the soil and not in the plant. Thus, their effect is protectant rather than systemic, with the fungicide providing a pathogen-free zone around the seed. I have used thiabendazole as a seed dressing and achieved very good disease control. Seedlings from treated seed were almost completely disease free, whereas those from untreated seed carried 80–90% infection.

ROVIRA:

In our experiments with fungicides, the best control has been obtained with soil treatments of benomyl and PCNB. Thiabendazole gives effective control in pot trials but is phytotoxic. We had no success with seed treatments. The only success in pot experiments has been with soil incorporation of fungicides; of those tested, two were effective over a period of 4 weeks. At the rates used, they would not have been economical but might be useful as an experimental tool.

HUBER:

We have been getting away from the seed treatment approach and concentrating on getting better distribution of fungicide in the root zone. In our trials, we use anhydrous ammonia as a carrier for the sterol-inhibiting fungicides, injecting it at 25-cm intervals in the soil. This system provides better control for a longer period of time, but, again, the rates are not economic.

BALLINGER:

I support what Professor Huber has said. My work in Australia has indicated only marginal control of take-all with seed treatments, but excellent control has been achieved by applying fungicides as granules or as pellets banded with the seed at sowing. This may support the theory that the fungicides are acting as protectants rather than systemics.

PARKER:

I wonder if enough thought has been given to the possibility of using volatile fungicides. Volatile compounds such as pheromones can be effective with insects at extremely low levels; they can, for example, cause a developing insect to jump a complete instar, resulting in a sterile dwarf adult.

Question 4 (BATEMAN):

Assuming that fungicidal control of take-all is possible, can it be made practical and economic? Fungicide applications to soil can partially control take-all and increase yield of winter wheat, but the amount of fungicide needed may make such a treatment uneconomic. The problem therefore is to minimize fungicide use. One approach was to try to test the ability of localized applications to bring the fungicide into direct contact with the pathogen only at the most damaging infection sites. Results suggest, for the fungicides benomyl and nuarimol, that partial eradication and protection by application at the depth of crown formation may give the best control. Systemic protection from these fungicides did not seem to occur when they were placed at various depths. The effects of restricting the site of fungicide application, the practical means of doing this, and the most suitable fungicidal compounds and formulations are not yet completely clear. I am very interested in Mr. Ballinger's work with fungicide granules and pellets.

BALLINGER:

Granules can be easily applied to the seed bed with conventional sowing equipment to protect the seminal roots. I'm sure the system could be adapted to apply the granules just below the soil surface in the crown root area.

BATEMAN:

I have been broadcasting the fungicide on top of the soil and harrowing it in before sowing. This is an extra treatment, extra cultivation, and an extra expense that makes the technique unsatisfactory.

BALLINGER:

Application of chemicals, either as seed or soil treatments at sowing, appeals to most growers because they are familiar with the technique and it is easily incorporated in their normal routine.

RODGERS:

Could fungicide granules be applied through a small seeds box?

BALLINGER:

Small seeds boxes, familiar to growers in Australia, would be very suitable for granule application.

KOLLMORGEN:

Mr. Ballinger, in your work with granules, has there been any evidence of phytotoxicity?

BALLINGER:

Some of the fungicides I have worked with, particularly the sterol inhibitors, can be phytotoxic at relatively low levels. Granular application considerably reduced this problem because fungicides are not concentrated on the seed as with a seed treatment. My work has mainly involved two chemicals: triadimefon, which can be highly phytotoxic, and benomyl. Both are currently used as seed dressings for smut control in Australia (10 and 20 g/ha, respectively). At this rate, triadimefon can delay seedling emergence; but when applied as granules, the rate can be increased one hundred times with no increase in phytotoxicity.

HUBER:

All sterol-inhibiting fungicides are not equally phytotoxic. In our studies, triadimefon has an order of phytotoxicity probably eight times greater than triadimenol.

COOK:

Are you achieving a take-all response to the granules?

BALLINGER:

The drought last year ruined all field trials. However, this year, I have sown the fungicide trials on an irrigated site that had been sown to wheat the previous two seasons. The first sampling, 8 weeks after sowing, showed 90% of plants from untreated seed with take-all lesions. Triadimefon and benomyl granules significantly reduced infection on the seminal roots.

HUBER:

Expression of phytotoxicity can be environmentally sensitive.

POWELSON:

I am not sure that a little phytotoxicity is a bad thing, as there are many factors involved. A delayed seeding date can reduce the disease risk with most wheat diseases. If the initial onset is delayed, this may provide valuable protection.

Question 5 (HUBER):

What chemicals are available for delaying infection or pathogenesis?

HUBER:

We have covered most of this, but I wonder if there are others working with different materials.

BALLINGER:

We have discussed triadimefon, benomyl, and triadimenol. Imazalil has performed well in glasshouse trials.

ROVIRA:

Benomyl and PCNB were effective in pot experiments. However, when we used PCNB in the field, yield was reduced. We attributed this to the low soil-moisture content in the seedling stage.

HUBER:

The nitrification inhibitors nitrapyrin and etridiazole reduce take-all by stabilizing nitrogen in the ammonium form. This results in greater metabolic activity in root tissues and the rhizosphere. The growth regulator may aid in disease escape by increasing root growth.

Question 6 (WONG):

What recent information is there on the nontarget effects of herbicides, insecticides, and fungicides on *G. graminis* var. *tritici* and its antagonists?

WONG:

Direct drilling has increased in popularity in many parts of Australia; consequently, the use of pesticides has also increased. Is there any information on the effect of herbicides, insecticides, and nematicides on take-all and its antagonists? Could some of the conflicting effects of direct drilling on take-all be the result of the use of different pesticides?

SIVASITHAMPARAM:

In Western Australia, we have observed increased take-all severity in areas that have been treated with dicamba.

ROVIRA:

This has also been recorded in the United Kingdom. I have had reports in South Australia of increased incidence of whiteheads following dicamba treatment.

POWELSON:

Appleby and I have a graduate student who is looking at interactions between take-all and diclofop.

HUBER:

Several nematicides reduce take-all severity by inhibiting nitrification.

Question 7 (YARHAM):

How does herbicide practice influence take-all? To what extent can herbicides influence take-all—either directly by influencing pathogen growth or host root susceptibility, or indirectly through long-term effects on the soil microflora?

HORNBY:

I think this may refer to a report from New Zealand where glyphosate was connected with an increase in take-all.

POWELSON:

We have done some work with glyphosate, and results indicate that take-all was more severe where the herbicide had been used.

ROVIRA:

Our trials would support this. More take-all has been measured in wheat following a grassy pasture sprayed with glyphosate than a similar pasture sprayed with paraquat/diquat. This may occur because glyphosate does not have the sudden-death effect of paraquat/diquat. Thus, the roots survive longer.

Session 5: Disease Expression and Measurement

CHAIRPERSON: K. SIVASITHAMPARAM

Question 1 (PITTAWAY):

Are whiteheads the consequence of root damage at or beyond anthesis, or can they also be the clinical expression of latent seedling infection? The question relates to the developmentally associated capacity of a host to tolerate root damage and how the time of pathogen attack influences ultimate symptom development. Root-pruning experiments have shown that wheat can survive with a highly reduced root system, but the consequences of seminal and/or crown root reduction are largely dependent on developmental and environmental conditions after amputation. Both seminal root and leaf initiation are determinate in nature. Seminal root primordia are differentiated during germination, and leaf primordial production ceases when the apex commences reproductive growth. As a consequence, the vegetative phase represents the most active root and leaf initiation phases; during this phase, seedlings would be expected to show the greatest tolerance to root damage. Moreover, once the developing ear has emerged, adventitious root growth also ceases, at a time when demands from the expanding leaves and developing grain are very high. Therefore, unless a very high proportion of the seedling root system was damaged, would you expect to observe symptoms during the vegetative growth phase? Or would such early damage remain benign until the developmental and environmental demands on the reduced root system postanthesis proved too great?

PARKER:

Whiteheads are a symptom of the ultimate in disease—death. In Western Australia, on old land, there are usually thousands of seriously affected plants for every whitehead. On a farm we worked on for 9 years, we only once saw whiteheads, and this was only in a small area of one field. Yet *G. graminis* var. *tritici* was evident on this farm because most plants were infected every year. The seminal roots were almost always nonfunctional by the flowering stage. How much grain was lost was hard to assess; in our climate, where wheat matures in increasingly hot and dry conditions, I believe it can be around 35% without recognized clinical symptoms (blackened roots, blackened straw bases, and whiteheads) being visible.

COOK:

I would comment that the whitehead phase occurs under hot and dry conditions and is a more spectacular expression of the disease; it results from crown infection that restricts water movement to the foliage. Root infections alone will not lead to whiteheads. Invariably, we find that the fungus has grown into the crown of plants with whiteheads. Root infections result in smaller plants. The plants compensate for root infections, if signaled early enough, by producing fewer tillers or by not growing as tall as normal. Root damage by itself produces symptoms in the tops that are not unlike symptoms of inadequate nutrition.

MacNISH:

Under our conditions in Western Australia, I consider the seminal roots to be the most important. At the end of the season, when the plant is extracting moisture from deep in the soil profile, crown roots are not doing very much. If the take-all fungus penetrates the crown, the seminal system, even if functioning, becomes inoperative because of blockages in the crown region.

SKOU:

Take-all is common, whereas whiteheads occur infrequently. They may be caused by factors other than infection by *G. graminis* var. *tritici*, although there is some correlation between whiteheads and the take-all fungus. There are more reliable symptoms of infection by *G. graminis* var. *tritici* than whiteheads.

MacNISH:

I consider whiteheads to be a very poor indicator of take-all. They may reflect something that happens at anthesis, something unrelated to take-all.

COOK:

We count whiteheads in some cases and take other measurements in other cases. Obviously, there are several other causes of whiteheads. When we know that take-all is the only cause of crown or basal stem rot, then whiteheads can be a useful measure of disease.

ROVIRA:

I do believe that whiteheads are a significant indication of take-all attack and are correlated with a high incidence of infection under our conditions. A particular set of conditions is required for whitehead formation: we seldom see any in our plots unless we have more than 50% of the plants infected, and we have found a high correlation between disease incidence and disease severity. Requirements for whitehead formation are a high incidence of root infection, a moist spring that favors growth of the host and the pathogen, and a sharp cutoff in soil moisture in October. In this situation the take-all fungus is the most severe, because in blocking the xylem it restricts water supply to the tops.

PITTAWAY:

An observation by Simmonds in the 1940s that

prompted my question concerns results from root-pruning experiments. He found that if seminal roots were pruned after the seedling stage, plants matured early; if nodal roots were pruned, maturity was delayed. With take-all, seminal roots are usually the most affected, and whiteheads result. What is the importance, however, of nodal root attack, and do the same symptoms result?

COOK:

From my observations, if a pathogen mainly affects the seminal root system on young plants, maturity is delayed. If the pathogen infects the crown, then the plant will mature early.

PARKER:

I doubt whether seminal roots are as important to the maturing plant as most plant breeders would believe. Nodal roots are much thicker than seminals near the crown, where they serve both a feeding function and act as prop roots to keep the tillers erect. But as they penetrate deeper, they become finer and more like seminal roots; and I wonder if they can then take up subsoil water.

COOK:

You should remember that the wheat plant does not have crown roots before it has tillers. Crown roots emerge in pairs on the opposite sides of the tillers.

Question 2 (HORNBY):

What are the best designs for field experiments on take-all? It is difficult to predict take-all severity because of its dependence on weather conditions and previous cropping. It is also generally accepted that take-all has a patchy distribution and that in consecutive cereal crops the patches change in shape and intensity from year to year. Usually, patchiness is a problem early in runs of the host; but the onset of the take-all decline phenomenon may create less conspicuous patterns of disease intensity. The amount of sampling and disease assessment required to deal with the resulting variability may exceed normal resources. Because of these problems, the results of many field experiments have been inconclusive. Are there ways in which we can improve conventional layouts, or should we be looking for new approaches? Can we identify the strengths and weaknesses of small and large plots? Have there been any experiments to determine optimum plot sizes, and would such information have general application at all stages of crop growth or in different years and at different sites?

HORNBY:

When I started experimenting with naturally occurring take-all, I followed a tradition (which I think many people still follow) of working with large plots. Because of their size, it is unusual to have many replicates of such plots, and it is exceptional to sample and assess even 1% of plants within them. I had reason to question this kind of approach after testing a range of fertilizers on a site in which take-all was unevenly distributed and where the small number of replicates resulted in some treatments being by chance mostly in areas of much disease and others mostly in areas of little disease. It occurred to me that if the distribution of

inoculum in the field were known, experiments might be blocked according to inoculum density levels. A further idea, prompted by the single assessment/tiller technique used for some foliar diseases, was to have treatment and control plants in close proximity. These thoughts led to a compromise design. There are 10 replicated small plots, 35 cm × 35 cm, of each of six treatments at each of three sampling dates, and all the plants within these are assessed. The treatments are distributed among incomplete, balanced blocks of four plots, one of which is always a control. Blocks with complementary treatments are put down in pairs throughout the site. In this way, we hope to test treatments in small, relatively uniform areas of disease. Preliminary examination of data from first and second samplings indicated that variation in disease assessments among block pairs was up to 32 times greater than the variation among plots within blocks and 8 times greater than the variation within block pairs for a site at Woburn (U.K.), where take-all had become progressively worse in three consecutive crops of wheat. At a Rothamsted (U.K.) site, where take-all decline had occurred and disease was less, the variation among block pairs was only 11 times that among plots within blocks; at this site, it seems that plot size could have been increased to about 1 m × 1 m without experiencing increased variation.

WALLWORK:

I have a split-plot experimental design. Inoculum is scattered within subplots over a 40-cm^2 area on a plane below the seed. Grain of the test lines is scattered over a 20-cm^2 area within the inoculated zone. There are four subplots comprising three separate inoculum levels and a control (uninoculated ryegrass seed) that enables a nearest-neighbor comparison to be made between each dose rate and a control. Use of small plots and nearest-neighbor controls means that errors arising from variation in the soil or background inoculum can be minimized. Also, in using small plots, it is easier to increase the replicate number, which I consider to be an important consideration where local variation in the growing environment is large.

HOLLAMBY:

We are using a technique when selecting for tolerance to cereal cyst nematode (CCN) where we have plots of 5 m × six rows, and each plot is bordered by intolerant and tolerant genotypes. In the case of CCN, it is Festiguay (tolerant) and Egret (intolerant); with take-all, however, tolerant and intolerant plant species would be needed. With two border rows around each plot, we obtain from the tolerant genotype a measure of the site yield potential per se. From the difference between the tolerant and intolerant genotypes, we obtain an indication of the incidence of the disease; and we only take information from those patches of similar yield potential and those that have the greatest disease level. We use this information for tolerance determinations or single-plant selection.

In my experience, many cultivar evaluation trials are sown at sites at which there is excellent disease development. To obtain information from such sites, we need a means of using the data from the diseased patches and ignoring the rest, and vice versa. This is where we need the statisticians to help us. How do we utilize the information from such trials rather than "writing them off"?

WONG:

How many replicates would you have?

HOLLAMBY:

Yield plots are 5 m × six rows. Plots for single-plant selection are longer and may consist of single rows 60 m long, each with a tolerant and intolerant cultivar on each side. In our normal trials where we are assessing yields per se, there are usually six replicates.

WONG:

How would you discard the replicates?

HOLLAMBY:

We do not discard individual replicates but discard individual plots according to disease incidence. We therefore end up with a mean and a variance. The variance may be based on four to six plots.

SCOTT:

What do you do with the tolerant and intolerant values? Do you adjust the value of your experimental plot by the value of the neighboring one?

HOLLAMBY:

We use the difference between tolerant and intolerant entries. We treat them as indicators of whether the disease is present; if the disease is present, we use the plots in that area, and if not, we discard the data.

SCOTT:

There is a technique in use at the Plant Breeding Institute in which numerous control plots are planted in an experimental design. Values from control plots are used as covariates of values from neighboring test plots in the analysis of variance.

HOLLAMBY:

There is some interaction between the level of the disease with the genotype. The covariance analysis does not work because it assumes linear relationships.

POWELSON:

Suggestions we get from statisticians are to increase the number of samples per plot and, more important, to increase sampling frequency. This takes manpower; but, with proper analysis, the method will overcome the variation that occurs with single-time assessment.

Question 3 (ROVIRA, BRISBANE, and SIMON):

What is the best method of assessing take-all and linking disease levels with yield loss? There are a number of ways of assessing the level of take-all in a cereal crop. These include (a) incidence, i.e., percentage of plants with one or more stelar lesions; (b) percentage of roots with stelar lesions; (c) length of lesion per plant; and (d) severity, or a numerical rating system of 0–4 based upon the percentage of root systems affected by take-all. With natural levels of infestation in the field in Australia, there is not the severe blackened root system often encountered in glasshouse experiments with artificial inoculum and in wetter climates. There is a high correlation (0.96) between "incidence" and

"severity" over a wide range of disease levels. Field experiments in which increasing numbers of propagules have been added to soil have shown that at the levels of propagules that correspond to the natural field levels after grassy pasture, the yield loss is in the order of 25–30%—a loss that is difficult to measure without large plots and many replications. The yield loss that is going to occur for a given level of take-all in soil will be very dependent upon the soil conditions between tillering and maturity. Hence, one would expect the dose-response curves to vary greatly from season to season. The question is that unless we have data on the relationship between the disease level and yield over several seasons, will we ever be able to assess the economic losses caused by take-all?

SIVASITHAMPARAM:

The area of the relationship between disease level and yield is complex, and the general opinion tends to favor evaluations lasting through several seasons.

COOK:

Three years ago, I used an approach to assess losses from take-all that might be of interest. We began fumigating the soil in plots in a field that was about to be planted to wheat for the third year. Other plots of nonfumigated soil were marked out in a grid near the fumigated plots; each plot in the grid was approximately 5 m × 5 m. Plants from each plot were scored for disease at the dough stage. We removed the plants from the edges of the plots. We had 40 such plots, including the fumigated ones, and our assays were somewhat crude. We found that even the fumigated plots had plants that were infected. We harvested each plot by cutting rows down the middle. The correlation between disease and yield was −0.75. We tried various fits and found a straight-line fit to be the best. The grower harvested about 6.5 t/ha; but our best yield, which was in the fumigated plots, was about 7.5 t/ha. On this basis we estimated a loss of about 1 t/ha.

MacNISH:

My observations are certainly contrary to Dr. Rovira's. He states that under field conditions in Australia there is not the severe blackening of the root systems often encountered in the glasshouse. In an experiment at Mt. Barker, I had to include a super-severe take-all category in which 80% of plants had all the roots blackened and at least 1.5 cm of blackening up the stem.

ROVIRA:

How often did you observe that? I observed it one year in four when there were as many whiteheads as there were blackened roots and stem bases.

MacNISH:

Take-all must be more severe in Western Australia than in South Australia. I have also found that take-all incidence is not a good enough indicator of yield loss. I use a scale based on percentage of the root system discolored. With this system I often get moderate to severe take-all, giving me a better correlation with yield than just incidence. These categories are the ones I have used to evaluate disease-yield relationships.

PARKER:

A point that should be made is that little wheat is grown in the area where Dr. MacNish is working.

MacNISH:

The area is one where we should be able to grow a great deal of wheat; but because take-all is severe, farmers do not grow it.

PARKER:

We keep discussing take-all based on the severe symptoms described in textbooks. It is time we rewrote the field symptoms of take-all to include those often loosely called "subclinical." The subclinical symptoms are premature ripening of plants, fewer tillers, and heads containing little grain. The field symptoms may be confirmed or negated in the laboratory by plating discolored stelar tissues onto potato-dextrose agar with streptomycin. Here the method of surface treatment of tissues for the isolation of *G. graminis* var. *tritici* is most important! Davies' silver nitrate method, published in 1935, has been compared with mercuric chloride and sodium hypochlorite in our laboratory and is vastly superior to both.

HORNBY:

I would just like to take up the point that Dr. Rovira made about methods of assessment (which seem to be on some scale of increasing sophistication) all giving very similar results, and add that this has been remarked on in several published papers. In the work I was describing in the last question, we used simple disease ratings and also calculated the proportions of roots infected, which was very time-consuming and seemed not to lead to conclusions any different from those based on ratings. Why is it that for much work, there seems to be little return from assessing take-all by the more complex and sophisticated methods?

WALLWORK:

My inclination, as a researcher looking for resistance/tolerance in new cultivars, is to look for yield differences in the presence and absence of the disease. Other scores may not relate to economic loss resulting from infection. I wonder if others agree that this is an acceptable method of scoring experimental plots or whether estimates of root discoloration might be better.

SCOTT:

I would say it is unlikely that yield depression would be the most accurate way of assessing take-all. It would be unlikely to be as precise, in a statistical sense, as assessing the incidence of root lesions. Of course, another possible reason for assessing yield depression in addition to root lesions is that cultivars may differ in their yield response to a given amount of take-all; that is, they may differ in tolerance, although personally I am not attracted by that idea. But it would be another reason for assessing yield. If you wish to assess take-all, there is no better way than assessing root lesions.

HUBER:

There are several things that enter into our assessment of take-all. One of them is the time take-all damage is assessed. Tissue damage changes rapidly under some environmental conditions. Cultivar is also a factor. For example, an index in May may give a very good indicator one year; the next year, an index in June or April may be a better indicator of actual severity from the standpoint of yield. The other thing we find is that some cultivars do not show severe cortical damage. These are just some of the problems we have to accept. Although we use a root rot index in our own studies, at times we have found that this is not a true indication of yield loss because there is tolerance in some cultivars. We may not have any whiteheads at the time we take readings (at early milk stage), yet some varieties will not fill out completely. Thus, whiteheads are a good indicator for some cultivars, but not for others. In selecting our breeding lines, we make a yield determination as well as a root index and whitehead reading.

Question 4 (BALLINGER):

How is take-all severity best scored? Several different measurements are commonly made to determine the effect of *G. graminis* var. *tritici* on plant growth, including the amount of affected root tissue, number of tillers, plant height, plant weight, and number of deadheads. Which of these parameters provides the most reliable measure of take-all severity? Are the parameters suited to both field and glasshouse trials?

BALLINGER:

I rate my plants on a disease score of 0–5 depending on amount of infection on the roots.

MacNISH:

Statisticians do not like scoring systems because they are not continuous, whereas quantitative methods are just too time-consuming if you are assessing a large number of plants. Some compromise has to be made in such a situation.

Question 5 (MOORE):

As part of a broad project to develop adequate models to describe the effects of root and crown pathogens on wheat growth and to allow prediction of the effects of such pathogens on yield, an irrigated trial was conducted in 1982 on the Tamworth Research Centre, Australia. Varying levels of take-all were generated by sowing, with the wheat seed, different amounts of oat kernel inoculum of *G. graminis* var. *tritici*. Although excellent correlations existed between level of added inoculum and the parameters used to measure disease (whiteheads, incidence, and severity), neither level of inoculum nor the disease parameters were highly correlated with measurements of plant response to infection (tillers and heads per plant, dry matter, and grain yield) except at the highest level of added inoculum (2:1, oat:wheat). The poor relationship between disease and plant response can be explained best by the fact that at no stage were plants allowed to be moisture-stressed. The question I would like the workshop to address therefore is: what methods do other workers use to measure disease and evaluate losses, and how can we account for plant water relations in crop loss models?

BRISBANE:

Yield can be split into various components—number of plants, tillers per plant, heads per tiller, grains per head, and weight per grain. Yield depression due to disease can be attributed to particular yield components. Has anyone done this?

HUBER:

We have looked at the biochemical components of yield in relation to nitrogen movement and utilization, and we have it worked out fairly well for maize and sorghum. We are now in the process of working out the system for wheat based on grain storage proteins and disease severity. One of the factors we have identified is the effect of take-all on different types of wheat. Hard red wheat has a different level of tolerance than the soft red or soft white wheats, primarily because of the difference in nitrogen response and the saturation of storage protein sinks. The kernel protein sink of most hard red wheats saturates very rapidly, and they respond poorly to high levels of nitrogen. Take-all is thus much less damaging on a hard wheat with a limited nitrogen requirement than on one that has a higher requirement for nitrogen. Thus, the impact of take-all is quite dramatically different at the same levels of nitrogen with different cultivars because of their differing ability to saturate the kernel nitrogen sink.

Question 6 (POWELSON):

What growth and yield components best reflect the severity of attack by *G. graminis* var. *tritici* in field plots? Only a small portion of the root system can be examined when plants are dug and the soil washed from the roots. Many of the severely infected roots break and are lost in the digging and washing process. How does this influence the estimates of severity of attack? What is the relationship between seminal root and crown root attack? Top growth usually reflects the vigor and performance of the root system and is easy to measure. Thus, measurements of plant height, tiller number, grains per head, and fresh and dry weights should give valid estimates of root-attack stress that would be reflected in grain yields.

ROVIRA:

There is a reduction in grain size in plants that are suffering from take-all, and this has been used in a number of countries for screening for resistance rather than using yields.

COOK:

I think it will depend on the stages at which root infections began. We have done much work on yield components in relation to the fumigation response in wheat. If roots were infected early, plants responded by producing fewer tillers, which is important in terms of yield. Reductions in number of kernels per ear and especially reductions in thousand-kernel weights that we obtained were all the result of the fungus getting into the crown. In our experience, these reductions indicated that the plant was growing maximally until suddenly the fungus caused enough damage to the crown to impair water movement to the ears. Early infection could be more important, however, because it causes a reduction in the number of heads per unit area.

HUBER:

The yield components in wheat are kernel size and kernel number. Certainly, kernel numbers depend upon tiller numbers as well as number of grains per head. The other factor, kernel size, is a direct reflection of nitrogen and its efficient movement into the kernel. A soft red wheat that has a very dynamic grain storage interaction may move 18 carbon atoms for every nitrogen moved into the grain, in contrast to a hard red wheat where kernel saturation is achieved very rapidly. It is an excellent indicator of yield potential. The kernel sink relationship does not eliminate the kernel number interaction.

CUNNINGHAM:

In the case of spring-sown cereals, certainly grain size appears to be the factor most affected. Concurrent with this, less well-filled grain results in a substantial increase in grain nitrogen because of less dilution of the protein laid down earlier by starch. The effect of take-all in reducing grain size and increasing grain nitrogen was more pronounced for wheat than barley.

KOLLMORGEN:

Professor Powelson has demonstrated a strong correlation between foliage fresh weight and root damage.

MacNISH:

This is not surprising, because the plant is an integrator of what has gone on before. At any time in the season, there will be high correlation between plant growth and what has happened to plants in the past.

KOLLMORGEN:

I agree, but I believe that researchers have been reluctant to use such a simple measurement as foliage fresh weight.

MacNISH:

This is alright if you are sure you only have take-all. In any field situation, you cannot be sure that you only have take-all.

HUBER:

The concern I have with foliage fresh weight is that wheat is extremely poor at recycling nutrients. Because of that, there is a continuous requirement for nitrogen and other nutrients. If nitrogen is partitioned, 65% of the total nitrogen in mature plants is in the kernel; yet, if you look at the time of uptake of that nitrogen, 50% has probably been acquired after anthesis. Most wheats have a continuous uptake of nitrogen and some other nutrients, rather than extensive recycling from vegetative tissue. In fact, you can remove every leaf except the flag leaf from some wheat plants and only reduce yields 15%. You cannot do that with maize or sorghum and some other plants. Thus, wheat is a very inefficient recycler and requires a continuous uptake of nutrients.

BRISBANE:

So the implication of what you say is that if we protected the wheat plant for 4 weeks it would make no difference?

HUBER:

Essentially, that is what we have observed. With a favorable environment, *G. graminis* var. *tritici* continues to infect; however, if that 4 weeks is the only infection period, then it would be a factor.

SIVASITHAMPARAM:

One of the issues that concerns Professor Powelson is that many of the severely infected roots break and are lost in the digging and washing process. How does this influence the estimates of severity of attack?

KOLLMORGEN:

Professor Powelson's point is very valid. No matter whether you pull the plant from soil or dig it, you leave behind a portion of the root system, and the roots most likely to be left behind are those affected by *G. graminis* var. *tritici*.

Question 7 (BALLINGER):

Is it possible to predict the likely severity of take-all from the amount of infected debris in the soil before sowing? This would have practical significance in aiding farmers' management decisions regarding rotations and the possible use of control measures.

BALLINGER:

We attempted a bioassay to identify suitable naturally infected field sites for fungicide studies. Core samples (90 mm) were taken from field sites previously affected by take-all, placed in the glasshouse, and sown to wheat. After 6 weeks, the plants were assessed for disease. Of 11 sites assayed, 3 indicated high levels of take-all inoculum. Subsequent field trials showed a high correlation between the bioassay results and actual field infection.

HORNBY:

This problem has been investigated on several occasions in the United Kingdom. Certainly, about 20 years ago there was hope that soil bioassays could be used to identify risks and perhaps even predict take-all; but this faded as no reliable relationship could be found. Whether, because of differences in climate and farming systems, they would be more reliable in Australia remains to be seen. Some of the assumptions that are made in such discussions in Australia, however, are difficult for me to accept. In Britain, there are several known modifying factors that make it very difficult to use bioassay data for prediction. Take-all decline and cropping history are two of them, and if biological control agents such as *Phialophora graminicola* are present, these too may complicate interpretation. The weather is a modifier, operating between the time you assess the soil for its infectivity and the time you assess disease in the field. Such modifiers obscure any predictive value these techniques may have. Have any Australian workers tried these techniques?

ROVIRA:

Ridge developed a soil bioassay when we started our rotation × tillage trials. He worked on the basis that in the first 3 weeks of growth the root system of the wheat plant explores about 300 g of soil. He sieved the soil and broke up the root fragments and then grew wheat in 300 g of the soil at 15°C. After 3–4 weeks, he assessed plants for take-all lesions and found that he could distinguish between soils that had grown a pasture made up of a mixture of barley grass and *Medicago* spp. and ones that had grown *Medicago* spp., peas, or oats. Although such information could be useful for farming, it is probably too late to do anything about it. There are some problems with the system; for example, how much of that residue material do you include in your pots after sieving, and should a deal of the top material be sieved out to leave just the roots and crowns?

HORNBY:

What can you do if the prediction is available only after the farmer has sown his crop? There seems very little he can do—perhaps cut his losses by plowing in the crop? Slope and Gutteridge at Rothamsted (U.K.) also considered soil assays for assessing the risk of take-all in farmers' fields after they had successfully used such assays to measure inoculum buildup, but the labor involved and the dependence of disease severity on weather offered a gloomy prospect for general forecasting of take-all.

SIVASITHAMPARAM:

I have been assaying field inoculum using the most-probable-number technique and disease severity assays for the past 6 years in Western Australia. The numbers appear to peak at certain times of the year. In certain instances, the inoculum assay showed some relationship to disease levels of the crop sown at the same sites.

MacNISH:

We need predictions well in advance, not just a few weeks before sowing.

IVERS:

In my opinion, I would need information the previous year.

SIVASITHAMPARAM:

This is what we had in mind when we started our project. It was our intention to assay the inoculum in the previous year and relate it to crop disease. I have been doing this assay in various fields and will continue to do it for several years to see if I can establish reliable relationships.

BALLINGER:

When I first posed that question I did not know any answer, but now I am a little more enlightened. We did a bioassay test during summer, so we knew before sowing what the probable level of *G. graminis* var. *tritici* was. We used the information to select naturally infected sites for field trials. Six sites rated as having low take-all risk using this technique were found to have low levels of take-all in the following crop. A site at Numurkah, Victoria, Australia, however, that showed a high risk of take-all has produced a high level of disease. This technique may be valuable in identifying suitable sites for field trials, particularly in disease control studies.

SCOTT:

Even if you have a method of assessing inoculum accurately, you do not know what the weather is going to be.

SIVASITHAMPARAM:

If a predictive test is a viable proposition, it would be the outcome of an understanding of the inoculum and the biotic and abiotic soil environment.

Question 8 (HUBER):

What is the relationship of the time of infection to yield and quality losses, i.e., fall versus spring infection?

HUBER:

I raised that question after evaluating 10 years of data on the effect of rate of seeding on infection and 3 years of data on triadimenol seed treatments that provide a 6- to 8-week delay in infection. If we seed 2 weeks earlier than recommended, we can anticipate a 50% yield reduction from take-all in a second wheat crop. With delayed seeding of only 2 weeks, we reduced our disease severity (yield reduction and whiteheads) much greater than with the 6- to 8-week delayed infection provided by seed treatment. I do not know how to interpret these two situations.

GLENN:

Is it possible that the chemical has other effects on the plant, which make it more susceptible to disease as well as delaying emergence?

HUBER:

In 3 years of study, we observed a consistent delay in infection; but in two of those years, we had very severe infection later on. In those same years, we did not have such severe later infection where seeding was delayed. A subclinical effect of the chemical on the plant cannot be ruled out, and we can only conclude that similar delays in infection with the two different systems have different effects on disease and yield.

Question 9 (KOLLMORGEN):

What progress has been made toward control of take-all? Take-all has been a topic of scientific investigation for over a century. In which regions of the world is the disease presently under control as a result of these investigations, and what are the savings to wheat growers?

PARKER:

In Western Australia, a very successful rotation for controlling take-all was clover-wheat. On some farms this rotation was maintained for 12–14 years, and grain yields steadily increased over this period. The success of this rotation resulted from the use of subterranean clover, which buries its seed, and heavy grazing by sheep, which prevents seeding by gramineous weeds such as *Hordeum leporinum, Bromus* spp., *Vulpia*

myuros, and *Lolium rigidum* (both the subclover and grasses are annuals). We found soil infested with take-all along the fence lines, where annual grasses remained, but almost none in the field where grasses were virtually eliminated. A recent rotation showing great promise on the sandier soils is lupine-wheat, where wheat has yielded up to double that of wheat after wheat. Australian plant breeders are seeking alternative grain legume crops for the heavier soils.

HUBER:

There is no doubt about it. Crop rotation, recommended some years ago, is an effective control measure.

SCOTT:

I do not think it should be overlooked that great progress has been made. I wish I could say that there is a resistant wheat cultivar. I think the biggest single factor is probably the amount of inorganic nitrogen used. That is something Professor Garrett predicted in the late 1940s. The wheat plant is better able to replace diseased crown roots if it receives a generous supply of nitrogen, be it organic or inorganic. I completely agree with Professor Parker that subclinical take-all is still a serious problem, but it has been controlled to a huge degree compared with what could happen some years ago.

HUBER:

In the United States, crop rotation, date of seeding, and nitrogen are the three most important factors affecting take-all. If control recommendations were followed, take-all would be a minor problem.

SKOU:

In Denmark, take-all is not a very serious problem even though barley growing has increased during the 1960s and 1970s, with more than half of the Danish agricultural area under barley today. We have not told the farmers that they should monocrop with barley. The change came because the farmers were forced to grow barley for economic reasons, and they are willing to pay for the higher amount of nitrogen necessary for high yields. So many farmers in Denmark have been through a take-all decline following continuous barley. We have, of course, farms with serious take-all. These farmers have cattle, and they grow wheat. If they don't practice good rotations, they have serious trouble growing wheat.

Session 6: Grower Observations and Questions

CHAIRPERSON: I. RODGERS

Question 1 (RODGERS and HAYWARD):
Is it possible to determine levels of take-all inoculum in soils and thereby predict disease levels in particular fields?

ROVIRA:
As stated in Session 5, we developed a test similar to that used by Mr. Ballinger to predict several months in advance whether take-all would occur in a particular field. I felt that most growers would not be prepared to collect samples for such a test because they would probably not be convinced about potential losses due to take-all. Nevertheless, the technique worked and could be used to predict high-risk situations. Would the grower participants be keen to collect samples in January for bioassay?

MacNISH:
What is the latest a grower would need to know the result of the bioassay?

RODGERS:
Late February.

MANLEY:
If you thoroughly assess your pasture the year before you crop it, you should be able to predict the likelihood of a take-all problem. I agree that sampling for a bioassay may be a nuisance.

IVERS:
A predictive test to quantify levels of inoculum would be very valuable to me. I crop continuously on my farm and therefore cannot predict take-all levels from the incidence of grasses in a pasture phase.

ROVIRA:
Our test involved sieving out the larger organic fragments, sowing wheat, and, after 4 weeks, examining the roots for lesions. Using Dr. Hornby's methods, we estimated population densities.

SIVASITHAMPARAM:
We have also been investigating this matter. Dr. MacNish has produced a chart that can be used to predict losses due to take-all. The chart is based on factors such as cropping history and soil type. Dr. Hornby's most-probable-number method is only slightly more helpful than Dr. MacNish's chart.

TROLLDENIER:
I would like to comment on the preparation of samples for the bioassay technique. Sampling of soils and subsequent storage strongly affect the microbiological and, to a lesser extent, the chemical soil parameters. Storage conditions play an important role in how rapidly the microbial biomass and the composition of the microflora are being changed. Bioassays can only give adequate information on the natural state of a soil when the material is analyzed immediately after sampling. This is, however, often rather difficult in practice.

HORNBY:
Two points have cropped up in this discussion—risk and prediction. Australian researchers have spoken about estimating risk, but the original question was about prediction. I would like to ask a question of the Australian scientists: have they any indication of how frequently the situations they identify as risky actually develop serious disease when a susceptible crop is grown?

ROVIRA:
I can only speak from results in our experimental plots where the test had good predictive value over 3 years and related well to the amount of take-all we had on the roots at tillering.

HORNBY:
In the United Kingdom, there are complicating factors—take-all decline and such modifiers as weather—that affect prediction. Taking these into account, we believe that about 70% of the risk situations we identify on our experimental farm ultimately develop significant disease.

COOK:
In the Pacific Northwest of the United States, the key question facing a grower is not when he will have a take-all problem but when it will go away. Growers know that take-all is likely to be severe after a grassy alfalfa field or after sagebrush containing native grasses. What the growers need is a bioassay to determine the magnitude of suppression resulting from take-all decline so that they can predict losses.

Question 2 (RODGERS and HAYWARD):
How long does the take-all fungus survive in soil?

COOK:
In Washington State (U.S.A), only a small amount of inoculum would survive more than a year. One year's break to a nonsusceptible crop such as potatoes causes a very significant reduction in inoculum levels.

KOLLMORGEN:

In experiments at Walpeup (Victoria, Australia), the take-all fungus survived in infected wheat crowns on the soil surface for 12 months.

PARKER:

There are reports of survival for up to 6 years. Although that survival would be minimal, it may be sufficient to initiate a further outbreak.

Question 3 (RODGERS and HAYWARD):

What soil characteristics (texture, pH, etc.) favor the buildup of inoculum of *G. graminis* var. *tritici* and disease expression?

RODGERS:

Take-all does not appear to be as much of a problem in northern areas of Australia as it is in the south, and we wonder why this is so.

MacNISH:

I think this is because of the wetter conditions in the south rather than differences in soil type. In Western Australia, take-all occurs on both acid and alkaline soils.

PARKER:

I agree with Dr. MacNish.

COOK:

Professor Parker, in those areas where take-all occurs on acid soils, what is the effect of liming? In the Pacific Northwest, take-all may occur with acid soils, but inevitably liming increases its severity.

PARKER:

I have only done that experiment once; in that instance, liming did not increase disease. I have an isolate with a pH optimum of 5, and liming would not be expected to increase its disease potential.

POWELSON:

In Douglas County, Oregon (U.S.A.), where we have acid soils and a problem with aluminum toxicity, liming reduces the severity of take-all.

MacNISH:

We tested the effects of different levels of lime on take-all, but no differences were evident.

HUBER:

In the midwestern United States, take-all is not related to pH per se, but it is related to pH-dependent reactions involving the availability of other nutrients and rhizosphere microbial interactions.

TROLLDENIER:

We conducted a pot experiment with an acid sandy soil amended with different levels of lime. The effect of liming was almost nil with healthy plants, whereas liming reduced yields considerably with plants infected by *G. graminis* var. *tritici*. At the lowest pH tested (3.9), there was no significant difference between the yields of healthy and diseased plants. At pH 4.5, grain yield reduction due to take-all was 43%; at pH 5.3 it increased to 77%; and at pH 6.7 it was as high as 84%.

GLENN:

Isolates of *G. graminis* var. *tritici* with different pH optima may be selected at different pH values.

HORNBY:

I was interested in Professor Powelson's remarks concerning aluminum toxicity. In southern Brazil, where there are also aluminum toxicity problems, liming causes severe take-all. This is contrary to Professor Powelson's observations.

POWELSON:

The situation in Brazil is different because of the different soil profile.

Question 4 (RODGERS and HAYWARD):

Why does take-all often occur in patches within a field?

HUBER:

We observed that available manganese is often low in severe take-all patches because of high populations of manganese-oxidizing bacteria in the rhizosphere, although soil tests indicated that there were no innate differences.

TROLLDENIER:

Manganese-oxidizing bacteria are unlikely to be restricted to certain parts of a field. Are there any other factors involved?

HUBER:

I am sure that there are: pH and organic matter, for example, may be involved.

CUNNINGHAM:

We investigated these patches and established that a pH gradient occurred from the center, with lower pH levels at the margins. This may have been the result of uneven liming and may have induced manganese deficiency.

Question 5 (RODGERS and HAYWARD):

How does the take-all fungus spread within a field?

WONG:

Infected tissues may be spread by machinery and by wind. Ascospores are unlikely to be important because we seldom see perithecia in the field. Furthermore, *G. graminis* var. *avenae* is uncommon in oat and wheat crops even though it is prevalent in bowling or golf greens close to these crops.

CUNNINGHAM:

Ascospores are sometimes considered important in epidemics in Ireland in the first and second cereal crops on cutaway peatland.

Question 6 (RODGERS and HAYWARD):

How effective is fallowing in the control of take-all?

KOLLMORGEN:

In pot experiments I conducted, survival of *G. graminis* var. *tritici* was greater in unplanted (i.e., fallow) pots than in pots planted with either cereals or noncereals. This suggested that fallow may not be as effective as a grass-free break crop (lupine, peas, or rapeseed) in breaking the disease cycle. However, in the field a grass-free fallow was generally as effective as break crops of lupine, peas, and rapeseed.

HUBER:

In the western United States where fallowing is common, take-all is often severe in a wet year but is of little consequence in a dry year. We concluded, therefore, that a major effect of the fallow was to increase nitrogen levels.

Question 7 (MANLEY):

Is there any correlation between seeding rate and incidence of take-all?

MANLEY:

In an experimental site on my property, wheat buffer plots sown at 55 kg/ha had a much higher incidence of take-all than one of my fields sown at 35 kg/ha.

KOLLMORGEN:

Studies in Western Australia demonstrated an increase in take-all with an increase in seeding rate.

ROVIRA:

A possible explanation would be the increased chance of spread of *G. graminis* var. *tritici* from one plant to the next. In addition, at the higher seeding rate there would be more crowns to carry the fungus over to the next season.

HUBER:

At a high seeding rate, plants are more likely to be stressed for nitrogen and possibly other nutrients, and take-all will be more severe. A reduction in seeding rate can be an important control measure on nutrient-deficient soils.

COOK:

Clarke demonstrated the advantages of low seeding rate in a pot trial. He infested soil with *G. graminis* var. *tritici* and put the soil into pots. In one set of pots he grew one plant per pot, and in the other several plants per pot. Take-all was much more severe where there were several plants per pot.

KOLLMORGEN:

Mr. Griffiths, what is the recommended seeding rate in the Victorian Mallee?

GRIFFITHS:

Normal seeding rate is 40 kg/ha. Experimental work has established that within the range of 25 to 75 kg/ha, yield is virtually unaffected by seeding rate.

KOLLMORGEN:

Were the seeding rate experiments conducted on sites prone to take-all?

GRIFFITHS:

In the course of the studies, the plots would have been affected by take-all.

Question 8 (RODGERS and HAYWARD):

What are the effects of stubble retention on take-all?

CUNNINGHAM:

We did some experiments on different methods of stubble disposal and found no pronounced effects on take-all.

COOK:

In the Pacific Northwest, Dr. Moore established that take-all was more severe where wheat was direct drilled.

GRIFFITHS:

Our work in the Victorian Mallee would corroborate that.

HUBER:

When stubble is left on the soil surface, nutrient stress is intensified because there is a very poor recycling of nutrients into the soil profile.

HORNBY:

In experiments in the United Kingdom, the incidence of take-all has been similar in direct-drilled, conventionally cultivated, and chisel-plowed treatments.

Question 9 (MANLEY):

Is there any evidence to suggest that take-all is suppressed in soils with high salt levels or where sodium chloride has been applied as a control measure?

MANLEY:

I have heard of Professor Powelson's work in the United States and am wondering if sodium chloride could be applied at seeding to control take-all.

POWELSON:

Growers do not apply sodium chloride at seeding. Fertilizers applied at seeding are ammonium nitrogen, phosphate, and potassium chloride. Sodium chloride is sometimes used as a topdressing in spring, although ammonium chloride is more common.

ROVIRA:

Professor Powelson, are your fertilizer treatments affected by pH?

POWELSON:

Very definitely. Control is usually not achieved outside the pH range of 5.4 to 6.8.

TROLLDENIER:

Professor Powelson, have you compared sodium chloride with potassium chloride to determine if potassium is involved?

POWELSON:

Our experiments have shown that it is chloride and not potassium that is involved in suppression of take-all.

TROLLDENIER:

The chloride ion is very readily being leached from the topsoil.

POWELSON:

This is correct, and timing of the application is therefore very important.

Question 10 (RODGERS and HAYWARD):

Take-all is often most severe in those strips within a field where chaff and straw have been deposited during harvest of a previous crop. What are the possible explanations?

VENN:

We have also made this observation in South Australia on barley following a previous barley crop and confirmed that *G. graminis* var. *tritici* was the major pathogen present.

HUBER:

The effect may result from immobilization of nitrogen and other nutrients in the strips.

WONG:

Another explanation is that the straw layer reduced soil temperatures and that this promoted survival of *G. graminis* var. *tritici*.

HAYWARD:

We burned the straw so temperatures in the strips would have been high.

HORNBY:

In England, we have demonstrated that self-sown wheat will increase inoculum levels, and this may be a factor.

HAYWARD:

Where the straw was burned, there was little possibility of volunteer wheat. In one instance, the following crop was barley and there was no wheat in the barley.

PARKER:

Did the fires burn all the trash?

HAYWARD:

Yes. It is also interesting that the same effect occurred when the stubble was not burned.

MANLEY:

I would think that cultivation across the tracks would tend to disperse inoculum.

COOK:

Cultivation moves inoculum to a very limited extent.

HORNBY:

Where were the wheel tracks in relation to the strips?

HAYWARD:

On either side, so the effect was not caused by compaction.

Question 11 (IVERS):

Under what conditions and after how many years does take-all decline establish in wheat and barley?

MacNISH:

In an experiment I conducted near your property on a loamy gravel, it took 6 years for take-all decline to become established.

HORNBY:

In the United Kingdom, the time required depends very much on the previous cropping history. If a field has previously been in pasture, it may take 6 or more years of continuous cereals; but if it has had arable crops, it may occur after 3–4 years of cereals.

PARKER:

At Green Hills, Western Australia, although take-all decline was evident after 7 years, there was still considerable yield loss due to take-all.

COOK:

I have observed take-all decline in different soil types. At one location where the soil pH was 7.5, take-all decline did not occur until the 10th or 11th wheat crop, whereas at another location on an acid soil take-all decline occurred after only two crops. Where lime had been added to the acid soil, the decline was delayed by 3–4 years.

Question 12 (IVERS):

How effective are oats, lupine, peas, linseed, and rapeseed in breaking the disease cycle of take-all in a system of continuous cropping?

ROVIRA:

A single crop of oats, peas, or *Medicago* spp. will reduce take-all to an almost negligible level in a following wheat crop.

MacNISH:

In Western Australia, lupine is the most effective break crop provided it is free from grasses. Oats are not so effective because they will host *G. graminis* var. *tritici*.

Session 7: Nutrition and Fertilizers

CHAIRPERSON: R. POWELSON

Question 1 (YARHAM):

How does the timing of nitrogen application affect the ability of plants to cope with infection? Recent work at Rothamsted Experimental Station (U.K.) and by the U.K. Agricultural Development and Advisory Service has pointed to the benefits of later winter (February) applications of nitrogen in reducing the effects of take-all on yield. Will such treatments be beneficial whatever the season and whatever the pattern of the take-all epidemic? And what of autumn nitrogen application? Normally we would regard this as dangerous, favoring the saprophytic survival of the pathogen. Nowadays, however, with even earlier sowing, we are finding more and more evidence of sometimes quite high levels of infection by early winter. Given that the fungus has to be regarded as an active parasite in early sown crops in the autumn, should we be giving autumn nitrogen to promote root development?

HUBER:

We are encouraging farmers to move away from a split nitrogen application and have a single autumn application by stabilizing that nitrogen to avoid denitrification losses. There are two reasons for this: one is to reduce autumn infection by the ammoniacal nitrogen; the other is that early spring topdressing increases losses caused by *Rhizoctonia* and *Pseudocercosporella*. We try to apply all fertilizer before sowing provided we can maintain adequate nitrogen nutrition throughout the growth of the crop. Although we can reduce yield losses with nitrate nitrogen, ammoniacal nitrogen not only increases yield but also reduces disease.

Question 2 (HUBER):

Are there interactions between nutrients and take-all?

HUBER:

Many of the effects we have observed between form of nitrogen and take-all are related to local farming practices. We are trying to understand what nutritional and biological interactions are involved. For instance, where take-all tolerance and response to different nutrients are concerned, our hard red wheats generally suffer less yield loss from take-all than soft white wheat.

The form of nitrogen becomes a very critical factor for us in reducing take-all. With the ammoniacal form of nitrogen, incidence of take-all is reduced. We are trying to understand why this happens. For a long time, people said that most plants cannot utilize ammonium nitrogen and that they only utilize nitrate nitrogen. Now we realize that plants use either form equally well. However, there is a distinct difference in metabolism between the two different types of nitrogen, which has a dramatic effect on root pathogens.

We find that with the ammonium form of nitrogen, the normal defense reaction of the plant is enhanced. In addition, availability of phosphorus, zinc, and manganese is increased.

TROLLDENIER:

With nitrate nutrition, bacterial activity in the rhizosphere is lower than with ammonium nutrition. We studied the effect of the nitrogen form on bacterial numbers in the rhizosphere and on root respiration of wheat. Both were higher with ammonium than with nitrate nutrition. This may also be of significance with regard to antagonistic bacteria, which might become more abundant with ammonium rather than with nitrate nutrition, and also with regard to the oxygen status of the root zone. As a consequence, greater quantities of manganese may become available to the plants at higher root respiration, which may lower the oxygen partial pressure.

SMILEY:

I did a considerable amount of work with different nitrogen sources when I was investigating the influences of rhizosphere pH. When using soil at pH greater than 6, I obtained very good disease suppression but could eliminate the effect by pasteurizing the soil by steaming or with fumigation. Therefore, it was not a direct effect of the nutritional status of the plant. At very low levels of soil pH, we could not eliminate the suppressing effect by pasteurization of the soil. When I started looking at the microbial populations in the root zone, I was able to detect the changes in the population not in terms of total numbers but between different groups.

GRAHAM:

The chemistry of manganese in soil is exceedingly complex. Dissolution as Mn^{2+} of insoluble Mn^{III} and Mn^{IV} oxides and hydroxides in soils is promoted by H^+ (low pH), electrons or reducing power, and low oxygen tension. Such transformations are favored by ammonium nitrogen, superphosphate fertilizers, and organic matter, which also decrease take-all. Although manganese transformations in soil are complex, involving both chemical and microbial oxidations and reductions, pH is a dominant factor; and pH is also an important factor affecting the severity of take-all. Further, uptake of manganese is low from cool, wet soils, or conditions that also favor infection by take-all. Level of pH should not affect take-all in soils devoid of Mn^{III}

and Mn^{IV} oxides; take-all would, however, be responsive to applied fertilizer in adequate amounts. Acid soils on which take-all is serious have been regarded as anomalous in the etiology of the disease. It is likely that the few such acid soils with serious take-all problems do not contain significant amounts of Mn^{III} and Mn^{IV} and therefore are approaching an absolute deficiency of the element. The literature suggests that a similar story operated with *Streptomyces* of potato.

We have been looking at cultivars for resistance to take-all based on the efficiency of the cultivars to extract manganese from marginal or manganese-deficient soil. Our cultivars that are most tolerant or resistant to take-all are efficient users of manganese. Dr. Scott has said that he does not see any major sources of variability in wheats for resistance to take-all in England. I am surprised, but perhaps we are seeing it because we are looking at wheat in a background of much more manganese-deficient soil.

MacNISH:
Would you expect the manganese to be higher in plant tissue?

GRAHAM:
No; much of our manganese work was done by mixing manganese with superphosphate and ammonium fertilizer, which lowers soil pH and slows the manganese immobilization. The roots that contact the fertilizer in the band can supply the leaves, and the crop looks as though it has recovered completely from manganese deficiency. However, since manganese moves poorly in the soil and is not translocated downward in the plant, most of the roots grow in manganese-deficient soil, and they can be susceptible to take-all.

MacNISH:
We have measured manganese levels in tops and grains following fertilization with different sources of nitrogen. Although there were large differences in levels of take-all, there were no significant differences in the levels of manganese.

GRAHAM:
Manganese deficiency can be caused by low temperatures during the winter months, and this favors attack by *G. graminis* var. *tritici*. However, as the weather warms up, manganese is mobilized and levels in plant tissues may rise. Thus, grain manganese does not reflect the manganese status during the critical winter months.

Question 3 (SIVASITHAMPARAM):
What is the current state of understanding on the effect of copper, chloride, and manganese on the expression of disease by the host?

GRAHAM:
Robson and students have done some work on the effects of copper and zinc, which decrease take-all in deficient soils. Other work suggests that copper is involved in lignin synthesis. Manganese is also involved in lignin synthesis in the secondary cell wall. The latest information on zinc is that it is involved in membrane integrity. There is evidence that amino acids,

sugars, and other nutrients leak out of zinc-deficient root cells, and this of course must change the microbiology of the rhizosphere.

COOK:
The manganese aspect is extremely interesting, but I was most pleased to hear reference to the other trace nutrients. Reis's work indicated that manganese, copper, or zinc, if withheld from the plant singly, could result in the roots becoming intrinsically much more susceptible to attack by the take-all fungus. I am sure that in the Pacific Northwest (U.S.A.), we have areas where zinc is deficient subclinically. Although manganese is obviously very important, I do not think we should overlook the possibility that in some areas other trace nutrients might be the critical ones.

HUBER:
I would just mention for the growers' benefit that there are several ways they can supply trace elements. I notice you use blade plows and scarifiers. This type of equipment works very well for applying anhydrous ammonia, as well as trace elements in solution. The ammonium form of nitrogen enhances favorable biological interactions and helps maintain the availability of micronutrients.

GRAHAM:
Other trace elements can be important within particular soils. Getting back to manganese, I would like to make the point that it is easily the most mercurial in its availability in the soil, depending as it does on several physical, chemical, and biological factors.

HUBER:
With respect to manganese, we have investigated the soil physical factors that Dr. Graham has mentioned. We found that the biological interactions in the rhizosphere were critical factors governing availability. We have soils that have a relatively high manganese level but where the element is not available to the plant because of oxidation.

POWELSON:
I would like to make one comment on chloride. We are not thinking of chloride as having anything to do directly with plant nutrition. Chloride is known to regulate uptake of other anions and cations and can inhibit nitrifying bacteria. Our emphasis on chloride is that it enhances the biological suppression of take-all that is induced by ammonium nitrogen. When you pasteurize soil for 30 min at 60° C or fumigate it, there is no suppression. You have to have a natural soil. You can show the biological effect of reintroducing a small proportion of native soil, then you see an ammonium nitrogen suppression enhanced by chloride. With spring topdress applications of chloride, there is a direct relationship with how much spring chloride you put on and how much ends up in the plant tissue; and there is additional suppression of take-all. The thing that compounds the take-all disease syndrome in Oregon (U.S.A.) is that take-all rarely occurs alone; and we have seen disease suppression of stripe rust and *Septoria*, especially with spring-topdressed chloride. We have not yet been able to separate out all the components of the yield response.

Question 4 (COTTERILL):

What can be recommended as a nitrogen fertilizer for pot trials where yield of wheat is used as a measure of the effects of take-all? In a pot trial where artificial *G. graminis* var. *tritici* inoculum is added to the soil and its effect measured by the yield, the test soil is severely deficient in nutrients. It is essential that nutrients be added to achieve a reasonable yield. What nitrogen form can be added to the soil so that there is neither a beneficial nor a detrimental effect on the *G. graminis* var. *tritici* inoculum?

MacNISH:

I would use sodium nitrate.

HUBER:

Soils that do not have a take-all problem at all develop one when calcium nitrate is used. Potassium nitrate and sodium nitrate will give a similar disease response as calcium nitrate. We do know that manganese-oxidizing populations increase markedly with the addition of nitrate nitrogen.

Session 8: Environmental Factors

CHAIRPERSON: O. GLENN

Question 1 (YARHAM):

How does pH influence disease development and symptom expression? Take-all has long been regarded as a disease of alkaline soils and is certainly suppressed at pH values of less than 5.4. Yet increasingly we are finding severe attacks occurring in acid areas of fields (pH 5.6–6.0). The term "acid-patch take-all" has been coined to describe this situation. What is the explanation for this apparently anomalous phenomenon? Can a fall in pH weaken plant roots (and thus make them more vulnerable to attack) before the soil becomes so acid that the pathogen is inhibited?

PARKER:

In Western Australia, we have isolates of *G. graminis* var. *tritici* adapted to grow optimally at pH 5.

Question 2 (NICOLL):

Which geographical areas of the Australian cereal belt have a problem with take-all? We are aware of Mayfield's survey of wheat diseases in South Australia. Have any other surveys been conducted in other states, and if so, with what result? Is it possible to describe some cereal areas as prone to take-all?

MacNISH:

In Western Australia, high levels of take-all occur in a band that lies between the 450- to 525-mm rainfall isohyets. In this zone, the area planted with wheat is restricted because of take-all.

HUBER:

What size is the area prone to a high level of take-all?

ROVIRA:

In southern Australia, a high proportion of the 10 million hectares sown to wheat each year would be affected to some extent by take-all.

WONG:

In New South Wales, take-all is less common in the northwest. One reason may be the wet and warm summer conditions, which reduce survival of the fungus. Another reason may be the greater range of summer crops for rotation with winter cereals.

WILDERMUTH:

In Queensland, take-all is not a serious problem, probably for the same reasons as in northwestern New South Wales. Even in wheat monoculture areas, we have not yet detected a serious take-all problem.

SMILEY:

I would like to interject an opinion about survival. Many of the parameters seem to be the same as for the plant litter decomposition. This makes sense because this pathogen survives predominantly in infected plant tissue. Therefore, any cultivation program that favors a decrease in decomposition rate of plant litter is likely to increase survival of the pathogen. Optimum plant litter decomposition occurs at about −100 to −500 kPa and about pH 6.5, and it declines rapidly on either side of those values. As with pathogen survival, it is markedly affected by temperature, nitrogen levels, and contact of the litter with the soil.

PARKER:

At 25–26° C, the fungus does not grow well because it cannot compete with other microbes (which are decomposing plant litter); but at 10–18° C, the fungus not only survives but grows out in soil and has, in our laboratory, colonized fresh straw pieces.

KOLLMORGEN:

Take-all is a serious disease in the Victorian Mallee, especially on alkaline calcareous soils. The disease also occurs to some extent on the clay soils in the Wimmera. Recently, take-all has been severe on acid soils in the southern Wimmera.

Question 3 (GARRETT):

What is the mechanism of the effect of neutral to alkaline soils in promoting growth of the take-all fungus along cereal roots?

GLENN:

Professor Garrett asks for comments on his idea that it is likely that the effect of pH is due to intolerance of acidity by *G. graminis* var. *tritici* in its metabolism and growth.

SMILEY:

In my experiments, I was able to distinguish between the direct effects of very low rhizosphere pH and the indirect effects of higher rhizosphere pH. In that work, bulk soil pH seemed to have little bearing on the expression of take-all.

I find a fair amount of take-all in acid soils, especially when rhizosphere pH is buffered toward neutrality. I feel the best perspective for examining the effects of pH is an interrelated package that has to include rainfall, soil texture and type, and chemical factors.

GRAHAM:

It seems to me important to separate the pH tolerance

330

of the organism from the pH tolerance of its infectivity. Its invasiveness is much more pH dependent than its growth.

GLENN:

Your comment emphasizes the importance of where pH is measured. Growth can occur in soil where bulk soil pH may be more relevant; but infection is initiated in the rhizosphere, so pH must be measured there.

HUBER:

A series of studies on the effect of form of nitrogen on pH indicated two distinct factors. The greatest effect on acidification is caused by nitrification and not plant uptake. Since we can inhibit nitrification, we can control that part of the system.

SMILEY:

If the pH drop resulting from use of ammonium nitrogen is caused by nitrification alone, there would be no need to add the nitrification inhibitor nitrapyrin. Because of the necessity to add nitrapyrin to suppress nitrification in many areas, I feel there is another effect in the rhizosphere.

HUBER:

Without nitrapyrin, there is nitrate uptake. In less than 6 days, 90% of the ammoniacal nitrogen is nitrified in many of our soils.

SMILEY:

You are suggesting that the acidification through nitrification is not as important as the effect in the rhizosphere of ammonium versus nitrate nitrogen uptake by the plant. To get a marked effect of nitrification, there would be no need to inhibit it.

HUBER:

We can separate these two mechanisms, and acidification of the bulk soil is not involved in the process.

MacNISH:

We do not need nitrapyrin to get ammonium nitrogen to work in Western Australia.

HUBER:

Is that because of your low soil pH?

MacNISH:

I have assumed it has something to do with our dry conditions.

Question 4 (COTTERILL):

What effect, if any, does waterlogging of the soil have on *G. graminis* var. *tritici* once it is within the living root system? In our current work, we are sampling soils from the Esperance area in Western Australia, which receives heavy rainfall. Is constant waterlogging of the soil likely to affect development and expression of the disease once the fungus is within the living root system?

HUBER:

I have no information on direct effects of waterlogging on the take-all fungus, but the asparagine content of the plant goes up manyfold when it is waterlogged.

COOK:

G. graminis var. *tritici* is highly aerobic and therefore the main controlling factor—whether inside or outside the root—in waterlogged soil would be oxygen supply. Possibly oxygen could be supplied by conduction down through the plant. There are data showing that oxygen can be supplied to roots as far as 7–10 cm below the soil surface. Beyond this, without oxygen supplied externally, growth will be halted quickly.

Session 9: Host-Parasite Interactions

CHAIRPERSON: L. PENROSE

Question 1 (SIVASITHAMPARAM):
What is the present situation with respect to breeding for resistance to take-all?

PARKER:
Plants have vast amounts of genetic information that is not expressed, much of it being evolutionary memory, which may in part be exploited. I had wondered whether there could be selection in the field for resistance, using this genetic material. For example, a farmer plants 500–800 million seeds a year, harvests, and replants in conditions where take-all is influencing yield though it is not necessarily apparent. Under these conditions, there may be natural selection for resistance to take-all. Indeed, data were presented yesterday that suggest that there is within-cultivar selection for take-all resistance. This is not to imply that this resistance will solve the take-all problem, though a lot of work remains to take this idea further.

To some degree, *G. graminis* var. *tritici* is a successful pathogen because it can go from the outer cortex into the stele before the host has time to mobilize its defenses. If host plants had the ability to mobilize their defenses more quickly, they might be more resistant. Such resistance would most likely be found among those plants escaping early infection. In our program, such plants were selected from a crop and further selection was performed under conditions where soil temperatures and water were controlled. From an initial 100 lines, we now have 5 with some resistance.

SCOTT:
Why did you decide to look within a cultivar for differences in resistance, since wheat lines are essentially genetically homogeneous? Why not look across a wider range of genetic material?

PARKER:
For two reasons. First, other people with greater resources than I were looking across cultivars, and it appeared pointless to repeat their work. Second, although genetically homogeneous in theory, there is evidence that many cultivars are not homogeneous to the degree we think. The heterogeneity is mostly concealed; but in bacteria, for instance, it may suddenly appear and be confused with a new mutation. Similarly for plants: for example, some legumes (*Acacia*) have juvenile pinnate leaves that are only seen when the plant first germinates, whereas subsequent "leaves" are really phyllodes. The juvenile leaf is a vestigial structure from the past that only develops upon germination. I believe similar variations may exist in wheat.

PENROSE:
Has anyone gone further in breeding for resistance?

HUBER:
We have screened for resistance for the past 12 years. At first we examined the world wheat collection and decided that there was no resistance to take-all present. We then started looking for a response to nitrogen under severe take-all conditions and identified lines that suffered less severe take-all with nitrogen applications. Generally, the world wheat collection showed a graded response to nitrogen with severe take-all, and a few lines appeared promising.

SCOTT:
Yesterday I said I was skeptical about the existence of resistance. By "resistance" I mean the host's ability to resist infection by the pathogen, and thus the extent of take-all lesions. I have seen very few data that convince me that there are differences between cultivars in the extent of lesioning.

PARKER:
We have lines selected for reduced take-all lesioning.

SCOTT:
I have reservations about accepting isolated observations of resistance because of the difficulty in reproducing results over a number of environments and particularly over a number of years. There are almost no published reports of resistance that have been substantiated by the original or another worker. One of the strongest environmental interactions is with season. Resistance that exists in 1983 might not exist in 1984.

Question 2 (MacNISH):
What realistic hope is there for resistance to take-all? Where should we be looking for sources of resistance? If sources of resistance are developed, do we expect them to be stable?

SCOTT:
As an example of the likelihood of resistance being stable, one can consider oats. The resistance of oats is based on the production of avenacin. This has been overcome by *G. graminis* var. *avenae*, which produces avenacinase, thus rendering oats susceptible. If it were possible to introduce oat resistance into wheat, I would predict with some confidence that it would break down. As far as resistance within wheat is concerned, I don't think one can attempt to address the question of stability until we have established that there is

332

resistance within wheat.

Question 3 (HUBER):

Is genetic resistance real? We have released several cultivars of winter wheat with fairly good levels of tolerance to take-all under field conditions and have preliminary indications of mechanisms responsible.

HUBER:

We selected for resistance in wheat and saw a reduction in lesion development in cultivar Auburn. Among the differences between this and other cultivars is the correlation of resistance with populations of manganese-oxidizing microorganisms in the rhizosphere. The induced rhizosphere activity with manganese may be under genetic control. We have a similar relationship with manganese-oxidizing populations and resistance to gray-speck in oats, which is under genetic control.

KOLLMORGEN:

Professor Huber, what are your breeders doing with the resistance you have identified? Is it possible to compound it?

HUBER:

We are continually crossing those lines that appear better than current cultivars and are moving the resistance into our adapted local cultivars. We have released one cultivar, which has been well received. We have tested it at 22 locations throughout Indiana (U.S.A.), and its yield has held up in all these trials with or without take-all. We hope to release a better line within 2 years and to have even better lines available for release in the near future.

SCOTT:

I am interested in the selection procedure for resistance in Professor Huber's breeding program. In addition, how can the take-all resistance of the released cultivar be verified over 22 locations on the basis of yield trials?

HUBER:

Its resistance was not based on yield trials but from studies in disease tests under severe infection. In many of the statewide trials, we had take-all to various degrees. In addition, we have two selection nurseries every year: one at the agronomy farm where infection is initiated the year before testing by inoculation, and the other in a farmer's field selected for high take-all incidence. The same replicated nursery is sown at both locations, so selection is under commercial farm conditions as well as where take-all is induced.

SCOTT:

How do you select?

HUBER:

Selection is performed using a root index, whiteheads, and yield. Roots tend to be indexed on farmers' sites, since the agronomy farm has a silt-loam soil that is difficult to remove from roots. Roots are dug, washed, and scored for percentage of root infection.

SCOTT:

What breeding generations are screened?

HUBER:

The fourth to the sixth generations.

Question 4 (SKOU):

What are the prospects in breeding for take-all resistance in wheat and barley? It appears very difficult to breed for resistance to take-all because extensive screening of wheat and barley has not disclosed promising sources of resistance. The possibilities are thus either an accumulation of the small resistances found in some cultivars or lines within the two species, or a transfer of resistance from more resistant lines in related genera. The question is, however, have we utilized all that we know about the way the take-all fungus attacks? We know that:

(a) Differences in resistance between cultivars or lines within the two species are small, whereas lines within related species or genera have more resistance (e.g., in rye). These facts must have fundamental reasons, whatever they might be.

(b) Experiments with chromosome addition lines with the single rye chromosomes in wheat have shown the rye resistance to be dispersed over the seven rye chromosomes. The resistance in rye is thus most likely polygenic.

(c) The ability to regenerate new roots in place of those killed by the fungus may vary much more from one genus to another than from one species to another within a genus, and the rate of root regeneration is correlated with a higher resistance. Also, this ability is most likely polygenic. It is, however, very dependent on the environmental conditions, including edaphic factors.

Will it—on the basis of (a)–(c)—be possible and reasonable to use the rate of root regeneration as the basis for a take-all breeding program?

SKOU:

We have already heard about problems encountered when breeding for resistance. However, I would be particularly interested in comments on selection for new root production as compensation for roots killed by the fungus.

Selecting for new root production can be conducted in the absence of take-all by cutting a number of plant roots. However, this introduces complications since the stressed plant then produces lateral roots to compensate. Moreover, this approach is only valid if take-all simply cuts roots at a point. This does not simply occur when a host is infected with take-all; hence, I don't think such a method is useful.

SIMON:

Is tolerance possibly the result of a more vigorous root system before infection?

ROVIRA:

I think root vigor is very important, certainly in the early stages of a crop. An apparently resistant Pitic cross from an early screening of cultivars from the Waite Institute, South Australia, was vigorous in both root and top growth.

SKOU:

Is there a difference between a more vigorous root system and the ability to produce new roots?

ROVIRA:

There may be no difference; but, I ask, does early root vigor contribute to tolerance to take-all?

SKOU:

There are many problems in studying the influence of root vigor on take-all severity. A barley line with only one seminal root and minimal nodal roots has been produced by mutation at my institute. Although this line yielded proportionally less than normal varieties, it was not easy to kill with take-all, though there was little early root vigor.

Question 5 (SCOTT):

How useful is the resistance of rye to take-all? Rye is much more resistant to take-all than wheat, and rye crops probably seldom suffer severe damage from the disease. Triticale, the amphiploid of wheat and rye, is widely reported to be only slightly more resistant than wheat. However, this is based chiefly on data from seedlings; at the adult plant stage in field trials, triticale appears to stand halfway between wheat and rye, both in resistance to root infection and in yield response to inoculation. Can this resistance be agriculturally useful beyond the cultivation of rye itself? Can it be exploited by growing triticale, or by genetic transfer to wheat? Will it be durable, or do rye-adapted isolates of the take-all pathogen pose a threat to it?

SCOTT:

There are three considerations in this question. First, rye crops are seldom troubled by take-all. Second, in our experience, rye resistance is very useful in triticale. This has not been universally recognized in the literature because tests have often been conducted on seedlings. Yesterday, however, Dr. Wong claimed to find triticale and wheat equally susceptible at the adult stage of development. Over numerous trials, we find triticales to have adult plant resistance midway between that of wheat and rye. Third, in England we have clear evidence for rye-adapted isolates of *G. graminis* that behave in a predictable way in repeated tests. Some isolates are adapted to rye and others are clearly not. We don't know how serious a threat these isolates pose to rye crops, or to triticale crops. I would be interested to know if anyone has experience with such isolates or the use of rye resistance.

POWELSON:

We have been operating a take-all nursery for 6 years. Our experience has been that two or three triticale lines were the only material appearing consistently more resistant over that time.

KOLLMORGEN:

Is there much variation between triticale lines? Dr. Scott sent us two lines, Cambridge 1 and Cambridge 2, which seemed more resistant in terms of lesion development than our local triticale cultivar.

SCOTT:

Wheat cultivars appear susceptible whereas rye cultivars appear more resistant, yet there appears to be little variation within these species. Oddly, it is within the triticale species that we find clear evidence of varietal variation. We initially thought some triticales were incomplete, that is to say they didn't possess all seven rye chromosomes. Upon checking, this did not account for the varietal differences in resistance. Possibly the triticales tested in Britain are a little more resistant than those in Australia.

KOLLMORGEN:

Is there evidence for different races of *G. graminis* var. *tritici*? Can triticale cultivars be used to test for different races of this pathogen?

SCOTT:

I don't know of evidence showing that races adapted to different triticale cultivars exist.

PARKER:

I refer to unpublished work by Yeates where *G. graminis* var. *tritici* isolates were found that were more or less pathogenic on barley and a range of triticales, and were pathogenic on oats. This was checked with Mr. Walker. There are variants of this pathogen that can attack wheat, oats, and barley. I have seen a rye crop and a ryegrass stand heavily damaged by take-all. Like Dr. Scott, I am left somewhat skeptical of achieving my ambitions.

SCOTT:

Do you still have the isolate?

PARKER:

I think so.

Question 6 (ROVIRA, BRISBANE, and SIMON):

Is reduction in yield with added take-all inoculum the most important criterion in screening wheat cultivars for resistance and/or tolerance to take-all? Grain yield of wheat depends on many interacting factors at different stages of growth. Is it therefore better to measure resistance by assessing incidence or severity of disease on the roots or top weight at tillering? The latter must also be questionable because in South Australia take-all is rarely severe enough to cause a measurable or significant reduction in the number of tillers.

SIMON:

Dr. Scott has already questioned evidence for resistance to take-all in wheat. We are convinced that there is no resistance in the wheat lines in our screening program, but rather tolerance to the pathogen. In screening for tolerance, then, should not yield be examined rather than assessing pathogen attack on host roots?

I will present data collected last year from small field plots in which 33 cultivars were tested for take-all tolerance. These cultivars have been selected over 7 years from 4,000 entries in the Australian Wheat Collection. Also included are commercial cultivars from Chile. This work began in conjunction with Dr. Rathjen from the Waite Institute, South Australia. We became involved through our techniques for inoculating soil with the take-all fungus. In the present experiment, ryegrass propagules colonized by take-all were used to infect plants at rates of 16 and 64 propagules per kilogram of soil. At the low rate of inoculum, there was no effect on host growth during the season or on final yield. We use a tolerance index to assess cultivars. This was the ratio between grain yield at the high inoculation rate and mean grain yield at the zero and low inoculation rates. Our standard cultivar is Condor, a commercial cultivar that consistently appears at the bottom of our tolerance index. We find a range of tolerance among wheat cultivars. However, a problem we encounter is the variation in yield between replicates in a season. Nevertheless, the cultivars presented appear consistently tolerant over a number of seasons, and in 1 year at two sites.

SCOTT:

To the question, should you select for yield when searching for tolerance differences, the answer is yes. However, although Mr. Simon's data appear very interesting, I would like to see the extent to which his cultivar rankings are consistent over several years. Many of us could present data like this, but in our experience it is difficult to obtain consistent cultivar rankings over 3, 4, or 5 years. Many people have looked for resistance, but they have finally had to conclude that there is only slight variation within wheat.

Question 7 (MacNISH):

What level of infection in the root system is necessary to make it inoperative? The wheat plant has two root systems (i.e., seminal and nodal). The seminal roots, of which there are usually five, have only one xylem vessel each whereas the nodal roots (no set number) have five or six xylem vessels each. How important is damage to either system? How much penetration or colonization of a xylem vessel is necessary to make it inoperative? What techniques are available for assessing internal damage to the root system?

MacNISH:

I am particularly interested in the seminal roots, which are important in Australia where the end of the growing season is very dry. I would like to know how much damage must be done to the xylem vessels to make them inoperative.

HORNBY:

We don't think blockage of the xylem vessels is initially as important as restriction of translocate flow. If there is any flow of translocates, the root is functional.

MacNISH:

I am considering Australian conditions where moisture uptake is most important.

ROVIRA:

The first thing seen in an infected root is a slowing of root growth and a halt in lateral and fine root production. This must in turn have an effect on water and nutrient uptake.

HORNBY:

Work by Clarkson et al in England showed that translocation and root elongation ceased following phloem disintegration and, within 2-3 days, ion uptake and transport to the shoot ceased, although there was no extensive plugging of the xylem.

MacNISH:

It is possible to find roots that may not be functional, yet appear perfectly healthy under close examination. Have those roots died below the lesion?

Question 8 (SKOU):

The number of lignitubers formed in different root tissues is correlated with the activity of the tissue, but why is their formation most pronounced in the root and endodermis and around developing roots—for example, lateral roots? Is the formation of lignitubers just a result of the common activity in the root tissue or may there be other reasons? Why is the root endodermis of maize able to stop the fungus, whereas in the other cereals it only slows down the development of the attack for a while?

SKOU:

The production of lignitubers requires much energy, not only to synthesize these structures but to increase the size of the plasmalemma that must fold around them.

HORNBY:

Has Dr. Skou any evidence that lignituber formation slows down the pathogen, or that a pathogen above a region containing many lignitubers is a less virulent strain?

SKOU:

When scanning across an infected root, few lignitubers are found in the outer cells, but deeper in the root they become more frequent. Lignitubers can only be produced by living cells. With more extensive colonization, the root may be killed and no more lignitubers are formed. However, if the root is still living, even more lignitubers are formed until, in the case of barley, wheat, and rye, the endodermis is reached. Here up to a hundred lignitubers can form in a single cell.

SCOTT:

Lignitubers appear to be resistance mechanisms, but do they actually stop fungal penetration of the cell wall?

SKOU:

Lignitubers temporarily inhibit the pathogen in wheat, barley, and rye. If it is not lignitubers that are stopping the fungus, what is?

SCOTT:

Do lignitubers form at the endodermis in maize, the point at which fungal growth stops?

SKOU:

Yes.

Question 9 (PITT):

Has any work been carried out on the enzymology of the infection process over the last 5 years? As with other root-infecting fungi, infection and ramification of the take-all pathogen through wheat roots must depend heavily upon the complement and sequential activities of its pectinases and cellulases. Indeed, one reason for different infection potential between take-all isolates may be that different isolates possess different pectinase and cellulase activities and specificities that may reflect their particular isozyme pattern. How much is known about the intraspecies diversity of these exocellular carbohydrases, and can these patterns be used to type given isolates into particular subspecies that may closely correlate with pathogenicity?

What is known about natural (i.e., from within the wheat plant itself) or synthetic inhibitors of these enzymes, and could these find field application? Does anyone know whether phytoalexins or similar compounds inhibit the action of these enzymes and so delimit spread of tissue degradation? Lastly, could feedback inhibition of exocellular carbohydrases during growth in laboratory media partially account for attenuation of pathogenicity during extended subculturing?

SKOU:

The take-all fungus has all the enzymes necessary for the degradation of lignitubers, which not only consist of lignin but other related substances. It can be seen under the microscope that lignitubers are degraded from within. This must be the site where all these enzymes are operating. Thus, do lignitubers have the ability to slow the pathogen, despite the pathogen having the ability to degrade lignitubers?

Session 10: Microbial Interactions

CHAIRPERSON: P. COTTERILL

Question 1 (SIVASITHAMPARAM):
Do the same roots that get infected by take-all get further infected by other pathogens, such as *Rhizoctonia* and *Fusarium* species? Is it likely that these fungi could interact, with the success of each species depending on the time of infection and soil environment?

SIVASITHAMPARAM:
Studies by Lal (1939) and Zogg (1976) suggest an interference phenomenon, with combinations of pathogens resulting in a reduction in disease.

KOLLMORGEN:
R. solani and *G. graminis* var. *tritici* inoculated together are devastating, because *G. graminis* var. *tritici* infects the seminal roots and *R. solani* infects the crown roots.

ROVIRA:
If *R. solani* infects first it causes cortical rotting, which can limit the progress of *G. graminis* var. *tritici*.

HORNBY:
Orange discolorations, attributed to a *Cylindrocarpon* species, and take-all lesions have occurred occasionally on the same root system in the United Kingdom.

PITTAWAY:
At one site in South Australia known to be high in *G. graminis* var. *tritici*, *Pythium graminicola* was isolated from roots with black lesions both by itself and in combination with *G. graminis* var. *tritici*. Such interactions in disease development may be of pathological significance and should be investigated.

Question 2 (GARRETT):
Is biological control of take-all through prior colonization of wheat roots by avirulent fungi of similar infection habit an effective control measure?

WONG:
This can be successful if the roots are sufficiently colonized by these fungi, as is the case with *Phialophora graminicola* in the United Kingdom. In Australia, *G. graminis* var. *graminis* and a lobed hyphopodiate *Phialophora* sp. gave a 10–20% increase in yield when they were added in the seed furrow at the time of sowing in soil infested with take-all. Since the mechanism of suppression appears to be an induced host response resulting in a more rapid lignification and suberization of the stele, large populations of the fungal antagonists have to be present in the rooting zone. If this can be achieved, control should be possible. Moreover, if "compatible" isolates of avirulent fungi and fluorescent pseudomonads can be used, a "double-barreled" protective effect may be possible, with the avirulent fungi colonizing the cortex and giving a host response and the bacteria inhibiting the take-all fungus on the rhizoplane.

Question 3 (WONG):
Has the role of viruses in *G. graminis* var. *tritici* and related fungi been clarified?

WONG:
Do viruses play a role in hypovirulence or take-all decline?

COOK:
For Pacific Northwest (U.S.A.) conditions, there is no evidence to support the hypovirulence theory to account for our observations of take-all decline. This supports the conclusions already reached in the United Kingdom. In fields that had undergone take-all decline, 90% of the isolates were highly virulent. If hypovirulence accounts for take-all decline, why is there so much virulence in the population? Secondly, hypovirulence does not explain why, in pot tests, a small amount of soil introduced into a fumigated rooting medium suppresses fresh, virulent inoculum that is introduced. That would require a background of hypovirulent strains added with the 1% of soil, which would then need to quickly match up with the introduced fungus and transfer hypovirulence. There is no evidence that this occurs, and the role of viruses in the fungus has not been clarified. There is no question that the fungus carries viruses. There are many serologic types, and they appear latent with the fungus.

PARKER:
More physiological plant pathology work is needed on this fungus, and this will perhaps enlighten us on the role of viruses.

SCOTT:
In France, there is commercial exploitation of hypovirulent strains. Is the virus still implicated in this hypovirulence?

HORNBY:
The French described the hypovirulent fungus as having virus in it, but I don't think it is claimed that the viruses are the cause of hypovirulence.

COOK:
In the French work the hypovirulent isolates are not recovered directly from the field, but isolates from the field are kept in vegetative culture until they become "hypovirulent." This conversion to a weakly pathogenic form may be caused by something other than viruses.

Session 11: Disease Management

CHAIRPERSON: G. MacNISH

Question 1 (YARHAM):

How does type and duration of break crop affect the subsequent development of take-all and the development of take-all decline? In recent years, many English farmers have noted that take-all attacks are worse in second than in third wheat crops, and yet all the original work on take-all decline suggested that we should expect the "peak" in the third or fourth crops. If a very severe attack occurs in the second year, then presumably this could of itself trigger the early development of take-all decline. But what of those cases where third-year wheat appears less affected than nearby second-year wheat, even when the preceding second-year wheat appeared to suffer very little from the disease? This often seems to occur where single-year breaks are used in intensive cereal situations. Are such breaks insufficient to eliminate take-all decline completely, and do we therefore start "halfway up" the antagonist buildup and therefore reach the peak after only 2 years of cropping following a 1-year break? If this is so, do we know enough about the relative effects of 1- and 2-year breaks in this context? And what of different types of break crops?

COOK:

I commenced experiments at Lind, Washington (U.S.A.), in 1967 on a formerly dryland site, but where wheat is now produced under irrigation. Initially, when we introduced break crops after 7 years of continuous wheat, they totally offset the decline phenomenon—or at least made the soil conducive again. However, more recently, when break crops were introduced, suppressiveness revived very rapidly when wheat was again planted. Where a field had gone far enough into the decline phase, suppressiveness could be restored quickly after a break crop. It did not matter whether it was peas, soybeans, or common beans—none of these break crops destroyed suppressiveness.

HORNBY:

I support Dr. Cook. The old idea—that if you break a cereal monoculture exhibiting take-all decline for one season, take-all suppressiveness is lost—is in fact incorrect. We have been able to grow break crops for 2–3 years without eliminating the suppressive principle.

PARKER:

Soybeans are in doubt as a break crop after Professor Huber's comments earlier about take-all after 20 consecutive years of soybeans. Does the fungus actually adapt itself to rhizosphere growth or something like that?

COOK:

I would support Professor Huber's observations about soybeans. In the experiment I described above, we had 7 years of wheat followed by 3 years of break crops (common beans and 2 years of soybeans), and then in the 11th year back into wheat. Following soybeans, every wheat plant in the noninoculated, nonfumigated plot developed take-all. In contrast, the wheat plants were essentially all disease free after 3 years of oats or alfalfa.

MacNISH:

Professor Huber, do you think the take-all fungus has become adapted to living or surviving on soybean?

HUBER:

Apparently, since take-all is not suppressed by soybeans.

Question 2 (CUNNINGHAM):

Is there scope for exploitation of host specificity in *G. graminis* var. *tritici* through strategic alternation of wheat and barley crops?

CUNNINGHAM:

Although we have observed host specificity with monocereal cropping, there are also other factors involved. Barley tends to be more tolerant of the disease, even with the same amount of infection as wheat. The actual reduction in barley yield is between 30 and 40% of the loss in yield with wheat. During the take-all decline years, Irish farmers tend to grow continuous barley; but many growers would prefer to take the occasional wheat crop. Such a practice often does not work because during the decline phase, at least on certain soils, the minor peak seasons may give rise to take-all insufficient to affect barley greatly, but having a considerable effect on wheat. Have others had a similar experience in their country?

HUBER:

Barley exudates and residues may inhibit nitrification. This is one of the reasons why there is sometimes a reduction in take-all in wheat following barley. In Australia, where you use organic or mineralizable nitrogen as the source of nitrogen, I would anticipate that you would see a more pronounced effect than you see where large amounts of inorganic nitrogen are added.

KOLLMORGEN:

Professor Huber, did you say that take-all would be less severe in wheat following barley?

HUBER:

I would anticipate it being less severe.

KOLLMORGEN:

That is certainly not the case in Australia.

SKOU:

In Denmark, if we grow wheat after barley, we may get a severe attack.

KOLLMORGEN:

I think the take-all fungus can multiply and spread very well on barley roots without actually killing the plant, and the same applies to triticale.

HORNBY:

I have an impression that when this strategy (i.e., growing barley in the risk years) was used in England, the results were inconclusive.

KOLLMORGEN:

An effective rotation in Victoria, Australia, is to grow wheat, barley, peas. Barley can manage take-all much better than a second wheat crop, so barley is sown and then a cleaning crop of peas.

Question 3 (ROVIRA, BRISBANE, and SIMON):
Does tillage affect take-all decline?

SIMON:

At Avon, South Australia, rotation and tillage trials have been conducted since 1977. Plots have been continuously cropped to wheat with two tillage treatments since then. The treatments are conventional tillage practice of the district (CC) and direct drilling with a narrow-tined implement that gives minimal soil disturbance (DD). Take-all incidence under CC declined from 70% of plants infected in 1980 to 6% in 1982. Even though there was a decline in take-all incidence in both tillage methods between 1980 and 1982, due to severe drought in 1982, there was significantly less take-all under CC than DD. Other factors may be involved. For example, there was more available nitrogen at seeding in the CC than the DD. There are also differences in soil structure, but there was no measurable difference in the amount of total organic matter between the two tillage practices. What may be different is the distribution of that organic matter in the soil.

COOK:

If the factors or principles responsible for take-all decline operate on the root or in the rhizosphere of wheat, then cultivation by fragmenting and blending the infested residue with the surface soil is forcing the fungus to become a root-infecting fungus; and it is then more vulnerable to the decline principle. On the other hand, when the fungus occupies large fragments of intact crown tissues, direct drilling results in the fungus being able to grow from a large food base (with more energy) into the crowns of the next crop, escaping any interaction in the rhizosphere. Our experience with direct drilling is that infection begins very quickly in the crown of the next crop and is the main cause of serious losses with direct drilling. So, I am suggesting that blending the inoculum into the topsoil forces the fungus to become a root-infecting fungus, where it is then more subject to the mechanism of take-all decline or take-all suppression.

ROVIRA:

Cultivation could also be blending the suppressive organisms in the soil so that the fungus is not only starting from a poorer food base, but it has a more hazardous route to the root past suppressive organisms dropped around the soil during cultivation.

SIMON:

It may be that soil structure and availability of oxygen in the two types of tillage practices have a bearing on expression of suppression. Some of Dr. Cook's work would point to the smaller pore sizes in soil under direct drilling inhibiting the expression of suppression.

PARKER:

Is the take-all fungus such an aerobe that it couldn't grow at, say, 1% oxygen? I have seen the take-all fungus growing on oat kernels in bottles with metal caps where the conditions would be nearly anaerobic.

COOK:

An experiment that Dr. Rovira and I did 10 years ago showed that the factor responsible for suppressiveness was more sensitive to reduced aeration than the take-all fungus. So, at a certain point in the aerobic/anaerobic regime, the take-all fungus was still able to operate but the suppressive factor was not; and in that case we had more severe take-all.

Question 4 (KOLLMORGEN):
Is it possible to predict the likely effects of reduced tillage practices and stubble retention on take-all?

Question 5 (YARHAM):
How does cultivation method influence take-all? In Washington, a change from moldboard plowing to direct drilling has been found to aggravate take-all substantially. In the United Kingdom, such an effect has sometimes been noted on peat soils; but it is the exception rather than the rule, and on light mineral soil direct drilling can sometimes reduce the severity of the disease. What is the reason for such differences? Can we now predict, for a given area and soil type, whether the disease will be aggravated or alleviated by a change from plowing to minimal cultivation or direct drilling.

KOLLMORGEN:

It seems that there are conflicting reports on the effects of direct drilling on take-all. Some reports from the United Kingdom suggest that take-all is less severe following direct drilling and others that it is more severe. Dr. Moore's work in Washington indicated that it is more severe; work that I have conducted last year would indicate that it makes little difference whether you direct drill or cultivate conventionally.

MacNISH:

In Western Australia, experiments established in 1977 tested zero till (sown with a triple-disk drill) against increasing levels of cultivation. In a continuous cropping experiment at Esperance, there was distinctly less take-all as we moved from cultivation toward zero till. In experiments conducted at three locations, there were 11 times out of the 29 when zero till had significantly less take-all than the cultivation and 18 times when it was not significantly different. These results are in conflict with those of Drs. Rovira, Cook, and others.

ROVIRA:

From the first experiments we conducted at Avon, we have consistently found a higher level of infection in wheat sown by direct drilling. This year at another site we had treble the incidence of take-all in the direct drill wheat compared with the conventionally cultivated wheat. In 1979 the yields were consistently lower by 0.8 t/ha, regardless of treatment, than the conventional cultivation. In that year we had a correlation between incidence and yield of −0.69. Our direct drilling is an absolute minimum soil disturbance using a coulter wheel that cuts through the trash, followed by a narrow tine, followed by a press wheel. In 1982 there was again an excellent correlation of −0.75 between the incidence and yield. This is in contrast to what Dr. MacNish has obtained in another part of Australia. With our data there is a good correlation between yield and mineral nitrogen at seeding (a correlation of 0.62). We have found consistently in our direct drilling that there is less mineral nitrogen at seeding; this may affect the ability of the plants to respond to take-all infection. It would also affect early vigor and yield.

POWELSON:

I've met people working on direct drilling who show that a correlation exists between yield and soil moisture.

ROVIRA:

We checked our soil moisture at the Avon site and found no significant difference between tillage treatments. In 1982, take-all accounted for 43% of the yield variability, nitrate at seeding 21%, *Rhizoctonia* for 5%, which leaves 31% for the soil chemists and physicists.

PARKER:

Could it be that what Drs. MacNish and Rovira are both calling "direct drilling" are different operations?

MacNISH:

What I call "zero tilling" is sowing with a three-disk drill that disturbs the soil even less than Dr. Rovira's machine. What I call a "direct drill" is sowing with a tine combine drill that disturbs the soil slightly more than Dr. Rovira's machine. We still get the same trend with the direct drilling.

HORNBY:

In the mid-1970s, I looked at some large experiments set up in conjunction with the National Institute of Agricultural Engineering and the Agricultural Development and Advisory Service. The results showed clearly that take-all differed little in conventionally plowed (turning over the soil to a depth of about 20–23 cm), chisel-plowed (pulling the chisel through the soil to a depth of about 10–15 cm), and direct-drilled plots. These different cultivations produced different organic residue profiles in the soil, and at times throughout the season these seemed to be associated with infectivity profiles. However, the net result was no difference at all in the disease during three consecutive years.

COOK:

The work done by Dr. Moore can help to explain some of the issues that have been raised. First, he worked with a seeder with a very sharp double disk that sliced the seed into the soil, with virtually no soil disturbance. Second, his experiments were done at three different locations: (a) on acid soils in a high rainfall area where take-all can be quite severe; (b) at Lind, Washington, with irrigated wheat on fairly alkaline soil; and (c) at Pullman, Washington, where take-all is normally marginal, using natural rainfall and rich loam soil. In all cases, over a period of 3 years, the incidence of take-all (or the severity, whichever he chose to measure) was greater with direct drilling than conventional cultivation. Moreover, using our quadrat technique of fumigated versus nonfumigated and inoculated versus noninoculated, he demonstrated in at least three experiments that when the soil was fumigated and the fungus reintroduced, there was the same amount of disease whether soil was tilled or not. We felt this ruled out the environment as a factor. Also, if the soil was fumigated, yield was the same whether or not the soil was tilled. So zero-tilled soil is satisfactory for wheat growing if the soil is free from pathogens. His conclusion was that there was more inoculum available for infection of the crop with the direct drilling. Stirring the soil and restirring it accelerated the breakdown and made it very difficult for the pathogen to remain within large food bases. Thus, conventional tillage provides smaller and smaller fragments and, consequently, some disease control results.

HUBER:

A lot of literature indicates that to reduce take-all we should have a firm seedbed following conventional tillage. I wonder how you envisage no-till or direct drill relating to the firm seedbed effect.

MacNISH:

The seedbed of the no-till treatments is firm.

SIMON:

The work of physicists on the site at Avon has shown that, once through the top 10 cm of the seedbed, there is no difference in water penetration or in penetrometer readings.

ROVIRA:

Is there any chance that in the Western Australian environment in which Dr. MacNish has been working there is a faster breakdown of the undisturbed crown compared with our situation where the crowns just sit in that soil for 4 months?

MacNISH:

There could be; the Esperance site is fairly wet and the Mt. Barker site is certainly very wet during winter, but it is dry in summer.

ROVIRA:

In Western Australia, the break in the season (i.e., onset of autumn rains) is quite sharp and is followed by consistently wet weather. In South Australia there is a very unpredictable break in the season, so we may have less opportunity for decomposition as a result of direct drilling than there would be in Western Australia.

SMILEY:

In soil moisture characteristics curves, the dry end of the curve is dominated by textural considerations, whereas the wet end of the curve is dominated by structure as well as texture. Structural differences, as they would be affected by tillage, can certainly affect the water content and water potential relationship. These may have some bearing, especially in some of the wetter soils compared with some of the drier soils in Australia.

Question 6 (YARHAM):

How does straw burning influence take-all? Should we, as some say, burn the straw, as to plow it in would give too "puffy" a seedbed? Or ought we, as has been advocated in South Africa, plow in the straw so that the mopping up of the nitrogen during its microbial breakdown may help to starve out the pathogen?

KOLLMORGEN:

In Victoria, it has been accepted that straw burning has little effect on take-all. In an experiment at Rutherglen (Victoria, Australia), there was more take-all where the stubble was burned than where the stubble was not burned. I find it difficult to explain that result, but certainly we cannot advocate stubble burning as a means of reducing take-all.

MacNISH:

That would be my impression in Western Australia also.

ROVIRA:

In the Wagga, New South Wales, Australia, region, they believe that a "hot burn" is beneficial for reducing take-all. By a hot burn they mean something that penetrated 2–4 cm into the soil and burned out a fair bit of the crown. A cool burn is regarded as unsatisfactory by growers.

PARKER:

I don't believe that the heat from any stubble fire gets down even a centimeter.

MacNISH:

Some work that Stynes did suggests that the temperature hardly changed below the surface of the soil in a normal grass fire.

Session 12: Suppressive Soils and Take-all Decline

CHAIRPERSON: G. WILDERMUTH

Question 1 (CUNNINGHAM):

Should the status of pathogen and environment in studies of take-all decline be always contrasted with that during the take-all peak?

CUNNINGHAM:

In Europe more so than in Australia, when a sequence of cereals follows grassland there is a phase of increase of the disease up to a peak that is followed by a decline phase in the field. These are two very distinct phases during which various activities are taking place, and we need to have a reference point. It seems that the disease peak is the ideal reference point to contrast changes occurring up to and subsequent to that time. A lot of evidence has accumulated of changes, particularly the degree of suppressiveness of the soil. In Europe there are two series of checks restricting the pathogen, one following grassland and the other during the established decline phase; the peak is an interval between these two, when a lack of overlap exists between the two sets of constraints. I would like to hear what people have to say in this part of the world on the take-all pattern, although I am aware that there does not seem to be a pronounced peak in Australia.

ROVIRA:

I think that Dr. Cunningham has mentioned the two constraints that occur in Europe, the first one following the grassland and the next one in the decline phase after continuous wheat. We do have a real basic difference in the Australian environment, and that is that after grassland here any suppression is outweighed by the buildup of take-all inoculum on grass roots. There is some suppression, but *Phialophora* spp. do not seem to play a role in that suppression. In the first year of wheat after grassy pasture, take-all is probably going to be highest. Take-all does not start off at a low level and gradually build up if wheat is planted into a grassy pasture. On the other hand, if wheat is planted into a medic or clover-rich pasture, 10% of plants may have take-all in the first year, 40–50% in the second year, and 60–70% in the third year.

POWELSON:

In Oregon (U.S.A.), it is recognized that areas that produce annual and perennial ryegrass seed will produce a very fine wheat crop after grass. They can have 2 years of wheat without any significant yield reduction by take-all. It is only in the third or fourth wheat crop that take-all becomes severe.

MacNISH:

We have one experiment where we looked at the grass composition of the pasture 2 years prior to the crop. There was a good correlation between level of grass in the pasture and level of take-all in the second year after the grass.

CUNNINGHAM:

Is there any evidence of a shift in virulence of isolates in the sequences from different rotations?

MacNISH:

We have a take-all decline situation where we were unable to show that there were actual differences in virulence of the fungus.

ROVIRA:

Why don't *Phialophora* spp. function in Australia in a similar fashion to that in Europe?

WONG:

I think the reason why we are not getting a large buildup of *Phialophora* spp. under these grasses is that during the wheat phase there is such a low population of these fungi. It may be necessary to go back to a permanent pasture for a long period before we get a buildup of *Phialophora* spp. so that take-all is controlled. The Australian environments may not be favorable for this rapid buildup of *Phialophora* spp.

HORNBY:

Is *P. graminicola* absent from Australian soils?

WONG:

No, it is quite common in bowling and golf greens and lawns. It is not very common on grasses or in wheat because I think it is reduced by cultivation. *Phialophora* spp. have a preference for grasses rather than wheat. Many of my isolates have been from either introduced grasses or wild grasses, and we don't see a great deal of *Phialophora* spp. in cultivated crops. I have been doing some work to control Ophiobolus patch in bowling greens by using *Phialophora* spp.

ROVIRA:

We have tried Australian isolates of *G. graminis* var. *tritici* of low virulence with no measurable control of take-all.

Question 2 (POWELSON):

What can be done to facilitate the more rapid establishment of take-all decline?

POWELSON:

Growers in Oregon do not have a serious take-all

problem after the fourth to sixth year of wheat monoculture, and take-all decline is recognized as an effective management practice. We believe that ammonium fertilizers encourage the onset of take-all decline. We also believe that there are similarities in the mechanisms of reduction of take-all by ammonium fertilizers and take-all decline. Pseudomonads appear to be associated with reductions in take-all in both cases.

SMILEY:

Pseudomonads have been shown to be a component of the systems, and we can often get our best form of nitrogen response where we know that we have good pseudomonad populations in the soil. The suppressive soils tend to give a response of greater magnitude than the nonsuppressive soils.

WONG:

I believe the British have found very little correlation between pseudomonads and take-all decline.

HORNBY:

In Britain, about 2% of bacteria recovered from wheat rhizospheres in soils where take-all decline was operating were pseudomonads, and we do not believe they were the explanation for take-all decline.

HUBER:

Two things occur with wheat monoculture. One is a reduction in the rate of nitrification, which would affect the form of nitrogen; the second is a reduction in the population of manganese-oxidizing bacteria in the rhizosphere of the wheat crop. Those two systems would correlate very well with what we see with the effect of nutrition on take-all.

MacNISH:

I would like to ask Dr. Hornby whether we have to stick to a strict definition of take-all decline. Does the peak have to be high?

HORNBY:

I think if you don't have that situation, you should be looking for another description of that form of suppressiveness. In the literature, people have described take-all suppression arising in six or seven different ways. We need to ask: Are they all the same kind of suppression? Although there is no evidence that they are, I think it is probably wise and prudent to keep them separated. After all, the original description of the take-all decline phenomenon was based on observations of continuous winter wheat in which a peak of disease occurred, and therefore the term should not be applied uncritically to suppressive soils that have no record of a significant attack of take-all.

Suppression of take-all with ammonium fertilizers may not be the same as take-all decline achieved by wheat monoculture. So I think it is important to distinguish them until a little more is known. It seems to me that we are likely to find many organisms more or less able to control take-all, and one single organism is unlikely to be the cause of take-all decline.

MacNISH:

Should "suppressive soils" be the name given to all soils that do not follow the traditional pattern for take-all decline?

CUNNINGHAM:

I would not agree totally with Dr. MacNish, because at different sites we have obtained different type peaks and different decline patterns. On some sites, the incidence of the take-all peak can be very high (80–90%) and can subsequently recede to a series of pronounced minor peaks and troughs during the decline phase. At other sites, there is not a pronounced peak at all, with an incidence of 35–40% for 2 or 3 years around the third to sixth crop and then a drop to a uniformly low level. However, I do agree with Dr. MacNish that there should be some term to describe soils that are inherently not prone to take-all.

HORNBY:

My poster points out the danger of taking things for granted. Inoculum in a continuous spring barley site was monitored for more than 14 years, commencing at a time we considered to be a peak of disease. In the succeeding years, there was a considerable drop in disease and in the level of inoculum, and it seemed the criteria for take-all decline were fulfilled. However, when the soil was tested for suppressiveness, it was not suppressive. So we appear to have a take-all decline site with nonsuppressive soil. Where do we go from here?

MacNISH:

Was soil suppressiveness proved when take-all decline was first described?

HORNBY:

Take-all decline was defined in terms of disease reduction, but there have been studies showing that take-all decline soils were also suppressive. The soil I have been describing apparently shows take-all decline, but it appears not to be suppressive.

WILDERMUTH:

You have actually tested that?

HORNBY:

Yes.

Question 3 (WELLER):

Will biocontrol of take-all by introduced microorganisms be more successful in fields where the soil is conducive and with high levels of inoculum of *G. graminis* var. *tritici* or in fields where take-all decline is developing?

WELLER:

We need to consider introducing a biological control organism into a wheat crop at a time when it will be most effective. For example, if it is introduced into a very conducive field, there may be too much of the take-all fungus present at that particular time for the introduced agent to be effective. If the biological control agent is added during a period when take-all decline is commencing, suppression of disease may be more evident. In a test where we treated seed with suppressive bacteria and planted it in a wheat field that had a high level of inoculum, there was little response in the first year. In the second year, the bacterial treatment gave a better response in the same field. This year, the third year, we will perhaps see a significant level of suppression by the bacteria. Three years of the bacterial treatment were needed for bacteria to become incorporated into the inoculum and inhibit the take-all fungus during the parasitic and saprophytic phases. I would like to comment on what Dr. Hornby has been saying about all the various mechanisms that have been presented to explain take-all decline. An organism can provide biological control of take-all without being the agent responsible for the take-all decline phenomenon.

WILDERMUTH:

Do we have a comment on the original question?

COOK:

I would just like to reemphasize the importance of follow-up. In 1969, Shipton and I transferred soil from one site where wheat had been grown for many years to plots at a site where wheat had not been grown for many years; in doing so, we initiated take-all decline 1 year earlier than occurred naturally at the plot site. In the first year, we introduced the soil to a depth of 15 cm. The take-all fungus was introduced as inoculum at the time of seeding, and there was no benefit from the introduced soil. None of the six soils compared in that experiment gave improved wheat growth by the end of the first season. We planted again, with no new inoculum and no new suppressive soil, and the second crop was healthy where we had introduced this one soil. It had come from a field that had been in wheat for 12 years and where take-all was no longer a problem; we don't know that take-all decline had gone on in that field because we weren't able to observe the rise and fall of the disease earlier.

WONG:

I would like to ask Dr. Hornby whether *Phialophora* spp. are also suppressed in take-all decline soils, since they are somewhat related?

HORNBY:

I cannot think of any data that throw light on this. There is often an inverse association between levels of the take-all fungus and *Phialophora* spp. during monoculture.

WILDERMUTH:

I have done some work on this, but only in fumigated soil using artificially induced suppressive soil. There was a reduction in *Phialophora* spp. and *G. graminis* var. *graminis* with suppressive soils.

WONG:

I would like to add to what Dr. Weller said about the swamping effect of inoculum on the biocontrol agent. I find this is the same with the avirulent fungi that I have been working with. The control is only significant if there is a low population of the take-all fungus. There is no control when there is severe take-all. I suggest that the way to manage take-all using biocontrol agents would be to introduce antagonists at the beginning of a series of cereal crops, after a break. The antagonists would lengthen the period in which cereals can be grown without serious take-all, and take-all decline may then occur.

WELLER:

If we can accelerate the onset of take-all decline, it would be important to growers. Reducing the number of years during which take-all would be severe from three or four to one or two by the application of a biocontrol agent would be an example of successful biological control. Farmers need to be educated about realistic expectations for biological control. In Washington State (U.S.A.), biological control is going to be integrated with several other practices to produce the best wheat crop.

Question 4 (HORNBY):
What techniques are there for testing for soil suppressiveness?

HORNBY:

Henis, Ghaffar, and Baker, working with *Rhizoctonia* and radish seedlings, measured "soil conduciveness," i.e., the incidence of disease in a given plant population as expressed by proportions of healthy seedlings developing in the infested and noninfested treatments. They described this conduciveness in terms of a linear function, called the "conduciveness index." The French have used a similar approach, a host-infection technique for measuring soil resistance to *Pythium* spp. The receptivity of a test soil to natural inoculum was determined by comparisons with a sterile soil and one with maximum induced resistance. Bouhot expressed receptivity as soil infectivity potential at some time after time zero (when the pathogen was introduced) divided by soil infectivity potential immediately after introduction. Both techniques depend on added inoculum.

SIVASITHAMPARAM:

Parker and Fang Chang-Sha have been using a technique for the past 10 years to look at saprophytic competence of isolates. They dilute yellow sand with field soil in different proportions and look at the saprophytic and parasitic competence of isolates; the technique is based on the Cambridge test for saprophytic competence and could be used to indicate the suppressiveness of a soil by assessing the competence of an isolate with increasing proportions of natural soil.

PARKER:

We compared the number of strikes or the number of actual infection points in whole soil with those in a series of dilutions of yellow sand. When soil was diluted to 1/10 in this way, we frequently saw more infections in the 1/10 soil than in the original soil; very occasionally, there was even more in the 1/100. It did occur to me that this might be a means whereby we could look at suppressiveness.

SIVASITHAMPARAM:

Where I took soil from under wheat and compared it with soil from under oats, there was more infection in the soil diluted 1/10 than in the original soil.

PITTAWAY:

In a study unrelated to wheat, I compared the growth response of both sensitive and tolerant species to a fungal pathogen using naturally infected and uninfected soils of similar type for dilution-dose mixes. I observed a dose response in the tolerant but not the sensitive species. Further analysis indicated that there was a soil factor, independent of the pathogen, influencing the tolerant species. I think the results of dilution-dose studies using fungal pathogens should interpret both the influence of the dilution series on pathogen activity and on plant growth and sensitivity to infection, preferably comparing known tolerant and sensitive species and cultivars.

COOK:

The pot test that Shipton and I developed and that Dr. Rovira and I used with success grew out of the successful transfer of the suppressive soils from one field to another. We introduced a small amount of soil into plots, and by the time we got it blended in with the original field soil it amounted to only a half of 1% by weight to 15-cm depth. That soil gave suppression of take-all in the second sowing. We subsequently used that method in the greenhouse with considerable success, but it does not always work. In fact, we have had quite a few cases where we were unable to demonstrate any suppression by the method. One percent of old wheat field soil blended in with 99% of fumigated soil plus the take-all fungus demonstrates suppression. It would be nice to use 100% natural soil; but soils vary in texture, pH, and in many other factors, all of which affect the severity of disease.

When chemical and physical differences are reduced by maintaining 99% of the background soil the same, with the only difference being the 1% soil added, then it can be seen that the suppressive factor can be transferred. Once that is determined, I think the points Dr. Hornby is raising here would be very appropriate. But how is it indexed? I believe a crucial need is to have a technique for testing soil suppressiveness. Farmers in Washington are not interested in how much inoculum is in their soil but whether the soil is suppressive. If we had such a method that was really reliable, I think it would have practical as well as scientific value.

ROVIRA:

It is a difficult but important question because, although the technique of Cook and Shipton works well, it does not always work. To date, we have not been able to demonstrate transferable (or "specific") suppression in a calcareous soil where take-all in continuous wheat was less with cultivation than with direct drilling. Is this a take-all decline situation or take-all suppression? There is some general suppression in that when inoculum was introduced into the soil before planting wheat, the added inoculum produced less take-all in soil from the cultivated continuous wheat than in soil from direct-drilled wheat.

HORNBY:

As far as I can make out, other methods have been tried; one is to add inoculum to soils to see how much reduction in take-all occurs compared with some conducive control. I was hoping that somebody might have a novel approach, and I was wondering whether I had struck a chord when Professor Parker talked about soil dilutions. I seem to recall that when we did dilutions of naturally infested soils for bioassays, I also noticed this occasional unexpected increase in infectivity. I think that might be worth exploration.

WILSON:

When soils are diluted, changes in nutrients occur and these may affect infection. Inoculation should be such that the nutrient levels in the bulk soils are not affected. This can be achieved by using only a small volume of inoculum (e.g., a core) so that inoculum dilution has an insignificant effect on nutrient levels.

WILDERMUTH:

Has anybody got an answer to what is the best test?

SIVASITHAMPARAM:

Can someone suggest a standard test until such time as we get something that is really better?

ROVIRA:

Would you say a standard test could be the incorporation of a range of inoculum levels into natural soil compared with a soil diluted 1/10?

SIVASITHAMPARAM:

That is right. If one calls a soil suppressive, people in another country can work out what it means.

ROVIRA:

I think there are other areas of standardization: one is what do you do with bulk soil and another would be the form of inoculum you put into the soil; another question would be the length of time that you leave the inoculum in the soil before planting.

HORNBY:

That is an interesting point. We have talked about kinds of inoculum in terms of how they might affect other organisms. We use colonized oats as added inoculum, and yesterday we heard this criticized as possibly providing nutrient sources for antagonistic organisms. So the question of what an added inoculum should be is important.

PARKER:

To overcome the problem of nutrition, to which of course my yellow sand method was obviously open to criticism, I would suggest this possibility: take a particular soil in which suppressiveness is to be tested and treat it at 60° C with aerated steam. This would not seriously affect the chemical status of the soil. This could be used as the diluent for mixing with the standard soil under test. This could be a way of avoiding nutritional differences.

WILDERMUTH:

Wouldn't there be ammonium nitrogen released at that temperature?

PARKER:

Yes, there would be, but perhaps you could make up for that by adding that ion.

SIMON:

The suppression of growth of take-all from different-sized fractions of ground oat inoculum may vary. The use of small seeds of uniform size and infectivity would assist in the standardization of a test for soil suppressiveness.

WILDERMUTH:

Is there any more comment on the source of inoculum—or perhaps the time we should conduct these tests? I think 28 days is standard for the Cook and Rovira test.

ROVIRA:

It could be the time factor before you even seed. I would certainly appreciate some views on that. In natural soil under Australian conditions, inoculum is dormant through the summer; then, 4–6 weeks before seeding, rain occurs and the soils become moist. This gives several weeks for suppression to occur.

SIMON:

Another point about standardization of suppressive tests is the temperature at which the plants in the bioassay are grown. The Cook-Shipton method states 15°C, but some say 15°C is too high. Is there any comment from other researchers as to the optimum temperature?

HORNBY:

If our prime concern is to compare results from different countries, we should concentrate on developing a standard test applicable to all regions. As long ago as 1967, Lester and Shipton published the method in which the soil was mixed with added inoculum in plastic bags, allowed to stand for a month at 18–20°C, and then tested by growing wheat seedlings for 3 weeks at 15°C. We have adopted this method but use 19°C for the bioassay phase.

Session 13: Bacterization and Biological Control

CHAIRPERSON: P. WONG

Question 1 (YARHAM):

Is biological control of take-all nearer to becoming a practical reality? Numerous organisms are known to be antagonistic to *G. graminis* var. *tritici*, but we still seem to be a long way from a generally accepted system for using them commercially. Do seed treatments with antagonists place antagonists where they are needed? Is there any consensus of opinion as to which organism is best?

WELLER:

My impression is that control of take-all by bacterial treatments is close to becoming a reality. Work on biocontrol of take-all and other diseases and the use of plant-growth-promoting rhizobacteria (PGPR) show that beneficial organisms can be introduced onto the plant root system and are able to suppress major pathogens. Our work in Washington State (U.S.A.) is at a stage where we are close to commercialization. We have conducted a series of tests in the glasshouse and in small plots in the field in which we have demonstrated significant suppression of take-all. We are now testing the treatments in large commercial field plots where we simply coat the seed with the bacteria and hand them to the grower, who plants the coated seed in fields where take-all is a problem. I think that seed treatment is the best way to apply the bacteria, as opposed to soil treatments in the rhizosphere and rhizoplane. The suppressive bacteria need to be positioned in the rhizoplane and rhizosphere so as to protect the root in the first year and then become associated with root and crown tissues infected with *G. graminis* var. *tritici* that provide the inoculum in the following year. Soil treatments have been effective in glasshouse studies, but in the field I do not think the soil treatments are very practical.

As far as the best organism to use, fluorescent pseudomonads are popular perhaps because they are easy to work with. However, beyond this, their ecological and biological characteristics make them suitable for use in biocontrol.

PARKER:

Can I ask what are those characteristics?

WELLER:

First, the fluorescent pseudomonads are normally found in the rhizosphere. They survive in organic material, and when the root comes in contact with the organic material the pseudomonads quickly colonize it. A second factor is that they have a faster growth rate in the rhizosphere compared with other organisms there. Bowen and Rovira have shown that in the rhizosphere

pseudomonads had a generation time that was shorter than *Bacillus* spp. and other bacteria. Another factor is that the pseudomonads are very nutritionally versatile and are able to use a large number of nutrients in root exudates. Finally, they are producers of antibiotics and siderophores. Because of these characteristics, I think they are well suited for biological control of take-all and other diseases.

SIVASITHAMPARAM:

In 1973, when I was working with pseudomonads, Griffin warned me that with the soils and climate we have in Western Australia we may be expecting too much from these bacteria because the moisture conditions there may not be as suitable for bacteria as for fungi. For this reason he was not very enthusiastic about bacteria having much potential as control agents. What is your opinion on that?

WELLER:

It is commonly thought that pseudomonads require high soil moisture to provide biological control. I have been surprised by work done at Pullman, Washington, and at Adelaide, South Australia, that demonstrates that introduced pseudomonads colonize roots under fairly dry conditions. In fact, where the soil profile is too wet, oxygen becomes limiting and bacterial activity ceases. In work that is going on at CSIRO, pseudomonads suppressive to take-all that have been introduced onto wheat seeds into fairly dry soils are able to move down the roots 11–13 cm. We need to know a lot more about the physical and chemical factors that affect root colonization.

COOK:

A point that I think should be brought out with regard to the isolates that Dr. Weller is now working with is that we stack the deck in favor of finding good strains by starting with a suppressive soil where take-all decline has occurred and by isolating from roots of wheat that were relatively healthy while growing in the soil, even though the take-all fungus was present.

WONG:

I beg to differ with that concept because I have found that fluorescent pseudomonads from a soil suppressive of Fusarium wilt from Salinas Valley, California (U.S.A.), were as effective against take-all as against Fusarium wilt; and these bacteria came from a soil that had never seen a crop of wheat. So, it does not appear to have much to do with take-all decline.

WELLER:

I agree that you will find isolates that are effective against a variety of diseases. But across a broad spectrum of pseudomonads, I think you have a better chance of selecting the very best strains if you isolate them from the same system that you expect them to operate in.

CUNNINGHAM:

Is there any evidence of differential sensitivity to the pseudomonads of different isolates of *G. graminis* var. *graminis*?

WELLER:

If antibiotics or siderophores are involved, you may find this.

ROVIRA:

The same organism that has been effective in Washington against *Gaeumannomyces* is looking good against the South Australian *G. graminis* var. *tritici* isolate 500. It indicates to me that there is no specificity against different *G. graminis* var. *tritici* cultures.

WONG:

I have shown some differences in in vitro antibiotic tests between various isolates of bacteria and isolates of take-all, but I believe that in vitro tests do not correlate well with suppression in soil.

BRISBANE:

One factor we must take into account is that no matter whether we suppress *G. graminis* var. *tritici*, *Fusarium*, or *Rhizoctonia* from a farming viewpoint, if we can get an organism that has a wide spectrum of activity, it will be far superior to something that is only suppressive to take-all.

WILDERMUTH:

Working with suppressive soils but not directly with bacteria, we found that the fungus itself was important. We were able to artificially induce suppressiveness to take-all by using plants inoculated with *F. graminearum*. This suppressiveness was as effective in the system we were using as the system where we just used take-all inoculum. So, it seemed to me that the take-all fungus per se was not as important as other factors.

WONG:

Could I swing the discussion back to the question about biological control becoming a practical reality? Would anyone like to talk about fungal antagonists?

PARKER:

We have done quite a lot of work on fungal antagonists. First, following Lemaire's work on hypovirulent take-all fungi, we obtained very similar results to those of Lemaire. At that time, we rather fancied his idea of some immune response on the part of the plant because the organisms we were working with were closely related. Then, Dr. Sivasithamparam showed in some of his tests that *F. oxysporum* had considerable promise. This was an unrelated fungus, and it turned out to be better than any of the others. Following this, we tested a number of root-inhabiting fungi, some of which also protect against *G. graminis*

var. *tritici*. This control, however, was only effective when the roots were preinoculated with the fungi. The procedure does not lend itself to large-scale farming and so there has been no further progress.

There may be effects on the plants from bacteria or fungi, which could induce plant resistance from the rhizosphere. I am thinking of possible elicitors, and we should not forget the volatiles!

WONG:

I demonstrated a 10–20% increase in yield when I added either *G. graminis* var. *graminis* or a *Phialophora* sp. with lobed hyphopodia to soil infested with take-all. I was not able to get any control with *P. graminicola* by adding inoculum in the form of colonized oat grains in the same furrow as the wheat. Since then, I have not taken it any further because of lack of commercial interest. Very recently, some biotechnology companies have shown some interest, and I am now looking at seed pelleting—getting a mycelial preparation and pelleting it onto wheat seed to see whether I can get a similar level of control. I believe, like Dr. Weller, that we have to get the fungus as close to the seed as possible so that the seminal roots will be colonized before they are attacked by take-all.

Question 2a (HUBER):

What is the role of rhizosphere bacteria in take-all? We have 3 years of data on bacterial-variety-nutrition interactions in different soils, indicating both direct and indirect effects that are reversible by environment.

Question 2b (ROVIRA, BRISBANE, and SIMON):

What are the mechanisms by which pseudomonads inhibit *G. graminis*? On agar, pseudomonads inhibit *G. graminis* because of the production of various antibiotics. As the inhibition varies with the media, it may be possible to identify the nature of these antibiotics—for example, are they siderophores? The inhibition of *G. graminis* by pseudomonads in pot trials does not correlate well with the production of antibiotics on agar. Is there any possibility of improving this correlation, perhaps even extending it to field trials?

HUBER:

We are seeing an indirect interaction on rhizosphere bacteria rather than a direct effect—in other words, an effect on the manganese-oxidizing bacteria in the rhizosphere, rather than a direct effect on the take-all fungus. Another concern is that if we add one of these other organisms to the rhizosphere of nutritionally stressed plants, we actually increase the severity of take-all rather than reduce it. Have others seen that interaction?

WELLER:

In the rhizosphere, some bacteria are beneficial, some are deleterious, and some have no effect whatsoever. We hope to be able to replace the deleterious microorganisms with beneficial ones.

HUBER:

Do you see the same organism having different effects depending on the status of the host?

WELLER:

Some of our most suppressive bacteria may be pathogens either of wheat or of the take-all fungus itself; if the population of bacteria on the seed is high enough, the bacteria may have deleterious effects on seed germination.

TROLLDENIER:

I would like to come back to the involvement of several organisms that counteract *Gaeumannomyces* in the rhizosphere. When you take soil suspensions of different origins, you find that the total rhizosphere population is more or less involved. We compared take-all suppression by soil microflora of different origins. Sterilized wheat plants were grown in tubes containing sterilized soil under axenic conditions, or inoculated with *G. graminis* var. *tritici* alone or with *G. graminis* var. *tritici* plus a small amount of soil suspension. We obtained growth comparable to that of healthy control plants when a soil suspension from a bean field was added together with the pathogen. A suspension from a wheat soil, however, did not improve the growth of infected plants. This shows clearly that many organisms are probably involved in controlling *Gaeumannomyces* in the rhizosphere.

COOK:

I would like to say something about the value of antibiotic-producing ability in a control agent. First, keep in mind that in soil ecology the ability of antibiotic production is more a weapon of defense than one of offense, although it would appear that we are using it as a weapon of offense against the take-all fungus, which might not work. Antibiotic-producing organisms will not be totally effective in excluding the take-all fungus from roots. What we need is an organism that will serve as a secondary colonist of the lesion by establishing and multiplying in the lesion. The lesion is a very nutritious habitat, and antibiotic production could then be a very valuable weapon of defense in inhibiting the take-all fungus.

WELLER:

I think that many individuals believe that because there is not an absolute relationship between antibiotic production by bacteria in vitro and suppression of the fungus in the soil, antibiotics have no role. Antibiotic or siderophore production is just one of probably several characteristics that are necessary for a bacterium to possess to be effective at disease suppression. We find that antibiotic or siderophore production is important to certain strains; if they lose these characteristics, they lose suppressiveness. The work at Berkeley, California, demonstrates clearly the importance of siderophore production to some PGPR strains and shows that the loss of siderophore production by the bacteria correlates with the loss of their ability to enhance plant growth. So I think that in certain strains antibiotic production is very important.

WONG:

Could you please elaborate on siderophores?

WELLER:

Siderophores are iron-chelating compounds produced by pseudomonads and many other organisms that help to make iron in the soil or the rhizosphere more available to bacteria. It is thought that the siderophores tie up the iron that is present in the rhizosphere, and the pathogens—be they the take-all fungus or *Fusarium* or deleterious

bacteria—are starved for iron. Evidence in support of this is that PGPR mutants that no longer produce siderophores lose their ability to stimulate plant growth and that synthetic iron chelators that mimic the effect of siderophores give the same disease-suppressive effect.

Question 3a (PITT):

What are the current screening strategies for obtaining microbial take-all antagonists? Apart from direct field isolation of suppressive bacteria from wheat-root rhizosphere or enrichment for such bacteria in systems comprising plant, soil, and pathogen, the only other approach of which I am aware is the mass screening of soil isolates in the hope of obtaining inhibition zones between pathogen and isolate on agar media. Because considerable information is available on the saprophytic and parasitic life-cycle phases of *G. graminis* var. *tritici*, as well as on the infection process per se, are other techniques now available to isolate take-all antagonists?

Question 3b (WELLER):

How can the selection of microorganisms that provide control of take-all in the field be improved? A glasshouse test is necessary to select potential field-effective strains and eliminate ineffective strains before field testing. The assay should permit rapid screening of many strains and should provide results that correlate to the field. At Pullman, pseudomonads are being screened for the ability to protect wheat against take-all when the bacteria are applied to seed. Strains of *Pseudomonas* spp. are screened in plastic cones (2.5 cm in diameter and 17 cm long, with 200 cones suspended in a rack) in which bacteria-coated seeds are placed on a layer of soil infested with *G. graminis* var. *tritici* that is sandwiched between two layers of vermiculite. The level of suppressiveness of a strain depends on the soil used in the test, the particle size, concentration of the take-all inoculum, and the source of the bacterial strain (from either a suppressive or conducive soil). Generally, strains that are highly effective in the glasshouse test also show some suppressiveness in the field. What other screening methods are currently in use, and do they predict the field effectiveness of a strain?

Question 3c (WELLER):

How effective in the field are the microorganisms that are currently being tested for biological control of take-all? It seems unlikely that any microorganism that is applied to control take-all will be able to provide absolute control of the disease. However, when biocontrol agents are used in conjunction with other cultural practices, substantial suppression of take-all can be achieved. In the state of Washington, strains of fluorescent *Pseudomonas* spp., when applied to wheat seed, increase by 10–27% the grain yields of wheat that is grown in fields where *G. graminis* var. *tritici* is introduced or naturally present compared with nontreated wheat. What microorganisms are currently being tested for control of take-all, and how many are being considered for commercial production? What are the best methods for applying the biological treatments? Are combinations of microorganisms better than single treatments?

WELLER:

I would like to present our selection process. One of the most important areas in biological control is the selection of antagonists. I think everyone is faced with the same problem of how to select the best isolates. We have developed a system that is effective in selecting the best pseudomonads for use against take-all in the field. First, we isolate bacteria from the wheat roots or from the rhizosphere. We also isolate from wheat roots that are growing in take-all suppressive soil to which the take-all fungus has been added. This step enriches for the most suppressive organism. Next we test the pseudomonads for in vitro inhibition and production of antibiotics and siderophores. Then we run candidate organisms through a cone test that allows us to screen many organisms at once. In the cone test, organisms are tested in both a fumigated and nonfumigated soil at several concentrations or levels of inoculum. Finally, we select the best strains and test those in our field tests. By this process, we eliminate large groups of organisms and pick out only the best.

SIVASITHAMPARAM:

What media have you been using?

WELLER:

King's medium B for siderophore production and homemade quarter-strength potato-dextrose agar for antibiotic production.

ROVIRA:

I must act as devil's advocate here because, for the life of me, I cannot see that the number of pseudomonads and their concentrations on the root surface are sufficient to produce either enough antibiotics or siderophores to affect the fungus several micrometers away. If we look at a colony of bacteria on an agar plate, there are millions of cells in that one colony; but, in contrast, there is no way in the rhizosphere that the bacteria are going to be that well fed. Maybe we need some lateral thinking about other mechanisms that may operate.

WONG:

Fungi and bacteria tend to be found along the grooves of epidermal cells of the roots. I believe that because both organisms are concentrated in those areas, there is a possibility of a sufficient accumulation of antibiotics or other chemicals that could be significant.

Question 4 (SIVASITHAMPARAM):

Pseudomonads appear not to survive for long periods on seed surfaces. Has a method of extending their survival on seed surfaces been found?

WELLER:

I guess it is a matter of what is considered a long period of time. With the methods that we use, the bacteria survive for about 3 weeks at 5°C with very little decline in the initial populations. This is using methyl cellulose as a sticker. Even though pseudomonads do not produce resting spores, they can survive in a desiccated state for quite a long time. A xanthan gum treatment with talc is one of the best ways of ensuring the survival of the bacteria.

WONG:

I have used the techniques that *Rhizobium* researchers have been using, that is, using irradiated peat and preparing a bacterial inoculum within the irradiated peat. We can then keep the peat for probably 6 months or a year in the refrigerator with little loss in viability. I think there is no problem also if you want to use it on the farm because the farmer will only pellet the seed almost immediately before sowing. The time interval we are talking about is probably a few days. So, I do not think survival on the seed is going to be a big problem.

ROVIRA:

When Ridge was working with these pseudomonads, he found that once he got the organism into the peat it survived for some months.

BRISBANE:

There is a whole field of work on additives, but, of course, some of these pseudomonads produce their own gums. I suspect that by adjusting the cultural conditions, you can alter the amount of gum around the bacterium and increase its survival. The problem is, you then increase its resistance to settling or centrifuging; so if you want to harvest it, it is much more difficult.

WONG:

Yes, I found this with some of my isolates. In fact, we have the added advantage that we do not need a sticker for putting the bacteria onto the seed because there is this natural gum from the bacteria that serves as a sticker.

HUBER:

Gaeumannomyces frequently penetrates through the root and travels for long distances without ever having any surface contact between the stelar and cortical tissues. If we are looking at biological control only as a surface phenomenon, we are going to miss some areas. An organism like *G. graminis* var. *tritici* is also ubiquitous; maybe we ought to look at suppression in the saprophytic phase rather than the parasitic phase.

SKOU:

It is interesting to discuss all the differences in bacteria and fungi and their potential in controlling take-all, but if they cannot compete in soil they will be without any practical interest.

WONG:

I think that the commercial companies would not be too worried about this because they would be quite happy to put the seed coating on every year rather than have the antagonists survive in the soil. I think, ideally, we probably would like the bacteria to survive in the soil between crops. Is that what you mean?

SKOU:

But even if you use them for the present crop, they have to compete with all the other organisms in the soil.

WONG:

Well, that is one of the important criteria of selection.

WILSON:

Yesterday, Dr. Cook made some very interesting

comments and suggested that there might be a genetic factor that suppresses pathogenicity. I was just wondering whether he had ever considered the possibilities of this phenomenon in terms of biological control.

COOK:

I will certainly consider it and hope that we can look into it. However, as to how it might be exploited, you can just let your imagination run wild. Perhaps one way would be to do some genetic engineering and transfer the genetic material from the fungus and establish it in the plant. Another way might be through some cultural practice, to put the selection pressure on the fungus, such that this phenomenon that now seems to be expressed spontaneously in the agar plate could in fact be induced to occur more frequently in field soil.

WELLER:

Getting back to Professor Huber's comment that we should study suppression more in the saprophytic phase, I think the suppressive pseudomonad that is introduced to the seed in one year is inhibiting the fungus the following year as it grows out from infected root material.

ROVIRA:

Dr. Wildermuth showed that a great deal of the suppression was occurring as the mycelium was coming out of the inoculum into the suppressive soil. If the fluorescent pseudomonads and other organisms proliferated around the lesion, then when that root dried out during the summer months, the fungus would normally be protected; but as it emerged, it would encounter some of the suppressive organisms. I see this as a very effective way of building up suppression over time.

TRUTMANN:

Seed dressing with biocontrol agents appears to be a very effective method of maximizing the chances of contact between antagonist and the plant pathogen during its parasitic phase without expenditure of large resources. As mentioned, this method probably also has a residual effect on *G. graminis* var. *tritici* in its saprophytic state. An alternative strategy of applying biocontrol agents during the saprophytic state of *G. graminis* var. *tritici* does not appear to be a feasible proposition because large amounts of inoculum would need to be incorporated into the soil or left on the soil surface, and chances of contact between the biocontrol agent and the target are reduced.

WELLER:

In the Pacific Northwest (U.S.A.), planting the bacteria-treated seed into fields with take-all resulted in a statistically significant increase (20%) in yield.

WONG:

I have also been able to show at Tamworth, Australia, about 20% increase in yield in small plot experiments using fluorescent pseudomonads.

Question 5 (WONG):

What are the guidelines in the various countries for the registration of biocontrol agents for commercial use?

COOK:

In the United States, the Environmental Protection Agency (which regulates all agents for control of pests) has come up with some softened guidelines for what they call biorationals. It includes pheromones and biocontrol agents, living or their products, but there has been a very definite softening of requirements for the registration of biorationals compared with more traditional synthetic chemical pesticides.

PITT:

Could I ask Dr. Cook to be more specific about the guidelines for biorationals?

COOK:

Apparently some of the mammalian toxicity data are not required compared with what would be required with the synthetic chemicals. I think there may be fewer tests required on mammalian toxicity and those general areas where chemicals were certainly more important.

Index

pathogenic variability, 181
Phialophora, effect of, 337
preservation techniques, 299
pseudomonads, effect of, 145, 152
remote sensing for, 28
Rhizoctonia solani, inoculation with, 337
rotation and tillage effects, 255
ryegrass roots, 67
saprophytic survival, 36, 240
silica gel use, 299
soil fungicides, 259
spread in field, 324
triadimenol seed treatment, 232
variety and forma specialis, 297
viruses in, 337
on wheat: roots, 78; yield, effect on, 129, 218
Gahnia radula
Phytophthora cinnamomi on, 177, 178
rhizoplane microflora, 92
Phytophthora effect on tissue
permeability, 162
Gel (see also Pelgel)
root secretion, 75
Geococcus
mycophagous amoeba, 15
Geotrichum candidum
specific growth rate, 189
Gephyramoeba
deliculata, mycophagy by, 14
mycophagous, take-all affected by, 107
Gibberella zeae
maize stalk rot, 39
Gilmaniella humicola
lethal temperature, 193
Gliocladium
on clover and ryegrass roots, 67
Glucose
suppressive soil role, with *Fusarium*, 102, 103
Glyphosate
take-all affected by, 314, 315
Gossypium hirsutum (see also Cotton)
soil insects in fields, 20
Graphium spp.
specific growth rates, 189
Growth rates
specific, selected bacteria and fungi, 189

Haynaldia spp.
take-all resistance in, 158
Helicotylenchus dihystera
on sugarcane, 18
Herbicides
crop and weed resistance, 229
plant disease affected by, 227
root disease affected by, eight crops, 268
soil microflora, factors affecting, 4
Heterodera
avenae: fumigation, 259; relation to
take-all, 260; tillage, effect of, 252
schachtii, herbicide effect on larval
emergence, 228
trifolii, clover roots, fungal relation, 66
Hordeum (see also Barley)
leporinum, take-all host, 240
vulgare, take-all host, 240
Host-parasite interaction
Gaeumannomyces graminis, 332
Hyperparasitism
Trichoderma on *Thanatephorus*, 119
Hyphopodium
Gaeumannomyces graminis, function, 298
Phialophora, 298
Hypochnus
cinnamomi, binucleate species, 58
sasakii, *Rhizoctonia* fusion tried, 57
setariae, binucleate species, 58
Hypomyces solani (see *Nectria
haematococca*)
Hypovirulence
take-all decline, status, 337

Imazalil
take-all control, 314

Insects
fungus interaction, 20
Integrated control
Phytophthora root rot of soybean, 263
Integrated pest management
advocated, 5
discussion, 229-230
fumigation and *Trichoderma* seed
treatment, 110
sclerotial pathogens, 124
Intensification cropping
cause and effects, 127
International Society for Plant Pathology
Soilborne Plant Pathogens Committee, 5
Iodophanus carneus
in Australian wheat fields, 55
Iprodione
onion white rot, 124, 126
wheat *Rhizoctonia*, 65
Irrigation
clover root rot, 271
potato scab, timing, 198
root disease affected by, nine crops, 267
Isopogon ceratophyllus
rhizoplane mycoflora, 92

Juncus bufonius
Phytophthora effect on tissue
permeability, 162

Klebsiella sp., 107
Koch's postulates
infrequent use, 5

Lactuca sativa
Rhizoctonia host, 55
Lagenocystis sp.
on clover roots, 66
Laimosphere
wheat, antagonistic bacteria, role, 146
Lectin
rhizosphere hyphae, *Trichoderma*
relation, 111
Lepidosperma laterale
Phytophthora cinnamomi on, 177
Leptomyxa flabellata
mycophagy, 14, 15
Lignitubers
cereals, infection role, 335
Lolium perenne
root fungi, pasture, 66
Loss
cereal root disease assessment, 29
wheat yield from take-all, effect of
NPK, 218
Lucerne (see Alfalfa)
Lupinus angustifolius
Rhizoctonia host, 55
Lycopersicon esculentum (see also Tomato)
Rhizoctonia host, 55

Macadamia
Phytophthora cinnamomi on, 71
Macrophomina phaseolina
Collembola effect on, 21
in wheat root disease complex, 48
Macroposthonia
on sugarcane, 18
Maize (see also *Zea mays*)
biological seed treatments, 141
stalk rot nature, 38
Manganese
relation to take-all, 212, 327, 328
wheat rhizosphere, effect of, 333
Mayorella
mycophagous amoeba, 14
Medic
rotation for wheat take-all, 240, 256
Medicago
crop sequence, take-all affected by, 240
sativa, *Rhizoctonia* on, 55
Meloidogyne javanica
soil solarization, effect on, 282
on sugarcane, 18

Metalaxyl
apple crown rot control, 244
clover root rot control, 234, 235, 271, 272
Phytophthora root rot of soybean, 263
Methyl bromide
biological seed treatment compared
with, 111
potato yield affected by, causes, 128
soil solarization compared with, 281
Microdochium bolleyi
Coprinus-like fungus association, 50
Fusarium association, rye and wheat
stem base, 146
Pythium affected by, in wheat cortex, 146
root disease complex in wheat, 48
Microscopy
differential interface, mycoparasite, 111
Microtis unifolia
Rhizoctonia on, Australia, 69
Minirhizotrons
usefulness, 4
Mites
fungus interactions, 20
in rhizosphere, 75
Model
clover mycorrhizae, 88
usefulness, 5
Monoculture
potato, *Verticillium* resistance affected
by, 165
take-all, and disease management, 338
Mortierella spp.
on *Gahnia* and *Isopogon*: rhizoplane,
93-95; roots, 97
Mucor
on *Gahnia* and *Isopogon* rhizoplane,
93-95
hiemalis, specific growth rate, 189
Mulching
polyethylene, pathogen affected by, 286
Multiple cropping
root disease affected by, field crops and
vegetables, 268
Mycoparasitism
Gaeumannomyces by *Pythium*, 146
Mycorrhizae
vesicular-arbuscular, clover model, 88
Mycostasis
status, 4

Nectandra
Phytophthora cinnamomi on, 71
Nectria haematococca
on passion fruit, collar rot, 41-44
Nematicides
sugarcane root rot control, 17
Nematodes
cyst, trifluralin effect on hatching, 228
disease identification by aerial
photography, 27
Neocosmospora vasinfecta
Fusarium antagonist, 131
Neurospora crassa
nutrient uptake systems, 188
specific growth rate, 189
Niche
ecological, concept, 187
Nitrification inhibitors
take-all control, 314, 331
Nitrogen
application, timing of, 327
bacterial activity related to, 327
eyespot affected by, 11
pea diseases, 204
root disease affected by, 269
rotation and yield related to, 128
seeding rate related to, 325
take-all affected by, 10, 86, 128, 218, 246,
322, 329, 331
Nothofagus
Phytophthora cinnamomi on, 71
Nuarimol
take-all control, 232, 313

355